CLASSICAL MODERN ALGEBRA

CLASSICAL
MODERN ALGEBRA

SETH WARNER

Professor of Mathematics
Duke University

Prentice-Hall, Inc., Englewood Cliffs, New Jersey

Current printing (first digit): 1 2 3 4 5 6 7 8 9 10

13-136069-8

Library of Congress catalog card number 78-12003

Printed in the United States of America

Prentice-Hall International, Inc., London
Prentice-Hall of Australia, Pty. Ltd., Sydney
Prentice-Hall of Canada, Ltd., Toronto
Prentice-Hall of India Private Ltd., New Delhi
Prentice-Hall of Japan, Inc., Tokyo

To three Susan Emilys

PREFACE

To more than a generation of mathematics students, "modern algebra" has come to be identified with the contents of B. L. van der Waerden's classic text bearing that name. To refer now to subject matter that was already standard in 1930 as "modern" is somewhat anomalous, but it is difficult to abandon the terminology we have received from our mathematical forebears. For this reason, "classical" has been added to the traditional "modern algebra" in the title of this text not only to indicate the historical place of the subject matter but also to suggest its continuing, decisive role in all of mathematics.

Although this text is intended primarily as an introduction to abstract algebra for advanced undergraduates, it may prove useful to anyone seeking an introduction to the subject. Many instructors will find more material here than can be covered conveniently in an academic year. Consequently, this text may serve a variety of different courses. Instructors may choose, for example, to assume as "known" the natural numbers and integers and therefore to omit sections 11, 12, and 14 of Chapter II; or instructors may wish to omit sections 25 and 26 of Chapter IV, sections 31 and 32 of Chapter V, all of Chapter VII, or sections 44 and 47 in Chapter VIII.

The contents of this text are mostly drawn from the first eight chapters of the author's *Modern Algebra*. At many points the exposition has been simplified and made less general to facilitate the use of the text in the classroom. Material on finite groups, including the Sylow theorems, has been added.

Readers will find all the linear algebra needed for a reading of Chapters VI–VIII in Chapter V, plus some additional material on duality that is designed to relate concepts of analytic geometry with those of linear algebra. The text does not contain enough material for a thorough course in linear algebra, however.

The author is convinced that the only way a beginner can really absorb algebra is by solving problems. For this reason a large number of exercises is included, and it is hoped that the text when read is sufficiently clear so that the instructor and his students may devote most of class time to the exercises. They range in difficulty from the routine to the very advanced, but hints are supplied for those of even modest difficulty so that they will be amenable to many students. Exercises that are not routine are starred. Instructors will perceive that often exercises on the same topic are grouped together; such a group of exercises may serve as the basis for papers by students particularly interested in the topic considered. Errors undoubtedly persist, especially in the exercises, and the author will welcome corrections from readers. Many exercises are based on contributions to *The American Mathematical Monthly*, and the author wishes to express his thanks to the many contributors to that journal who will find their contribution (sometimes slightly modified) among the exercises.

SETH WARNER

CONTENTS

CLASSICAL MODERN ALGEBRA

CHAPTER I

ALGEBRAIC STRUCTURES

The best way of learning what modern algebra is all about is, of course, to study it. But some preliminary insight may be obtained by comparing it with the mathematics taught in many secondary schools. Beginning algebra is concerned with the operations of addition and multiplication of real or complex numbers and with the related operations of subtraction and division. Emphasis is on manipulative techniques (such as rearranging parentheses, solving quadratic equations) and on translating problems into the language of algebra whose solutions may then be worked out.

Emphasis in modern algebra, in contrast, is on deriving properties of algebraic systems (such as the system of integers, or of real numbers, or of complex numbers) in a formal, rigorous fashion, rather than on evolving techniques for solving certain kinds of equations. Modern algebra is thus more abstract in nature than elementary algebra. In contrast with elementary algebra, modern algebra studies operations on objects that need not be numbers at all but are assumed to satisfy certain laws. Such operations are already implicit in elementary mathematics. Thus, addition and multiplication of polynomials are best thought of as operations on polynomials considered as objects themselves, rather than on numbers. In the theorem "the derivative of the sum of two differentiable functions is the sum of their derivatives" of calculus, "sum" refers essentially to an operation not on numbers but on differentiable functions.

In spirit, modern algebra is more like plane geometry than elementary algebra. In plane geometry one deduces properties of plane geometrical figures from a given set of postulates. Similarly, in modern algebra we deduce properties of algebraic systems from given sets of postulates. Some statements proved, such as "$-(-a) = a$ and $a \cdot 0 = 0$ for any number a," are so familiar that it is sometimes hard to realize that they need proof. Others, such as "there are polynomial equations of fifth degree that cannot be solved by

1

radicals," are solutions to problems mathematicians worked on for centuries. In plane geometry, however, only one set of postulates is considered, and these postulates are applicable only to the plane. In contrast, many sets of postulates are considered in modern algebra, some of which are satisfied by many different algebraic systems. This generality, of course, greatly increases the applicability of the theorems proved.

But increased generality is not the only benefit of the abstract approach in algebra. In addition, the abstract approach exposes the essential ideas at work in a theorem and clears away what is merely fortuitous. Consider, for example, the following two assertions:

(1) There are at most three real numbers x satisfying the polynomial equation
$$x^3 - 3x^2 + 2x - 1 = 0,$$

for the degree of the polynomial is 3.

(2) There is at least one real number x between 2 and 3 satisfying

$$x^3 - 3x^2 + 2x - 1 = 0,$$

for

$$2^3 - 3 \cdot 2^2 + 2 \cdot 2 - 1 = -1 < 0$$

and

$$3^3 - 3 \cdot 3^2 + 2 \cdot 3 - 1 = 5 > 0.$$

The truth of (1) depends only on relatively few properties of the real numbers, and the statement obtained by replacing "real" with "rational" or "complex" is also valid. In contrast, the truth of (2) depends on much deeper properties of the real numbers, and, indeed, the statement obtained by replacing "real" with "rational" is false. The theorem underlying (1) is actually valid for many algebraic systems besides the real numbers, but for relatively few of such systems is the theorem underlying (2) valid. When we examine these two theorems in their natural abstract setting, we shall see at once what properties of the real numbers are essential for the truth of (2) but incidental for the truth of (1).

1. The Language of Set Theory

Any serious discussion must employ terms that are presumed understood without definition. Any attempt to define all terms used is futile, since it can lead only to circular definitions. The terms we shall discuss but not attempt to define formally are *object, equals, set, element, is an element of, function, ordered pair*.

The English verb **equals,** symbolized by $=$, means for us "is the same as," or "is identical with." Thus, if a is an object and if b is an object, the expression

$$a = b$$

means

"a" and "b" are both names for the same object.

For example, the sixteenth president of the United States = the author of the Gettysburg address.

By a **set** we mean simply a collection of objects, which may be finite or infinite in number and need not bear any obvious relationship to each other. For variety of expression, the words **class, collection,** and occasionally **space** are sometimes used as synonyms for "set." An object belonging to a set is called an **element** of the set or a **member** of the set. A set E and a set F are considered identical if and only if the elements of E are precisely those of F. The symbol $\in$ means "is an element of" or "belongs to" or grammatical variants of these expressions, such as "belonging to." The assertion

x is an element of E

is thus symbolized by

$$x \in E.$$

Similarly, the symbol $\notin$ means "is not an element of."

For example, if P is the set of presidents of the United States before 1900, then Hayes $\in P$, Tilden $\notin P$, 17 $\notin P$. If H is the set of all heads of state in 1863 together with the number 1863, then Lincoln $\in H$, Wilson $\notin H$, Victoria $\in H$, 1776 $\notin H$, Napoleon III $\in H$, 1863 $\in H$.

Five mathematical sets, with which the reader is already acquainted, are so important that we shall introduce symbols for them now. The set of **natural numbers** consisting of the whole numbers 0, 1, 2, 3, etc., we denote by N. The set of **integers** consisting of all natural numbers and their negatives, we denote by Z. Thus, -3 is an integer but not a natural number. The set of all **rational numbers** consisting of all fractions whose numerators are integers and whose denominators are nonzero integers is denoted by Q. Every integer n is also the fraction $n/1$ and so is an element of Q.

The set of **real numbers** is denoted by R and may be pictured as the set of all points on an infinitely long line, e.g., the X-axis of the plane of analytic geometry. Every rational number is a real number, but not conversely; for example, $\sqrt{2}$ and π are real numbers but not rational numbers. We shall say that a real number x is **positive** if $x \geq 0$ and that x is **strictly positive** if $x > 0$. It is simply for convenience that we deviate in this way from the ordinary meaning of "positive." Similarly, a real number x is **negative** if $x \leq 0$, and x is **strictly negative** if $x < 0$. With this terminology, zero is both a positive and a negative number and, indeed, is the only number that is both positive and negative.

The set of **complex numbers** is denoted by C and may be pictured as the set of all points in the plane of analytic geometry, the point (x, y) corresponding to the complex number $x + iy$. Complex numbers may also be written in trigonometric form: If z is a nonzero complex number and if

$$z = x + iy,$$

then also

$$z = r(\cos \theta + i \sin \theta)$$

where

$$r = \sqrt{x^2 + y^2},$$

the length of the line segment joining the origin to (x, y), and where θ is the angle whose vertex is the origin, whose initial side is the positive half of the X-axis, and whose terminal side passes through (x, y). If

$$z_1 = x_1 + iy_1 = r_1(\cos \theta_1 + i \sin \theta_1),$$

$$z_2 = x_2 + iy_2 = r_2(\cos \theta_2 + i \sin \theta_2),$$

then

$$z_1 + z_2 = (x_1 + x_2) + i(y_1 + y_2)$$

and

$$z_1 z_2 = (x_1 x_2 - y_1 y_2) + i(x_1 y_2 + x_2 y_1)$$
$$= r_1 r_2 [\cos(\theta_1 + \theta_2) + i \sin(\theta_1 + \theta_2)].$$

In this and subsequent chapters our discussion will frequently proceed on two levels. In §11 we shall formally state postulates for addition and the ordering on the natural numbers, prove theorems about them, and go on later to define and derive properties of the integers, the rational numbers, the real numbers, and the complex numbers. On an informal level, however, we shall constantly use these sets and the familiar algebraic operations on them in illustrations and exercises, since the reader is already acquainted with many of their properties. It is important to keep these levels distinct. In our formal development we cannot, of course, use without justification those properties of N, Z, Q, R, or C that we tacitly assume known on the informal level.

We shall say that a set E is **contained** in a set F or is a **subset** of F if every element of E is also an element of F. The symbol $\subseteq$ means "is contained in" and its grammatical variants, such as "contained in," so that

$$E \subseteq F$$

means

$$E \text{ is contained in } F.$$

Similarly, we shall say that E **contains** F and write

$$E \supseteq F$$

if F is contained in E; the symbol $\supseteq$ thus means "contains" and its grammatical variants. Clearly, $E \subseteq E$, and if $E \subseteq F$ and $F \subseteq G$, then $E \subseteq G$. Since E and F are identical if they have the same elements,

$$E = F \text{ if and only if } E \subseteq F \text{ and } F \subseteq E.$$

For example, $N \subseteq Z$, $Z \subseteq Q$, $Q \subseteq R$, and $R \subseteq C$, or more succinctly, $N \subseteq Z \subseteq Q \subseteq R \subseteq C$.

One particularly important set is the **empty set** or **null set,** which contains no members at all. It is denoted by $\emptyset$. One practical advantage in admitting $\emptyset$ as a set is that we may wish to talk about a set without knowing *a priori* whether it has any members. For any set E, we have $\emptyset \subseteq E$; indeed, every member of $\emptyset$ is also a member of E, simply because there are no members at all of $\emptyset$.

A set having just a few elements is usually denoted by putting braces around a list of symbols denoting its elements. No significance is attached to the order in which the symbols of the list are written down, and several symbols in the list may denote the same object. For example, {the hero of Tippecanoe, President Wilson, the paternal grandfather of President Benjamin Harrison} = {the author of the League of Nations, President William Henry Harrison}.

Very often, a set is defined as consisting of all those objects having some given property. In this case, braces and a colon (some authors use a semicolon or a vertical bar in place of a colon) are often used to denote the set, as in the following illustrations. For example,

$$\{x \colon x \text{ has property} Q\}$$

denotes the set of all objects x such that x has property Q (the colon should be read "such that"); the set H discussed on page 3 may thus be denoted by

$$\{x \colon \text{either } x \text{ was a head of state in 1863, or } x = 1863\}.$$

Often, a set is described as the subset of those elements of some set possessing a certain property. Thus,

$$\{x \in E: x \text{ has property } Q\}$$

denotes the set of all objects x that belong to the set E and have property Q. Many different properties may serve in this way to define the same set. For example, the sets

$$\{2, 3, 5\},$$

$$\{x \in N: x \text{ is a prime number and } x < 7\},$$

$$\{x \in N: 2x^4 - 21x^3 + 72x^2 - 91x + 30 = 0\}$$

are identical.

A **function** is a rule which associates to each element of a given set, called the **domain** of the function, one and only one element of a given set, called the **codomain** of the function. Thus, to specify a function, we must specify two sets, the domain and codomain (they may be the same set) and a rule which associates to each element of the domain a unique element of the codomain. If f is a function with domain E and codomain F, for each $x \in E$ the unique element of F that is associated to x by f is called the **value** of f at x, or the **image** of x under f, and is usually denoted by $f(x)$. The expression "f is a function from E into F" means "f is a function whose domain is E and whose codomain is F" and is symbolized by

$$f: E \to F.$$

If b is the image under a function f of an element a, then

$$b = f(a)$$

by the definition of the expression "$f(a)$"; another frequently used way of expressing this is to write

$$f: a \to b.$$

For example, let us determine all functions from E into F where $E = \{0, 1\}$ and $F = \{4, 5\}$. Clearly, there are precisely four such functions, which we denote by f, g, h, k, whose rules are given as follows:

$$f: 0 \to 4, 1 \to 4 \qquad h: 0 \to 5, 1 \to 4$$

$$g: 0 \to 4, 1 \to 5 \qquad k: 0 \to 5, 1 \to 5.$$

Sometimes, a function may be defined by means of a formula, in which case the formula and an arrow or equality sign are often used in specifying the function. For example, $f: \mathbf{R} \to \mathbf{R}$, where

$$f: x \to x \sin x$$

means that f is the function with domain $\mathbf{R}$ and codomain $\mathbf{R}$ that associates to each number x the number $x \sin x$. Similarly, $g: \mathbf{R} \to \mathbf{R}$, where

$$g(x) = \begin{cases} -1 \text{ if } x \leq 2, \\ x^2 + 1 \text{ if } x > 2 \end{cases}$$

means that g is the function with domain $\mathbf{R}$ and codomain $\mathbf{R}$ that associates -1 to x if $x \leq 2$ and $x^2 + 1$ to x if $x > 2$.

If f and g are functions, then f and g are regarded as identical if they have the same domain, the same codomain, and if to each element of their domain they associate the same element of their codomain. Thus, if $f: \mathbf{R} \to \mathbf{R}$ and $g: \mathbf{R} \to \mathbf{R}$ are given by

$$f: x \to \sqrt{x^2}$$

$$g: x \to |x|,$$

then f and g are the same function, a fact we express, of course, by writing $f = g$. On the other hand, if h is the function with domain $\mathbf{R}$ and codomain the set $\mathbf{R}_+$ of positive numbers that is given by

$$h: x \to |x|,$$

then $h \neq g$, since h and g have different codomains.

The last of our undefined terms is **ordered pair**. To every object a and every object b (which may possibly be identical with a) is associated an object, denoted by (a, b) and called **the ordered pair whose first term is a and whose second term is b,** in such a way that for any objects a, b, c, d,

(OP)   if $(a, b) = (c, d)$, then $a = c$ and $b = d$.

This concept is familiar from analytic geometry. Indeed, the points of the plane of analytic geometry are precisely those ordered pairs both of whose terms are real numbers. It is possible to give a definition of "ordered pair" that yields (OP) solely in terms of the set-theoretic concepts already introduced (Exercise 1.11). Therefore, it is not really necessary to take "ordered pair" as an undefined term. However, the precise nature of any definition of "ordered pair" is unimportant; all that matters is that (OP) follow from any proposed definition.

Using "ordered pair," we wish to define the **ordered triad** (a, b, c) of objects a, b, c in such a way that if $(a, b, c) = (d, e, f)$, then $a = d$, $b = e$, and $c = f$. We accomplish this by defining (a, b, c) to be the ordered pair $((a, b), c)$. Indeed, if

$$((a, b), c) = ((d, e), f),$$

then

$$(a, b) = (d, e)$$

and

$$c = f$$

by (OP), whence again by (OP),

$$a = d,$$
$$b = e.$$

Similarly, if a, b, c, and d are objects, we define (a, b, c, d) to be $((a, b, c), d)$, and so forth.

We could, of course, have defined (a, b, c) to be $(a, (b, c))$ and again derived the result that if $(a, b, c) = (d, e, f)$, then $a = d$, $b = e$, and $c = f$. Thus, we have simply chosen for our definition of (a, b, c) one of several equally plausible possibilities.

Definition. If E and F are sets, the **cartesian product** of E and F is the set $E \times F$ defined by

$$E \times F = \{(x, y): x \in E \text{ and } y \in F\}.$$

Similarly, if E, F, and G are sets, the **cartesian product** of E, F, and G is the set $E \times F \times G$ defined by

$$E \times F \times G = \{(x, y, z): x \in E, y \in F, \text{ and } z \in G\}.$$

For example, $R \times R$ is the plane of analytic geometry, $R \times \{0\}$ is the X-axis and $\{0\} \times R$ the Y-axis of that plane, $R \times \{1, 2\}$ is the set of points lying on either of two certain lines parallel to the X-axis, and $\{0, 1\} \times \{0, 2\}$ is the set of corners of a certain rectangle. Similarly, $R \times R \times R$ is space of solid analytic geometry, $R \times R \times \{0\}$ is the XY-plane, and $R \times \{0\} \times R$ is the XZ-plane.

The concept of the cartesian product of two sets enables us to define in a natural way the graph of a function.

If f is a function from E into F, the **graph** of f is the subset $\text{Gr}(f)$ of $E \times F$ defined by

$$\text{Gr}(f) = \{(x, y) \in E \times F : y = f(x)\}.$$

When is a subset G of $E \times F$ the graph of a function from E into F? Suppose that $G = \text{Gr}(f)$, where f is a function from E into F. As f associates to each element of E an element of F,

(Gr 1)   for each $x \in E$, there exists at least one element $y \in F$ such that $(x, y) \in G$,

and since to each $x \in E$ is associated only one element of F,

(Gr 2)   if $(x, y) \in G$ and if $(x, z) \in G$, then $y = z$.

Conversely, if G is a subset of $E \times F$ satisfying (Gr 1) and (Gr 2), then G is the graph of exactly one function whose domain is E and whose codomain is F, namely, the function that associates to each $x \in E$ the unique [by (Gr 2)] element y of F such that $(x, y) \in G$.

If G is a subset of $E \times F$ satisfying (Gr 2) [but not necessarily (Gr 1)], then G is the graph of exactly one function whose domain is a subset of E and whose codomain is F, namely, the function f whose domain D is

$$\{x \in E : \text{for some } y \in F, (x, y) \in G\}$$

that associates to each $x \in D$ the unique [by (Gr 2)] element y of F such that $(x, y) \in G$.

For example, the graph G of a function f whose domain is a subset of R and whose codomain is R is a certain subset of $R \times R$. By (Gr 2), every line parallel to the Y-axis contains at most one point (i.e., ordered pair of real numbers) belonging to G; conversely, every subset of the plane having this property is the graph of exactly one function whose domain is a subset of R and whose codomain is R. Thus, the circle of radius 1 about the origin is not the graph of a function, but the semicircles G_1 and G_2 defined by

$$G_1 = \{(x, y) \in R \times R : x^2 + y^2 = 1 \text{ and } y \geq 0\}$$
$$G_2 = \{(x, y) \in R \times R : x^2 + y^2 = 1 \text{ and } y \leq 0\}$$

are graphs of functions.

A function f is completely determined by its domain, codomain, and graph, for the rule of f associates to each element x of its domain the unique element y of its codomain such that (x, y) belongs to its graph. (If we had wished to avoid introducing "function" as a primitive term, we could have done so by *defining* a function to be any ordered triad (E, F, G) of sets such that G is a subset of $E \times F$ satisfying (Gr 1) and (Gr 2); if f were such a triad, its first, second, and third terms would, of course, be called respectively the domain, codomain, and graph of f. The elegance of this alternative method of introducing the notion of a function is outweighed by its lack of intuitiveness.)

From given sets we may form new sets in a natural way. The collection of all subsets of a given set E is again a set which we shall denote by $\mathfrak{P}(E)$. The collection of all functions from a given set E into a given set F is a very important set, which we shall denote by F^E. Thus

$$\mathfrak{P}(E) = \{X : X \subseteq E\}$$

and

$$F^E = \{f : f \text{ is a function from } E \text{ into } F\}.$$

We have already listed all the members of the set F^E for the case where $E = \{0, 1\}$ and $F = \{4, 5\}$.

EXERCISES

1.1. Let A, B, C, and D be the sets defined by

$A = \{x \in \mathbf{R} : 1 < x \le 2\}$, $\quad B = \{x \in \mathbf{R} : \text{either } 1 \le x < 2 \text{ or } x = 3\}$,
$C = \{0, 3, 5\}$, $\quad D = \{x \in \mathbf{R} : \text{either } 2 \le x \le 3 \text{ or } x = 5\}$.

In the plane of analytic geometry, draw each of the sixteen sets $X \times Y$ where X and Y are among A, B, C, and D. When does $X \times Y = Y \times X$?

1.2. If either $E = \emptyset$ or $F = \emptyset$, what is $E \times F$? If $E \times F = \emptyset$, what can you say about E and F? If E has m elements and if F has n elements, how many elements does $E \times F$ have?

1.3. Prove that if $E \subseteq G$ and if $F \subseteq H$, then $E \times F \subseteq G \times H$. If $E \times F \subseteq G \times H$, does it necessarily follow that $E \subseteq G$ and $F \subseteq H$? Under what circumstances does it follow?

1.4. Let M be the set of all American men now alive, and let W be the set of all American women now alive. Determine whether G is the graph of a function whose domain is a subset of M and whose codomain is W (and, if so, what the domain of the function is), where G is the set of all $(x, y) \in M \times W$ such that

(a) y loves x.   (b) y is taller than x.
(c) x is the husband of y.   (d) y is the wife of x.
(e) y is the wife of x and x has curly hair.
(f) x and y have the same parents.
(g) y is the youngest sister of x.
(h) y's father has the same given name as x.
(i) x voted for y's husband for president in 1968.
(j) y is your instructor's wife. (How does your answer depend on his marital status?)

1.5. Let E be the set of all real numbers x satisfying $0 \le x \le 1$. Determine whether G is the graph of a function from E into E, where G is the set of all $(x, y) \in E \times E$ such that

(a) $y = x^3$.   (b) $x = y^3$.
(c) $y = e^x$.   (d) $y = e^{x-1}$.
(e) $y = \sin x$.   (f) $x = \sin y$.
(g) $x = \sin(\pi/2)y$.   (h) $(x - \frac{1}{2})^2 + y^2 = \frac{1}{4}$.
(i) $(x - \frac{1}{2})^2 + (y - \frac{1}{2})^2 = \frac{1}{4}$.   (j) $y = 1$ if $x \in Q$, and $y = 0$ if $x \notin Q$.

1.6. Determine whether G is the graph of a function whose domain is a subset of $R \times R$ and whose codomain is R (and, if so, what the domain of the function is), where G is the set of all $((x, y), z) \in (R \times R) \times R$ such that

(a) $z = x + y$.   (b) $x = 1$.   (c) $x^2 + y^2 + z^2 = 1$.
(d) $x^2 + y^2 + z = 1$. (e) $y = e^{x+z}$.   (f) $z = 1$.
(g) $z \ne y$, and x is the area of the triangle with vertices $(y, 0)$, $(z, 0)$, and $(0, 1)$.
(h) $z > y$, and x is the area of the triangle with vertices $(y, 0)$, $(z, 0)$, and $(0, 1)$.
(i) $x \ge 0$, $z \ge 0$, and y is the sine of the angle whose vertex is at $(0, 0)$, whose initial side passes through $(x, 1)$, and whose terminal side passes through $(z, 1)$.
(j) $x \ge 0$, $z \ge 0$, and y is the cosine of the angle whose vertex is at $(0, 0)$, whose initial side passes through $(x, 1)$, and whose terminal side passes through $(z, 1)$.

1.7. How many elements are there in each of the sets $\mathfrak{P}(E)$, $\mathfrak{P}(F)$, F^E, and E^F if $E = \{1803, \text{Jefferson}, \text{Louisiana}\}$ and $F = \{$the first odd prime, the first successful Republican presidential candidate, the number of Persons in the Trinity, the first assassinated president, the commander-in-chief of the Grand Army of the Republic, the number of sons of Adam whose names are given in Genesis$\}$? List all the elements in each of those four sets.

1.8. How many elements does $\mathfrak{P}(\emptyset)$ have? If E is a finite set of n elements, how many elements does $\mathfrak{P}(E)$ have?

1.9. Let E be a finite set having m members, and let F be a finite set having n members. How many members does F^E have if

(a) $m > 0$ and $n > 0$?   (b) $m = 0$ and $n > 0$?

(c) $m > 0$ and $n = 0$?   (d) $m = 0$ and $n = 0$?

*1.10. Prove that $\sqrt{2} \notin Q$. [Arrive at a contradiction from the assumption that $\sqrt{2} \in Q$ by use of the fact that every rational number may be expressed as the quotient of integers at least one of which is odd.]

*1.11. (a) For each object a and each object b, define (a, b) to be $\{\{a\}, \{a, b\}\}$. Prove that if $(a, b) = (c, d)$, then $a = c$ and $b = d$. [Consider separately the two cases $\{a\} = \{a, b\}$ and $\{a\} \neq \{a, b\}$.]

(b) For each object a and each object b, define $]a, b[$ to be $\{\{b\}, \{a, b\}\}$. Prove that if $]a, b[=]c, d[$, then $a = c$ and $b = d$. [Use (a).]

2. Compositions

Addition and multiplication, the two basic operations of arithmetic, are examples of the fundamental objects of study in modern algebra:

Definition. A **composition** (or a **binary operation**) on a set E is a function from $E \times E$ into E.

The symbols most frequently used for compositions are $\cdot$ and $+$. We shall at first often use other symbols, however, to lessen the risk of assuming without justification that a given composition has properties possessed by addition or multiplication on the set of real numbers. If $\triangle$ is a composition on E, for all $x, y \in E$ we denote by

$$x \triangle y$$

the value of $\triangle$ at (x, y).

Just as in elementary calculus attention is limited to certain special classes of functions (e.g., differentiable functions in differential calculus, continuous functions in integral calculus), so also in algebra we shall consider only compositions having particularly important properties. If $\triangle$ is a composition on E and if x, y, and z are elements of E, $(x \triangle y) \triangle z$ may very well be different from $x \triangle (y \triangle z)$.

Definition. A composition $\triangle$ on E is **associative** if

$$(x \triangle y) \triangle z = x \triangle (y \triangle z)$$

for all $x, y, z \in E$.

If $\triangle$ is an associative composition, we shall write simply $x\triangle y\triangle z$ for $(x\triangle y)\triangle z$. It is intuitively clear that if $\triangle$ is associative, all possible groupings of a finite number of elements yield the same element; for example,

$$(x\triangle(y\triangle z))\triangle(u\triangle v) = x\triangle(y\triangle((z\triangle u)\triangle v)).$$

We shall formulate precisely and prove this in Appendix A.

Although algebraists have intensively studied certain nonassociative compositions, we shall investigate associative compositions only. We shall, however, encounter both commutative and noncommutative compositions:

Definition. Let $\triangle$ be a composition on E. Elements x and y of E **commute** (or **permute**) for $\triangle$ if

$$x\triangle y = y\triangle x.$$

The composition $\triangle$ is **commutative** if $x\triangle y = y\triangle x$ for all $x, y \in E$.

If E is a finite set of n elements, a composition $\triangle$ on E may be completely described by a table of n rows and n columns. Symbols denoting the n elements of E head the n columns and, in the same order, the n rows of the table. For all $a, b \in E$, the entry in the row headed by a and the column headed by b is the value of $\triangle$ at (a, b). It is easy to tell from its table whether $\triangle$ is commutative; indeed, $\triangle$ is commutative if and only if its table is symmetric with respect to the diagonal joining the upper left and lower right corners. However, usually it is not possible to determine by a quick inspection of its table whether a composition is associative.

Example 2.1. Let E be one of the sets Z, Q, R, C. Then addition and multiplication (the functions $(x, y) \to x + y$ and $(x, y) \to xy$ respectively) are both associative commutative compositions on E. Subtraction is also a composition on E, but it is neither associative nor commutative since, for example, $6 - (3 - 2) \neq (6 - 3) - 2$ and $2 - 3 \neq 3 - 2$.

Example 2.2. Let E consist of the English words "odd" and "even." We define two compositions $\oplus$ and $\odot$ on E by the following tables:

$\oplus$	even	odd
even	even	odd
odd	odd	even

$\odot$	even	odd
even	even	even
odd	even	odd

These compositions mirror the rules for determining the parity of the sum and product of two integers (e.g., the sum of an even integer and an odd integer is odd, and the product of an even integer and an odd integer is even). It is easy to verify that $\oplus$ and $\odot$ are both associative and commutative.

Example 2.3. For each positive integer m, we define N_m to be the set $\{0, 1, \ldots, m-1\}$ of the first m natural numbers. Two very important compositions on N_m, denoted by $+_m$ and $\cdot_m$ and called *addition modulo m* and *multiplication modulo m*, are defined as follows: $x +_m y$ is the remainder after $x + y$ has been divided by m, and $x \cdot_m y$ is the remainder after xy has been divided by m. In other words, $x +_m y = x + y - jm$ where j is the largest integer such that $jm \leq x + y$, and $x \cdot_m y = xy - km$ where k is the largest integer such that $km \leq xy$. The tables for addition modulo 6 and multiplication modulo 6 are given below.

$+_6$	0	1	2	3	4	5
0	0	1	2	3	4	5
1	1	2	3	4	5	0
2	2	3	4	5	0	1
3	3	4	5	0	1	2
4	4	5	0	1	2	3
5	5	0	1	2	3	4

$\cdot_6$	0	1	2	3	4	5
0	0	0	0	0	0	0
1	0	1	2	3	4	5
2	0	2	4	0	2	4
3	0	3	0	3	0	3
4	0	4	2	0	4	2
5	0	5	4	3	2	1

Manipulations with these compositions are in some respects easier than the corresponding manipulations with ordinary addition and multiplication on Z. For example, to find all $x \in N_6$ such that

$$(x \cdot_6 x) +_6 x +_6 4 = 0,$$

we need only calculate the expression involved for each of the six possible choices of x to conclude that 1 and 4 are the desired numbers. No analogue of the quadratic formula is needed.

The modulo m compositions arise in various ways: In computing time in hours, for example, addition modulo 12 is used (3 hours after 10 o'clock is 1 o'clock, i.e., $3 +_{12} 10 = 1$).

It is easy to infer the associativity and commutativity of $+_m$ and

$\cdot_m$ from the associativity and commutativity of addition and multiplication on $\mathbf{Z}$. Let us prove, for example, that

$$(x \cdot_m y) \cdot_m z = x \cdot_m (y \cdot_m z).$$

Let j be the largest integer such that $jm \leq xy$, and let p be the largest integer such that $pm \leq yz$. By definition,

$$x \cdot_m y = xy - jm,$$
$$y \cdot_m z = yz - pm.$$

Let k be the largest integer such that $km \leq (xy - jm)z$, and let q be the largest integer such that $qm \leq x(yz - pm)$. Then $(jz + k)m \leq (xy)z$ and

$$(q + xp)m \leq x(yz),$$

and by definition

$$(x \cdot_m y) \cdot_m z = (xy - jm)z - km,$$
$$x \cdot_m (y \cdot_m z) = x(yz - pm) - qm.$$

But $jz + k$ is the largest of those integers i such that $im \leq (xy)z$; if not, $(jz + k + 1)m \leq (xy)z$, whence $(k + 1)m \leq (xy - jm)z$, a contradiction of the definition of k. Similarly, $q + xp$ is the largest of those integers i such that $im \leq x(yz)$. As $(xy)z = x(yz)$, we have $jz + k = q + xp$, and thus

$$(x \cdot_m y) \cdot_m z = (xy - jm)z - km = xyz - (jz + k)m$$
$$= xyz - (q + xp)m = x(yz - pm) - qm$$
$$= x \cdot_m (y \cdot_m z).$$

Example 2.4. On any set E we define the compositions $\leftarrow$ and $\rightarrow$ by

$$x \leftarrow y = x,$$
$$x \rightarrow y = y$$

for all $x, y \in E$. Clearly, $\leftarrow$ and $\rightarrow$ are associative compositions, but, if E contains more than one element, neither is commutative.

Example 2.5. Let P be a plane geometrical figure. A *symmetry* of P, or a *rigid motion of P into itself*, is a motion of P such that the center of

same class, of which one defines an associative (commutative) composition but the other a nonassociative (noncommutative) composition? Can you make precise the assertion that the compositions defined by two tables in the same class are "just like" each other? [Rewrite one of the tables by heading its first row and column by "b" and its second row and column by "a".]

2.3. If E is a finite set of n elements, how many compositions are there on E? Of these, how many are commutative?

2.4. Find all $x \in N_6$ such that

(a) $3 \cdot_6 x = 3$.
(b) $2 \cdot_6 x = 5$.
(c) $(5 \cdot_6 x) +_6 3 = 4$.
(d) $x \cdot_6 x = 1$.
(e) $x \cdot_6 x = 5$.
(f) $(x \cdot_6 x) +_6 (3 \cdot_6 x) = 4$.
(g) $2 \cdot_6 x = 4 \cdot_6 x$.
(h) $x \cdot_6 x \cdot_6 x = (5 \cdot_6 x) +_6 5$.
(i) $x \cdot_6 x \cdot_6 x = 4 \cdot_6 x$.
(j) $(x +_6 x +_6 x) +_6 (x +_6 x +_6 x) = 0$.

What facts that you remember from elementary algebra concerning the solution of equations no longer hold when addition and multiplication modulo 6 replace ordinary addition and multiplication of real numbers?

2.5. Let $m > 2$. Prove that $(m - 1) \cdot_m (m - 1) = 1$. Infer that there exists $p \in N_m$ such that $x \cdot_m x \neq p$ for all $x \in N_m$; i.e., there exists an element of N_m that is not a square for multiplication modulo m.

2.6. Prove that $+_m$ is associative.

2.7. Prove that $+_m$ and $\cdot_m$ are commutative.

*2.8. Let $m \geq 0, n > 0$. Let $+_{m,n}$ be the composition on N_{m+n} defined as follows: if $x + y < m$, $x +_{m,n} y$ is defined to be $x + y$; if $x + y \geq m$, then $x +_{m,n} y$ is defined to be $x + y - kn$ where k is the largest integer satisfying $m + kn \leq x + y$.

(a) Write out the table for $+_{3,4}$. Show how to label the stars of the Big Dipper $0, 1, \ldots, 6$ so that the sequence $0, 1, 1 +_{3,4} 1, 1 +_{3,4} 1 +_{3,4} 1$, etc. traces out first the handle and then the bowl infinitely many times in a clockwise fashion. What is the analogue of this model for $+_{m,n}$? What does the model become if $m = 0$? if $n = 1$?
(b) Prove that $+_{m,n}$ is associative and commutative.
(c) Find all $x \in N_7$ such that:

(1) $x +_{3,4} 2 = 3$.   (2) $x +_{3,4} x = 4$.   (3) $x +_{3,4} x = x$.

2.9. Prove that $\leftarrow$ and $\rightarrow$ are associative compositions. How can you recognize from the table of a composition on a finite set whether it is $\leftarrow$ or $\rightarrow$?

2.10. Complete the table for the composition $\circ$ of Example 2.5.

2.11. Construct the table for the composition analogous to that of Example 2.5 on the set of (the six) symmetries of an equilateral triangle. Do the same for (the four) symmetries of a rectangle that is not a square.

2.12. Determine whether $\triangle$ is an associative composition on R, where for all $x, y \in R$, $x \triangle y =$

(a) $\max\{x, y\}$.

(b) $x + y + x^2 y$.

(c) $\min\{x, 2\}$.

(d) $x + \log(10^{y-x} + 1)$.

(e) $2x + 2y$.

(f) $x + y - 3$.

(g) the largest integer $\leq x + y$.

(h) $x + y - xy$.

(i) $\sqrt{x^2 + y^2 + 1}$.

(j) $x + \log(1 + 10^{y-x} + 10^y)$.

(k) $\tan\left[\dfrac{2}{\pi}(\arctan x)(\arctan y)\right]$.

(l) $\min\{x, y\}$ if $\min\{x, y\} < 13$, and $\max\{x, y\}$ if $\min\{x, y\} \geq 13$.

(Logarithms are to base 10. "Max" stands for "the maximum of," and "min" stands for "the minimum of.")

2.13. Determine whether $\triangle$ is an associative composition on the set R_+^* of all strictly positive real numbers, where for all $x, y \in R_+^*$, $x \triangle y =$

(a) $3xy$.

(b) $x \log(1 + y)$.

(c) $x^{\log y}$.

(d) $x + 2\sqrt{x} + y + 2\sqrt{y} + 2\sqrt{xy} + 1$.

(e) x^y.

(f) $\dfrac{xy}{x + y}$.

(g) $xy + 1$.

(h) $\dfrac{x + y}{1 + xy}$.

(i) $yx^{1-\log y}$.

(j) $\dfrac{x + y + 2xy + 2}{2x + 2y + xy + 1}$.

(k) 17.

(l) $\dfrac{xy}{x + y + 1}$.

(Logarithms are to base 10.)

2.14. Which compositions of Exercises 2.12 and 2.13 are commutative?

2.15. If $\triangle$ is an associative composition on E and if $a \in E$, then the composition $\triangledown$ on E defined by $x \triangledown y = x \triangle a \triangle y$ is associative.

2.16. If $\triangle$ is an associative composition on E and if x commutes with y and with z for $\triangle$, then x commutes with $y \triangle z$.

*2.17. Let $\triangle$ be a composition on E. An element $a \in E$ is **idempotent** for $\triangle$ if $a \triangle a = a$. The composition $\triangle$ is an **idempotent** composition if every element of E is idempotent for $\triangle$. The composition $\triangle$ is an **anticommutative** composition if for all $x, y \in E$, if $x \triangle y = y \triangle x$, then $x = y$.

(a) The compositions $\rightarrow$ and $\leftarrow$ are idempotent anticommutative compositions on E.

(b) If $\triangle$ is associative, then $\triangle$ is anticommutative if and only if $\triangle$ is idempotent and $x \triangle y \triangle x = x$ for all $x, y \in E$.

(c) If $\triangle$ is associative and anticommutative, then $x \triangle y \triangle z = x \triangle z$ for all $x, y, z \in E$. [Consider $x \triangle y \triangle z \triangle x \triangle z$.]

3. Unions and Intersections of Sets

Two fundamental compositions on the set $\mathfrak{P}(E)$ of all subsets of E are defined as follows:

Definition. Let A and B be subsets of E. The **union** of A and B is the set $A \cup B$ defined by

$$A \cup B = \{x \in E: \text{either } x \in A \text{ or } x \in B\}.$$

The **intersection** of A and B is the set $A \cap B$ defined by

$$A \cap B = \{x \in E: x \in A \text{ and } x \in B\}.$$

If no elements belong to both A and B, that is if $A \cap B = \emptyset$, we shall say that A and B are **disjoint** sets.

Let us picture E as the area bounded by a rectangle, A and B as overlapping discs lying in that area (a disc consists of all points lying inside and on a given circle). The rectangle is then divided into four mutually disjoint pieces, and Figure 1 gives a pictorial representation of $A \cup B$ and of $A \cap B$. These diagrams are examples of *Venn diagrams*, which may be used to picture new sets formed from given ones or to illustrate relations subsisting between sets. For example, three overlapping discs A, B, and C in a rectangle divide it into eight mutually disjoint pieces, and sets formed from A, B, and C by taking unions and intersections have pictorial representations (Venn diagrams) in the rectangle. Thus, Figure 2 gives the Venn diagram of $A \cap (B \cup C)$, and Figure 3 gives the Venn diagrams of $A \cap B$ and $A \cap C$. The diagram for $(A \cap B) \cup (A \cap C)$ is then formed by shading all areas that are shaded in either of the diagrams of Figure 3, and consequently, the Venn diagram for $(A \cap B) \cup (A \cap C)$ is identical with that

$A \cup B$

$A \cap B$

Figure 1

$A \cap (B \cup C)$

Figure 2

$A \cap B$

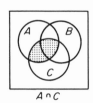
$A \cap C$

Figure 3

for $A \cap (B \cup C)$. This illustrates one of the equalities of the following theorem, which gives basic properties of the compositions $\cup$ and $\cap$ on $\mathfrak{P}(E)$.

Theorem 3.1. If A, B, and C are subsets of E, then

$$(A \cup B) \cup C = A \cup (B \cup C), \qquad A \cup B = B \cup A,$$
$$(A \cap B) \cap C = A \cap (B \cap C), \qquad A \cap B = B \cap A,$$
$$A \cap (B \cup C) = (A \cap B) \cup (A \cap C), \qquad A \cup \emptyset = A,$$
$$A \cup (B \cap C) = (A \cup B) \cap (A \cup C), \qquad A \cap E = A.$$

Proof. We shall prove, for example, that $A \cup (B \cap C) = (A \cup B) \cap (A \cup C)$. If $x \in A \cup (B \cap C)$, then either $x \in A$, in which case x belongs both to $A \cup B$ and to $A \cup C$ and consequently to $(A \cup B) \cap (A \cup C)$, or else $x \in B \cap C$, in which case x again belongs both to $A \cup B$ and to $A \cup C$ and hence to $(A \cup B) \cap (A \cup C)$. Therefore,

$$A \cup (B \cap C) \subseteq (A \cup B) \cap (A \cup C).$$

Conversely, if $x \in (A \cup B) \cap (A \cup C)$ but $x \notin A$, then since $x \in A \cup B$, we have $x \in B$, and since $x \in A \cup C$, we have $x \in C$, whence $x \in B \cap C$. Thus, if $x \in (A \cup B) \cap (A \cup C)$, then either $x \in A$ or $x \in B \cap C$, and consequently, $x \in A \cup (B \cap C)$. Therefore,

$$(A \cup B) \cap (A \cup C) \subseteq A \cup (B \cap C).$$

The compositions $\cup$ and $\cap$ on $\mathfrak{P}(E)$ are thus both associative and commutative.

Definition. If A and B are subsets of E, the **relative complement** of B in A is the set $A - B$ defined by

$$A - B = \{x \in E : x \in A \text{ and } x \notin B\}.$$

The **complement** of B is the set B^c defined by

$$B^c = \{x \in E : x \notin B\}.$$

$A - B$

B^c

Figure 4

It makes no sense to speak of the complement of a set, of course, unless it is clearly understood in the context that all sets considered are subsets of a certain given set, with respect to which all complements are taken.

The Venn diagrams for $A - B$ and for B^c are given in Figure 4.

Theorem 3.2. If A and B are subsets of E, then

$$A \cup A^c = E, \qquad\qquad \emptyset^c = E,$$

$$A \cap A^c = \emptyset, \qquad\qquad E^c = \emptyset,$$

$$(A \cap B)^c = A^c \cup B^c, \qquad (B^c)^c = B.$$

$$(A \cup B)^c = A^c \cap B^c.$$

Proof. We shall prove, for example, that $(A \cup B)^c = A^c \cap B^c$. If $x \in (A \cup B)^c$, then $x \notin A \cup B$, so $x \notin A$ and $x \notin B$, whence $x \in A^c$ and $x \in B^c$, and therefore, $x \in A^c \cap B^c$. Thus,

$$(A \cup B)^c \subseteq A^c \cap B^c.$$

On the other hand, if $x \in A^c \cap B^c$, then $x \notin A$ and $x \notin B$, whence $x \notin A \cup B$, and therefore, $x \in (A \cup B)^c$. Thus,

$$A^c \cap B^c \subseteq (A \cup B)^c.$$

Venn diagrams are useful in analyzing data about subsets of a given set. There is no need to represent subsets in a Venn diagram by discs, although it is convenient to do so if the number of initially given subsets does not exceed three.

To illustrate the use of Venn diagrams, let us consider two studies on the ownership of homes, cars, and TV sets by industrial workers, one study made on a sample of 1,000 workers in Muskegon, the other on a sample of 1,000 workers in Muskogee. The data reported are summarized below.

	Muskegon	Muskogee
Home owners	158	128
Car owners	333	323
TV owners	693	692
Home and car owners	23	13
Home and TV owners	73	25
Car and TV owners	103	49
Home, car, and TV owners	13	3

Let H, C, and T be respectively the set of home owners, car owners, and TV owners in one of the samples, and for any subset X of the sample, let $n(X)$ be the number of its members. We form the Venn diagram for three subsets and determine the number of elements in each of the eight resulting subsets for the Muskegon study as follows:

As

$$n(H \cap T) = n(H \cap T \cap C) + n(H \cap T \cap C^c),$$

we have

$$n(H \cap T \cap C^c) = 73 - 13 = 60;$$

as

$$n(H \cap C) = n(H \cap T \cap C) + n(H \cap T^c \cap C),$$

we have

$$n(H \cap T^c \cap C) = 23 - 13 = 10;$$

as

$$n(H) = n(H \cap T \cap C) + n(H \cap T^c \cap C) + n(H \cap T \cap C^c)$$
$$+ n(H \cap T^c \cap C^c),$$

we have

$$n(H \cap T^c \cap C^c) = 158 - 60 - 10 - 13 = 75.$$

Similarly, we find that

$$n(C \cap T \cap H^c) = 90,$$
$$n(C \cap T^c \cap H^c) = 220,$$
$$n(T \cap C^c \cap H^c) = 530.$$

Thus,

$$n(H \cup T \cup C) = 60 + 10 + 75 + 90 + 220 + 530 + 13 = 998,$$

so

$$n(H^c \cap T^c \cap C^c) = 1,000 - n(H \cup T \cup C) = 2.$$

The Venn diagram for the Muskegon study is given in Figure 5. From the diagram we may read off many facts. For example, 923 workers own either a car or a TV set, 620 own a TV set but not a home, 690 own a home if they own a car, 865 own a home only if they own a car, 555 own a home if and only if they own a car.

Proceeding similarly with the Muskogee report, however, we find that $n(H \cup C \cup T) = 1,059$, which is impossible since the sample contained only 1,000 workers. The data are therefore inconsistent, and the report is false.

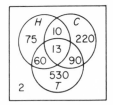

Figure 5

We may easily extend the notion of the union and intersection of two sets to any class whatever of sets:

Definition. Let $\mathscr{A}$ be a class of subsets of E. The **union** of $\mathscr{A}$ is the set $\cup\mathscr{A}$ defined by

$$\cup\mathscr{A} = \{x \in E : \text{there exists } A \in \mathscr{A} \text{ such that } x \in A\}.$$

The **intersection** of $\mathscr{A}$ is the set $\cap\mathscr{A}$ defined by

$$\cap\mathscr{A} = \{x \in E : x \in A \text{ for every member } A \text{ of } \mathscr{A}\}.$$

For example, if $\mathscr{A}$ is the class of all discs in the plane of analytic geometry that are tangent to the X-axis and have radius 1 (the points on the boundary of a disc as well as in its interior are considered elements of the disc), then $\cup\mathscr{A}$ is the infinite strip consisting of all points (x, y) such that $-2 \le y \le 2$. If $\mathscr{B}$ is the class of all discs of radius 1 lying inside and tangent to the circle of radius 2 and center $(0, 0)$, then $\cap\mathscr{B} = \{(0, 0)\}$.

EXERCISES

3.1. Let A be the set of words occurring in the first sentence of Chapter I, B the set of words occurring in the first sentence of the second paragraph of Chapter I. What is $A \cup B$? $A \cap B$? $A - B$? $B - A$?

3.2. Prove that $A \cup (B \cap C) = (A \cup B) \cap (A \cup C)$ and that $(A \cap B)^c = A^c \cup B^c$, and draw the pertinent Venn diagrams.

3.3. Prove that the following five statements concerning subsets A and B of E are equivalent:

(a) $A \subseteq B$.
(b) $A \cap B^c = \emptyset$.
(c) $A \cap B = A$.
(d) $A \cup B = B$.
(e) $A^c \supseteq B^c$.

3.4. Prove the following identities concerning subsets A, B, and C of E, and draw the pertinent Venn diagrams:

(a) $A - (B \cup C) = (A - B) \cap (A - C)$.
(b) $(A - C) \cap (B - C) = (A \cap B) - C$.
(c) $(A - B) - C = A - (B \cup C)$.
(d) $A - (B - C) = (A - B) \cup (A \cap C)$.

3.5. What is $\cup \mathscr{A}$ if $\mathscr{A}$ is the class of all discs of radius 1 lying inside and tangent to the circle of radius 3 and center $(0, 0)$? What is $\cap \mathscr{B}$ if $\mathscr{B}$ is the class of all discs of radius 2 lying inside and tangent to that circle?

3.6. Let $\mathscr{A}$ be a class of subsets of E. For each subset B of E let $B \wedge \mathscr{A}$ be the class of all subsets of E of the form $B \cap A$ where $A \in \mathscr{A}$, and let $B \vee \mathscr{A}$ be the class of all subsets of E of the form $B \cup A$ where $A \in \mathscr{A}$. Let $\mathscr{A}'$ be the class of all subsets of E of the form A^c where $A \in \mathscr{A}$. Prove the following statements:

(a) $B \cap (\cup \mathscr{A}) = \cup(B \wedge \mathscr{A})$.
(b) $B \cup (\cap \mathscr{A}) = \cap(B \vee \mathscr{A})$.
(c) $\cup \mathscr{A}' = (\cap \mathscr{A})^c$.
(d) $\cap \mathscr{A}' = (\cup \mathscr{A})^c$.
(e) If $B \subseteq A$ for all $A \in \mathscr{A}$, then $B \subseteq \cap \mathscr{A}$.
(f) If $B \supseteq A$ for all $A \in \mathscr{A}$, then $B \supseteq \cup \mathscr{A}$.

3.7. If $\mathscr{A}$ is the empty class of subsets of E, what is $\cup \mathscr{A}$? $\cap \mathscr{A}$?

3.8. If D and E are the domains of functions f and g respectively and if $D \cap E = \emptyset$, then $\mathrm{Gr}(f) \cup \mathrm{Gr}(g)$ is the graph of a function whose domain is $D \cup E$.

3.9. Complete the analysis of the Muskogee report.

3.10. Four hundred students in a class of 800 are studying either French, German, or Russian. No student studies all three languages, but 11 are studying both French and Russian. Of the 242 students studying French, 211 are studying no other foreign language. Sixty-eight students study Russian. How many study German only? either Russian or German? French or German but not Russian? Of the 800 students, how many study Russian only if they study French? Russian if they study French? Russian if and only if they study both French and German?

*3.11. Special seminars concerning the life, work, and times of a single prominent man are offered by various departments of a certain university. One year seminars are offered on Kant, Pope, Bach, and Molière respectively by the philosophy, English, music, and French departments. A total of 110 students is enrolled in these seminars. Thirty-seven are enrolled in the Kant seminar; of these, 21 are taking no other seminar, but 7 are taking the Molière seminar, 8 the Pope seminar, and 6 the Bach seminar in addition. Thirty-seven are enrolled in the Pope seminar; of these, 20 are taking no other seminar, but 6 are taking the Molière seminar and 10 the Bach seminar in addition. Thirty-nine are enrolled in the Bach seminar; of these, 22 are taking no other seminar, but 7 are taking the Molière seminar in addition. Thirty-three are enrolled in the Molière seminar, of whom 18 are taking no other seminar.

(a) How many are taking all four seminars? How many the Kant and Bach seminars only? How many are taking the Molière and Bach seminars but not the Pope seminar? Denote the class of those taking the Kant, Pope,

Bach, and Molière seminar respectively by K, P, B, and M. Insert appropriate numbers on a Venn diagram for four sets (a model is given in Figure 6) from which these answers may easily be obtained. [First find $n(K \cap P \cap B)$, $n(K \cap P \cap M)$, $n(K \cap B \cap M)$, and $n(P \cap B \cap M)$; then find $n(K \cap P \cap B \cap M)$.]

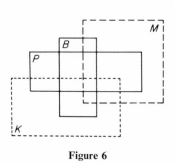

Figure 6

(b) By an error the number of students enrolled in the Pope seminar is recorded as 38 instead of 37, but the other numbers given above are recorded correctly. Show that the thus altered figures are inconsistent, and hence that the existence of the error can be detected internally from the figures themselves.

4. Neutral Elements and Inverses

The integers 0 and 1 play similar roles for addition and multiplication: 0 added to any number yields that number, and 1 multiplied by any number yields that number. Such elements are called neutral elements:

Definition. Let $\triangle$ be a composition on E. An element $e \in E$ is a **neutral** element (or **identity** element or **unity** element) for $\triangle$ if

$$x \triangle e = x,$$

$$e \triangle x = x$$

for all $x \in E$.

Theorem 4.1. There exists at most one neutral element for a composition $\triangle$ on E.

Proof. Suppose that e and e' are neutral elements. As e is a neutral element,

$$e \triangle e' = e',$$

and as e' is a neutral element,

$$e \triangle e' = e.$$

Hence, $e' = e$.

Therefore, if there is a neutral element e for $\triangle$, we may use the definite article and call e *the* neutral element for $\triangle$. If a composition denoted by a symbol similar to $+$ admits a neutral element, that element is often also called the **zero** element and is usually denoted by 0, so that

$$x + 0 = x = 0 + x$$

for all x. A neutral element for a composition denoted by a symbol similar to $\cdot$ is usually called an **identity** element and is often denoted by 1, so that

$$1x = x = x1$$

for all x (as in elementary algebra, if a composition is denoted by $\cdot$, we often write simply xy for $x \cdot y$). Unless otherwise indicated, *if there is a neutral element for a composition denoted by $\triangle$, we shall denote it by e.*

Addition and multiplication on each of the sets N, Z, Q, R, and C have neutral elements 0 and 1 respectively. Zero is the neutral element for addition modulo m, and if $m > 1$, 1 is the neutral element for multiplication modulo m. The composition $\circ$ defined in Example 2.5 on the set of all symmetries of the square admits r_0, the symmetry leaving each point of the square fixed, as neutral element. On the set $\mathfrak{P}(E)$ of all subsets of E, $\emptyset$ is the neutral element for $\cup$ and E is the neutral element for $\cap$.

If x is a nonzero real number, $-x$ and x^{-1} (or $1/x$) play similar roles for addition and multiplication, for $-x$ added to x yields the neutral element 0 for addition, and x^{-1} multiplied by x yields the neutral element 1 for multiplication. For this reason, $-x$ and x^{-1} are called inverses of x for addition and multiplication respectively:

Definition. Let $\triangle$ be a composition on E. An element $x \in E$ is **invertible** for $\triangle$ if there is a neutral element e for $\triangle$ and if there exists $y \in E$ such that

$$x \triangle y = e = y \triangle x.$$

An element y satisfying $x \triangle y = e = y \triangle x$ is called an **inverse** of x for $\triangle$.

Theorem 4.2. If $\triangle$ is an associative composition on E, an element $x \in E$ admits at most one inverse for $\triangle$.

Proof. If

$$x \triangle y = e = y \triangle x,$$

$$x \triangle z = e = z \triangle x,$$

then

$$y = y \triangle e = y \triangle (x \triangle z)$$
$$= (y \triangle x) \triangle z = e \triangle z$$
$$= z.$$

Therefore, if $\triangle$ is associative and if x is invertible for $\triangle$, we may use the definite article and speak of *the* inverse of x. The inverse of an element x invertible for an associative composition denoted by a symbol similar to $+$ is denoted by $-x$, so that by definition,

$$x + (-x) = 0 = (-x) + x.$$

The inverse of an element x invertible for an associative composition denoted by a symbol similar to $\cdot$ is often denoted by x^{-1}, so that by definition,

$$xx^{-1} = 1 = x^{-1}x.$$

Unless otherwise indicated, *if x is invertible for an associative composition denoted by $\triangle$, we shall denote its inverse by x^*.*

If E is either Q, R, or C, every nonzero element of E is invertible for multiplication on E; however, 1 and -1 are the only integers invertible for multiplication on Z. By inspection of the tables, every element of N_6 is invertible for addition modulo 6, but only 1 and 5 are invertible for multiplication modulo 6. Every symmetry of the square is invertible for the composition defined in Example 2.5; for example, $r_1^{-1} = r_3$ and $d_1^{-1} = d_1$.

The conclusion of Theorem 4.2 need not hold for nonassociative compositions. For example, if $\triangle$ is the composition on N_3 defined by the following table, then 0 is the neutral element, and both 1 and 2 are inverses of 1 (and also of 2) for $\triangle$.

$\triangle$	0	1	2
0	0	1	2
1	1	0	0
2	2	0	0

Theorem 4.3. If y is an inverse of x for a composition $\triangle$ on E, then x is an inverse of y for $\triangle$. Thus, an inverse of an invertible element is itself invertible. In particular, if x is invertible for an associative composition $\triangle$, then x^* is invertible, and

$$x^{**} = x.$$

Proof. The first two assertions follow at once from the equalities

$$x \triangle y = e = y \triangle x.$$

In particular, if x is invertible for an associative composition $\triangle$, then x^* is invertible, and the inverse of x^*, which is unique by Theorem 4.2, is x; thus, $x^{**} = x$.

If $+$ is an associative composition and if x is invertible for $+$, then by Theorem 4.3, $-x$ is invertible and

$$-(-x) = x.$$

Similarly, if $\cdot$ is an associative composition and if x is invertible for $\cdot$, then by Theorem 4.3, x^{-1} is invertible and

$$(x^{-1})^{-1} = x.$$

Theorem 4.4. If x and y are invertible elements for an associative composition $\triangle$ on E, then $x \triangle y$ is invertible for $\triangle$, and

$$(x \triangle y)^* = y^* \triangle x^*.$$

Proof. We have

$$\begin{aligned}
(x \triangle y) \triangle (y^* \triangle x^*) &= ((x \triangle y) \triangle y^*) \triangle x^* \\
&= (x \triangle (y \triangle y^*)) \triangle x^* \\
&= (x \triangle e) \triangle x^* \\
&= x \triangle x^* = e
\end{aligned}$$

and similarly

$$\begin{aligned}
(y^* \triangle x^*) \triangle (x \triangle y) &= ((y^* \triangle x^*) \triangle x) \triangle y \\
&= (y^* \triangle (x^* \triangle x)) \triangle y \\
&= (y^* \triangle e) \triangle y \\
&= y^* \triangle y = e.
\end{aligned}$$

Thus, if x and y are invertible for an associative composition $+$, then so is $x + y$, and

$$-(x + y) = (-y) + (-x).$$

If x and y are invertible for an associative composition $\cdot$, then so is xy, and

$$(xy)^{-1} = y^{-1}x^{-1}.$$

Of course, if $\cdot$ is commutative, then we also have $(xy)^{-1} = x^{-1}y^{-1}$, but in the contrary case, $(xy)^{-1}$ need not be $x^{-1}y^{-1}$. In Example 2.5, for instance,

$$(r_1 \circ h)^{-1} = d_1^{-1} = d_2,$$

but

$$r_1^{-1} \circ h^{-1} = r_3 \circ h = d_1.$$

Similarly, to undo the result of putting on first a sweater and then a coat, one does not first remove the sweater and then the coat, but rather one removes first the coat and then the sweater.

The conclusion of Theorem 4.4 need not hold for a nonassociative composition. For example, if $\triangle$ is the composition on N_3 given by the table below, then 0 is the neutral element for $\triangle$, and each element x admits a unique inverse x^*, but $(1 \triangle 1)^* \neq 1^* \triangle 1^*$.

$\triangle$	0	1	2
0	0	1	2
1	1	1	0
2	2	0	1

Theorem 4.5. Let $\triangle$ be an associative composition on E, and let x, y, and z be elements of E.

1° If both x and y commute with z, then $x \triangle y$ also commutes with z.
2° If x commutes with y and if y is invertible, then x commutes with y^*.
3° If x commutes with y and if both x and y are invertible, then x^* commutes with y^*.

Proof. If $x \triangle z = z \triangle x$ and $y \triangle z = z \triangle y$, then

$$(x \triangle y) \triangle z = x \triangle (y \triangle z) = x \triangle (z \triangle y)$$
$$= (x \triangle z) \triangle y = (z \triangle x) \triangle y$$
$$= z \triangle (x \triangle y).$$

If $x \triangle y = y \triangle x$ and if y is invertible, then

$$y^* \triangle x = y^* \triangle (x \triangle (y \triangle y^*)) = y^* \triangle ((x \triangle y) \triangle y^*)$$
$$= y^* \triangle ((y \triangle x) \triangle y^*) = (y^* \triangle (y \triangle x)) \triangle y^*$$
$$= ((y^* \triangle y) \triangle x) \triangle y^* = x \triangle y^*.$$

Finally, if both x and y are invertible and if $x \triangle y = y \triangle x$, then

$$x^* \triangle y^* = (y \triangle x)^* = (x \triangle y)^* = y^* \triangle x^*$$

by Theorem 4.4.

Now that we have begun to present some theorems of algebra, the reader may ask why certain theorems are chosen for presentation in the text while others are, presumably, omitted. This is a perfectly reasonable question, and it will recur frequently. The answer, often, is simply that they are chosen because of their usefulness in later developments—the discussion of axioms for the natural numbers and the construction of the integers, the rationals, the real and complex numbers, for example, to mention just a few of the basic concerns of algebra—that the beginner in algebra cannot know about at this point. Consequently, the reader should hold these questions in abeyance, since the answers will be easier to give later. Rereading earlier sections after later chapters have been studied is often rewarding, because the reader will then be able to add to his logical understanding of the statements and proofs of theorems an insight into the reasons they were chosen for presentation.

EXERCISES

4.1. Rewrite the statement and proof of Theorem 4.4 if the composition is
   (a) denoted by $+$ rather than by $\triangle$,
   (b) denoted by $\cdot$ rather than by $\triangle$.

4.2. Rewrite the statement and proof of Theorem 4.5 if the composition is
   (a) denoted by $+$ rather than by $\triangle$,
   (b) denoted by $\cdot$ rather than by $\triangle$.

4.3. (a) Let M be the set of married men now alive, W the set of married women now alive. Is the statement "For every $w \in W$ there exists $m \in M$ such that m is the husband of w" equivalent to the statement "There exists $m \in M$ such that for every $w \in W$, m is the husband of w"?
   (b) Is the statement "For every $x \in E$ there exists $e \in E$ such that $e \triangle x = x = x \triangle e$" equivalent to the statement "There exists a neutral element for $\triangle$"?
   (c) Let E be a set containing more than one element. Prove that for every $x \in E$ there exists $e \in E$ such that $e \leftarrow x = x = x \leftarrow e$. Is there a neutral element for $\leftarrow$?

4.4. Let E be a finite set of n elements. If $a \in E$, for how many compositions on E is a the neutral element? Of these, how many are commutative? How many compositions on E admit a neutral element? Of these, how many are commutative? How many compositions on E admit no neutral element?

4.5. An element $e \in E$ is a **left neutral element** for a composition $\triangle$ on E if

$$e \triangle x = x$$

for all $x \in E$, and e is a **right neutral element** for $\triangle$ if

$$x \triangle e = x$$

for all $x \in E$.

(a) If there exist a left neutral element and a right neutral element for $\triangle$, then there exists a neutral element e for $\triangle$, and furthermore, e is the only left neutral element and the only right neutral element for $\triangle$.

(b) If E is a set containing more than one element, which elements of E are left neutral elements for $\leftarrow$? right neutral elements for $\leftarrow$? left neutral elements for $\rightarrow$? right neutral elements for $\rightarrow$?

4.6. Which subsets of E are invertible elements of $\mathfrak{P}(E)$ for $\cup$? for $\cap$?

4.7. For each $x \in N_{12}$, find its inverse $-x$ for addition modulo 12 and, if it is invertible for multiplication modulo 12, find its inverse for that composition.

4.8. In Exercise 2.2, which tables determine compositions admitting a neutral element? Of these compositions, for which ones is exactly one element invertible? For which are both elements invertible? Are there two tables in the same class determining compositions for one and only one of which is there a neutral element? is exactly one element invertible? are both elements invertible?

4.9. Let $E = \{e, a, b\}$ be a set having three elements. Write down tables for all commutative compositions on E for which e is the neutral element (let the rows and columns be headed by e, a, and b in that order). For each table, determine whether the composition defined is associative, and determine which elements are invertible. Can you divide the tables into classes by a principle similar to that of Exercise 2.2, so that tables belonging to the same class define compositions "just like" each other?

4.10. Determine which of the compositions defined in Exercises 2.12 and 2.13 admit a neutral element. For each such composition, exhibit the neutral element and determine which elements are invertible.

*4.11. Let $\triangle$ be a composition on E for which there is a neutral element e. An element y is a **left inverse** of x for $\triangle$ if

$$y \triangle x = e,$$

and z is a **right inverse** of x for $\triangle$ if

$$x \triangle z = e.$$

(a) If $\triangle$ is associative and if x has both a left inverse and a right inverse for $\triangle$, then x has an inverse x^* for $\triangle$, and furthermore, x^* is the only left inverse and the only right inverse of x for $\triangle$.

(b) If $\triangle$ is associative, if $x \triangle y$ has a left inverse for $\triangle$, and if $y \triangle x$ has a right inverse for $\triangle$, then x and y are invertible for $\triangle$.

(c) Let $\triangle$ be the composition on R defined by

$$x \triangle y = x + y + x^2 y.$$

Prove that $\triangle$ admits a neutral element and that every real number has a unique right inverse for $\triangle$, but that there exist numbers that have no left inverse for $\triangle$. Show also that the neutral element is the only invertible element.

*4.12. If $\triangle$ is an associative composition on E and if there is an element u of E such that for every $a \in E$ there exist $x, y \in E$ satisfying

$$u \triangle x = a = y \triangle u,$$

then there is a neutral element for $\triangle$.

5. Composites and Inverses of Functions

Let f be a function from E into F, g a function from F into G. In particular, for each $x \in E$, g associates to $f(x)$ the element $g(f(x))$ of G. We introduce the following notation for the rule which, roughly speaking, consists of applying first f to each element of E and then g to the resulting element of F.

Definition. Let f be a function from E into F, and let g be a function from F into G. The **composite** of g and f is the function $g \circ f$ with domain E and codomain G defined by

$$(g \circ f)(x) = g(f(x))$$

for all $x \in E$.

Note that $g \circ f$ *is defined only if the codomain of f is the domain of g.* Note also that $g \circ f$ is the function obtained *first* by applying f to elements of E and *then* g to the result; in this sense, we read the notation for the composite of two functions from right to left. For example, if f and g are the functions from R into R defined by

$$f(x) = 2x,$$

$$g(x) = \sin x$$

for all $x \in R$, then

$$(f \circ g)(x) = 2 \sin x,$$

$$(g \circ f)(x) = \sin 2x$$

for all $x \in R$.

Let $f: E \to F$, $g: F \to G$, and $h: G \to H$ be three functions. The composite functions $g \circ f$ and $h \circ g$ are then defined, as are $h \circ (g \circ f)$ and $(h \circ g) \circ f$. Taking apart the meaning of the composite of two functions, we see that both $h \circ (g \circ f)$ and $(h \circ g) \circ f$ are, roughly speaking, the function obtained by first applying f to each element x of E, then applying g to the resulting element $f(x)$ of F, and finally applying h to the resulting element $g(f(x))$ of G. Consequently, we may conjecture that $(h \circ g) \circ f$ and $h \circ (g \circ f)$ are really the same function, namely, the function that associates to each $x \in E$ the element $h(g(f(x)))$ of H. The following theorem is a formal verification of this conjecture.

Theorem 5.1. If f is a function from E into F, if g is a function from F into G, and if h is a function from G into H, then

$$(h \circ g) \circ f = h \circ (g \circ f).$$

Proof. The domain of both $(h \circ g) \circ f$ and $h \circ (g \circ f)$ is the domain E of f, and the codomain of both functions is the codomain H of h. For every $x \in E$,

$$[(h \circ g) \circ f](x) = (h \circ g)(f(x)) = h(g(f(x)))$$
$$= h((g \circ f)(x)) = [h \circ (g \circ f)](x).$$

Hence, $(h \circ g) \circ f = h \circ (g \circ f)$.

Figure 7 is a pictorial representation of Theorem 5.1; all ways indicated of going from E to H are identical.

In view of Theorem 5.1, we shall write simply $h \circ g \circ f$ for $(h \circ g) \circ f$.

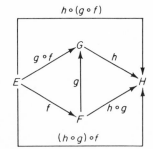

Figure 7

Definition. For any set E, the **identity function** on E is the function 1_E from E into E defined by

$$1_E(x) = x$$

for all $x \in E$.

The graph of 1_E is the set of all ordered pairs (x, x) where $x \in E$; this set is also called the **diagonal subset** of $E \times E$.

If f is a function from E into F, clearly

$$f \circ 1_E = f,$$
$$1_F \circ f = f.$$

If f is a function, there may exist elements u, x of its domain such that $u \neq x$ but $f(u) = f(x)$. For example, if $f: x \mapsto x^2$ from R into R, $f(2) = f(-2)$. Functions for which this does not happen, that is, functions f for which $u \neq x$ implies that $f(u) \neq f(x)$, are sufficiently important to warrant a special name:

Definition. A function f from E into F is an **injection** if the following condition holds:

(Inj)   For all u, $x \in E$, if $u \neq x$, then $f(u) \neq f(x)$.

A function f is **injective** or **one-to-one** if f is an injection.

An equivalent formulation of (Inj) is the following: For all u, $x \in E$, if $f(u) = f(x)$, then $u = x$.

Definition. A function f from E into F is a **surjection** if for every element y in the codomain F of f there exists at least one element x in the domain E of f such that $y = f(x)$. The function f is **surjective** or a function **onto** F if f is a surjection.

For example, the function $f: x \mapsto x^2$ from R into R is not a surjection, but the function $g: x \mapsto x^2$ from R into the set R_+ of positive numbers is a surjection.

Definition. A function f is a **bijection** or a **one-to-one onto** function if f is both an injection and a surjection. A function if **bijective** if it is a bijection. A **permutation** of a set E is a bijection whose domain and codomain are both E.

Thus, $f: E \to F$ is bijective if and only if for each $y \in F$ there is exactly one element $x \in E$ such that $y = f(x)$. Consequently, if f is a bijection from E onto F, we may define a function $f^{\leftarrow}$ from F into E by associating to each $y \in F$ the unique $x \in E$ such that $f(x) = y$.

Definition. Let f be a bijection with domain E and codomain F. The **inverse** of f is the function $f^{\leftarrow}$ with domain F and codomain E that associates to each $y \in F$ the unique $x \in E$ such that $f(x) = y$.

Thus, if f is a bijection from E onto F, then

$$y = f(x)$$

if and only if

$$f^{\leftarrow}(y) = x.$$

Customarily, the inverse of a bijection is denoted by f^{-1} rather than f. That notation, however, may lead to ambiguities in certain contexts, and consequently, we shall avoid denoting the inverse of a bijection f by f^{-1}. The following table gives some familiar examples of bijections and their inverses.

Domain f	Codomain f	$f(x) =$	$f^{\leftarrow}(y) =$
$x \geq 0$	$y \geq 3$	$2x + 3$	$\frac{1}{2}(y - 3)$
$1 \leq x \leq 3$	$1 \leq y \leq 9$	x^2	$\sqrt{y}$
$1 \leq x \leq 2$	$\frac{1}{2} \leq y \leq 1$	x^{-1}	y^{-1}
$0 \leq x \leq \pi/2$	$0 \leq y \leq 1$	$\sin x$	$\arcsin y$
$x \leq 0$	$0 < y \leq 1$	10^x	$\log_{10} y$

Often properties of a function are deduced from properties of its composite with another function. The following theorem is an example of this.

Theorem 5.2. Let f be a function from E into F, and let g be a function from F into G.

1° If $g \circ f$ is injective, then f is injective.
2° If $g \circ f$ is surjective, then g is surjective.

Proof. If $g \circ f$ is injective and if $f(u) = f(x)$, then

$$(g \circ f)(u) = g(f(u)) = g(f(x)) = (g \circ f)(x),$$

so $u = x$. If $g \circ f$ is surjective and if $z \in G$, then there exists $x \in E$ such that $(g \circ f)(x) = z$, so if $y = f(x)$, we have

$$g(y) = g(f(x)) = (g \circ f)(x) = z.$$

The following theorem presents a very important criterion for bijectivity.

Theorem 5.3. Let f be a function from E into F. If there exist functions g and h from F into E such that

$$g \circ f = 1_E,$$
$$f \circ h = 1_F,$$

then f is a bijection from E onto F and

$$g = h = f^{\leftarrow}.$$

Proof. Since 1_E is injective, f is injective by $1°$ of Theorem 5.2, and since 1_F is surjective, f is surjective by $2°$ of Theorem 5.2. Thus, f is a bijection from E onto F.

Let $y \in F$, and let $x = f^{\leftarrow}(y)$. Then $y = f(x)$, so

$$f^{\leftarrow}(y) = x = 1_E(x) = g(f(x)) = g(y),$$

and also as

$$f(x) = y = 1_F(y) = f(h(y)),$$

we have

$$f^{\leftarrow}(y) = x = h(y)$$

as f is injective. Thus, $g = h = f^{\leftarrow}$.

Theorem 5.4. If f is a bijection from E onto F, then $f^{\leftarrow}$ is a bijection from F onto E,

$$f^{\leftarrow} \circ f = 1_E,$$
$$\circ f^{\leftarrow} = 1_F,$$

and

$$f^{\leftarrow\leftarrow} = f.$$

Proof. If $x \in E$ and if $y = f(x)$, then $x = f^{\leftarrow}(y)$, so

$$(f^{\leftarrow} \circ f)(x) = f^{\leftarrow}(f(x)) = f^{\leftarrow}(y) = x.$$

Thus, $f^{\leftarrow} \circ f = 1_E$. If $y \in F$ and if $x = f^{\leftarrow}(y)$, then $y = f(x)$, so

$$(f \circ f^{\leftarrow})(y) = f(f^{\leftarrow}(y)) = f(x) = y.$$

Thus, $f \circ f^{\leftarrow} = 1_F$. From these equalities and Theorem 5.3, applied to $f^{\leftarrow}$, we conclude that $f^{\leftarrow}$ is a bijection from F onto E and that $f^{\leftarrow\leftarrow} = f$.

The inverse of a bijection f may be thought of as the rule which "undoes" the effect of f. If $f: E \to F$ and $g: F \to G$ are bijections, then $g \circ f$ is the function obtained by applying first f to elements of E and then g to the resulting elements of F; clearly, therefore, to undo the effect of $g \circ f$, one should first undo the effect of g by applying $g^{\leftarrow}$ and then undo the effect of f by applying $f^{\leftarrow}$. Consequently, it is reasonable to guess that $(g \circ f)^{\leftarrow} = f^{\leftarrow} \circ g^{\leftarrow}$, a conjecture that we formally verify in the following theorem.

Theorem 5.5. If f is a bijection from E onto F and if g is a bijection from F onto G, then $g \circ f$ is a bijection from E onto G, and

$$(g \circ f)^{\leftarrow} = f^{\leftarrow} \circ g^{\leftarrow}.$$

Proof. If u, $x \in E$ and if $u \neq x$, then $f(u) \neq f(x)$ as f is injective, whence $(g \circ f)(u) = g(f(u)) \neq g(f(x)) = (g \circ f)(x)$ as g is injective. Thus, $g \circ f$ is injective. If $z \in G$, there exists $y \in F$ such that $g(y) = z$ as g is surjective, and there exists $x \in E$ such that $f(x) = y$ as f is surjective, whence $(g \circ f)(x) = g(f(x)) = g(y) = z$. Thus, $g \circ f$ is surjective. The domain of both $(g \circ f)^{\leftarrow}$ and $f^{\leftarrow} \circ g^{\leftarrow}$ is G, and the codomain of both functions is E. To show that they are the same function, therefore, we need to show that for each $z \in G$, $(f^{\leftarrow} \circ g^{\leftarrow})(z) = (g \circ f)^{\leftarrow}(z)$. Let $y = g^{\leftarrow}(z)$, and let $x = f^{\leftarrow}(y)$. Then

$$(f^{\leftarrow} \circ g^{\leftarrow})(z) = f^{\leftarrow}(g^{\leftarrow}(z)) = f^{\leftarrow}(y) = x,$$

and as $f(x) = y$ and $g(y) = z$, $(g \circ f)(x) = g(f(x)) = g(y) = z$, whence

$$(g \circ f)^{\leftarrow}(z) = x = (f^{\leftarrow} \circ g^{\leftarrow})(z).$$

Figure 8 is a pictorial representation of Theorem 5.5; all ways indicated of going from G to E are identical.

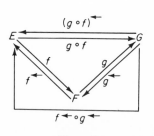

Figure 8

Let us interpret these results for the special case where $F = E$. If f and g are functions from E into E, then $g \circ f$ is also a function from E into E, so

$$(g, f) \mapsto g \circ f$$

is a composition on E^E, which we shall denote by $\circ$. By Theorem 5.1, $\circ$ is associative, and the identity function 1_E is the neutral element for $\circ$. By Theorems 5.3 and 5.4, a function f from E into E is invertible for $\circ$ if and only if f is a permutation of E, in which case the inverse of f for $\circ$ is $f^{\leftarrow}$. The final assertion of Theorem 5.4 and the assertion of Theorem 5.5 for the case where $F = E$ are then respectively the assertions of Theorems 4.3 and 4.4 for the associative composition $\circ$ on E^E.

EXERCISES

5.1. Let f, g, and h be the functions from R into R defined by

$$f(x) = 2x - 3,$$
$$g(x) = 7 \sin 5x,$$
$$h(x) = 10^{x^2}.$$

Write expressions giving the value at x for each of the following functions:

$$
\begin{array}{lllll}
g \circ f & h \circ g & f \circ f & f \circ g \circ h & g \circ h \circ f \\
f \circ g & f \circ h & g \circ g & g \circ f \circ h & f \circ h \circ g \\
g \circ h & h \circ f & h \circ h & h \circ g \circ f & h \circ f \circ g.
\end{array}
$$

5.2. Determine the codomain of f and write an expression for $f^{\leftarrow}(y)$, if f is a bijection such that $f(x) =$

(a) $5x + 7$, $x \in R$.

(b) $\dfrac{3x - 4}{2x - 6}$, $x \neq 3$.

(c) $-x^2 + 2$, $x \geq 0$.

(d) $3 \cos 2x$, $0 \leq x \leq \dfrac{\pi}{2}$.

(e) $2 \log_5(x + 3)$, $x > -3$.

(f) 7^{3x+5}, $x \in R$.

(g) $x^4 + 5x^2 + 3$, $x \geq 0$.

(h) $\arcsin(\tan x)$, $\dfrac{-\pi}{4} \leq x \leq \dfrac{\pi}{4}$.

(i) $\arccos(\cos x)$, $3\pi \leq x \leq 4\pi$.

(j) $\cos(\pi^x)$, $x \leq 1$.

5.3. Let $J = \{x \in R : 0 \leq x \leq 1\}$, let $\mathscr{C}$ be the set of all continuous functions from J into R, and let $\mathscr{C}'_0$ be the set of all differentiable functions v from J into R such that $v(0) = 0$ and the derivative Dv of v is continuous.

(a) Let S be the function from $\mathscr{C}$ into $\mathscr{C}'_0$ defined by

$$[S(f)](x) = \int_0^x f(t)\, dt$$

for all $x \in J$ and all $f \in \mathscr{C}$. Cite theorems from calculus to show that S is a bijection from $\mathscr{C}$ onto $\mathscr{C}'_0$, and give an expression for $S^{\leftarrow}$.

(b) Cite theorems from calculus to show that the function $D : v \mapsto Dv$ is a bijection from $\mathscr{C}'_0$ onto $\mathscr{C}$, and give an expression for $D^{\leftarrow}$.

*5.4. If m and n are natural numbers, $\dbinom{n}{m}$ is defined to be the number of subsets having m elements of a set having n elements. Determine $\dbinom{n}{m}$ if $0 \leq m \leq n$.

$$\left[\text{Obtain the answer first for } m = 1, 2, \text{ and } 3, \text{ and then determine a formula} \right.$$

for $\binom{n}{m}$ in terms of $\left. \binom{n}{m-1}. \right]$ Express your answer as a quotient whose numerator is a factorial and whose denominator is a product of two factorials.

5.5. How many injections are there from a set having m elements into a set having n elements if $m = n$? if $m > n$? if $m < n$? [Use Exercise 5.4.]

5.6. If E is a nonempty set and if f is an injection from E into F, then there is a function g from F into E such that $g \circ f = 1_E$. [Use Exercise 3.8.]

5.7. (a) If f is the function from N into N defined by

$$f(n) = n + 1,$$

then f has infinitely many left inverses (Exercise 4.11) for the composition $\circ$ on N^N.

(b) If g is the function from N into N defined by

$$g(n) = \begin{cases} \dfrac{n}{2} & \text{if } n \text{ is even,} \\[2mm] \dfrac{n-1}{2} & \text{if } n \text{ is odd,} \end{cases}$$

then g has infinitely many right inverses (Exercise 4.11) for the composition $\circ$ on N^N.

5.8. Let E, F, G, and H be sets. For each subset A of $E \times F$ and each subset B of $F \times G$, the **composite** of B and A is the set $B \circ A$ defined by

$$B \circ A = \{(x, z) \in E \times G : \text{for some } y \in F, (x, y) \in A \text{ and } (y, z) \in B\},$$

and the set $A^{\leftarrow}$ is defined by

$$A^{\leftarrow} = \{(y, x) \in F \times E : (x, y) \in A\}.$$

(a) If $f: E \to F$ and if $g: F \to G$, then $\mathrm{Gr}(g) \circ \mathrm{Gr}(f) = \mathrm{Gr}(g \circ f)$.

(b) If A, B, and C are subsets respectively of $E \times F$, $F \times G$, and $G \times H$, then $(C \circ B) \circ A = C \circ (B \circ A)$.

(c) If A and B are subsets respectively of $E \times F$ and $F \times G$, then $(B \circ A)^{\leftarrow} = A^{\leftarrow} \circ B^{\leftarrow}$.

(d) If A is a subset of $E \times F$, then $A \circ \mathrm{Gr}(1_E) = A$ and $\mathrm{Gr}(1_E) \circ A = A$.

(e) If A is a subset of $E \times F$, then $A \circ A^{\leftarrow} = \mathrm{Gr}(1_F)$ if and only if A is the graph of a surjection whose domain is a subset of E and whose codomain is F.

(f) If A is a subset of $E \times F$, then $A \circ A^{\leftarrow} = \mathrm{Gr}(1_F)$ and $A^{\leftarrow} \circ A = \mathrm{Gr}(1_E)$ if and only if A is the graph of a bijection from E onto F.

(g) The composition $(B, A) \mapsto B \circ A$ on $\mathfrak{P}(E \times E)$, which we denote by $\circ$, is associative with neutral element $\mathrm{Gr}(1_E)$. A subset A of $E \times E$ is invertible for $\circ$ if and only if A is the graph of a permutation of E. If A is invertible for $\circ$, its inverse for $\circ$ is $A^{\leftarrow}$.

6. Isomorphisms of Algebraic Structures

In mathematics, ordered pairs or ordered triads are frequently used formally to create a new entity from two or three component parts. Similarly, ordinary language often suggests the origins of something new in naming it by joining together in some fashion the names of its component parts, e.g., "bacon, lettuce, and tomato sandwich," "A.F.L.-C.I.O.," "State of Rhode Island and Providence Plantations." We shall define an algebraic structure essentially to be a nonempty set together with one or two compositions on that set:

Definition. An **algebraic structure with one composition** is an ordered pair $(E, \triangle)$ where E is a nonempty set and where $\triangle$ is a composition on E. An **algebraic structure with two compositions** is an ordered triad $(E, \triangle, \triangledown)$ where E is a nonempty set and where $\triangle$ and $\triangledown$ are compositions on E. An **algebraic structure** is simply an algebraic structure with either one or two compositions.

This definition is quite artificial in two respects. First, on the set E we consider only binary operations. In a more general definition, ternary operations (functions from $E \times E \times E$ into E), unary operations (functions from E into E), and, in general, n-ary operations for any integer $n > 0$ would be allowed. Second, we have limited ourselves to at most two compositions. A more general definition would allow any number of compositions, even infinitely many. However, algebraic structures with more than two binary operations are rarely encountered in practice.

The only reason we have for imposing these limitations is that it is convenient to do so. The reader is invited to consider a more general definition of algebraic structure and to modify correspondingly the concepts subsequently introduced for algebraic structures.

It is customary to use expressions such as "the algebraic structure E under $\triangle$," for example, instead of the more formal "the algebraic structure $(E, \triangle)$."

Two algebraic structures, though distinct, may be "just like" each other. But before we explore this concept, let us consider some nonmathematical examples of situations just like each other.

In checker game (1) of Figure 9 white is to move, and in game (2) black is to move (the initial position of the black checkers is always the lower-numbered squares).

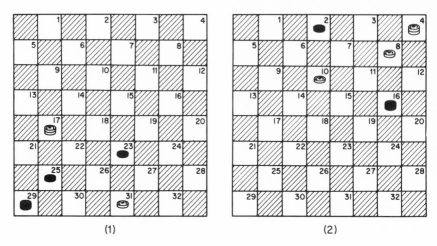

Figure 9

Strategically, situation (1) for white is just like (2) for black. A white man (respectively, white king, black man, black king) occupies square k in (1) if and only if a black man (respectively, black king, white man, white king) occupies square $33 - k$ in (2), and the move of a white piece from square k to square m in (1) has the same strategic value as the move of a black piece from square $33 - k$ to square $33 - m$ in (2). Though not identical, the two situations strategically have the same form, or to use a word whose Greek roots mean just that, they are "isomorphic" situations. A winning strategy for white in (1), for example, is the following: white moves $31 \to 26$, black must jump $23 \to 30$, white moves $17 \to 21$, black must move $30 \to 26$, white double-jumps $21 \to 30 \to 23$, black must move $29 \to 25$, white moves $23 \to 26$ and $26 \to 22$ on his next move and wins. The corresponding strategy in (2) is obtained mechanically by interchanging black and white and replacing k by $33 - k$. Thus, black moves $2 \to 7$, white must move $10 \to 3$, black moves $16 \to 12$, etc. Similarly, any situation in checkers has a strategically isomorphic counterpart, obtained by corresponding to a man (king) on square k of the given situation a man (king) of opposite color on square $33 - k$.

It is easy to describe situations in other games that are strategically just like each other. If two decks of cards are stacked so that the nth card of one deck has the same denomination and suit as the nth card of the other for $1 \leq n \leq 52$, then corresponding hands dealt from the two decks for any card game whatever will have the same strategic value, even though the backs of cards from different decks have different decorative patterns. For a less trivial example, consider any card game in which the strategic value of a card depends only on its color and denomination and not on its specific suit. If two decks are such that the nth card of one deck is a spade (respectively, heart, diamond, club) of a certain denomination if and only if the nth card of the other deck is a club (respectively, diamond, heart, spade) of the same denomination, corresponding hands dealt from the two decks will clearly have the same strategic value, regardless of whether the entire deck is dealt. Thus, the two decks are "isomorphic" for any such game, though not for the game of contract bridge, in contrast.

In each of these examples there is a bijective correspondence f between the elements of one situation and those of the other that preserves all strategic relationships [in the checkers situation, f(black king on square 29) = white king on square 4, for example, and in a card game of the type just described, f(ten of hearts) = ten of diamonds]. Similarly, we shall say that two algebraic structures are isomorphic if there is a bijective correspondence between elements of their sets that preserves all algebraic relationships:

Definition. Let $(E, \triangle)$ and $(F, \triangledown)$ be algebraic structures with one composition. An **isomorphism** from $(E, \triangle)$ onto $(F, \triangledown)$ is a bijection f from E onto F such that

$$(1) \qquad f(x \triangle y) = f(x) \triangledown f(y)$$

for all $x, y \in E$. Let $(E, \triangle, \wedge)$ and $(F, \triangledown, \vee)$ be algebraic structures with two compositions. An **isomorphism** from $(E, \triangle, \wedge)$ onto $(F, \triangledown, \vee)$ is a bijection f from E onto F such that (1) holds and similarly,

$$(2) \qquad f(x \wedge y) = f(x) \vee f(y)$$

for all $x, y \in E$. If there exists an isomorphism from one algebraic structure onto another, we shall say that they are **isomorphic** algebraic structures. An **automorphism** of an algebraic structure is an isomorphism from itself onto itself.

If E and F are finite sets having the same number of elements and if $\triangle$ and $\triangledown$ are compositions on E and F respectively, then $(E, \triangle)$ and $(F, \triangledown)$ are isomorphic if and only if $\triangle$ and $\triangledown$ have tables that are "just like" each other. More precisely, let f be a bijection from E onto F, and for notational convenience, let a' denote $f(a)$ for each $a \in E$. Let T be a table for $\triangle$, and let T' be the table for $\triangledown$ such that if the jth row and column of Table T are headed by a, then the jth row and column of Table T' are headed by a'. If a and b are elements of E, the entry in the row headed by a and the column headed by b in Table T is $a \triangle b$, and the entry in the row headed by a' and the column headed by b' in Table T' is $a' \triangledown b'$. By definition, f is an isomorphism if and only if $a' \triangledown b' = (a \triangle b)'$ for all $a, b \in E$. Consequently, *f is an isomorphism if and only if for all $a, b \in E$, the entry in the row headed by a' and the column* *headed by b' in Table T' is the image under f of the entry in the row headed by a and the column headed by b in Table T,* or more concisely, *if and only if each entry in Table T' is the image under f of the entry in the corresponding square of Table T.*

It may well happen that $(E, \triangle)$ is isomorphic to $(F, \triangledown)$ and that $(E, \wedge)$ is isomorphic to $(F, \vee)$, but that $(E, \triangle, \wedge)$ is not isomorphic to $(F, \triangledown, \vee)$ (Exercise 6.6).

Theorem 6.1. *Let $(E, \triangle)$, $(F, \triangledown)$, and $(G, \vee)$ be algebraic structures, let f be a bijection from E onto F, and let g be a bijection from F onto G.*

1° *The identity function 1_E is an automorphism of $(E, \triangle)$.*

2° *The bijection f is an isomorphism from $(E, \triangle)$ onto $(F, \triangledown)$ if and only if $f^{\leftarrow}$ is an isomorphism from $(F, \triangledown)$ onto $(E, \triangle)$.*

3° *If f is an isomorphism from $(E, \triangle)$ onto $(F, \triangledown)$ and if g is an isomorphism from $(F, \triangledown)$ onto $(G, \vee)$, then $g \circ f$ is an isomorphism from $(E, \triangle)$ onto $(G, \vee)$.*

Proof. We shall prove 2°. Necessity: Let $z, w \in F$. Then there exist $x, y \in E$ such that $z = f(x)$ and $w = f(y)$ as f is surjective. Hence,

$$f^{\leftarrow}(z \triangledown w) = f^{\leftarrow}(f(x) \triangledown f(y)) = f^{\leftarrow}(f(x \triangle y))$$
$$= x \triangle y = f^{\leftarrow}(z) \triangle f^{\leftarrow}(w).$$

Sufficiency: If $f^{\leftarrow}$ is an isomorphism from $(F, \triangledown)$ onto $(E, \triangle)$, then by what we have just proved, $f^{\leftarrow \leftarrow}$ is an isomorphism from $(E, \triangle)$ onto $(F, \triangledown)$. But $f^{\leftarrow \leftarrow} = f$ by Theorem 5.4.

It is obvious how to formulate and prove the analogue of Theorem 6.1 for algebraic structures with two compositions. By Theorem 6.1, every algebraic structure is isomorphic to itself; if one algebraic structure is isomorphic to a second, the second is isomorphic to the first; if one algebraic structure is isomorphic to a second and if the second is isomorphic to a third, then the first is isomorphic to the third.

Two isomorphic algebraic structures possess exactly the same algebraic properties, and, indeed, one may derive the properties of one from those of the other mechanically by use of any given isomorphism. We shall illustrate this procedure in the proof of the following theorem.

Theorem 6.2. *Let f be an isomorphism from $(E, \triangle)$ onto $(F, \triangledown)$.*

1° *The composition $\triangle$ is associative if and only if $\triangledown$ is associative.*

2° *The composition $\triangle$ is commutative if and only if $\triangledown$ is commutative.*

3° An element $e \in E$ is a neutral element for $\triangle$ if and only if $f(e)$ is a neutral element for ∇.

4° An element $y \in E$ is an inverse of $x \in E$ for $\triangle$ if and only if $f(y)$ is an inverse of $f(x)$ for ∇.

Proof. To prove the condition of 1° is necessary, let u, v, $w \in F$. Then there exist x, y, $z \in E$ such that $f(x) = u$, $f(y) = v$, $f(z) = w$. Consequently, as $\triangle$ is associative,

$$
\begin{aligned}
(u\nabla v)\nabla w &= (f(x)\nabla f(y))\nabla f(z) = f(x\triangle y)\nabla f(z) \\
&= f((x\triangle y)\triangle z) = f(x\triangle (y\triangle z)) \\
&= f(x)\nabla f(y\triangle z) = f(x)\nabla (f(y)\nabla f(z)) \\
&= u\nabla (v\nabla w).
\end{aligned}
$$

Conversely, if ∇ is associative, then as $f^{\leftarrow}$ is an isomorphism from (F, ∇) onto $(E, \triangle)$ by Theorem 6.1, we conclude from what we have just proved that $\triangle$ is also associative.

We shall omit the proofs of 2° and 3°. To prove 4°, let e be the neutral element for $\triangle$, whence by 3°, $f(e)$ is the neutral element for ∇. If

$$x\triangle y = e = y\triangle x,$$

then

$$f(x)\nabla f(y) = f(x\triangle y) = f(e) = f(y\triangle x) = f(y)\nabla f(x),$$

so $f(y)$ is an inverse of $f(x)$. Conversely, if $f(y)$ is an inverse of $f(x)$ for ∇, then as $f^{\leftarrow}$ is an isomorphism from (F, ∇) onto $(E, \triangle)$, we conclude from what we have just proved that $f^{\leftarrow}(f(y))$ is an inverse of $f^{\leftarrow}(f(x))$ for $\triangle$. But by Theorem 5.4, $f^{\leftarrow}(f(y)) = y$ and $f^{\leftarrow}(f(x)) = x$.

Example 6.1. The algebraic structure $(E, \oplus, \odot)$ of Example 2.2 is isomorphic to the algebraic structure $(N_2, +_2 \cdot_2)$. Indeed, it is easy to verify that the function f from E into N_2 defined by

$$f(\text{even}) = 0,$$

$$f(\text{odd}) = 1$$

satisfies

$$f(x \oplus y) = f(x) +_2 f(y),$$

$$f(x \odot y) = f(x) \cdot_2 f(y)$$

for all $x, y \in E$. Notice that the tables for $\oplus$ and $\odot$ on page 13 are just like the tables given below for $+_2$ and $\cdot_2$ respectively.

$+_2$	0	1
0	0	1
1	1	0

$\cdot_2$	0	1
0	0	0
1	0	1

Example 6.2. Let R_n be the set of all complex nth roots of unity, that is, let

$$R_n = \{z \in \mathbf{C} \colon z^n = 1\}.$$

Thus, R_n is the set of all the complex numbers

$$\cos \frac{2\pi k}{n} + i \sin \frac{2\pi k}{n}$$

where $0 \leq k \leq n - 1$. If $z, w \in R_n$, then

$$(zw)^n = z^n w^n = 1,$$

so $zw \in R_n$. Therefore, multiplication is a composition on R_n. The bijection f from N_n onto R_n defined by

$$f(k) = \cos \frac{2\pi k}{n} + i \sin \frac{2\pi k}{n}$$

for all $k \in N_n$ is an isomorphism from the algebraic structure $(N_n, +_n)$ onto the algebraic structure $(R_n, \cdot)$. Indeed,

$$f(j)f(k) = \left(\cos \frac{2\pi j}{n} + i \sin \frac{2\pi j}{n}\right)\left(\cos \frac{2\pi k}{n} + i \sin \frac{2\pi k}{n}\right)$$

$$= \cos \frac{2\pi j}{n} \cos \frac{2\pi k}{n} - \sin \frac{2\pi j}{n} \sin \frac{2\pi k}{n}$$

$$+ i\left(\sin \frac{2\pi j}{n} \cos \frac{2\pi k}{n} + \cos \frac{2\pi j}{n} \sin \frac{2\pi k}{n}\right)$$

$$= \cos \frac{2\pi(j + k)}{n} + i \sin \frac{2\pi(j + k)}{n}.$$

Hence, if $j + k < n$, then

$$f(j +_n k) = f(j + k) = f(j)f(k),$$

and if $j + k \geq n$, then

$$f(j +_n k) = f(j + k - n)$$

$$= \cos \frac{2\pi(j + k - n)}{n} + i \sin \frac{2\pi(j + k - n)}{n}$$

$$= \cos \frac{2\pi(j + k)}{n} + i \sin \frac{2\pi(j + k)}{n}$$

$$= f(j)f(k).$$

Notice that the tables given below for multiplication on R_4, which is the set $\{1, i, -1, -i\}$, and addition modulo 4 are just like each other.

$+_4$	0	1	2	3
0	0	1	2	3
1	1	2	3	0
2	2	3	0	1
3	3	0	1	2

$\cdot$	1	i	-1	$-i$
1	1	i	-1	$-i$
i	i	-1	$-i$	1
-1	-1	$-i$	1	i
$-i$	$-i$	1	i	-1

Example 6.3. Let $\triangle$ be a composition on a set $E = \{a, b\}$ of two elements. What compositions $\triangledown$ are there on E for which $(E, \triangle)$ is isomorphic to $(E, \triangledown)$? Since the only permutations of E are the identity permutation and the permutation J defined by

$$J(a) = b,$$

$$J(b) = a,$$

there are at most two such compositions, one of which is $\triangle$ itself, of course, since 1_E is an automorphism of $(E, \triangle)$. The composition $\triangledown$ on E such that J is an isomorphism from $(E, \triangle)$ onto $(E, \triangledown)$ is easily determined: as $a \triangledown a = J(b) \triangledown J(b)$, we must have $a \triangledown a = J(b \triangle b)$; similarly, $a \triangledown b = J(b \triangle a)$, $b \triangledown a = J(a \triangle b)$, and $b \triangledown b = J(a \triangle a)$. Thus, if the first row and column of the tables for $\triangle$ and $\triangledown$ are headed by a

and the second row and column by b, then each entry in the table for $\triangledown$ is *not* the same as the diagonally opposite entry in the table for $\triangle$. Let us divide the compositions on E into classes which we shall call "isomorphic classes," so that two compositions belong to the same class if and only if they define isomorphic algebraic structures. The isomorphic class determined by $\triangle$ is thus $\{\triangle, \triangledown\}$ and hence has one member if $\triangledown = \triangle$ [equivalently, if J is an automorphism of $(E, \triangle)$] and has two members if $\triangledown \neq \triangle$ [equivalently, if J is not an automorphism of $(E, \triangle)$]. There are four compositions $\triangle$ such that $\triangledown = \triangle$; their tables are given below.

	a	b
a	a	a
b	b	b

	a	b
a	b	b
b	a	a

	a	b
a	a	b
b	a	b

	a	b
a	b	a
b	b	a

Thus, four isomorphic classes contain one member each and the remaining six isomorphic classes contain two members each. Taking one composition from each isomorphic class, we obtain a list of ten compositions, no two of which define isomorphic algebraic structures, but such that every composition on E defines an algebraic structure isomorphic with that defined by one of the compositions in our list. One sometimes expresses this by saying, "There are to within isomorphism ten compositions on E."

Example 6.4. Let E be the set of even integers. The sum of two even integers is even; thus, addition is a composition on E. The bijection $f: n \mapsto 2n$ from $\mathbf{Z}$ onto E is an isomorphism from $(\mathbf{Z}, +)$ onto $(E, +)$; indeed,

$$f(n + m) = 2(n + m) = 2n + 2m = f(n) + f(m).$$

It may seem strange that the algebraic structures $\mathbf{Z}$ and E under addition are "just like" each other in the sense of being isomorphic, because we do not usually think of the integers as being just like the even integers. The paradox is explained by the fact that we have carefully ignored multiplication in this example. When we refer to the integers or to the even integers, we usually have the algebraic structures $\mathbf{Z}$ or E under addition *and* multiplication in mind. But $(\mathbf{Z}, +, \cdot)$ and $(E, +, \cdot)$ are *not* isomorphic even though $(\mathbf{Z}, +)$ and $(E, +)$ are; indeed, $\mathbf{Z}$ has a neutral element for multiplication, but E does not.

Example 6.5. Let R_+^* be the set of strictly positive real numbers. The bijection $f: x \mapsto 10^x$ from R onto R_+^* is an isomorphism from the algebraic structure $(R, +)$ onto $(R_+^*, \cdot)$, for, by the law of exponents,

$$f(x + y) = 10^{x+y} = 10^x 10^y = f(x)f(y).$$

The isomorphism $f^\leftarrow$ from $(R_+^*, \cdot)$ onto $(R, +)$ is, of course, the logarithmic function $x \mapsto \log_{10} x$. That problems involving multiplication of strictly positive numbers can be transformed into problems involving addition by means of the logarithmic function is essentially just a restatement of the fact that $f^\leftarrow$ is an isomorphism of these two algebraic structures. Although $(R, +)$ and $(R_+^*, \cdot)$ are isomorphic, $(R, +, \cdot)$ and $(R_+^*, \cdot, +)$ are not, since R has a neutral element for multiplication but R_+^* has no neutral element for addition.

Our final theorem will assist us in making certain constructions in later chapters. Occasionally in algebra we are given an algebraic structure $(E, \triangle)$ and a bijection f from E onto a set F, and we wish to "transplant" $\triangle$ from E to F by means of f, that is, we wish to construct a composition $\triangledown$ on F such that f is an isomorphism from $(E, \triangle)$ onto $(F, \triangledown)$. This can always be done, and moreover, it can be done in only one way.

Theorem 6.3. (Transplanting Theorem) Let $(E, \triangle)$ be an algebraic structure, and let f be a bijection from E onto a set F. There is one and only one composition $\triangledown$ on F such that f is an isomorphism from $(E, \triangle)$ onto $(F, \triangledown)$, namely, the composition $\triangledown$ defined by

$$(3) \qquad u \triangledown v = f(f^\leftarrow(u) \triangle f^\leftarrow(v))$$

for all $u, v \in F$.

Proof. To show that there is at most one composition having the desired properties, we shall show that if $\triangledown$ is a composition on F such that f is an isomorphism from $(E, \triangle)$ onto $(F, \triangledown)$, then $\triangledown$ satisfies (3). Indeed, as f is an isomorphism and as $f \circ f^\leftarrow = 1_F$,

$$f(f^\leftarrow(u) \triangle f^\leftarrow(v)) = f(f^\leftarrow(u)) \triangledown f(f^\leftarrow(v)) = u \triangledown v.$$

It remains for us to show that the composition $\triangledown$ defined by (3) has the desired properties. Let $x, y \in E$, and let $u = f(x)$, $v = f(y)$. Then $x = f^\leftarrow(u)$ and $y = f^\leftarrow(v)$, so

$$f(x \triangle y) = f(f^\leftarrow(u) \triangle f^\leftarrow(v)) = u \triangledown v = f(x) \triangledown f(y).$$

Thus, f is an isomorphism from $(E, \triangle)$ onto $(F, \triangledown)$.

It is natural to call the composition $\triangledown$ defined by (3) the **transplant** of $\triangle$ under f. If $F = E$ and if f is an automorphism of $(E, \triangle)$, then the transplant of $\triangle$ under f is, of course, $\triangle$ itself.

In Example 6.3 we calculated the transplant of a composition on $\{a, b\}$ under J. The transplant $\triangledown$ of multiplication on Z under the bijection $f: n \mapsto 2n$ from Z onto the set E of even integers is given by

$$n \triangledown m = f(f^{\leftarrow}(n) \cdot f^{\leftarrow}(m)) = f\left(\frac{n}{2} \cdot \frac{m}{2}\right)$$

$$= 2 \cdot \frac{nm}{4} = \frac{1}{2} nm.$$

The transplant $\triangledown$ of multiplication under the bijection $f: x \mapsto 10^x$ from R onto the set R_+^* of strictly positive real numbers is given by

$$u \triangledown v = f(f^{\leftarrow}(u) \cdot f^{\leftarrow}(v)) = f((\log u)(\log v))$$

$$= 10^{(\log u)(\log v)} = (10^{\log u})^{\log v}$$

$$= u^{\log v}.$$

If two algebraic structures are isomorphic, it is sometimes easy to find a "natural" isomorphism between them. But if we suspect that two algebraic structures are not isomorphic, how should we go about trying to prove that they are not? If the sets involved are infinite, it may be impossible to examine all possible bijections from one onto the other, and if they are both finite with a large number of elements, it may be too arduous to show that none of the bijections is an isomorphism. The best procedure is to try to find some property of one structure not possessed by the other that would have to be preserved by any isomorphism. We have already seen, for example, that neither the even integers under multiplication nor the strictly positive real numbers under addition are isomorphic either to the integers or to the real numbers under multiplication, because the latter two algebraic structures have neutral elements while the former two do not. A more difficult problem is to prove that the nonzero real numbers R^* under multiplication and the real numbers R under addition are not isomorphic. Both compositions are associative, commutative, possess neutral elements, and all elements of either set are invertible for the composition in question, so we must look further for some algebraic property that differentiates the two. Such a property is the following: There are two elements of R^* that are their own inverses for multiplication, namely, -1 and 1, but there is only one element of R that is its own inverse for addition, namely, 0. If f were an isomorphism from $(R^*, \cdot)$ onto $(R, +)$, then

$$0 = f(1) = f((-1)(-1))$$
$$= f(-1) + f(-1) = 2f(-1),$$

whence

$$f(-1) = 0 = f(1),$$

so f would not be injective, a contradiction.

EXERCISES

If $E \subseteq C$, the set of nonzero members of E is denoted by E^*, and if $E \subseteq R$, the set of strictly positive members of E is denoted by E^*_+.

6.1. Given a deck of cards to be used to play a game in which the strategic value of a card depends on its denomination only and not on its suit (e.g., certain forms of rummy), describe $(4!)^{13}$ decks of cards that strategically have the same form as the given deck. How many of these have the same strategic form as the given deck for a game in which the strategic value of a card depends on its denomination and color, but not on its specific suit? How many decks "isomorphic" to the given one can you describe for canasta? [Special attention must be given the 3's.]

6.2. Complete the proof of Theorem 6.1.

6.3. Complete the proof of Theorem 6.2. State and prove assertions similar to those of Theorem 6.2 concerning left neutral elements, right neutral elements (Exercise 4.5), left inverses, right inverses (Exercise 4.11), and the commutativity of two elements.

6.4. For what real numbers α is the function $f : x \mapsto \alpha x$ an automorphism of $(R, +)$? For what strictly positive real numbers α is the function $g : x \mapsto \alpha^x$ an isomorphism from $(R, +)$ onto $(R^*_+, \cdot)$?

*6.5. Show that multiplication modulo 5 is a composition on N^*_5. Exhibit two isomorphisms from $(N_4, +_4)$ onto $(N^*_5, \cdot_5)$. Show that these two are the only ones.

*6.6. Let $\vee$ and $\wedge$ be the compositions on R defined by

$$x \vee y = \max\{x, y\},$$
$$x \wedge y = \min\{x, y\}.$$

(a) Exhibit an isomorphism from $(R, \vee)$ onto $(R, \wedge)$ and one from $(R, \cdot)$ onto $(R, \cdot)$.

(b) Prove that $(R, \vee, \cdot)$ is not isomorphic to $(R, \wedge, \cdot)$.

[If f were an isomorphism, show first that $x > f^{\leftarrow}(0)$ would imply that $f(x) < 0$, and then use the fact that every square is positive.]

6.7. Let $Z[\sqrt{3}] = \{m + n\sqrt{3}: m, n \in Z\}$.

(a) The sum and product of two elements of $Z[\sqrt{3}]$ belong to $Z[\sqrt{3}]$; consequently, addition and multiplication are compositions on $Z[\sqrt{3}]$.

(b) The function $f: m + n\sqrt{3} \mapsto m - n\sqrt{3}$ is an automorphism of $(Z[\sqrt{3}], +, \cdot)$.

(c) Multiplication is a composition on the set $E = \{2^m 3^n: m, n \in Z\}$. Exhibit an isomorphism from $(Z[\sqrt{3}], +)$ onto $(E, \cdot)$.

6.8. Determine the transplant of

(a) multiplication on R under the permutation $f: x \mapsto 1 - x$ of R;

(b) addition on R_+^* under the permutation $f: x \mapsto x^2$ of R_+^*;

(c) addition on R_+^* under the bijection $f: x \mapsto \log_{10} x$ from R_+^* onto R.

6.9. Prove that $(C^*, \cdot)$ is not isomorphic with $(R^*, \cdot)$, $(C, +)$, or $(R, +)$.

*6.10. Prove that $(Q, +)$ is not isomorphic with $(Q_+^*, \cdot)$. [Use Exercise 1.10.]

*6.11. (a) Every permutation of E is an automorphism of $(E, \leftarrow)$ and of $(E, \rightarrow)$.

(b) If E contains at least two elements, then $(E, \leftarrow)$ and $(E, \rightarrow)$ are not isomorphic.

(c) If E contains at least three elements and if $\triangle$ is a composition on E such that every permutation of E is an automorphism of $(E, \triangle)$, then $x \triangle x = x$ for all $x \in E$.

(d) If E contains at least four elements and if $\triangle$ is a composition on E such that every permutation of E is an automorphism of $(E, \triangle)$, then $\triangle$ is either $\leftarrow$ or $\rightarrow$. [First prove that for all $x, y \in E$, $x \triangle y$ is either x or y.]

7. Semigroups and Groups

Algebraic structures whose compositions satisfy particularly important properties are given special names.

Definition. A **semigroup** is an algebraic structure with one associative composition. A semigroup is **commutative** or **abelian** if its composition is commutative.

Definition. A **group** is a semigroup $(E, \triangle)$ such that there exists a neutral element for $\triangle$ and every element of E is invertible for $\triangle$.

If $(E, \triangle)$ (or $(E, \triangle, \triangledown)$) is a specially named algebraic structure, it is customary to call the set E itself by that name, when it is understood or clear from the context what the composition in question is. The convenience resulting from this convention outweighs the fact that it is, of course, an abuse of language. For example, if $(E, \triangle)$ is a group, the set E itself is often called a group when it is clearly understood that $\triangle$ is the composition under consideration; if we speak of an element or a subset of the group $(E, \triangle)$, we mean an element or a subset of E. Thus, in suitable contexts, the words "semigroup" and "group" refer not only to certain algebraic structures but also to the sets from which they are formed.

Example 7.1. Under addition, Z, Q, R, and C are abelian groups. Under addition, N is an abelian semigroup but is not a group. Under addition modulo m, N_m is an abelian group, for if $n \in N_m$ and if $n \neq 0$, then $m - n$ is the inverse of n for addition modulo m, and 0 is, of course, its own inverse for addition modulo m.

Example 7.2. If $E \subseteq C$, we shall denote by E^* the set of all nonzero elements of E. None of N, Z, Q, C, and C is a group under multiplication, since 0 is not invertible for multiplication. However, Q^*, R^*, and C^* are abelian groups under multiplication. Under multiplication, N^* and Z^* are abelian semigroups but are not groups.

Example 7.3. The set of all symmetries of a square is a nonabelian group under the composition defined in Example 2.5. This group is also called the **group of symmetries of the square,** the **octic group,** or the **dihedral group of order 8.** The set of symmetries of any geometrical figure forms a group in a similar way. If the figure is a regular polygon of n sides, the resulting group has $2n$ elements, as we shall see in §25.

Example 7.4. The **symmetric difference** of two subsets X and Y of E is the set $X \triangle Y$ defined by

$$X \triangle Y = (X - Y) \cup (Y - X).$$

Thus, $\triangle$ is a composition on $\mathfrak{P}(E)$, and one may show that $(\mathfrak{P}(E), \triangle)$ is an abelian group (Exercise 7.1).

Example 7.5. The set of all permutations of a set E is often denoted by $\mathfrak{S}_E$. By Theorem 5.5, the composite of two permutations of E is a permutation of E; thus,

$$(g, f) \mapsto g \circ f$$

is a composition on $\mathfrak{S}_E$, which we shall denote by $\circ$. The identity function 1_E is a permutation of E and clearly is the neutral element for $\circ$. By Theorems 5.1 and 5.5, therefore, $(\mathfrak{S}_E, \circ)$ is a group, called the **symmetric group** on E. If $E = \{1, 2, \ldots, m\}$, it is customary to write $\mathfrak{S}_m$ for $\mathfrak{S}_E$. The group $\mathfrak{S}_m$ is called the symmetric group on m objects (or on m letters). If E has n elements, then $\mathfrak{S}_E$ contains $n!$ permutations.

Here is an informal proof. If $a_1, \ldots, a_n$ are the elements of E, a permutation of E is constructed first by choosing a value for the permutation to have at a_1, which can be done in n ways, then a value for the permutation to have at a_2, which can be done in $n - 1$ ways as one element has already been chosen for a_1, then a value for the permutation to have at a_3, which can be done in $n - 2$ ways, etc. Consequently, there are $n(n - 1)(n - 2) \ldots 2 \cdot 1 = n!$ permutations.

By Theorem 6.2, *an algebraic structure isomorphic to a group (semigroup) is itself a group (semigroup)*.

A basic technique for solving equations in elementary algebra is cancellation. From

$$2x = 4 = 2 \cdot 2,$$

for example, we infer that $x = 2$ by "cancelling out" a "2" in "$2x$" and in "$2 \cdot 2$"; similarly, from

$$x + 3 = 8 = 5 + 3$$

we conclude that $x = 5$ by cancelling out a "3." This procedure is not valid for all compositions. For example, from

$$2 \cdot_6 x = 4 = 2 \cdot_6 2$$

we cannot infer that $x = 2$; indeed, x could be either 2 or 5. On the other hand, from

$$2 \cdot_5 x = 2 \cdot_5 y$$

we may conclude that $x = y$. Elements that can always be cancelled out in this fashion are called cancellable for the composition in question:

Definition. An element $a \in E$ is **cancellable** for a composition $\triangle$ on E if the following condition holds:

(Canc)   For all $x, y \in E$, if either $a \triangle x = a \triangle y$ or $x \triangle a = y \triangle a$, then $x = y$.

For example, 1 and 5 are the only elements of N_6 cancellable for multiplication modulo 6, but every nonzero element of N_5 is cancellable for multiplication modulo 5.

The word "regular" is often used in this context. Some authors call cancellable elements regular, but others call invertible elements regular. Consequently, we shall not employ the term.

If $\triangle$ is a composition on a finite set E, then an element a of E is cancellable for $\triangle$ if and only if no entry is repeated in the row headed by a nor in the column headed by a in the table of $\triangle$.

Clearly, a is cancellable for a composition $\triangle$ on E if and only if the functions $L_a: x \mapsto a\triangle x$ and $R_a: x \mapsto x\triangle a$ from E into E are injective.

We shall repeatedly use the fact, proved in the following theorem, that *an element invertible for an associative composition is always cancellable for that composition.*

Theorem 7.1. If a is invertible for an associative composition $\triangle$ on E, then the functions L_a and R_a from E into E defined by

$$L_a(x) = a\triangle x,$$
$$R_a(x) = x\triangle a$$

are permutations of E. In particular, an element invertible for an associative composition is cancellable for that composition.

Proof. We denote the inverse of a by a^*. If

$$a\triangle x = a\triangle y,$$

then

$$a^*\triangle(a\triangle x) = a^*\triangle(a\triangle y),$$

so

$$(a^*\triangle a)\triangle x = (a^*\triangle a)\triangle y,$$

that is,

$$e\triangle x = e\triangle y,$$

whence

$$x = y.$$

Thus, L_a is injective. For any $y \in E$,

$$y = (a\triangle a^*)\triangle y = a\triangle(a^*\triangle y)$$
$$= L_a(a^*\triangle y).$$

Thus, L_a is surjective. Similarly, R_a is a permutation of E.

Corollary. Every element of a group is cancellable for the composition of the group.

The conclusion of Theorem 7.1 actually characterizes groups in the class of semigroups; indeed, if $(E, \triangle)$ is a semigroup such that L_a is a permutation for all $a \in E$ and R_b is a permutation for some $b \in E$, then $(E, \triangle)$ is a group (Exercise 7.14).

EXERCISES

*7.1. (a) Draw a Venn diagram for the symmetric difference of two sets.
(b) Prove that $(\mathfrak{P}(E), \triangle)$ is a group. [The Venn diagram for $(X \triangle Y) \triangle Z$ may suggest a proof of associativity.]

7.2. Is the set of all complex numbers of absolute value 1 a group under multiplication?

7.3. If E has more than two elements, then the symmetric group $\mathfrak{S}_E$ is not abelian.

7.4. Every group containing exactly two elements is isomorphic to $(N_2, +_2)$. Every group containing exactly three elements is isomorphic to $(N_3, +_3)$. How many automorphisms are there of $(N_2, +_2)$? of $(N_3, +_3)$?

7.5. If f is a bijection from E onto F, then the function $\Phi: u \mapsto f \circ u \circ f^{\leftarrow}$ is an isomorphism from $\mathfrak{S}_E$ onto $\mathfrak{S}_F$.

7.6. An element a of E is **left cancellable** (respectively, **right cancellable**) for a composition $\triangle$ on E if the function $L_a: x \mapsto a \triangle x$ (respectively, $R_a: x \mapsto x \triangle a$) is injective. State and prove a theorem similar to Theorem 6.2 for cancellable, left cancellable, and right cancellable elements.

*7.7. (a) Let $(E, \cdot)$ be a group, let e be the neutral element of E, and let $\div$ be the composition on E defined by

$$x \div y = xy^{-1}.$$

For all $x, y, z \in E$, the following equalities hold:
$1°$ $x \div x = e$.
$2°$ $x \div e = x$.
$3°$ $(x \div z) \div (y \div z) = x \div y$.
(b) Let $\div$ be a composition on E and e and element of E such that for all $x, y, z \in E$, equalities $1°$–$3°$ hold. Let $\cdot$ be the composition on E defined by

$$xy = x \div (e \div y).$$

Prove that $(E, \cdot)$ is a group. [Show first that e is the neutral element; then show that $e \div (x \div y) = y \div x$ and conclude that each element is invertible; for associativity, show successively that $xz \div yz = x \div y, (x \div z)(z \div y) = x \div y, x \div y = e$ implies that $x = y, x \div z = y \div z$ implies that $x = y, xy \div y = x, \cdot$ is associative.]

7.8. Let $(E, \cdot)$ be a commutative semigroup satisfying the following two properties:

1° For every $x \in E$, there exists $y \in E$ such that $yx = x$.
2° For all $x, y \in E$, if $yx = x$, then there exists $z \in E$ such that $zx = y$.

(a) If $yx = x = y'x$, then $y = y'$.
(b) If $yx = x$, then $yy = y$.
(c) If $yx = x$ and $zw = w$, then $y = z$. [Consider $y(yw)$ and $z(yw)$.]
(d) $(E, \cdot)$ is a group.
(e) Is the conclusion of (d) necessarily correct if we omit the hypothesis of commutativity? [Consider $(E, \leftarrow)$.]

7.9. A composition $\cdot$ on E is the composition $\rightarrow$ (respectively, $\leftarrow$) if and only if $\cdot$ is an associative, anticommutative composition for which there is a left (right) cancellable element. [Use Exercise 2.17.]

In the remaining exercises, $(E, \cdot)$ is a semigroup.

*7.10. If E has a neutral element e and if $xx = e$ for all $x \in E$, then E is an abelian group.

7.11. Let $a \in E$. If there exist $e, b \in E$ such that $ex = x$ for all $x \in E$ and $ba = e$, then a is left cancellable.

*7.12. If there exists $e \in E$ such that for all $x \in E$, $ex = x$ and $x^*x = e$ for some $x^* \in E$, then E is a group. [Use Exercise 7.11 to show that e is the neutral element by considering x^*xx^*x.]

*7.13. If a is an element of E such that the functions $L_a: x \mapsto ax$ and $R_a: x \mapsto xa$ are permutations of E, then there is a neutral element for $\cdot$, and a is invertible.

*7.14. If for every $a \in E$, $L_a: x \mapsto ax$ is a permutation of E, and if there exists $b \in E$ such that $R_b: x \mapsto xb$ is a permutation, then E is a group. [Use Exercise 7.13.]

*7.15. If there exists an idempotent $e \in E$ (Exercise 2.17), and if for every $a \in E$, there exists at least one element $x \in E$ satisfying $xa = e$ and at most one element y satisfying $ay = e$, then E is a group. [Use Exercise 7.12.]

8. Subgroups

The familiar compositions addition on R and addition on Q, though very similar, are not the same composition even though we use the same word "addition" to denote them both. Their similarity results from the fact that addition on Q is simply the "restriction" to $Q \times Q$ of addition on R.

Definition. Let $f: E \to F$, and let D be a subset of E. The **function obtained by restricting the domain** of f to D (or simply, the **restriction** of f to D) is the function

$$x \mapsto f(x)$$

from D into F.

Often we denote the function obtained by restricting the domain of f to D by f_D. Thus, f_D has domain D, the same codomain as f, and associates to each element of its domain the same element that f does. Clearly,

$$\text{Gr}(f_D) = \text{Gr}(f) \cap (D \times F).$$

Definition. Let $f: E \to F$. The **range** of f is the subset

$$\{y \in F: y = f(x) \text{ for some } x \in E\}$$

of F. Let G be a subset of F that contains the range of f. The **function obtained by restricting the codomain** of f to G is the function

$$x \mapsto f(x)$$

with domain E and codomain G.

Thus, the function obtained by restricting the codomain of a function f has the same domain and graph as f but, in general, has a smaller codomain. The function obtained by restricting the codomain of a function to its range is clearly a surjection; the function obtained by restricting the codomain of an injection to its range is clearly a bijection.

Definition. Let $f: E \to F$, $g: D \to G$. We shall say that f is an **extension** of g to a function from E into F, or that g is a **restriction** of f to a function from D into G, if $D \subseteq E$, $G \subseteq F$, and $g(x) = f(x)$ for all $x \in D$.

Thus, if $D \subseteq E$ and $G \subseteq F$, then there is a (unique) restriction of f to a function from D into G if and only if $f(x) \in G$ for all $x \in D$. Clearly, if $g: D \to G$ is a restriction of $f: E \to F$, then

$$\text{Gr}(g) = \text{Gr}(f) \cap (D \times G).$$

In particular, let $\triangle$ be a composition on E, and let $A \subseteq E$. If there is a restriction of $\triangle$ to a function from $A \times A$ into A, that restriction will be a composition on A; such a restriction exists if and only if $x \triangle y \in A$ for all $x, y \in A$.

Definition. Let $\triangle$ be a composition on E. A subset A of E is **stable** for $\triangle$, or **closed** under $\triangle$, if $x \triangle y \in A$ for all $x, y \in A$.

If A is stable for a composition $\triangle$ on E, we shall denote the restriction of $\triangle$ to $A \times A$ by $\triangle_A$ when it is necessary to emphasize that it is not the same as the given composition on E; but when no confusion would result, we shall simply drop the subscript and use the same symbol to denote both the given composition and its restriction. The composition $\triangle_A$ is called the **composition induced on A by $\triangle$**.

Example 8.1. If m is a positive integer, the set of all integral multiples of m is a subset of Z stable for both addition and multiplication on Z. If m and p are positive integers, the set of all integral multiples of m that are greater than p is another subset of Z stable for both addition and multiplication.

Example 8.2. The sets of nonzero integers, of nonzero rationals, of nonzero real numbers, and of nonzero complex numbers are all stable for multiplication. When is the set N_m^* of all nonzero elements of N_m stable for multiplication modulo m? If $m > 1$ and if N_m^* is stable, then m is a prime number; otherwise, $m = rs$ where r and s are strictly positive integers less than m, and consequently, r and s belong to N_m^*, but, since $rs = m$, $r \cdot_m s = 0$. Later we shall prove the converse (*cf.* Theorems 19.9 and 19.11): If m is prime, then N_m^* is stable for multiplication modulo m.

Example 8.3. The sets E and $\emptyset$ are stable for every composition $\triangle$ on E. If e is a neutral element for $\triangle$, then $\{e\}$ is stable for $\triangle$. More generally, if x is any element of E satisfying $x \triangle x = x$, then $\{x\}$ is stable for $\triangle$.

Example 8.4. Let $\vee$ be the composition on N defined by

$$x \vee y = \max\{x, y\}.$$

Then $(N, \vee)$ is a semigroup whose neutral element is zero, and every subset of N is stable for $\vee$.

Example 8.5. It is easy to enumerate all the stable subsets of the group G of symmetries of the square: They are $\emptyset$, $\{r_0\}$, $\{r_0, r_2\}$, $\{r_0, h\}$, $\{r_0, v\}$, $\{r_0, d_1\}$, $\{r_0, d_2\}$, $\{r_0, r_1, r_2, r_3\}$, $\{r_0, r_2, h, v\}$, $\{r_0, r_2, d_1 d_2\}$, and G.

Let $(E, \triangle)$ be an algebraic structure, and let A be a subset of E stable for $\triangle$. What properties does $\triangle_A$ inherit from $\triangle$? Surely if $\triangle$ is associative or commutative, so also is $\triangle_A$. In addition, every element of A cancellable for $\triangle$ is again cancellable for $\triangle_A$. If E has a neutral element e for $\triangle$, three possibilities arise: (1) A contains e, in which case e is, of course, the neutral element for $\triangle_A$; (2) there is no neutral element for $\triangle_A$ (for example, if A is the set of all strictly positive integers, then A is stable for addition, but there is no neutral element for addition on A); (3) A does not contain e, but there is, nevertheless, a neutral element for $\triangle_A$ (in Example 8.4, for instance, the set A of all strictly positive integers is stable for $\vee$ and 1 is the neutral element for $\vee_A$, although 0 is the neutral element for $\vee$). The third possibility cannot arise if every element of E is cancellable, for then e is the only element x satisfying $x \triangle x = x$:

Theorem 8.1. If e is the neutral element for a composition $\triangle$ on E, then e is the only cancellable element $x \in E$ satisfying $x \triangle x = x$.

Proof. If x is a cancellable element satisfying $x \triangle x = x$, then $x = e$ since $x \triangle x = x = x \triangle e$.

Definition. Let H be a nonempty stable subset of an algebraic structure $(E, \triangle)$. If $(H, \triangle_H)$ is a semigroup, we shall say that $(H, \triangle_H)$ is a **subsemigroup** of $(E, \triangle)$, and if $(H, \triangle_H)$ is a group, we shall say that $(H, \triangle_H)$ is a **subgroup** of $(E, \triangle)$.

If $(H, \triangle_H)$ is a subsemigroup (subgroup) of $(E, \triangle)$, the subset H of E is also called a subsemigroup (subgroup) of $(E, \triangle)$. Depending on the context, therefore, the words "subsemigroup" and "subgroup" may refer either to a set or to an algebraic structure. Each of the nonempty stable subsets of the group of symmetries of the square is, for example, a subgroup.

Theorem 8.2. If a subsemigroup of a group has a neutral element, that element is the neutral element of the group.

The assertion follows at once from Theorem 8.1 and the corollary of Theorem 7.1.

A nonempty subset of a semigroup is a subsemigroup if and only if it is stable. However, a nonempty stable subset of a group need not be a subgroup; for example, N is a subsemigroup but not a subgroup of the group $(Z, +)$.

We shall frequently apply either criterion 2° or 3° of the following theorem to determine whether a given nonempty subset of a group is a subgroup.

Theorem 8.3. Let H be a nonempty subset of a group $(G, \triangle)$. The following statements are equivalent:

1° H is a subgroup of G.
2° For all $x, y \in G$, if H contains x and y, then H contains $x \triangle y$ and y^*.
3° For all $x, y \in G$, if H contains x and y, then H contains $x \triangle y^*$.

Proof. Statement 1° implies 2°, for by Theorem 8.2, the neutral element for $\triangle_H$ is also the neutral element e for $\triangle$, and consequently, for every $y \in H$, the inverse of y for $\triangle_H$ is the unique inverse y^* of y for $\triangle$. Statement 2° implies 3°, for if x and y belong to H, then x and y^* belong to H by 2°, whence $x \triangle y^* \in H$ again by 2°. Finally, 3° implies 1°: There exists an element a in H by hypothesis, so $e \in H$ by 3° since $e = a \triangle a^*$. Also if $y \in H$, then $y^* \in H$ by 3° since $y^* = e \triangle y^*$. Moreover, H is stable for $\triangle$, for if $x, y \in H$, then by what we have just proved, x and y^* belong to H, so as

$$x \triangle (y^*)^* = x \triangle y,$$

$x \triangle y$ also belongs to H by 3°. Thus, H is stable for $\triangle$, H contains a neutral element for $\triangle_H$, every element of H has an inverse in H for $\triangle_H$, and $\triangle_H$ is, of course, associative since $\triangle$ is; therefore, $(H, \triangle_H)$ is a group.

If the composition of the group G is denoted by $+$ rather than $\triangle$, 2° and 3° become respectively:

For all $x, y \in G$, if H contains x and y, then H contains $x + y$ and $-y$.

For all $x, y \in G$, if H contains x and y, then H contains $x + (-y)$.

If the composition of G is denoted by $\cdot$ rather than $\triangle$, 2° and 3° become respectively:

For all $x, y \in G$, if H contains x and y, then H contains xy and y^{-1}.

For all $x, y \in G$, if H contains x and y, then H contains xy^{-1}.

Definition. Let $(G, \triangle)$ be a group. The **center** of G is the set $Z(G)$ consisting of all $x \in G$ such that $x \triangle y = y \triangle x$ for every $y \in G$.

A group G is abelian, of course, if and only if $Z(G) = G$.

Theorem 8.4. The center of a group $(G, \triangle)$ is a subgroup.

Proof. Clearly the neutral element of G belongs to $Z(G)$. If $x, y \in Z(G)$, then $x \triangle y \in Z(G)$ and $x^* \in Z(G)$ by 1° and 2° of Theorem 4.5. Thus, by Theorem 8.3, $Z(G)$ is a subgroup.

Theorem 8.5. Let $(E, \triangle)$ be a semigroup with neutral element e. The set G of all invertible elements of E is a subgroup of E.

Proof. If x and y are invertible, so is $x \triangle y$ by Theorem 4.4. Thus, G is a stable subset of E. Clearly, $e \in G$, and if $x \in G$, then $x^* \in G$ by Theorem 4.3. Also $\triangle_G$ is clearly associative as $\triangle$ is. Thus, G is a group.

From our discussion in §5, the subgroup of invertible elements of the semigroup $(E^E, \circ)$ of all functions from E into E is the group $\mathfrak{S}_E$ of all permutations of E.

Theorem 8.6. The set of all automorphisms of an algebraic structure E is a subgroup of $\mathfrak{S}_E$.

The assertion follows at once from 2° of Theorem 8.3 and Theorem 6.1.

We shall denote by $\text{Aut}(E)$ the group of all automorphisms of an algebraic structure E.

Definition. A group $\mathfrak{G}$ is a **permutation group** on E if $\mathfrak{G}$ is a subgroup of the group $\mathfrak{S}_E$.

Permutation groups were studied before the general definition of a group was formulated. When groups first began to be studied, mathematicians wondered if the concept of a group were really more general than that of a permutation group, that is, if there existed a group not isomorphic to any permutation group. The answer, furnished by Cayley, was negative; every group is isomorphic to a permutation group.

Theorem 8.7. (Cayley) Let $(G, \triangle)$ be a group, and for each $a \in G$, let L_a be the permutation of G defined by

$$L_a(x) = a \triangle x.$$

Then the function

$$L: a \mapsto L_a$$

is an isomorphism from $(G, \triangle)$ onto a permutation group on G.

Proof. By Theorem 7.1, L_a is indeed a permutation of G. If $L_a = L_b$, then

$$a = a \triangle e = L_a(e) = L_b(e) = b \triangle e = b.$$

Therefore, L is injective. For every $x \in G$,

$$(L_a \circ L_b)(x) = L_a(L_b(x)) = L_a(b \triangle x)$$
$$= a \triangle (b \triangle x) = (a \triangle b) \triangle x$$
$$= L_{a \triangle b}(x).$$

Therefore,

$$L_a \circ L_b = L_{a \triangle b}$$

for all $a, b \in G$. Thus, the set $\mathfrak{G} = \{L_a : a \in G\}$ is a stable subset of $\mathfrak{S}_G$, and L is an isomorphism from G onto $\mathfrak{G}$. Consequently, $\mathfrak{G}$ is a group and, therefore, is a permutation group on G.

The isomorphism L of Theorem 8.7 is called the **left regular representation** of $(G, \triangle)$.

Let $a_1, a_2, \ldots, a_r$ be elements of a set E such that $a_i \neq a_j$ whenever $i \neq j$. We shall denote by $(a_1, a_2, \ldots, a_r)$ the permutation σ of E defined as follows:

$$\sigma(a_i) = a_{i+1} \text{ if } 1 \leq i < r,$$
$$\sigma(a_r) = a_1,$$
$$\sigma(x) = x \text{ for all } x \notin \{a_1, \ldots, a_r\}.$$

Such permutations are called **cycles**. With this notation, for example,

$$(1, 2, 4) \circ (3, 4, 2) = (1, 2, 3).$$

Certain permutation groups have interesting geometrical models. For example, let G be the group of symmetries of the square. For each $u \in G$, let σ_u be the permutation of $\{1, 2, 3, 4\}$ describing the effect of u on the corners of the square. Thus,

$$\sigma_v = (1, 2) \circ (3, 4)$$

since v sends corner 1 into corner 2, corner 2 into corner 1, corner 3 into corner 4, and corner 4 into corner 3, and since $(1, 2) \circ (3, 4)$ is the permutation taking 1 into 2, 2 into 1, 3 into 4, and 4 into 3. Similarly,

$$\sigma_{r_1} = (1, 2, 3, 4), \qquad \sigma_h = (1, 4) \circ (2, 3),$$
$$\sigma_{r_2} = (1, 3) \circ (2, 4), \qquad \sigma_{d_1} = (1, 3),$$
$$\sigma_{r_3} = (1, 4, 3, 2), \qquad \sigma_{d_2} = (2, 4),$$

and σ_{r_0} is, of course, the identity permutation. It is easy to verify that

$$\sigma : u \mapsto \sigma_u$$

is an isomorphism from G onto a subgroup $\mathfrak{G}$ of $\mathfrak{S}_4$. Indeed, instead of proving directly that the composition of G is associative, it is easier to prove that σ is an isomorphism from G onto $\mathfrak{G}$; since G is, therefore, isomorphic to a group, we conclude that the composition of G is associative. Similarly, the group of symmetries of any polygon of n sides is isomorphic to a subgroup of $\mathfrak{S}_n$.

EXERCISES

8.1. Which subsets of E are stable for the compositions $\leftarrow$ and $\rightarrow$ on E?

8.2. Let E be a finite set of n elements. For how many compositions on E is every subset of E stable? Of these, how many are commutative?

8.3. If $F \subseteq E$, is $\mathfrak{P}(F)$ a stable subset of $\mathfrak{P}(E)$ for $\cup$? for $\cap$? for $\triangle$ (Example 7.4)? For which of those compositions is the set of all finite subsets of E stable? For which is the set of complements of all finite subsets of E stable?

8.4. Show that $H = \{3, 4, 5, 6\}$ is a subgroup of the semigroup $(N_7, +_{3,4})$ (Exercise 2.8). Is the neutral element of H the neutral element of N_7? Is every element of H invertible for the composition $+_{3,4}$ on N_7?

8.5. Write out the table for the group of (the ten) symmetries of a regular pentagon. Exhibit an isomorphism from it onto a subgroup of $\mathfrak{S}_5$.

8.6. Denote each element of $\mathfrak{S}_3$ other than the identity permutation as a cycle, and write out the table for the composition of the group $\mathfrak{S}_3$. Exhibit an isomorphism from the group of (the six) symmetries of an equilateral triangle onto $\mathfrak{S}_3$. Which subgroups of $\mathfrak{S}_3$ are isomorphic to the group of symmetries of an isosceles nonequilateral triangle?

8.7. Let R^* be the set of all nonzero real numbers, and let $E = R^* \times R$.

(a) Let $\triangle$ be the composition on E defined by

$$(a, b) \triangle (c, d) = (ac, ad + b)$$

for all (a, b), $(c, d) \in E$. Show that $(E, \triangle)$ is a group.

(b) For each $(a, b) \in E$ let $f_{a,b}$ be the function from R into R defined by

$$f_{a,b}(x) = ax + b,$$

and let

$$\mathfrak{G} = \{f_{a,b} \colon (a, b) \in E\}.$$

Show that $\mathfrak{G}$ is a permutation group on R isomorphic with the group $(E, \triangle)$ described in (a).

8.8. Construct a table for each composition on $\{a, b, c\}$ satisfying the following conditions: b is the neutral element, c is cancellable, a is invertible but not cancellable, and $\{a\}$ is not stable.

8.9. Let $\triangle$ be a composition on E, and let

$$A = \{x \in E \colon (x \triangle y) \triangle z = x \triangle (y \triangle z) \text{ for all } y, z \in E\}.$$

If $A \neq \emptyset$, then A is a subsemigroup of $(E, \triangle)$.

8.10. If $(G, \triangle)$ is a group and if $a \in G$, then the set of all elements of G commuting with a is a subgroup.

8.11. Let f be a function from E into E.
(a) The function f is left cancellable (Exercise 7.6) for the composition $\circ$ on E^E if and only if f is injective, and f is right cancellable if and only if f is surjective. Conclude that the subsemigroup of cancellable members of E^E coincides with the subgroup of its invertible members.
(b) The function f is idempotent (Exercise 2.17) for $\circ$ if and only if the function obtained by restricting both the domain and codomain of f to its range F is the identity function on F.

8.12. If E is a commutative semigroup, is the set of all noncancellable elements of E a subsemigroup?

*8.13. If H and K are subgroups of a group G neither of which contains the other, then there exists an element of G belonging neither to H nor to K.

*8.14. Let $E = \{a, b, c\}$ be a set of three elements. Let $\mathscr{A}$, $\mathscr{B}$, $\mathscr{C}_1$, $\mathscr{C}_2$, $\mathscr{C}_3$, and $\mathscr{D}$ be respectively the set of all compositions $\triangle$ on E such that the group of automorphisms of $(E, \triangle)$ is $\mathfrak{S}_E$, $\{1_E, (a, b, c), (a, c, b)\}$, $\{1_E, (a, b)\}$, $\{1_E, (a, c)\}$, $\{1_E, (b, c)\}$, and $\{1_E\}$ respectively, and let $\mathscr{C} = \mathscr{C}_1 \cup \mathscr{C}_2 \cup \mathscr{C}_3$.
(a) Show that $\mathscr{A}$ contains three members. [Use Exercise 6.11(c).] Write out the table for the unfamiliar composition belonging to $\mathscr{A}$. Show that $\mathscr{B}$ contains $3^3 - 3$ members and that each of $\mathscr{C}_1$, $\mathscr{C}_2$, and $\mathscr{C}_3$ contains $3^4 - 3$ members. Conclude that $\mathscr{D}$ has 19,422 members.
(b) Show that the isomorphic class determined by a composition in $\mathscr{A}$ consists of that composition alone, that the class determined by a composition in $\mathscr{B}$ contains one other composition, which is also from $\mathscr{B}$, that the class determined by a composition in $\mathscr{C}$ contains three compositions, one each from $\mathscr{C}_1$, $\mathscr{C}_2$, and $\mathscr{C}_3$, and that the class determined by a composition in $\mathscr{D}$ contains five other compositions, all from $\mathscr{D}$. Conclude that there are

3,330 isomorphic classes; that is, there are to within isomorphism 3,330 compositions on E.

(c) For how many compositions in each of the six classes is there a neutral element? Conclude that there are to within isomorphism 45 compositions on E admitting a neutral element.

(d) How many compositions in each of the six classes are commutative? Conclude that there are to within isomorphism 129 commutative compositions on E.

CHAPTER II

THREE BASIC NUMBER SYSTEMS

In this chapter we shall axiomatize the natural number system and then construct in a formal manner two other familiar number systems, the integers and the rational numbers. In plane geometry certain of the most self-evident statements about lines, points, angles, and circles are chosen as postulates; here also we shall select as postulates certain statements about the natural numbers that are particularly self-evident. Whether one mathematical statement is more self-evident than another is a psychological, not a mathematical, matter, and there are several alternative sets of postulates one may choose, all approximately equally self-evident. Incidentally, there is no mathematically compelling reason for preferring as postulates self-evident rather than unfamiliar though true statements about the natural numbers; indeed, some mathematicians delight in choosing as postulates for some familiar mathematical system certain true but unfamiliar statements and then deriving as theorems the more self-evident statements.

Our discussion of the natural numbers will not present any information that the reader does not already know, but rather it will serve to show how familiar facts about the natural numbers and finite sets may be derived from certain axioms. Consequently, readers who are willing to accept as "known" facts familiar to everyone about the natural numbers and finite sets may omit most of §§11–12, and similarly, readers willing to accept the integers as known and who are uninterested in their construction from the natural numbers may omit most of §14. (For a more precise statement, see the footnote at the beginning of each of those sections.)

In §11 we shall postulate the existence of a set N, a composition $+$ on N, and an "ordering" $\leq$ on N that satisfy certain conditions. Since the concept of an ordering is fundamental in all mathematics, we shall first consider it separately.

9. Orderings

The concept of a "relation" as used in everyday discourse is too fundamental to be defined in simpler terms, although, as we shall shortly see, a very simple definition may be given that is adequate for mathematics. In any event, before we can talk sensibly about a relation on a set E, we must be sure that for any elements x and y of E, either the relation holds between x and y or else it does not hold between x and y. Seizing upon this fact, we make the following informal definition of a relation:

A **relation** on a set E is a linguistic expression that may contain ____-blanks and . . .-blanks but contains blanks of no other kind, such that for every $x \in E$ and every $y \in E$, if "x" is inserted in every ____-blank and "y" in every . . .-blank, the resulting expression is a sentence that is either true or false. The following, for example, are relations on the set of all people now alive:

    (a) ____ loves . . . .
    (b) . . . is the husband of ____.
    (c) ____ and . . . have the same parents.
    (d) . . . and ____ are first cousins.
    (e) ____ and . . . have at least two grandparents in common.
    (f) —— is no taller than . . . .
    (g) ____ has the same parents and siblings as . . . .
    (h) ____ is the mother of . . . .
    (i) ____ is either an ancestor of . . . or the same person as . . . .
    (j) ____ is either a parent of . . . or the same person as . . . .
    (k) ____ and . . . are descendents of George III.
    (l) ____ is either at least as tall as . . . or at least as heavy as . . . .
    (m) ____ loves cheese.
    (n) The Sistine Chapel is in Rome.

No claim is made that the informal definition of "relation" given here corresponds exactly with the intuitive concept one may have. For example, it may seem contrary to ordinary usage to call (m) and (n) relations on the set of all people now alive, even though they satisfy the informal definition given.

If R is a relation, we shall say that x **bears** R to y if the sentence obtained by putting "x" in every ____-blank and "y" in every . . .-blank is true. If x bears R to y, we shall write $x \mathrel{R} y$, and if not, we shall write $x \not\mathrel{R} y$.

A relation R on E is **reflexive** if $x \mathrel{R} x$ for all $x \in E$, **symmetric** if for all $x, y \in E$, $x \mathrel{R} y$ implies that $y \mathrel{R} x$, **antisymmetric** if for all $x, y \in E$, $x \mathrel{R} y$ and $y \mathrel{R} x$ together imply that $x = y$, **transitive** if for all $x, y, z \in E$, $x \mathrel{R} y$ and $y \mathrel{R} z$ together imply $x \mathrel{R} z$.

Our human condition is such that relation (a), alas, is neither reflexive, symmetric, nor transitive, but at least it is not antisymmetric. A relation that is both symmetric and antisymmetric is also transitive (Exercise 9.1); consequently, there are at most $2^4 - 2$ possible combinations of the four properties defined that a given relation may possess; that there are, in fact, 14 possibilities is shown by the examples listed above. Thus, (m) is transitive but neither reflexive, symmetric, nor antisymmetric. Since (n) is a false statement, relation (n) is symmetric, antisymmetric, and transitive, but not reflexive.

To each relation R on E we may associate the subset of $E \times E$ consisting of all ordered pairs (x, y) such that $x \, R \, y$; this set is called the **truth set** of R. Every subset A of $E \times E$ is the truth set of at least one relation on E, namely, the relation

$$(\_\_\_, \ldots) \in A.$$

Knowing how to determine whether any given ordered pair in $E \times E$ belongs to the truth set of a given relation on E is certainly a long advance towards understanding the full meaning of the relation, and it is an epistemological question whether such knowledge should be regarded as the same as complete understanding of the meaning. Happily, the issue may be avoided in mathematics: Just as a set is completely determined by its elements (that is, E and F are identical sets if they have the same elements), so also two relations on a set of mathematical objects are regarded as the same relation if they have the same truth sets. We may, therefore, regard a relation on E simply as a certain subset of $E \times E$, namely, its truth set. Consequently, we make the following formal definition:

Definition. A **relation** on E is a subset of $E \times E$. A relation R on E is a **reflexive** relation on E if $(x, x) \in R$ for all $x \in E$; R is a **symmetric** relation on E if $(x, y) \in R$ implies that $(y, x) \in R$ for all $x, y \in E$; R is an **antisymmetric** relation on E if $(x, y) \in R$ and $(y, x) \in R$ together imply that $x = y$ for all x, $y \in E$; R is a **transitive** relation on E if $(x, y) \in R$ and $(y, z) \in R$ together imply that $(x, z) \in R$ for all $x, y, z \in E$.

As before, we shall write $x \, R \, y$ and say that x **bears** R to y if $(x, y) \in R$. In discussing relations we shall freely use both formal and informal definitions. The relation

$$\_\_\_ = \ldots$$

on E is, for example, simply the diagonal subset of $E \times E$.

Definition. A relation R on a set E is an **ordering** on E if R is reflexive, antisymmetric, and transitive. An ordering R on E is a **total ordering** if for all x, $y \in E$, either $x \, R \, y$ or $y \, R \, x$.

On the set of all British monarchs, the relation

$$M: \underline{\quad} \text{ was monarch after or at the same time as } \dots$$

is a total ordering, but the relation

$$\underline{\quad} \text{ was president of the United States after or at the same time as } \dots$$

on the set of all presidents of the United States is not antisymmetric because of President Cleveland's nonconsecutive terms. On the set of all persons who have ever lived, the relation

$$D: \underline{\quad} \text{ is a descendant of or the same person as } \dots$$

is an ordering that is not total, and on the set of all straight lines in the plane of analytic geometry, the relation

$L:$ _____ is parallel to . . . , and if . . . is not parallel to the Y-axis, then _____ coincides with or lies below . . . , but if . . . is parallel to the Y-axis, then _____ coincides with or lies to the right of . . .

is also an ordering that is not total (a line is considered parallel to itself). If F is a set having more than one element, the relation

$$\subseteq: \underline{\quad} \text{ is contained in } \dots$$

on the set $\mathfrak{P}(F)$ is an ordering that is not total.

Definition. Let R be a relation on a set E, and let A be a subset of E. The **restriction** of R to A, or the **relation induced on** A by R, is the relation R_A satisfying

$$x \, R_A \, y \text{ if and only if } x \, R \, y$$

for all $x, y \in A$. If S is a relation on subset A of E, relation R on E is an **extension** of S if $R_A = S$.

Thus, by definition,

$$R_A = R \cap (A \times A).$$

If R is respectively reflexive, symmetric, transitive, or antisymmetric, then R_A is also; consequently, if R is an ordering or a total ordering on E, then

R_A is also on A. To avoid cumbersome notation, we shall use the symbol denoting a given relation on E also to denote the relation it induces on A unless confusion results.

Figure 10 shows how an ordering R on a finite set may be represented diagrammatically.

Each small circle represents an element of the set, and the line segments connecting the circles are so drawn that $x \, R \, y$ if and only if either $x = y$ or

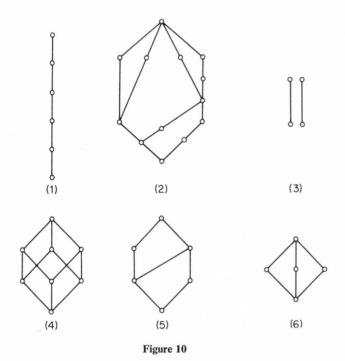

Figure 10

there is an ascending path of line segments joining the circle representing x to the circle representing y.

Thus, diagram (1) is the diagram for the restriction of M to the set of all British monarchs x such that $x \, M$ Victoria and Elizabeth II $M \, x$. Diagram (2) is the diagram for the restriction of D to the set of all persons x such that $x \, D$ Terah and Joseph $D \, x$ (we assume that Abraham is truthful in Genesis 20:12). Diagram (3) is the diagram for the restriction of L to the set of all lines parallel to and one unit away from either the X-axis or the Y-axis.

Definition. A **relational structure** is an ordered pair (E, R) where E is a set and R a relation on E. If R is an ordering on E, the relational structure (E, R) is also called an **ordered structure.**

Corresponding to the concept of an isomorphism between algebraic structures, an isomorphism of two relational structures is a bijection preserving everything in sight:

Definition. Let (E, R) and (F, S) be relational structures. An **isomorphism** from (E, R) onto (F, S) is a bijection f from E onto F such that for all x, $y \in E$,

$$x \, R \, y \text{ if and only if } f(x) \, S f(y).$$

Relational structures (E, R) and (F, S) are **isomorphic** if there is an isomorphism from one onto the other.

Theorem 9.1. Let (E, R), (F, S), and (G, T) be relational structures.

1° The identity function is an isomorphism from (E, R) onto itself.
2° If f is a bijection from E onto F, then f is an isomorphism from (E, R) onto (F, S) if and only if $f^{\leftarrow}$ is an isomorphism from (F, S) onto (E, R).
3° If f is an isomorphism from (E, R) onto (F, S) and if g is an isomorphism from (F, S) onto (G, T), then $g \circ f$ is an isomorphism from (E, R) onto (G, T).

Definition. Let R be a relation on E. We define $R^{\leftarrow}$ to be the relation on E satisfying

$$x \, R^{\leftarrow} y \text{ if and only if } y \, R \, x$$

for all $x, y \in E$.

If R is defined by a linguistic expression involving ____-blanks and . . .-blanks, $R^{\leftarrow}$ is obtained either by replacing every ____-blank by a . . .-blank and every . . .-blank by a ____-blank or by replacing certain crucial words and phases by their opposites. For example, $M^{\leftarrow}$ is the relation

____ was monarch before or at the same time as . . .

on the set of British monarchs. Clearly,

$$R^{\leftarrow\leftarrow} = R$$

for any relation R.

Theorem 9.2. If R is a reflexive (respectively, symmetric, antisymmetric transitive) relation on E, then $R^{\leftarrow}$ is also. If f is an isomorphism from a relational structure (E, R) onto a relational structure (F, S), then f is also an isomorphism from $(E, R^{\leftarrow})$ onto $(F, S^{\leftarrow})$.

If R is an ordering on a finite set, the diagram for $R^{\leftarrow}$ is obtained simply by turning the diagram for R upside down.

Symbols similar to $\leq$ and $\geq$ (e.g., $\subseteq$ and $\supseteq$, $\leqslant$ and $\geqslant$) are most frequently used to denote orderings. If $\leq$ is an ordering, then $\leq^{\leftarrow}$ is denoted by $\geq$. Similar, if $\geq$ is an ordering, then $\geq^{\leftarrow}$ is denoted by $\leq$. If $\leq$ is an ordering on E, for all $x, y \in E$ we write $x < y$ if $x \leq y$ and $x \neq y$, and similarly, we write $x > y$ if $x \geq y$ and $x \neq y$. It is customary to use such words as "less," "smaller," "greater," and "larger" and the corresponding superlatives when discussing orderings denoted by these symbols. However, the meaning of such words depends not only on the ordering involved but also on the particular symbol chosen to denote it. For example, if R is an ordering and if $x\ R\ y$, one would say "x is less than or equal to y" if $\leq$ were chosen to denote R, but in contrast "x is greater than or equal to y" if $\geq$ were chosen. Consequently, the use of such words for orderings not denoted by symbols similar to $\leq$ or $\geq$ could easily lead to confusion.

An ordering $\leq$ on E is total if and only if for all $x, y \in E$, either $x < y$ or $x = y$ or $x > y$, a condition known as the **trichotomy law.**

Definition. A **well-ordering** on E is a total ordering $\leq$ such that every nonempty subset A of E possesses a smallest element, i.e., such that every nonempty subset A of E contains an element a satisfying $a \leq x$ for all $x \in A$.

The familiar ordering $\leq$ on $\boldsymbol{R}$ is not a well-ordering; indeed, if $A = \{x \in \boldsymbol{R}: x > 1\}$, then A has no smallest member, for if $a \in A$, then $\frac{1}{2}(a + 1)$ is a yet smaller element of A.

It is intuitively evident that a total ordering on a finite set is a well-ordering; we shall formally prove this in §12. Clearly, if $\leq$ is a well-ordering on E and if A is a subset of E, the induced ordering $\leq_A$ on A is a well-ordering.

Definition. Let $(E, \leq)$ and $(F, \leqslant)$ be ordered structures. A function f from E into F is **increasing (decreasing)** if for all $x, y \in E$, if $x \leq y$, then $f(x) \leqslant f(y)$ $(f(x) \geqslant f(y))$; f is **strictly increasing (strictly decreasing)** if for all $x, y \in E$, if $x < y$, then $f(x) \prec f(y)$ $(f(x) \succ f(y))$.

For example, if $(E, \leq)$ is an ordered structure, 1_E is both increasing and decreasing, but 1_E is neither strictly increasing nor strictly decreasing unless $\leq$ is the equality relation.

Theorem 9.3. Let $(E, \leq)$ be a totally ordered structure, let $(F, \leqslant)$ be an ordered structure, and let f be a function from E into F.

1° If f is either strictly increasing or strictly decreasing, then f is injective.

2° The function f is an isomorphism from $(E, \leq)$ onto $(F, \leqslant)$ if and only if f is a strictly increasing surjection.

Proof. 1° If x, $y \in E$ and if $x \neq y$, then either $x < y$ or $y < x$, whence $f(x) \neq f(y)$ by hypothesis. 2° Necessity: If $x < y$, then $f(x) \leqslant f(y)$ and $f(x) \neq f(y)$ by hypothesis, so $f(x) \prec f(y)$. Sufficiency: Clearly, f is increasing, and f is injective by 1°. If $f(x) \leqslant f(y)$, then $x \leq y$, for otherwise, we would have $y < x$, whence $f(y) \prec f(x)$ as f is strictly increasing, a contradiction. Thus, f is a bijection satisfying $x \leq y$ is and only if $f(x) \leqslant f(y)$ for all x, $y \in E$.

EXERCISES

9.1. A relation R on E is both symmetric and antisymmetric if and only if R is contained in the diagonal subset of $E \times E$. In particular, if R is symmetric and antisymmetric, then R is transitive, and moreover, R is in addition reflexive if and only if R is the diagonal subset of $E \times E$ (i.e., the identity relation on E).

9.2. Determine whether relations (b)–(l) on page 68 are reflexive, symmetric, antisymmetric, or transitive.

9.3. Show explicitly that each of diagrams (1)–(3) of Figure 10 is the diagram of an ordered structure discussed in the text by labelling properly each small circle with the name of the element it represents.

9.4. Show that diagram (4) of Figure 10 is the diagram of the ordering $\subseteq$ on $\mathfrak{P}(F)$, where F is a set having three elements.

9.5. On the N^* of all strictly positive integers let $|$ be the relation defined by

$$x \mid y \text{ if and only if there exists } m \in N^* \text{ such that } mx = y.$$

(a) Show that $|$ is an ordering.

(b) Show that diagram (5) of Figure 10 is the diagram for the restriction of $|$ to $\{1, 2, 3, 4, 6, 12\}$.

(c) Draw the diagram for the restriction of $|$ to $\{1, 2, 3, 5, 6, 10, 30\}$.

9.6. (a) What subgroup of $\mathfrak{S}_4$ corresponds to the group H of symmetries of a rectangle that is not a square?

(b) Show that diagram (6) of Figure 10 is the diagram for the restriction of the ordering $\subseteq$ on $\mathfrak{P}(H)$ to the set of all subgroups of H.

9.7. Draw a diagram for

(a) the restriction of D to the set of all persons x such that $x \, D$ your paternal grandfather;

(b) the restriction of D to the set of all British monarchs x such that $x\, D$ Victoria;

(c) the set of all subsets of a set having four elements, ordered by $\subseteq$;

(d) the set of all subgroups of the group of symmetries of the square, ordered by $\subseteq$;

(e) the restriction of $|$ to $\{1, 2, 3, 4, 6, 8, 12, 24\}$;

(f) the restriction of $|$ to $\{1, 2, 3, 5, 6, 10, 30\}$.

9.8. Let E be a set having three elements. List all orderings of E. Divide them into isomorphic classes, so that orderings R and S belong to the same class if and only if (E, R) and (E, S) are isomorphic ordered structures. How many orderings are there? How many isomorphic classes? Draw the diagram for a representative of each isomorphic class.

9.9. Draw diagrams for all possible orderings on a set of four elements. How many orderings are there to within isomorphism on a set of four elements?

9.10. If E is a set having n elements, how many total orderings are there on E? How many isomorphic classes do the total orderings form?

9.11. Prove Theorem 9.1.

9.12. Let R be an ordering on E, and let R^* be the relation on E satisfying $x\, R^*\, y$ if and only if $x\, R\, y$ and $x \neq y$. If $S = R^*$, then the following conditions hold:

$1°$ For all $x \in E$, $x \not{S} x$.

$2°$ For all $x, y \in E$, if $x\, S\, y$, then $y \not{S} x$.

$3°$ For all $x, y, z \in E$, if $x\, S\, y$ and if $y\, S\, z$, then $x\, S\, z$.

Conversely, if S is a relation on E satisfying conditions $1°$–$3°$, then there is a unique ordering R on E such that $S = R^*$.

9.13. Prove Theorem 9.2.

9.14. Let $\leq_1$ and $\leq_2$ be orderings respectively on nonempty sets E_1 and E_2. Let $\leq$ be the relation on $E_1 \times E_2$ satisfying $(x_1, x_2) \leq (y_1, y_2)$ if and only if $x_1 \leq_1 y_1$ and $x_2 \leq_2 y_2$.

(a) The relation $\leq$ is an ordering on $E_1 \times E_2$.

(b) The ordering $\leq$ is a total ordering if and only if $\leq_1$ and $\leq_2$ are total orderings and either E_1 or E_2 has only one element.

9.15. Let E and F be sets, and let $\leq$ be an ordering on F. We shall denote also by $\leq$ the relation on F^E satisfying $f \leq g$ if and only if $f(x) \leq g(x)$ for all $x \in E$. Then $\leq$ is an ordering on F^E. Under what circumstances is the ordering $\leq$ on F^E a total ordering?

The terminology of the remaining exercises is that of Exercise 5.8. Let R be a relation on a set E. We denote by $\mathrm{Diag}(E)$ the diagonal subset $\{(x, x): x \in E\}$ of $E \times E$.

9.16. (a) R is reflexive if and only if $R \supseteq \mathrm{Diag}(E)$.

(b) R is symmetric if and only if $R^{\leftarrow} = R$.

(c) R is antisymmetric if and only if $R \cap R^{\leftarrow} \subseteq \mathrm{Diag}(E)$.

(d) R is transitive if and only if $R \circ R \subseteq R$.

(e) R is an ordering if and only if $R \circ R \subseteq R$ and $R \cap R^{\leftarrow} = \mathrm{Diag}(E)$.

(f) R is a total ordering if and only if $R \cap R^{\leftarrow} = \mathrm{Diag}(E)$, $R \circ R \subseteq R$, and $R \cup R^{\leftarrow} = E \times E$.

(g) If R and S are orderings on E, is $R \cap S$ necessarily an ordering? $R \cup S$? $R \circ S$?

9.17. An ordering R on E is **directed** if for all $x, y \in E$ there exists $z \in E$ such that $x\,R\,z$ and $y\,R\,z$. Prove that an ordering R is directed if and only if $R^{\leftarrow} \circ R = E \times E$.

10. Ordered Semigroups

An important property possessed by the familiar total ordering $\leq$ on the set of real numbers is that the addition of a number to both sides of an inequality preserves the inequality: if $x \leq y$, then $x + z \leq y + z$. Many important compositions and orderings are similarly related, and therefore, we make the following definition:

Definition. A composition $\triangle$ on E and an ordering $\leq$ on E are **compatible** with each other if for all $x, y, z \in E$,

(OS)   if $x \leq y$, then $x\triangle z \leq y\triangle z$ and $z\triangle x \leq z\triangle y$.

An **ordered semigroup** is an ordered triad $(E, \triangle, \leq)$ such that $(E, \triangle)$ is a semigroup and $\leq$ is an ordering on E compatible with $\triangle$. An **ordered group** is an ordered triad $(E, \triangle, \leq)$ such that $(E, \triangle)$ is a group and $\leq$ is an ordering on E compatible with $\triangle$.

For the most part, we shall henceforth denote compositions by symbols similar to $+$ and $\cdot$. If a composition is denoted by $+$, (OS) becomes

$$\text{if } x \leq y, \text{ then } x + z \leq y + z \text{ and } z + x \leq z + y,$$

and if it is denoted by $\cdot$ (in which case we usually write simply "ab" for "$a \cdot b$"), (OS) becomes

$$\text{if } x \leq y, \text{ then } xz \leq yz \text{ and } zx \leq zy.$$

If $(E, +, \leq)$ is an ordered semigroup and if H is a subsemigroup (subgroup) of $(E, +)$, then $(H, +_H, \leq_H)$ is clearly an ordered semigroup (ordered group). We shall say that an ordered semigroup $(H, \oplus, \preccurlyeq)$ is an **ordered subsemigroup** of an ordered semigroup $(E, +, \leq)$ if H is a subsemigroup of $(E, +)$, if $\oplus$ is the composition $+_H$ induced on H by $+$, and if $\preccurlyeq$ is the restriction $\leq_H$ of $\leq$ to H.

Definition. An **isomorphism** from an ordered semigroup $(E, +, \leq)$ onto an ordered semigroup $(F, \oplus, \preccurlyeq)$ is a bijection f from E onto F such that f is an isomorphism from the semigroup $(E, +)$ onto $(F, \oplus)$ and is also an isomorphism from the ordered structure $(E, \leq)$ onto $(F, \preccurlyeq)$. Two ordered semigroups are **isomorphic** if there is an isomorphism from one onto the other. An **automorphism** of an ordered semigroup is an isomorphism from itself onto itself.

It is easy to formulate and prove analogues of Theorems 6.1 and 9.1 for ordered semigroups.

In the following theorems we list some easy consequences of the definition of an ordered semigroup, which we shall repeatedly use in later sections.

Theorem 10.1. Let $(E, +, \leq)$ be an ordered semigroup.

1° If z is cancellable for $+$ and if $x < y$, then $x + z < y + z$ and $z + x < z + y$.

2° If z is invertible or if $\leq$ is a total ordering, and if either $x + z < y + z$ or $z + x < z + y$, then $x < y$.

Proof. If $x < y$ and if z is cancellable, then $x + z \leq y + z$ by (OS), and consequently, $x + z < y + z$, for otherwise, we would have $x + z = y + z$, and hence $x = y$, a contradiction. To prove 2°, suppose that $x + z < y + z$. If z is invertible, then by what we have just proved,

$$x = (x + z) + (-z) < (y + z) + (-z) = y$$

since $-z$ is cancellable by Theorem 7.1. If $\leq$ is a total ordering, then also $x < y$, for otherwise we would have $y \leq x$, and hence $y + z \leq x + z$, a contradiction.

Theorem 10.2. Let $(E, +, \leq)$ be an ordered semigroup possessing a neutral element 0. If x and y are invertible elements of E, then $x < y$ if and only if $-y < -x$. In particular, if x is invertible, then $x > 0$ if and only if $-x < 0$, and $x < 0$ if and only if $-x > 0$.

Proof. If $x < y$, then by Theorem 10.1,

$$0 = (-x) + x < (-x) + y,$$

whence

$$-y = 0 + (-y) < (-x) + y + (-y) = -x,$$

since both $-x$ and $-y$ are cancellable by Theorem 7.1. Conversely, by what we have just proved, if $-y < -x$, then

$$x = -(-x) < -(-y) = y.$$

From Theorems 10.1 and 10.2 we obtain the following theorem:

Theorem 10.3. Let $(G, +, \leq)$ be an ordered group, and let x, y, and z be elements of G. The following statements are equivalent:

$1°$ $x < y$.	$4°$ $-y < -x$.
$2°$ $x + z < y + z$.	$5°$ $y + (-x) > 0$.
$3°$ $z + x < z + y$.	$6°$ $(-x) + y > 0$.

The following theorem is an immediate consequence of Theorem 9.3.

Theorem 10.4. Let $(E, +, \leq)$ and $(F, \oplus, \leqslant)$ be ordered semigroups, and let f be a function from E into F. If $\leq$ is a total ordering, then f is an isomorphism from $(E, +, \leq)$ onto $(F, \oplus, \leqslant)$ if and only if f is a strictly increasing surjection satisfying $f(x + y) = f(x) \oplus f(y)$ for all $x, y \in E$.

EXERCISES

10.1. Formulate and prove the analogue for ordered semigroups of Theorems 6.1 and 9.1.

10.2. Let $\vee_n$ be the composition on N_n defined by

$$x \vee_n y = \max\{x, y\}$$

for all $x, y \in N_n$. Show that $(N_n, \vee_n, \leq)$ is an ordered semigroup (where $\leq$ is the restriction to N_n of the usual total ordering on N).

10.3. If $(E, +, \leq)$ is an ordered semigroup, then so is $(E, +, \geq)$. If $(G, +, \leq)$ is an ordered abelian group, then the function $g: x \mapsto -x$ from G into G is an isomorphism from $(G, +, \leq)$ onto $(G, +, \geq)$.

10.4. Let $(G, +, \leq)$ be an ordered abelian group, and let

$$G^+ = \{x \in G : x \geq 0\},$$
$$G^- = \{x \in G : x \leq 0\}.$$

Prove that G^+ and G^- are subsemigroups of G and that $g : x \mapsto -x$ is an isomorphism from $(G^+, +, \leq)$ onto $(G^-, +, \geq)$ and also from $(G^-, +, \leq)$ onto $(G^+, +, \geq)$.

10.5. For any set E, $(\mathfrak{P}(E), \cap, \subseteq)$, $(\mathfrak{P}(E), \cap, \supseteq)$, $(\mathfrak{P}(E), \cup, \subseteq)$, and $(\mathfrak{P}(E), \cup, \supseteq)$ are ordered semigroups. The function $C : X \mapsto X^c$ from $\mathfrak{P}(E)$ into $\mathfrak{P}(E)$ is an isomorphism from $(\mathfrak{P}(E), \cap, \subseteq)$ onto $(\mathfrak{P}(E), \cup, \supseteq)$ and is also an isomorphism from $(\mathfrak{P}(E), \cap, \supseteq)$ onto $(\mathfrak{P}(E), \cup, \subseteq)$.

10.6. Show that $(N^*, \cdot, \mid)$ (Exercise 9.5) is an ordered semigroup.

10.7. If $(G, +, \leq)$ is an ordered group, then $\leq$ is directed (Exercise 9.17) if and only if for every $x \in G$ there exist $y, z \in G$ such that $y \geq 0$, $z \geq 0$, and $x = y + (-z)$.

11. The Natural Numbers*

The first and most celebrated axiomatization of the strictly positive integers was made by the Italian mathematician G. Peano in 1889 (for a complete account, see E. Landau, "Foundations of Analysis," New York, Chelsea Publishing Company, 1951). Only a very slight modification of certain of his definitions is needed to yield an axiomatization of the natural numbers (for an account of Peano's axioms so modified, see B. Russell, "Introduction to Mathematical Philosophy," London, George Allen and Unwin, Ltd., 1956, or Exercise 11.8). Peano's postulates, modified for the natural numbers, concern a set N, a specific element 0 of N, and a "successor function" s from N into N. In the development, addition and the natural number 1 are defined in such a way that

$$s(n) = n + 1$$

for all $n \in N$.

Many alternative axiomatizations of the natural numbers may be chosen. For example, it is possible to give a purely algebraic axiomatization of N, that is, one concerning only addition on N (Exercise 11.7). We shall formally adopt as postulates certain statements concerning addition and the ordering of the natural numbers.

---

* This section, which is devoted to an axiomatic treatment of the natural numbers, may be omitted except for the definition of "distributive" on page 87 and pages 94–96, which introduce some new terminology.

Definition. A **naturally ordered semigroup** is a commutative ordered semigroup $(E, +, \le)$ satisfying the following conditions:

(NO 1)   The ordering $\le$ is a well-ordering.
(NO 2)   Every element of E is cancellable for $+$.
(NO 3)   For all m, $n \in E$, if $m \le n$, then there exists $p \in E$ such that $m + p = n$.
(NO 4)   There exist $m \in E$ and $n \in E$ such that $m \ne n$.

We adopt as a fundamental postulate the following assertion:

There exists a naturally ordered semigroup.

Later we shall prove that any two naturally ordered semigroups are isomorphic as ordered semigroups. Thus, to within isomorphism there is exactly one naturally ordered semigroup. There is, therefore, nothing arbitrary in our choice if we simply select some naturally ordered semigroup $(N, +, \le)$ and call the elements of N **natural numbers.**

For any one of the four defining properties of a naturally ordered semigroup, there exists a commutative ordered semigroup that satisfies the remaining three properties but is not naturally ordered (Exercises 11.1–11.3). For example, if E is a set having just one element, then $(E, +, \le)$ is a commutative ordered semigroup satisfying (NO 1), (NO 2), and (NO 3), but not (NO 4), where $+$ is the only possible composition on E and where $\le$ is the only possible ordering on E. Thus, the four conditions are "independent"; that is, no one of them may be deduced from the remaining three. On the other hand, we need not have assumed that $+$ is commutative, for if $(E, +, \le)$ is an ordered semigroup satisfying (NO 1)–(NO 3), then $+$ is necessarily a commutative composition (Exercise 11.5).

Since $\le$ is a well-ordering, there is a smallest element in N; this element we denote by 0, and we denote the complement of $\{0\}$ by N^*. By (NO 4), $N^* \ne \emptyset$ and so also has a smallest element; this element we denote by 1. For each $n \in N$, we define the set N_n by

$$N_n = \{m \in N : m < n\}.$$

Finally, if m and n are natural numbers, we define the set $[m, n]$ by

$$[m, n] = \{x \in N : m \le x \text{ and } x \le n\}.$$

Of course, if $m > n$, then $[m, n]$ is the empty set, and it is convenient to attach this meaning to $[m, n]$ if $m > n$ even though we shall most often consider the sets $[m, n]$ where $m \le n$. An **integer interval** is any set $[m, n]$ where m and n are natural numbers satisfying $m \le n$.

Theorem 11.1. The natural number 0 is the neutral element for $+$.

Proof. Since $0 \leq 0$, by (NO 3) there exists $p \in N$ such that $0 + p = 0$. As 0 is the smallest natural number, $0 \leq 0 + 0$. For the same reason, $0 \leq p$, so

$$0 + 0 \leq 0 + p = 0$$

by (OS). Therefore, $0 + 0 = 0$. For each natural number n,

$$(n + 0) + 0 = n + (0 + 0) = n + 0,$$

so $n + 0 = n$ by (NO 2). Since $+$ is commutative, 0 is thus the neutral element.

Theorem 11.2. If m and n are natural numbers, then $m \leq n$ if and only if there exists a natural number p such that $m + p = n$.

Proof. The condition is necessary by (NO 3). It is also sufficient, for if $m + p = n$, then

$$m = m + 0 \leq m + p = n$$

by Theorem 11.1 and (OS).

If $m \leq n$, there is a *unique* natural number p such that $m + p = n$, for if $m + q = n = m + p$, then $q = p$ by (NO 2). If $m \leq n$, we shall denote the unique natural number p such that $m + p = n$ by $n - m$. Thus, if $m \leq n$, then by definition

$$m + (n - m) = n = (n - m) + m.$$

Theorem 11.3. If m, n, and p are natural numbers, then $m < n$ if and only if $m + p < n + p$.

Proof. Since $\leq$ is a total ordering and since every natural number is cancellable for addition, the assertion follows from Theorem 10.1.

Theorem 11.4. If n and p are natural numbers, then $n < n + 1$, and $n < p$ if and only if $n + 1 \leq p$.

Proof. As $0 < 1$,

$$n = n + 0 < n + 1$$

by Theorem 11.3. If $n < p$, then $p - n \neq 0$ by Theorem 11.1, so $1 \leq p - n$ by the definition of 1, and hence,

$$n + 1 \leq n + (p - n) = p.$$

Conversely, if $n + 1 \leq p$, then $n < p$ since $n < n + 1$.

Corollary 11.4.1. For every natural number n, $N_{n+1} = N_n \cup \{n\}$.

Proof. For every natural number p, $p \notin N_{n+1}$ if and only if $n + 1 \leq p$, and $p \notin N_n \cup \{n\}$ if and only if $n < p$, since $\leq$ is a total ordering. Consequently, by Theorem 11.4, $(N_{n+1})^c = (N_n \cup \{n\})^c$, so by Theorem 3.2,

$$N_{n+1} = ((N_{n+1})^c)^c = ((N_n \cup \{n\})^c)^c = N_n \cup \{n\}.$$

Corollary 11.4.2. If m and n are natural numbers such that $m \leq n$, then $[m, n + 1] = [m, n] \cup \{n + 1\}$.

Proof. Since $m \leq n$, $x \in [m, n + 1]$ if and only if $x \geq m$ and either $x < n + 1$ or $x = n + 1$, and $x \in [m, n] \cup \{n + 1\}$ if and only if $x \geq m$ and either $x < n$ or $x = n$ or $x = n + 1$. Consequently, by Corollary 11.4.1,

$$[m, n + 1] = [m, n] \cup \{n + 1\}.$$

Theorem 11.5. (Principle of Mathematical Induction) Let S be a subset of N such that $0 \in S$ and for all natural numbers n, if $n \in S$, then $n + 1 \in S$. Then $S = N$.

Proof. Suppose that $S \neq N$. Then the complement S^c of S would be nonempty and hence by (NO 1) would contain a smallest element a. By hypothesis, $a \neq 0$, so $a \geq 1$. Consequently, $a - 1 < a$ by Theorem 11.4, since $(a - 1) + 1 = a$. Hence, $a - 1 \notin S^c$ as a is the smallest element of S^c, so $a - 1 \in (S^c)^c = S$, and therefore, by our hypothesis $a \in S$ since $a = (a - 1) + 1$, a contradiction. Consequently, $S = N$.

Corollary 11.5.1. Let p be a natural number, and let S be a set of natural numbers $\geq p$ such that $p \in S$ and for all natural numbers n, if $n \in S$, then $n + 1 \in S$. Then S is the set of all natural numbers $\geq p$.

Proof. Let $S' = S \cup N_p$. It is easy to verify that the hypotheses of Theorem 11.5 are satisfied by S'. Therefore, $S' = N$, so S is the set of all natural numbers $\geq p$.

Corollary 11.5.2. Let p and q be natural numbers such that $p \leq q$. Let S be a subset of $[p, q]$ such that $p \in S$ and for all natural numbers n, if $n \in S$ and if $n < q$, then $n + 1 \in S$. Then $S = [p, q]$.

Proof. Let $S' = S \cup \{m \in N : m > q\}$. Then S' satisfies the hypothesis of Corollary 11.5.1 and hence is the set of all natural numbers $\geq p$. Therefore, $S = [p, q]$.

Corollary 11.5.1 is also known as the *Principle of Mathematical Induction*. Arguments based on Theorem 11.5 or its corollaries are called arguments "by induction."

In elementary arithmetic one learns that the product nm of two strictly positive integers is simply the result of "adding m to itself n times." Our next goal is to formulate precisely a definition of multiplication on N that incorporates the basic idea just expressed concerning the relation of multiplication to addition. More generally, if $\triangle$ is a composition on a set E and if $a \in E$, we shall prove that there exists a unique function that associates to every strictly positive integer n the result of "forming the composite of a with itself n times." To do so, we shall use the following theorem, which justifies "definition by recursion."

Theorem 11.6. (Principle of Recursive Definition) Let p be a natural number, let E be a set, let $a \in E$, and let s be a function from E into E. There exists one and only one function f from the set of all natural numbers $\geq p$ into E such that
$$f(p) = a,$$
$$f(n + 1) = s(f(n))$$
for all $n \geq p$.

Proof. For each natural number $n \geq p$, we shall say that a function g is *admissible for n* if g is a function from $[p, n]$ into E such that

$$g(p) = a,$$
$$g(r + 1) = s(g(r))$$

for all natural numbers r satisfying $p \leq r < n$. Let S be the set of all natural numbers $\geq p$ for which there is one and only one admissible function. Clearly, $p \in S$, for the function g_p defined by

$$g_p(p) = a$$

with domain $\{p\}$ is the only admissible function for p. Suppose that $n \in S$, and let g be the unique admissible function for n. By Theorem 11.4, $n + 1 \notin [p, n]$, and by Corollary 11.4.2, $[p, n + 1] = [p, n] \cup \{n + 1\}$. Therefore, we may define a function h from $[p, n + 1]$ into E by

$$h(r) = g(r) \text{ if } r \in [p, n],$$
$$h(n + 1) = s(g(n)).$$

Clearly h is admissible for $n + 1$. If h' is any function admissible for $n + 1$, then the restriction of h' to $[p, n]$ is surely admissible for n and hence is g, since $n \in S$; therefore,

$$h'(r) = g(r) = h(r)$$

for all $r \in [p, n]$, and

$$h'(n + 1) = s(h'(n)) = s(g(n)) = h(n + 1).$$

Consequently, $h' = h$. Thus, $n + 1 \in S$, so by the Principle of Mathematical Induction, S is the set of all natural numbers $\geq p$.

For each natural number, $n \geq p$, let g_n be the unique admissible function for n. If $n \geq p$, the restriction of g_{n+1} to $[p, n]$ is admissible for n and hence is g_n, so $g_{n+1}(r) = g_n(r)$ for all $r \in [p, n]$, and in particular, $g_{n+1}(n) = g_n(n)$. Let f be the function with codomain E defined by

$$f(n) = g_n(n)$$

for all $n \geq p$. Then

$$f(p) = g_p(p) = a,$$

and for all $n \geq p$,

$$f(n + 1) = g_{n+1}(n + 1) = s(g_{n+1}(n))$$
$$= s(g_n(n)) = s(f(n)).$$

If f' is any function satisfying the desired conditions, then for each $n \geq p$ the restriction of f' to $[p, n]$ is clearly admissible for n and hence is g_n, so

$$f'(n) = g_n(n) = f(n).$$

Thus, there is exactly one function having the desired properties.

In trying to prove the existence of a function f having the properties given in Theorem 11.6, it is tempting to give a short but incorrect argument based on a misuse of the Principle of Mathematical Induction. The argument for the case $p = 0$ is the following: Define $f(0)$ to be a, and in general, if $f(n)$ is defined, define $f(n + 1)$ to be $s(f(n))$. Then the domain of the function f is N, for if

$$S = \{n \in N: f(n) \text{ is defined}\},$$

then $0 \in S$, and $n + 1 \in S$ whenever $n \in S$, so $S = N$ by the Principle of Mathematical Induction. Consequently, f is a function from N into E satisfying $f(0) = a$ and $f(n + 1) = s(f(n))$ for all $n \in N$.

Three objections may be made: First, in the above argument S is not really precisely defined. For a set to be defined, one ought to be able to reword the linguistic expression used in its definition so that only terms of logic and the basic undefined terms of set theory occur; it is not at all apparent how to do this for the expression "is defined" used in the definition of S. Indeed, ordinary usage suggests that "is defined" is more a psychological than a mathematical expression, for a term is usually considered defined only after a human agent has been willing to make the act of definition on it.

Second, in the argument f is not really defined either. To define a function, one must specify the domain and codomain of the function and the rule which associates to each element of the domain an element of the codomain. In the above argument, one specifies that 0 belongs to the domain S of f and that $n + 1$ belongs to S whenever n belongs to S. This condition implies that either S itself is used to define S, or else that the domain of f somehow changes from time to time during the act of definition. Neither possibility is admissible.

Third, the only property of the semigroup $(N, +)$ used in the argument is that it satisfies the Principle of Mathematical Induction. If the argument were valid, therefore, it could equally well be used to prove the following assertion: If 0 and 1 are elements of a commutative semigroup $(D, +)$ such that the only subset of D containing 0 and containing $x + 1$ whenever it contains x is D itself, then for any set E, any function s from E into E, and any element a of E, there exists a function f from D into E satisfying

$$f(0) = a,$$
$$f(x + 1) = s(f(x))$$

for all $x \in D$. This assertion is untrue, however; the elements 0 and 1 of the semigroup $(N_2, +_2)$ satisfy the hypothesis, but if s is the function from N into N defined by

$$s(n) = n + 1,$$

there is no function f from N_2 into N satisfying

$$f(0) = 0,$$
$$f(x +_2 1) = s(f(x))$$

for all $x \in N_2$.

Theorem 11.7. Let $\triangle$ be a composition on E, and let $a \in E$. There is one and only one function f_a from N^* into E such that

$$f_a(1) = a,$$
$$f_a(n + 1) = f_a(n)\triangle a$$

for all $n \in N^*$. If there is a neutral element e for $\triangle$, there is one and only one function g_a from N into E such that

$$g_a(0) = e,$$
$$g_a(n + 1) = g_a(n)\triangle a$$

for all $n \in N$; furthermore, $g_a(n) = f_a(n)$ for all $n \in N^*$.

Proof. Let s be the function from E into E defined by

$$s(x) = x\triangle a$$

for all $x \in E$. Applying the Principle of Recursive Definition to s, the natural number 1, and the element a, we obtain the desired function f_a. If there is a neutral element e, we obtain the desired function g_a by applying the Principle of Recursive Definition to s, the natural number 0, and the element e. In this case,

$$g_a(1) = g_a(0)\triangle a = e\triangle a = a,$$

so by the uniqueness of f_a, the restriction of g_a to N^* is f_a.

Definition. If n and m are natural numbers, the **product** of n and m is the natural number $n \cdot m$ (which we shall also denote by nm) defined by

$$n \cdot m = g_m(n),$$

where g_m is the unique function from N into N satisfying

$$g_m(0) = 0,$$
$$g_m(r + 1) = g_m(r) + m$$

for all $r \in N$. The composition $(n, m) \mapsto nm$ on N is called **multiplication.**

In view of the properties characterizing g_m, nm may be described as the result of adding m to itself n times. The definition of nm is a special case of the following definition:

Definition. Let $\triangle$ be a composition on E, and let $a \in E$. If there is a neutral element e for $\triangle$, we define $\triangle^n a$ by

$$\triangle^n a = g_a(n)$$

for all $n \in N$, where g_a is the unique function from N into E satisfying

$$g_a(0) = e.$$
$$g_a(r + 1) = g_a(r) \triangle a$$

for all $r \in N$. If there is no neutral element for $\triangle$, we define $\triangle^n a$ by

$$\triangle^n a = f_a(n)$$

for all $n \in N^*$, where f_a is the unique function from N^* into E satisfying

$$f_a(1) = a,$$
$$f_a(r + 1) = f_a(r) \triangle a$$

for all $r \in N^*$.

If a composition is denoted by a symbol similar to $+$ instead of $\triangle$, we write $n.a$ or simply na for $\triangle^n a$, and if it is denoted by a symbol similar to $\cdot$, we shall write a^n for $\triangle^n a$.

Thus, by definition,

$$\triangle^1 a = a,$$
$$\triangle^{n+1} a = (\triangle^n a) \triangle a$$

for all $n \in N^*$, and furthermore, if there is a neutral element e for $\triangle$,

$$\triangle^0 a = e.$$

A familiar rule of arithmetic is that multiplication is distributive over addition, that is, that

$$m(n + p) = mn + mp,$$
$$(m + n)p = mp + np$$

for all numbers m, n, and p.

Definition. A composition $\triangledown$ on E is **distributive** over a composition $\triangle$ on E if

$$x \triangledown (y \triangle z) = (x \triangledown y) \triangle (x \triangledown z),$$
$$(x \triangle y) \triangledown z = (x \triangledown z) \triangle (y \triangledown z),$$

for all $x, y, z \in E$.

Each of the compositions $\cup$ and $\cap$ on $\mathfrak{P}(E)$, for example, is distributive over the other and over itself by Theorem 3.1. The distributivity of multiplication over addition on N is one consequence of the following theorem.

Theorem 11.8. Let $\triangle$ be a composition on E, and let a and b be elements of E. If $\triangle$ is associative, then for all $n, m \in N^*$,

$$(1) \qquad \triangle^{n+m}a = (\triangle^n a)\triangle(\triangle^m a).$$

If $\triangle$ is associative and if $a\triangle b = b\triangle a$, then for all $n, m \in N^*$,

$$(2) \qquad (\triangle^m a)\triangle(\triangle^n b) = (\triangle^n b)\triangle(\triangle^m a),$$

$$(3) \qquad \triangle^n(a\triangle b) = (\triangle^n a)\triangle(\triangle^n b).$$

If there is a neutral element e for $\triangle$, then (1), (2), and (3) hold if either $n = 0$ or $m = 0$, and for all $m \in N$,

$$(4) \qquad \triangle^m e = e.$$

Proof. We shall assume that $\triangle$ is associative. Let $n \in N^*$, and let S be the set of all $m \in N^*$ for which (1) holds. Then $1 \in S$, for

$$\triangle^{n+1}a = (\triangle^n a)\triangle a = (\triangle^n a)\triangle(\triangle^1 a)$$

by definition. If $m \in S$, then $m + 1 \in S$, for

$$\triangle^{n+(m+1)}a = \triangle^{(n+m)+1}a = (\triangle^{n+m}a)\triangle a$$
$$= ((\triangle^n a)\triangle(\triangle^m a))\triangle a$$
$$= (\triangle^n a)\triangle((\triangle^m a)\triangle a)$$
$$= (\triangle^n a)\triangle(\triangle^{m+1}a).$$

Thus, by induction, $S = N^*$, so (1) is verified for all $n, m \in N^*$.

Henceforth, we shall assume in addition that $a\triangle b = b\triangle a$. Let T be the set of all $n \in N^*$ such that

$$(2') \qquad (\triangle^n a)\triangle b = b\triangle(\triangle^n a).$$

By our assumption, $1 \in T$. If $n \in T$, then $n + 1 \in T$, for

$$(\triangle^{n+1}a)\triangle b = ((\triangle^n a)\triangle a)\triangle b = (\triangle^n a)\triangle(a\triangle b)$$
$$= (\triangle^n a)\triangle(b\triangle a) = ((\triangle^n a)\triangle b)\triangle a$$
$$= (b\triangle(\triangle^n a))\triangle a = b\triangle((\triangle^n a)\triangle a)$$
$$= b\triangle(\triangle^{n+1}a).$$

Thus, by induction, $T = N^*$, so (2′) holds for all $n \in N^*$. Replacing a and b in (2′) respectively with b and $\triangle^m a$, which commute with each other as we have just seen, we obtain (2).

Let U be the set of all $n \in N^*$ such that (3) holds. Clearly, $1 \in U$. If $n \in U$, then $n + 1 \in U$, for by associativity and (2),

$$\triangle^{n+1}(a \triangle b) = (\triangle^n(a \triangle b)) \triangle (a \triangle b)$$
$$= ((\triangle^n a) \triangle (\triangle^n b)) \triangle (a \triangle b)$$
$$= (\triangle^n a) \triangle (((\triangle^n b) \triangle a) \triangle b)$$
$$= (\triangle^n a) \triangle ((a \triangle (\triangle^n b)) \triangle b)$$
$$= ((\triangle^n a) \triangle a) \triangle ((\triangle^n b) \triangle b)$$
$$= (\triangle^{n+1} a) \triangle (\triangle^{n+1} b).$$

Thus, by induction, $U = N^*$, so (3) holds for all $n \in N^*$.

The final assertion is proved by a simple inductive argument.

For a composition denoted by $+$, (1)–(4) become:

(5) $$(n + m).a = n.a + m.a$$

(6) $$m.a + n.b = n.b + m.a$$

(7) $$n.(a + b) = n.a + n.b$$

(8) $$m.0 = 0.$$

For a composition denoted by $\cdot$, they become:

(9) $$a^{n+m} = a^n a^m$$

(10) $$a^m b^n = b^n a^m$$

(11) $$(ab)^n = a^n b^n$$

(12) $$1^m = 1.$$

Theorem 11.9. Multiplication on N is distributive over addition.

Proof. By the definition of nm and $n.m$,

$$nm = n.m$$

for all n, $m \in N$. The assertion, therefore, follows from (5) and (7), since addition on N is associative and commutative.

Theorem 11.10. The natural number 1 is the identity element for multiplication on N, and multiplication is a commutative composition.

Proof. By the definition of multiplication, the function g_1 defined by

$$g_1(n) = n \cdot 1$$

is the unique function from N into N satisfying

$$g_1(0) = 0,$$
$$g_1(n + 1) = g_1(n) + 1$$

for all $n \in N$. But the identity function 1_N on N satisfies

$$1_N(0) = 0,$$
$$1_N(n + 1) = 1_N(n) + 1$$

for all $n \in N$. Hence, $1_N = g_1$, so $n = n \cdot 1$ for all $n \in N$. Also,

$$1 \cdot n = g_n(1) = g_n(0) + n$$
$$= 0 + n = n$$

for all $n \in N$ by the definition of multiplication. Hence, 1 is the identity element for multiplication.

To prove that multiplication is commutative, let $m \in N$, and let g_m and h_m be the functions from N into N defined by

$$g_m(n) = nm$$
$$h_m(n) = mn$$

for all $n \in N$. By the definition of multiplication, g_m is the unique function from N into N satisfying

$$g_m(0) = 0,$$
$$g_m(n + 1) = g_m(n) + m$$

for all $n \in N$. But

$$h_m(0) = m0 = 0$$

by (8), and

$$h_m(n + 1) = m(n + 1) = mn + m1$$
$$= mn + m = h_m(n) + m$$

by Theorem 11.9, since 1 is the identity element for multiplication. Hence, $h_m = g_m$, so multiplication is commutative.

Theorem 11.11. Let $\triangle$ be an associative composition on E, and let $a \in E$. For all $n, m \in N^*$,

$$(13) \qquad \triangle^{nm}a = \triangle^n(\triangle^m a) = \triangle^m(\triangle^n a),$$

and if in addition there is a neutral element for $\triangle$, then (13) holds for all $n, m \in N$.

Proof. Let $b = \triangle^m a$, and let h be the function from N^* into E defined by

$$h(n) = \triangle^{nm}a$$

for all $n \in N^*$. By definition, the function f_b from N^* into E defined by

$$f_b(n) = \triangle^n b$$

for all $n \in N^*$ is the unique function from N^* into E satisfying

$$f_b(1) = b,$$
$$f_b(n + 1) = f_b(n) \triangle b$$

for all $n \in N^*$. But

$$h(1) = \triangle^m a = b,$$

and by (1) and Theorems 11.9 and 11.10,

$$h(n + 1) = \triangle^{(n+1)m}a = \triangle^{nm+m}a$$
$$= (\triangle^{nm}a)\triangle(\triangle^m a) = h(n)\triangle b$$

for all $n \in N^*$. Hence, $h = f_b$, so

$$\triangle^{nm}a = \triangle^n(\triangle^m a)$$

for all $n, m \in N^*$. As multiplication on N is commutative, from what we have just proved we obtain

$$\triangle^m(\triangle^n a) = \triangle^{mn}a = \triangle^{nm}a$$

for all $m, n \in N^*$. Thus, (13) holds for all $n, m \in N^*$. If there is a neutral element e for $\triangle$, then (13) holds if either n or m is 0, for

$$\triangle^{n0}a = e = \triangle^{0m}a$$

since $n0 = 0 = 0m$, and

$$\triangle^n(\triangle^0 a) = e = \triangle^0(\triangle^m a),$$
$$\triangle^0(\triangle^n a) = e = \triangle^m(\triangle^0 a)$$

by (4).

For compositions denoted by $+$, (13) becomes

$$(14) \qquad\qquad (nm).a = n.(m.a) = m.(n.a),$$

and for compositions denoted by $\cdot$, (13) becomes

$$(15) \qquad\qquad a^{nm} = (a^m)^n = (a^n)^m.$$

Theorem 11.12. Multiplication on N is an associative composition.

Proof. Since $n.m = nm$ for all $n, m \in N$, the assertion follows from (14) and the associativity of addition on N.

Theorem 11.13. Let $m, n,$ and p be natural numbers.

1° $0 \cdot m = 0 = m \cdot 0$; conversely, if $mn = 0$, then either $m = 0$ or $n = 0$.
2° If $n > 0$, then $m < p$ if and only if $nm < np$.

Proof. 1° By definition, $0 \cdot m = 0$, and therefore $m \cdot 0 = 0$ as multiplication is commutative. If $mn = 0$ but $n \neq 0$, then $n \geq 1$, so

$$0 = mn = m((n - 1) + 1) = m(n - 1) + m \cdot 1 \geq m \geq 0,$$

whence $m = 0$. 2° Necessity: Since $m < p$ and $n > 0$, $n(p - m) > 0$ by 1°, so

$$np = n((p - m) + m) = n(p - m) + nm > nm$$

by Theorem 11.3. Sufficiency: As $nm < np$, clearly $m \neq p$; if $p < m$, then $np < nm$ by what we have just proved, a contradiction; hence, $m < p$.

Corollary. $(N, \cdot, \leq)$ is an ordered commutative semigroup. The sub-semigroup of cancellable elements of $(N, \cdot)$ is N^*. The only invertible element of $(N, \cdot)$ is 1.

Proof. The first two assertions follow at once from Theorem 11.13. Let m be invertible for multiplication, and let n be such that $mn = 1$. Then $m \neq 0$ and $n \neq 0$, so $m \geq 1$ and $n \geq 1$. If $m > 1$, then $mn > n \geq 1$ by Theorem 11.13, a contradiction. Hence, $m = 1$.

Definition. If $m \in N^*$ and if $n \in N$, we shall say that m **divides** n or that n is a **multiple** of m and write $m \mid n$ if there is a natural number p such that $mp = n$.

If $m \mid n$, there is exactly one natural number p such that $mp = n$, since m is cancellable for multiplication. If $m \mid n$, we shall denote the unique natural number p such that $mp = n$ by $\dfrac{n}{m}$ (or n/m), so that by definition,

$$m \cdot \frac{n}{m} = n = \frac{n}{m} \cdot m.$$

We may now prove that every naturally ordered semigroup is isomorphic to $(N, +, \leq)$. Although Theorems 11.1–11.5 were stated only for the specific naturally ordered semigroup $(N, +, \leq)$, they are, of course, valid for any naturally ordered semigroup.

Theorem 11.14. Let $(N', +', \leq')$ be a naturally ordered semigroup, let $0'$ be the smallest element of N', and let $1'$ be the smallest element of the complement of $\{0'\}$. The function g from N into N' defined by

$$g(n) = n.1'$$

for all $n \in N$ is an isomorphism from $(N, +, \leq)$ onto $(N', +', \leq')$.

Proof. Let S' be the range of g. By Theorem 11.1, the element $0'$ is the neutral element for $+'$, and by the definition of $0.1'$, therefore,

$$g(0) = 0.1' = 0'.$$

Thus, $0' \in S'$. If $x' \in S'$, then $x' +' 1' \in S'$, for if $g(n) = x'$, then

$$x' +' 1' = n.1' +' 1' = n.1' +' 1.1'$$
$$= (n + 1).1' = g(n + 1)$$

by (5). Hence, $S' = N'$ by Theorem 11.5 applied to N'. Therefore, g is surjective.

For each $p \in N$,
$$p.1' <' (p + 1).1',$$

for $p.1' <' p.1' +' 1'$ by Theorem 11.4 applied to N', and $p.1' +' 1' = p.1' +' 1.1' = (p + 1).1'$ by (5). Let

$$S = \{p \in N: n.1' <' p.1' \text{ for all } n \in N_p\}.$$

Clearly, $0 \in S$ as $N_0 = \emptyset$. If $p \in S$, then $p + 1 \in S$, for if $n < p + 1$, then by Corollary 11.4.1 either $n < p$, in which case,

$$n.1' <' p.1' <' (p + 1).1'$$

as $p \in S$, or $n = p$, in which case,

$$n.1' = p.1' <' (p + 1).1'.$$

Therefore, by induction, $S = N$, so if $n < p$, then $n.1' <' p.1'$. Thus, by (5) and Theorem 10.4, g is an isomorphism from $(N, +, \leq)$ onto $(N', +', \leq')$.

It is easy to see that g is the only isomorphism from the semigroup $(N, +)$ onto $(N', +')$ (Exercise 11.18).

An ordered pair may be thought of roughly as a list consisting of two entries written down in a definite order. For each strictly positive integer n, we may now introduce a concept corresponding analogously to a list consisting of n entries written down in a definite order.

Definition. Let $n \in N^*$. An **ordered n-tuple** is a surjection whose domain is the integer interval $[1, n]$. If f is an ordered n-tuple, $f(k)$ is called the **kth term** of the ordered n-tuple for each $k \in [1, n]$.

The reason for requiring that an ordered n-tuple be a surjection, rather than merely a function, is a technical one: If $a_1, \ldots, a_n$ are objects, there are infinitely many functions f with domain $[1, n]$ satisfying $f(1) = a_1, \ldots, f(n) = a_n$, since for each set E containing $\{a_1, \ldots, a_n\}$ there is such a function with codomain E. By requiring that the function be a surjection, however, we eliminate all but one of those functions, namely, that with codomain $\{a_1, \ldots, a_n\}$.

We shall, of course, employ the words "couple," "triple," "quadruple," etc., respectively for "2-tuple," "3-tuple," "4-tuple." If $a_1, \ldots, a_n$ are any objects, the ordered n-tuple f defined by $f(k) = a_k$ for each $k \in [1, n]$ is usually denoted by $(a_1, \ldots, a_n)$ or by $(a_k)_{1 \leq k \leq n}$. Frequently, an ordered n-tuple is denoted simply by putting parentheses around a list of its values written down in order. For example, $(3, 2, 5)$ is the ordered triple f defined by $f(1) = 3, f(2) = 2, f(3) = 5$, and (a_9, a_0, a_4, a_3) is the ordered quadruple g defined by $g(1) = a_9, g(2) = a_0, g(3) = a_4, g(4) = a_3$.

Clearly,

$$(a_1, \ldots, a_n) = (b_1, \ldots, b_m)$$

if and only if $n = m$ and $a_k = b_k$ for all $k \in [1, n]$. Consequently, we may "identify" ordered pairs with corresponding ordered couples and ordered triads with corresponding ordered triples; that is, we redefine every term previously defined so that every reference to an ordered pair is replaced by a reference to the corresponding ordered couple (i.e., the ordered couple having the same first term and the same second term as the ordered pair), and similarly, every reference to an ordered triad is replaced by a reference to the corresponding ordered triple. From now on, unless otherwise indicated, (a, b) denotes the ordered couple whose first term is a and whose second term is b, and similarly, (a, b, c) denotes the ordered triple whose first, second, and third terms are respectively a, b, and c. For example, if E, F, and G are sets, then $E \times F$ is by our redefinition the set of all ordered couples (x, y) such that $x \in E$ and $y \in F$, and $E \times F \times G$ is the set of all ordered triples (x, y, z) such that $x \in E$, $y \in F$, and $z \in G$. More generally, if $E_1, \ldots, E_n$ are sets, the **cartesian product** of $E_1, \ldots, E_n$ is the set $\prod_{k=1}^{n} E_k$ defined by

$$\prod_{k=1}^{n} E_k = \{(x_1, \ldots, x_n): x_k \in E_k \text{ for all } k \in [1, n]\}.$$

The cartesian product of $E_1, \ldots, E_n$ is often also denoted by $E_1 \times \ldots \times E_n$.

By means of the Principle of Recursive Definition it is possible to give a precise meaning to such expressions as $a_1 \triangle a_2 \triangle a_3 \triangle \ldots \triangle a_{n-1} \triangle a_n$, where $\triangle$ is a composition on E and $(a_1, a_2, a_3, \ldots, a_{n-1}, a_n)$ is any ordered n-tuple of elements of E. This is accomplished in a formal manner in Appendix A; here we shall give an informal summary of that appendix. Roughly speaking, we define $a_1 \triangle a_2 \triangle a_3 \triangle \ldots \triangle a_{n-1} \triangle a_n$ to be the element

$$(\ldots ((a_1 \triangle a_2) \triangle a_3) \triangle \ldots \triangle a_{n-1}) \triangle a_n,$$

where all the left parentheses occur at the beginning of the expression. This choice is, of course, arbitrary, but the General Associativity Theorem (Theorem A.5) assures us that if $\triangle$ is associative, any two ways of meaningfully inserting parentheses in the expression $a_1 \triangle a_2 \triangle a_3 \triangle \ldots \triangle a_{n-1} \triangle a_n$ yield the same element. If $\triangle$ is both commutative and associative, then no matter how we rearrange the terms of the n-tuple $(a_1, \ldots, a_n)$ and form the composite, the result is the same; more precisely (Theorem A.7), for every permutation σ of $[1, n]$,

$$a_{\sigma(1)} \triangle a_{\sigma(2)} \triangle a_{\sigma(3)} \triangle \ldots \triangle a_{\sigma(n-1)} \triangle a_{\sigma(n)} = a_1 \triangle a_2 \triangle a_3 \triangle \ldots \triangle a_{n-1} \triangle a_n.$$

Whenever $+$ is used to denote a composition, the sum $a_1 + a_2 + \ldots + a_n$ is often also denoted by $\sum_{k=1}^{n} a_k$, and similarly, when $\cdot$ is used to denote a composition, the product $a_1 a_2 \ldots a_n$ is often denoted by $\prod_{k=1}^{n} a_k$.

An inductive argument establishes that if $\triangle$ is associative and if b commutes with each of $a_1, \ldots, a_n$, then b commutes with $a_1 \triangle \ldots \triangle a_n$ (Theorem A.6). An inductive argument also establishes that if $+$ and $\cdot$ are compositions of E such that $\cdot$ is distributive over $+$, then

$$\left(\sum_{k=1}^{n} a_k \right) b = \sum_{k=1}^{n} a_k b,$$

$$b \left(\sum_{k=1}^{n} a_k \right) = \sum_{k=1}^{n} b a_k$$

for all $a_1, \ldots, a_n, b \in E$ (Theorem A.8). Henceforth, we shall assume the truth of these statements; formal definitions and proofs of the theorems mentioned are given in Appendix A.

EXERCISES

11.1. Let Q_+ be the set of all positive rational numbers. Then $(Q_+, +, \leq)$ is an ordered semigroup satisfying (NO 2), (NO 3), and (NO 4), and the neutral element of Q_+ for addition is its smallest element, but $(Q_+, +)$ is not isomorphic to $(N, +)$. [If f were an isomorphism from Q_+ into N, determine the parity of $f(q)$ by considering $f(q/2)$.]

11.2. Let β be an object not in N, and let $M = N \cup \{\beta\}$.
(a) We extend addition on N to a composition on M by defining

$$0 + \beta = \beta + 0 = \beta,$$

$$\beta + \beta = \beta,$$

$$n + \beta = \beta + n = n$$

for all $n \in N^*$. Then $(M, +)$ is a commutative semigroup, and 0 is the neutral element for addition on M.
(b) There is a unique total ordering $\leq$ on M inducing on N its given total ordering such that $0 < \beta < 1$.
(c) $(M, +, \leq)$ is an ordered semigroup satisfying (NO 1), (NO 3), and (NO 4).
(d) The semigroup $(M, +)$ is not isomorphic to $(N, +)$.

11.3. Let $E = N - \{1\}$. Then E is an ordered subsemigroup of $(N, +, \leq)$ satisfying (NO 1), (NO 2), and (NO 4), but $(N, +)$ is not isomorphic to $(E, +)$.

11.4. Let $(E, \triangle, \leq)$ be an ordered semigroup. If $a \leq b$, then $\triangle^n a \leq \triangle^n b$ for all $n \in N^*$. If $a < b$ and if either a or b is cancellable, then $\triangle^n a < \triangle^n b$ for all $n \in N^*$.

*11.5. If $(E, +, \leq)$ is an ordered semigroup satisfying (NO 1), (NO 2), and (NO 3), then $+$ is commutative. [Argue as in Theorem 11.1 to show that 0 is the neutral element, and then show that 1 commutes with every element.]

11.6. A semigroup $(E, +)$ is **inductive** if there exist α and β in E satisfying the following condition:

(Ind)   The only subset of E containing α and containing $x + \beta$ whenever it contains x is E itself.

If $(E, +)$ is a semigroup having elements α and β that commute and satisfy (Ind), then $+$ is a commutative composition. [Show that α commutes with every element, and then that β commutes with every element.]

*11.7. A semigroup $(E, +)$ is a **Peano semigroup** if (NO 2) holds and if there exist $\alpha, \beta \in E$ satisfying (Ind) and the following conditions:

(NO 5) α is the neutral element for $+$.

(PS) $x + \beta \neq \alpha$ for all $x \in E$.

Let $(E, +)$ be a Peano semigroup.

(a) If $x \neq \alpha$, then $x + y \neq \alpha$ for all $y \in E$.
(b) Let $\leq$ be the relation on E satisfying $x \leq y$ if and only if there exists $z \in E$ such that $x + z = y$. Then $(E, +, \leq)$ is a commutative ordered semigroup. [Use Exercise 11.6.]
(c) If $x < y$, then $x + \beta \leq y$.
(d) $(E, +, \leq)$ is a naturally ordered semigroup whose smallest element is α and whose smallest element distinct from α is β. [Use (Ind) in showing that $\leq$ is a total ordering. If B were a nonempty subset of E containing no smallest element, consider

$$\{x \in E: \text{for all } y \in E, \text{ if } y \leq x, \text{ then } y \notin B\}.]$$

Thus, there exists a Peano semigroup if and only if there exists a naturally ordered semigroup.

*11.8. An ordered triad (E, s, α) is a **Peano model** if E is a set, s is a function from E into E, α is an element of E, and the following conditions hold:

1° The range of s does not contain α.
2° The function s is injective.
3° The only subset of E containing α and containing $s(x)$ whenever it contains x is E itself.

Let (E, s, α) be a Peano model.

(a) A function h from E into E is *admissible* for an element x of E if $h(\alpha) = x$ and $h \circ s = s \circ h$. Prove that for each $x \in E$ there exists exactly one function admissible for x.
(b) Let $+$ be the composition on E defined by

$$x + y = A_x(y)$$

for all x, $y \in E$, where A_x is the unique admissible function for x. Prove that $(E, +)$ is a Peano semigroup (Exercise 11.7).

(c) There exists a Peano model if and only if there exists a naturally ordered semigroup.

11.9. (a) If $E = \{a + nd : n \in N\}$ where $a \in N$ and $d \in N^*$ and if s is the function from E into E defined by

$$s(x) = x + d$$

for all $x \in E$, then (E, s, a) is a Peano model (Exercise 11.8).

(b) If $E = \{ar^n : n \in N\}$ where a and r are strictly positive real numbers and $r \neq 1$ and if s is the function from E into E defined by

$$s(x) = xr$$

for all $x \in E$, then (E, s, a) is a Peano model.

(c) For each of the above two models, give an expression in terms of ordinary addition and multiplication for $x \oplus y$ and $x \odot y$, where $\oplus$ is the composition defined in Exercise 11.8(b) and where $\odot$ is the multiplication determined by the associated naturally ordered semigroup. [One method is to use Theorem 11.14.]

*11.10. Let $(E, +, \leq)$ be an ordered semigroup possessing elements α and β satisfying (NO 5), (Ind), and the following conditions:

(NO 6)   The ordering $\leq$ is a total ordering.

(NO 7) $\alpha < \beta$.

(a) If $x < y$, then $x + \beta \leq y$.

(b) $(E, +, \leq)$ satisfies (NO 1), (NO 3), and (NO 4).

(c) If $(E, +, \leq)$ is $(N_n, \vee_n, \leq)$ (Exercise 10.2) where $n > 1$, then $(E, +, \leq)$ satisfies (NO 5), (Ind), (NO 6), and (NO 7), and consequently also (NO 1), (NO 3), and (NO 4), but does not satisfy (NO 2).

*11.11. If $(E, +, \leq)$ is an ordered semigroup possessing elements α and β satisfying (Ind), (NO 6), and (NO 7), then one of the following four conditions holds:

1° $E = \{\alpha, \beta\}$, and $x + y = \beta$ for all $x, y \in E$.

2° $E = \{\alpha, \beta\}$, and $+$ is $\rightarrow$.

3° α is a neutral element for $+$.

4° There exists $\gamma \in E$ such that $\gamma > \beta$, $E = \{\alpha, \beta, \gamma\}$, and

$$x + y = \begin{cases} y & \text{if } x = \alpha, \\ \gamma & \text{if } x \neq \alpha. \end{cases}$$

for all $x, y \in E$.

[Prove successively that $\alpha < \alpha + \beta$, that α is the smallest element, that $x + y \geq y$, that either $x = \alpha$ or $x \geq \beta$, and that either $\alpha + \alpha = \alpha$ or $\alpha + \alpha = \beta$.]

11.12. Let $\triangledown$ be a composition on N. Then $\triangledown$ is distributive over addition if and only if there exists $k \in N$ such that

$$m \triangledown n = kmn$$

for all $m, n \in N$. Furthermore, $\triangledown$ is distributive over addition and there is a neutral element for $\triangledown$ if and only if $\triangledown$ is multiplication.

11.13. A composition $\triangledown$ on E is **left distributive** over a composition $\triangle$ on E if

$$x \triangledown (y \triangle z) = (x \triangledown y) \triangle (x \triangledown z)$$

for all $x, z \in E$, and is $\triangledown$ **right distributive** over $\triangle$ if

$$(x \triangle y) \triangledown z = (x \triangledown z) \triangle (y \triangledown z)$$

for all $x, y, z \in E$.

(a) The composition $\rightarrow$ on E is left distributive over every composition on E, and $\leftarrow$ is right distributive over every composition on E.

(b) The composition $\rightarrow$ (respectively, $\leftarrow$) is distributive over a composition $\triangle$ on E if and only if $\triangle$ is an idempotent composition (Exercise 2.17).

(c) If $\triangle$ is left (right) distributive over itself and if there is a right (left) neutral element for $\triangle$, then $\triangle$ is an idempotent composition.

(d) If $\triangle$ is left (right) distributive over itself and if a is an idempotent element of E, then $b \triangle a$ (respectively, $a \triangle b$) is idempotent for all $b \in E$.

11.14. For each $c \in E$, let $[c]$ be the composition on E defined by

$$x[c]y = c$$

for all $x, y \in E$. A composition $\triangledown$ on E is left distributive over $[c]$ if and only if $x \triangledown c = c$ for all $x \in E$. Also, $\triangledown$ is right distributive over $[c]$ if and only if $c \triangledown x = c$ for all $x \in E$. The composition $[c]$ is distributive over a composition $\triangle$ on E if and only if $c \triangle c = c$.

11.15. If $\triangledown$ is left distributive (respectively, right distributive) over every commutative associative composition on E, then $\triangledown$ is the composition $\rightarrow$ (respectively, $\leftarrow$). [Use Exercise 11.14.]

*11.16. (a) Every composition on E is distributive over $\rightarrow$ and $\leftarrow$.

(b) Let A be a nonempty subset of E distinct from E. Let $a \in A$, and let $b \in A^c$. The composition $\triangledown$ on E defined by

$$x \triangledown y = \begin{cases} a \text{ if } \{x, y\} \subseteq A, \\ b \text{ if } \{x, y\} \nsubseteq A \end{cases}$$

is associative and commutative.

(c) If $\triangle$ is a composition on E such that every commutative associative composition on E is distributive over $\triangle$, then $\triangle$ is either $\leftarrow$ or $\rightarrow$.

11.17. Let $m, n, p \in N$.

(a) Any two of the following three statements imply the third: $m \mid n$; $m \mid p$; $m \mid n + p$.

(b) If $n \leq p$, any two of the following three statements imply the third: $m \mid n$; $m \mid p$; $m \mid p - n$.

11.18. With the notation of Theorem 11.14, prove that g is the only isomorphism from $(N, +)$ onto $(N', +')$. [Show by induction that if f were an isomorphism such that $f(1) \neq 1'$, then $f(n) \neq 1'$ for all $n \in N$.]

12. Finite Sets*

Having developed the theory of the natural numbers, we may now formally define what it means for a set to be finite and prove rigorously certain familiar assertions about finite sets. Some of the theorems presented here are sophisticated expressions of facts obvious to any child encountering arithmetic for the first time. Indeed, they are so intuitively evident that it is sometimes hard to realize that they require proof. Yet in a formal development they do require proof, and their proofs demonstrate anew the importance of the Principle of Mathematical Induction.

Definition. If E and F are sets, we shall say that E is **equipotent** to F if there is a bijection from E onto F.

We shall write $E \sim F$ if E is equipotent to F, and $E \not\sim F$ if E is not equipotent to F.

Theorem 12.1. Let E, F, and G be sets.

$1°$ $E \sim E$.
$2°$ If $E \sim F$, then $F \sim E$.
$3°$ If $E \sim F$ and if $F \sim G$, then $E \sim G$.

Proof. Since 1_E is a permutation of E, $E \sim E$. From Theorems 5.4 and 5.5 we obtain $2°$ and $3°$.

Theorem 12.2. If $E \sim F$ and if a and b are elements of E and F respectively, then $E - \{a\} \sim F - \{b\}$.

---

* Except for the first definition and theorem, this section, which is devoted to proofs of familiar facts about finite sets, may be omitted.

Proof. By hypothesis, there is a bijection f from E onto F. It is easy to verify that the function g from $E - \{a\}$ into $F - \{b\}$ defined by

$$g(x) = \begin{cases} f(x) \text{ if } f(x) \neq b, \\ f(a) \text{ if } f(x) = b \end{cases}$$

for all $x \in E - \{a\}$ is a bijection from $E - \{a\}$ onto $F - \{b\}$.

In the sequel we shall use the following fact, which is a consequence of Corollary 11.4.1: For every natural number n, $n \in N_{n+1}$ and

$$N_n = N_{n+1} - \{n\}.$$

In particular, if $m > 0$, then $m - 1 \in N_m$ and

$$N_{m-1} = N_m - \{m - 1\}.$$

Theorem 12.3. If n and m are natural numbers, then $N_n \sim N_m$ if and only if $n = m$.

Proof. The condition is sufficient by $1°$ of Theorem 12.1.
Necessity: Let

$$S = \{n \in N : N_m \not\sim N_n \text{ for all } m \in N_n\}.$$

Clearly, $0 \in S$, for $N_0 = \emptyset$. Let $n \in S$. To prove that $n + 1 \in S$, let $m \in N_{n+1}$. If $m = 0$, then $N_m \not\sim N_{n+1}$ since $N_0 = \emptyset$ and $N_{n+1} \neq \emptyset$. If $m > 0$ and if $N_m \sim N_{n+1}$, then $N_{m-1} \sim N_n$ by Theorem 12.2 and $m - 1 < n$ by Theorem 11.3, in contradiction to our assumption that $n \in S$. Thus, $n + 1 \in S$. By induction, therefore, $S = N$, so $N_m \not\sim N_n$ whenever $m < n$. Consequently, by $2°$ of Theorem 12.1, $N_m \not\sim N_n$ whenever $m \neq n$.

Thus, by Theorem 12.3 and $3°$ of Theorem 12.1, if $E \sim N_n$, then $E \not\sim N_m$ for all $m \neq n$. In view of this fact, we may make the following definition.

Definition. A set E is **finite** if there exists a natural number n such that $E \sim N_n$. If E is finite, the unique natural number n such that $E \sim N_n$ is called the **number of elements** in E. A set E is **infinite** if E is not finite.

We shall also say that *E has n elements* if E is finite and the number of elements in E is n. By $3°$ of Theorem 12.1, to show that a set has n elements, it suffices to show that it is equipotent to a set having n elements.

Theorem 12.4. If E has $n + 1$ elements and if $a \in E$, then $E - \{a\}$ has n elements.

The assertion is an immediate consequence of Theorem 12.2.

Definition. A subset E of a set F is a **proper subset** of F if $E \neq F$. The set F **properly contains** E if E is a proper subset of F.

The symbol $\subset$ means "is a proper subset of," and $\supset$ means "properly contains." We, therefore, write $E \subset F$ or $F \supset E$ if E is a proper subset of F.

Theorem 12.5. If E is a proper subset of F and if F has n elements, then E is finite and has fewer than n elements. Consequently, every subset of a set having n elements is finite and has at most n elements.

Proof. Let S be the set of all natural numbers m such that every proper subset of any set having m elements is finite and has fewer than m elements. Clearly, $0 \in S$, for $\emptyset$ is the only set having zero elements, and there are no proper subsets of $\emptyset$. Let $m \in S$. To show that $m + 1 \in S$, let A be a proper subset of a set B having $m + 1$ elements. Then there exists $b \in B$ such that $b \notin A$, so $A \subseteq B - \{b\}$, which has m elements by Theorem 12.4. If $A = B - \{b\}$, then A has m elements; if $A \subset B - \{b\}$, then A is finite and has fewer than m elements, since $m \in S$; since $m < m + 1$, therefore, A has fewer than $m + 1$ elements. Hence, $m + 1 \in S$. By induction, therefore, $S = N$. In particular, $n \in S$, and the proof is complete.

Theorem 12.6. Let E be a set having n elements. If f is a surjection from E onto F, then F is finite and has at most n elements, and furthermore, F has exactly n elements if and only if f is a bijection.

Proof. Since E has n elements, there is a surjection from E onto F if and only if there is a surjection from N_n onto F. Therefore, we need only consider the case where $E = N_n$. For each $y \in F$ the set E_y defined by

$$E_y = \{x \in E : f(x) = y\}$$

is not empty as f is surjective. Therefore, by (NO 1) we may define a function g from F into E by

$$g(y) = \text{the smallest number in } E_y$$

for all $y \in F$. If $y \in F$, then $g(y) \in E_y$, whence $f(g(y)) = y$. Thus, $f \circ g = 1_F$, so g is injective by Theorem 5.2. The function obtained by restricting the codomain of g to its range E' is thus a bijection from F onto E'. By Theorem

12.5, E' is finite, and $m \leq n$ where m is the number of elements in E'. By Theorem 12.1, F has m elements. Suppose that $m = n$. Then by Theorem 12.5, $E' = E$, so g is a bijection from F onto E. Therefore,

$$f = f \circ 1_E = f \circ g \circ g^\leftarrow = 1_F \circ g^\leftarrow = g^\leftarrow.$$

In particular, f is a bijection from E onto F.

Theorem 12.7. Let E and F be finite sets having the same number of elements, and let f be a function from E into F. The following statements are equivalent:

    $1°$ f is bijective.
    $2°$ f is injective.
    $3°$ f is surjective.

Proof. By Theorem 12.5, $2°$ implies $3°$, for if f is injective, then the function obtained by restricting the codomain of f to its range F' is a bijection, so the subset F' of F has the same number of elements as F and consequently is F. By Theorem 12.6, $3°$ implies $1°$.

Theorem 12.8. The set N of natural numbers is infinite.

Proof. The function s defined by

$$s(n) = n + 1$$

for all $n \in N$ is an injection from N into N that is not a bijection, for

$$s(n) \geq 0 + 1 > 0$$

for all $n \in N$, and consequently, 0 is not in the range of s. Therefore, N is infinite by Theorem 12.7.

Theorem 12.9. If $(E, \leq)$ is a totally ordered structure, then every nonempty finite subset A of E has a greatest and a least member.

Proof. Let S be the set of all $n \in N^*$ such that every subset of E having n elements has a greatest and a least member. Clearly, $1 \in S$. Let $n \in S$. To show that $n + 1 \in S$, let A be a subset of E having $n + 1$ elements. Then there exists $b \in A$, and $A - \{b\}$ has n elements by Theorem 12.4. Consequently, $A - \{b\}$ has a greatest element c and a least element a as $n \in S$. The greater of c and b is then clearly the greatest element of A, and the lesser of a and b is the least element of A. Therefore, $n + 1 \in S$. Consequently, $S = N^*$ by induction, and the proof is complete.

EXERCISES

12.1. If every element of a finite semigroup E is cancellable, then E is a group. [Use Theorem 12.7 and Exercise 7.13.]

12.2. A set containing an infinite set is infinite.

12.3. If $E_1 \sim F_1$ and if $E_2 \sim F_2$, then $E_1 \times E_2 \sim F_1 \times F_2$.

12.4. If $E \sim G$ and if $F \sim H$, then $F^E \sim H^G$.

12.5. If E, F, and G are sets, then $(F \times G)^E \sim F^E \times G^E$.

12.6. If E, F, and G are sets, then $(G^E)^F \sim G^{E \times F}$.

*__12.7.__ A set E is **denumerable** if E is equipotent to N. A set is **countable** if it is either finite or denumerable. If E is a nonempty set, the following statements are equivalent:

1° E is countable.
2° There is a surjection from N onto E.
3° There is an injection from E into N.

[To show that 2° implies 3°, let f be a surjection from N onto E, and consider the function g from E into N defined by

$$g(y) = \text{the smallest number in } \{x \in N : f(x) = y\}$$

for all $y \in E$. To show that 3° implies 1°, show first that any infinite subset A of N is denumerable; for this, consider the function h from A into N defined by

$$h(x) = \text{the number of elements in } A \cap N_x$$

for all $x \in A$.]

12.8. Every subset of a countable set is countable, and every infinite subset of a denumerable set is denumerable.

*__12.9.__ Let E be a set. There is a proper subset F of E such that $E \sim F$ if and only if E contains a denumerable subset. [To show that the condition is necessary, apply the Principle of Recursive Definition to a bijection s from E onto F.]

12.10. (Cantor's Theorem) If E is a set, there is no surjection from E onto $\mathfrak{P}(E)$. [If f is a function from E into $\mathfrak{P}(E)$, show that $\{x \in E : x \notin f(x)\}$ is not in the range of f.] Infer that $\mathfrak{P}(N)$ is an uncountable set.

12.11. If $(E, \leq)$ and $(F, \leqslant)$ are well-ordered structures, then there exists at most one isomorphism from $(E, \leq)$ onto $(F, \leqslant)$.

*12.12. If $(E, +)$ is an infinite semigroup possessing distinct elements α and β satisfying (Ind) (Exercise 11.6), then $(E, +)$ is isomorphic to $(N, +)$. [Let $a_0 = \alpha$ and $a_k = \alpha + k.\beta$ for all $k \in N^*$; show that $f: k \mapsto a_k$ is a bijection from N onto E and that if $a_n = \beta$ and $a_m = \alpha + \alpha$, then $a_1 = a_{m+n}$.]

*12.13. A semigroup $(E, +)$ is **recursive** if there exist α, $\beta \in E$ such that for every function s from N into N, there is a unique function f from E into N satisfying

$$f(\alpha) = 0,$$

$$f(x + \beta) = s(f(x))$$

for all $x \in E$. A recursive semigroup is isomorphic either with $(N, +)$ or with $(N^*, +)$. [Show first that if $x \in E$, then either $x = \alpha$ or $x = y + \beta$ for some $y \in E$. Show then that E is an infinite inductive semigroup by considering the function $s: n \mapsto n + 1$ from N into N.]

*12.14. A finite semigroup $(E, \cdot)$ contains an idempotent element (Exercise 2.17). [Given $a \in E$, show that there exist n, $k \in N^*$ such that $a^{n+k} = a^n$. Then show that a^{2kn} is idempotent by considering a^{n+sk} where $s = 2n$.]

*12.15. (a) If f is an injection from A into A whose range is contained in a subset B of A, then $A \sim B$. [Show the existence of a sequence $(f_n)_{n \geq 0}$ in A^A such that $f_0 = 1_A$ and $f_{n+1} = f \circ f_n$ for all $n \geq 0$. (A formal definition of "sequence" is contained in Appendix A.) Let $C = \{y \in A: \text{for some } n \geq 0 \text{ and some } x \in A - B, y = f_n(x)\}$. Show that $g: A \to B$ defined by

$$g(x) = \begin{cases} x \text{ if } x \in A - C \\ f(x) \text{ if } x \in C \end{cases}$$

is the desired bijection.]
(b) (Cantor-Bernstein-Schröder Theorem) If there exist an injection u from E into F and an injection v from F into E, then $E \sim F$. [Apply (a) to $A = E$, $B =$ the range of v.]

13. The Division Algorithm

Long division, as taught in grammar school, is a technique for "dividing" a nonzero natural number b into a natural number a to obtain natural numbers q and r (called respectively the "quotient" and "remainder") satisfying $a = bq + r$ and $r < b$. The Division Algorithm is the assertion that natural numbers q and r satisfying these properties always exist and are unique. By use of the Division Algorithm we may justify the familiar "decimal" system of notation for natural numbers.

Theorem 13.1. (Division Algorithm) If a and b are natural numbers and if $b > 0$, then there exist unique natural numbers q and r satisfying

(1) $$a = bq + r,$$

(2) $$r < b.$$

Proof. To show that there exist natural numbers q and r satisfying (1) and (2), let $S = \{s \in N : b(s + 1) > a\}$. Then $S \neq \emptyset$, for as

$$b(a + 1) = ba + b > ba \geq 1 \cdot a = a,$$

$a \in S$. Thus, S has a smallest member q. If $q = 0$, then $b = b(0 + 1) > a$, so $q = 0$ and $r = a$ satisfy (1) and (2). Assume that $q > 0$. Then $q - 1 \notin S$ as $q - 1 < q$, so

$$bq = b((q - 1) + 1) \leq a < b(q + 1).$$

Let $r = a - bq$. Then (1) holds, and since

$$bq + r = a < b(q + 1) = bq + b,$$

(2) holds by Theorem 11.3.

To prove uniqueness, suppose that

$$a = bq + r = bq' + r',$$
$$r < b, \qquad r' < b.$$

Then

$$bq \leq bq + r = bq' + r' < bq' + b = b(q' + 1),$$

so $q < q' + 1$ by 2° of Theorem 11.13, whence $q + 1 \leq q' + 1$ by Theorem 11.4. Similarly,

$$bq' \leq bq' + r' = bq + r < bq + b = b(q + 1),$$

so $q' < q + 1$, whence $q' + 1 \leq q + 1$. Consequently, $q' + 1 = q + 1$, so $q' = q$. Therefore, $bq' = bq$, and since $bq' + r' = bq + r$, we conclude also that $r' = r$.

In our decimal system of notation for natural numbers, "0," "1," "2," ..., "9" denote the first ten natural numbers (they may formally be defined by $2 = 1 + 1, 3 = 2 + 1, \ldots, 9 = 8 + 1$), and all other natural numbers are denoted by a suitable juxtaposition of these numerals. Thus, "3407" denotes the natural number $3 \cdot 10^3 + 4 \cdot 10^2 + 0 \cdot 10^1 + 7 \cdot 10^0$ (where 10, of course, is defined to be $9 + 1$), and in general, if $r_1, \ldots, r_m \in [0, 9]$ and if $r_1 \neq 0$, then "$r_1 r_2 \ldots r_m$" denotes the natural number

$$\sum_{k=1}^{m} r_k \cdot 10^{m-k}.$$

The following theorem justifies this method of notation.

Theorem 13.2. Let b be a natural number such that $b > 1$. For every nonzero natural number a there exists one and only one sequence* $(r_k)_{1 \leq k \leq m}$ of natural numbers satisfying

(3) $$a = \sum_{k=1}^{m} r_k b^{m-k},$$

(4) $$0 \leq r_k < b \text{ for all } k \in [1, m],$$

(5) $$r_1 \neq 0.$$

Proof. Let us call a sequence $(r_k)_{1 \leq k \leq m}$ of natural numbers satisfying (3)–(5) a *development of a to base b*. We shall first prove that every nonzero natural number a has a development to base b. This is, indeed, the case if $a < b$, for then the sequence whose only term is a is a development of a to base b. Suppose that there were nonzero natural numbers having no development to base b; by (NO 1) there would exist a smallest such natural number c. By the Division Algorithm, there would exist natural numbers q and r satisfying $c = bq + r$ and $r < b$, and furthermore, $q > 0$, since otherwise we would have $c = r < b$. Hence, by the definition of c and since $0 < q < bq \leq c$, there would exist a development $(s_k)_{1 \leq k \leq n}$ of q to base b. But then

$$c = bq + r = \sum_{k=1}^{n} s_k b^{n+1-k} + r,$$

so $(r_k)_{1 \leq k \leq n+1}$ would be a development of c to base b where

$$r_k = \begin{cases} s_k \text{ if } k \in [1, n], \\ r \text{ if } k = n + 1, \end{cases}$$

a contradiction. Thus, every nonzero natural number possesses a development to base b.

---

* A formal definition of "sequence" is contained in Appendix A (page 487).

To show that each nonzero natural number has only one development to base b, let S be the set of all $m \in N^*$ such that for every nonzero natural number a, if a has a development to base b consisting of m terms, then a has only one development to base b. If a has a development to base b of one term (i.e., if $0 < a < b$) and if $(s_k)_{1 \leq k \leq n}$ is also a development of a to base b, then $n = 1$, for otherwise,

$$a = \sum_{k=1}^{n} s_k b^{n-k} \geq s_1 b^{n-1} \geq b^{n-1} \geq b > a,$$

a contradiction, and therefore, $a = s_1$. Consequently, $1 \in S$. Let $m \in S$. To show that $m + 1 \in S$, let a be a nonzero natural number having a development $(r_k)_{1 \leq k \leq m+1}$ to base b of $m + 1$ terms. Suppose that $(s_k)_{1 \leq k \leq n}$ is also a development of a to base b. Then $n > 1$, for otherwise, s_1 would have the development $(r_k)_{1 \leq k \leq m+1}$ to base b, a contradiction of the fact that $1 \in S$. Let

$$q = \sum_{k=1}^{m} r_k b^{m-k},$$

$$p = \sum_{k=1}^{n-1} s_k b^{n-1-k}.$$

Then

$$bq + r_{m+1} = bp + s_n,$$

so $q = p$ and $r_{m+1} = s_n$ by the Division Algorithm. As $m \in S$ and as $(r_k)_{1 \leq k \leq m}$ and $(s_k)_{1 \leq k \leq n-1}$ are developments to base b of p, we conclude that $n - 1 = m$ and that $r_k = s_k$ for all $k \in [1, m]$; therefore, $m + 1 \in S$. By induction, therefore, $S = N^*$, and the proof is complete.

Theorem 13.2 implies that not only 10 but any natural number b greater than 1 may be chosen as a "base" for a system of notation for the nonzero natural numbers. If $b > 1$ and if $(r_k)_{1 \leq k \leq m}$ satisfies (4) and (5), we shall denote by $[r_1 r_2 \ldots r_m]_b$ (or simply by $r_1 r_2 \ldots r_m$ if it is clear what base is being used) the natural number a defined by (3). For example,

$$[10101]_2 = 1 \cdot 2^4 + 0 \cdot 2^3 + 1 \cdot 2^2 + 0 \cdot 2^1 + 1 \cdot 2^0 = 21,$$

and if "t" and "e" denote ten and eleven respectively,

$$[3t0e7]_{12} = 3 \cdot 12^4 + 10 \cdot 12^3 + 0 \cdot 12^2 + 11 \cdot 12^1 + 7 \cdot 12^0$$

$$= 79,627.$$

The systems most frequently used are the "decimal" system to base 10 and the "binary" system to base 2, the latter of which is employed by many electronic computers.

The techniques learned in grammar school of "carrying over" and "borrowing" for calculating sums, differences, products, and quotients of integers expressed in the decimal system are equally valid if notation to a base other than 10 is used. For example, here are some calculations where notation to base 6 is employed:

```
   153        2431        1502          42
   345       -1545      ×   35      23)1501
 +  20         442       13114        140
 ─────        ────        5310        100
  1002                  110214         51
                        ──────         11
```

EXERCISES

13.1. Express each of the following in the decimal system of notation:

$$[1001001]_2 \qquad [5410]_6 \qquad [8888]_9 \qquad [2t5e9]_{12}$$

13.2. Express each of the following numbers in the systems of notation to base 2, 6, and 12: 3456, 1728, 1000, 1971.

13.3. Perform the indicated computations, where all the numbers are expressed to base 6:

```
   2451        3012        5043      25)5000
    501       -2453      ×   254
 + 3034
```

13.4. Let a and b be natural numbers such that $1 < a < b$, let $r_1, \ldots, r_m$ be natural numbers $<b$, and let $c = [r_1 r_2 \ldots r_m]_b$.
(a) If $a \mid b$, then $a \mid c - r_m$, and in particular, $a \mid c$ if and only if $a \mid r_m$.
(b) Which of 2, 3, 4, and 6 divide $[3t58]_{12}$? $[4t0e]_{12}$? $[et09]_{12}$? $[5096]_{12}$? $[tee]_{12}$?

13.5. Let a and b be natural numbers such that $1 < a < b$, let $r_1, \ldots, r_m$ be natural numbers $<b$, let $c = [r_1 r_2 \ldots r_m]_b$, and let $s_b(c) = r_1 + \ldots + r_m$.
(a) If $a \mid b - 1$, then $a \mid c - s_b(c)$, and in particular, $a \mid c$ if and only if $a \mid s_b(c)$.
(b) Which of 2, 4, and 8 divide $[4078]_9$? $[5173]_9$? $[5601]_9$?
(c) Which of 3 and 9 divide 12733? 31563? 40701?

14. The Integers†

We usually think of the set Z of integers as formed by adjoining to the set N of natural numbers an additive inverse for each number $n > 0$. Our formal construction of the integers will be closely modeled on this idea.

First, we need a set disjoint from N that is equipotent to N^*, since after addition is defined on the enlarged set Z, $n \mapsto -n$ will be a bijection from N^* onto $Z - N$, the set we have adjoined to N. For this purpose, we shall adopt as an axiom the following statement, which may be proved in any formal development of the theory of sets:

If E is a set, there is a bijection from E onto a set E' disjoint from E.

By virtue of this axiom, there exist a set N^{*-} disjoint from N and a bijection f from N onto N^{*-}; we define the set Z to be $N \cup N^{*-}$, and we shall call the members of Z **integers**. Our choice of N^{*-} insures that it is disjoint from N, so that $N^{*-} = Z - N$, but as noted earlier, we want a bijection (which later will be the bijection $n \mapsto -n$, after addition is suitably defined) from N^* onto N^{*-}, not from N onto N^{*-}. The function $g: n \mapsto n - 1$ with domain N^* and codomain N is a bijection; indeed, g is surjective, for if $n \in N$, then $n + 1 \in N^*$ and $g(n + 1) = n$ by the definition of $(n + 1) - 1$, since n is that unique natural number p satisfying $1 + p = n + 1$; g is injective, for if n, $m \in N^*$ and if $n - 1 = m - 1$, then by the definition of $n - 1$ and $m - 1$,

$$n = 1 + (n - 1) = 1 + (m - 1) = m.$$

Consequently, $f \circ g$ is a bijection from N^* onto N^{*-}; this is the bijection that will become the bijection $n \mapsto -n$ from N^* onto N^{*-} after we define addition; looking forward to this, we shall denote $(f \circ g)(n)$ by n^- for each $n \in N^*$. Thus, Z is the union of the disjoint sets N and $\{n^- : n \in N^*\}$.

To extend addition on N to a composition on Z, grammar school arithmetic suggests the following definition:

Definition. We extend addition $+$ on N to a composition, again called **addition,** on Z as follows: If $m \in N$, $n \in N^*$,

$$m + n^- = \begin{cases} m - n \text{ if } n \leq m, \\ (n - m)^- \text{ if } m < n. \end{cases}$$

† Except for Theorem 14.5, this section, which is devoted to the construction of the integers from the natural numbers, may be omitted.

If $m \in N^*$, $n \in N$,

$$m^- + n = \begin{cases} n - m \text{ if } m \leq n, \\ (m - n)^- \text{ if } n < m. \end{cases}$$

If $m, n \in N^*$,

$$m^- + n^- = (m + n)^-.$$

Clearly, addition on Z is commutative. Also, $0 + n^- = (n - 0)^- = n^-$ for each $n \in N^*$, so 0 is the neutral element for addition on Z. Moreover, for each $n \in N^*$,

$$n + n^- = n - n = 0,$$

since $n \leq n$ and since 0 is the unique natural number p such that $n + p = n$; thus, for each $n \in N^*$, n^- and n are inverses of each other for addition. Consequently, we shall have proved that $(Z, +)$ is a commutative group once we have proved that addition on Z is associative. To do so, we need the following theorem.

Theorem 14.1. Let m, n, and p be natural numbers.

  1° If $p \leq n$, then (a) $p \leq m + n$, and (b) $(m + n) - p = m + (n - p)$.
  2° If $n \leq p$ and $p \leq m + n$, then (a) $p - n \leq m$, and (b) $(m + n) - p = m - (p - n)$.
  3° If $m + n \leq p$, then (a) $n \leq p$ and $m \leq p - n$, and (b) $p - (m + n) = (p - n) - m$.

Proof. 1° Since $n = 0 + n \leq m + n$, (a) holds. By commutativity and the definition of $n - p$,

$$p + [m + (n - p)] = p + [(n - p) + m] = [p + (n - p)] + m$$
$$= n + m = m + n,$$

so $m + (n - p) = (m + n) - p$ by the definition of $(m + n) - p$. Similar proofs establish 2° and 3°.

To prove the associativity of addition on Z, we shall first show that

(1) $$(m + n) + p^- = m + (n + p^-)$$

for all $m, n \in N$ and all $p \in N^*$. If $p \leq n$, then $(m + n) + p^- = (m + n) - p$ by 1°(a) of Theorem 14.1, and $m + (n + p^-) = m + (n - p)$ by definition; hence, (1) holds by 1°(b) of Theorem 14.1. If $n < p$ and $p \leq m + n$, a similar appeal to 2° of Theorem 14.1 establishes (1), and if $m + n < p$, an appeal to 3° of Theorem 14.1 yields (1). Moreover,

$$(2) \qquad\qquad (n + p^-) + m = n + (p^- + m)$$

for all m, $n \in N$ and all $p \in N^*$, since

$$(n + p^-) + m = m + (n + p^-) = (m + n) + p^-$$
$$= (n + m) + p^- = n + (m + p^-)$$
$$= n + (p^- + m)$$

by commutativity and by (1) applied twice. Furthermore,

$$(3) \qquad\qquad (p^- + n) + m = p^- + (n + m)$$

for all m, $n \in N$ and all $p \in N^*$, since

$$(p^- + n) + m = m + (n + p^-) = (m + n) + p^- = p^- + (n + m)$$

by commutativity and (1).

Next, we shall show that

$$(4) \qquad\qquad (p + n^-) + m^- = p + (n^- + m^-)$$

for all $p \in N$ and all m, $n \in N^*$. If $p < n$, then

$$(p + n^-) + m^- = (n - p)^- + m^- = [(n - p) + m]^- = [m + (n - p)]^-$$

by definition, and

$$p + (n^- + m^-) = p + (n + m)^- = [(n + m) - p]^- = [(m + n) - p]^-$$

by definition and by 1°(a) of Theorem 14.1; hence, (4) holds by 1°(b) of Theorem 14.1. If $n \le p$ and $p < m + n$, a similar appeal to 2° of Theorem 14.1 establishes (4); and if $m + n < p$, an appeal to 3° of Theorem 14.1 yields (4). Moreover,

$$(5) \qquad\qquad (n^- + p) + m^- = n^- + (p + m^-)$$

for all $p \in N$ and all m, $n \in N^*$, since

$$(n^- + p) + m^- = (p + n^-) + m^- = p + (n^- + m^-) = p + (m^- + n^-)$$
$$= (p + m^-) + n^- = n^- + (p + m^-)$$

by commutativity and by (4) applied twice. Furthermore,

(6) $$(m^- + n^-) + p = m^- + (n^- + p)$$

for all $p \in N$ and all $m, n \in N^*$, since

$$(m^- + n^-) + p = p + (n^- + m^-) = (p + n^-) + m^- = m^- + (n^- + p)$$

by commutativity and (4). Finally,

(7) $$(m^- + n^-) + p^- = m^- + (n^- + p^-)$$

for all $m, n, p \in N^*$ by the associativity of addition on N, since

$$(m^- + n^-) + p^- = (m + n)^- + p^- = [(m + n) + p]^-,$$
$$m^- + (n^- + p^-) = m^- + (n + p)^- = [m + (n + p)]^-.$$

From (1)–(7) we conclude that addition on Z is associative. Consequently, for each $n \in N^*$, n^- is the unique additive inverse of n, which we are now justified in denoting by $-n$ rather than n^-. In sum, we have proved the following theorem:

Theorem 14.2. $(Z, +)$ is a commutative group. The complement of N in Z is $\{-n: n \in N^*\}$.

If n and m are natural numbers such that $n \leq m$, then by the definition of $m + n^-$,

$$m + (-n) = m + n^- = m - n.$$

Consequently, no ambiguity arises if we extend the definition of $m - n$ to all ordered couples (m, n) of integers by defining $m - n$ to be $m + (-n)$.

If m and n are any integers, then $m - n$ is the unique integer p satisfying $n + p = m$, since

$$n + (m - n) = n + (m + (-n)) = n + ((-n) + m)$$
$$= (n + (-n)) + m = m,$$

and since the cancellability of n insures that there is at most one integer p satisfying $n + p = m$.

The notation introduced above for integers is used more generally for elements of any semigroup whose composition is denoted by $+$. If x

and y are elements of a semigroup $(E, +)$ and if y is invertible, then $x + (-y)$ is usually denoted by $x - y$, $(-y) + x$ by $-y + x$, and if in addition x is invertible, $(-x) + (-y)$ is denoted by $-x - y$.

Theorem 14.3. If $(E, +)$ is a semigroup with neutral element 0 and if a and b are invertible elements of E, then $b - a$ is invertible, and

$$-(b - a) = a - b.$$

Proof. The assertion results from the identities

$$(b - a) + (a - b) = (b + (-a)) + (a + (-b)) = b + (-b) = 0,$$
$$(a - b) + (b - a) = (a + (-b)) + (b + (-a)) = a + (-a) = 0.$$

We next wish to extend the postulated total ordering $\leq$ on N to a total ordering $\leqslant$ on Z compatible with addition. To see how we should define $\leqslant$, let us first determine what conditions $\leqslant$ must satisfy if it is to have these properties. So let us assume for the moment that $\leqslant$ is an ordering on Z that extends $\leq$ and is compatible with addition. First, if $a \in N$, then we must have $0 \leqslant a$, since $0 \leq a$ and since $\leqslant$ is an extension of $\leq$. If $a \notin N$, then $a = -n$ for some $n \in N^*$, so $0 \prec n$ by the preceding and hence, $a = -n \prec 0$ by Theorem 10.3. Hence, if b, $c \in Z$ and if $b \leqslant c$, then $0 = b - b \leqslant c - b$, whence $c - b \in N$ since $a \prec 0$ for all $a \notin N$; conversely, if $c - b \in N$, then $0 \leqslant c - b$, whence $b = 0 + b \leqslant (c - b) + b = c$. In sum, if there is an ordering $\leqslant$ on Z that extends $\leq$ and is compatible with addition, then we must have

$$b \leqslant c \text{ if and only if } c - b \in N$$

for all b, $c \in Z$. This relation tells us how we must define a relation $\leqslant$ on Z if we are to have any hope that it will be an extension of $\leq$ compatible with addition.

Definition. We define a relation $\leqslant$ on Z by

$$b \leqslant c \text{ if and only if } c - b \in N.$$

Theorem 14.4. $(Z, +, \leqslant)$ is a totally ordered group, and $\leqslant$ induces on N its postulated ordering $\leq$.

Proof. Let a, b, $c \in Z$. Then $a \leqslant a$ since $a - a = 0 \in N$. If $a \leqslant b$ and $b \leqslant a$, then $b - a$, $a - b \in N$; if $a - b \in N^*$, then by Theorem 14.3, $b - a \in N \cap \{-n : n \in N^*\}$, a contradiction of Theorem 14.2; hence,

$a - b = 0$, so $a = b$. If $a \leqslant b$ and $b \leqslant c$, then $b - a$, $c - b \in N$, so $(c - b) + (b - a) \in N$; but

$$(c - b) + (b - a) = (c + (-b)) + (b + (-a)) = c + (-a) = c - a,$$

so $a \leqslant c$. Therefore, $\leqslant$ is an ordering on Z. If $a, b \in Z$, then either $b - a \in N$, or $b - a = -n$ where $n \in N^*$; in the first case, $a \leqslant b$, and in the second, $a - b = -(b - a) = n \in N^*$, whence $b \prec a$. Thus, $\leqslant$ is a total ordering.

If $a, b, c \in Z$ and if $a \leqslant b$, then $b - a \in N$, so

$$(b + c) - (a + c) = b + c + (-c) + (-a) = b - a \in N$$

by Theorem 4.4, whence $a + c \leqslant b + c$. Thus, $(Z, +, \leqslant)$ is a totally ordered group. To see that $\leqslant$ induces on N its given total ordering, let m, $n \in N$. If $m \leq n$, then $n + (-m) = n - m \in N$, so $m \leqslant n$. Conversely, if $m \leqslant n$, then $n + (-m) \in N$, so $0 \leq n + (-m)$, whence

$$m = 0 + m \leq [n + (-m)] + m = n.$$

Thus, we see that there is one and (by our earlier discussion) only one extension of $\leq$ to a total ordering on Z compatible with addition. Henceforth, we shall denote this extension by $\leq$ instead of $\leqslant$. Thus, if $b, c \in Z$, then $b \leq c$ if and only if $c - b \in N$. The set N of natural numbers may now be characterized as the set of positive integers. Indeed, for any integer a,

$$a \geq 0 \text{ if and only if } a \in N$$

by the definition of the ordering on Z, and since that ordering is total,

$$a < 0 \text{ if and only if } a \notin N.$$

To complete the construction of the integers, we need to extend multiplication to a composition on Z that will be distributive over addition, associative, and commutative. The following theorem suggests how to proceed.

Theorem 14.5. If $(G, +)$ is a group and if $\cdot$ is a composition on G that is distributive over $+$, then for all $x, y \in G$,

(8) $x0 = 0 = 0x$

(9) $x(-y) = -(xy) = (-x)y$

(10) $(-x)(-y) = xy$

and furthermore, if y is an inverse of x for $\cdot$, then $-y$ is an inverse of $-x$ for $\cdot$.

Proof. Since

$$x0 + x0 = x(0 + 0) = x0 = x0 + 0,$$

we conclude that $x0 = 0$ by the Corollary of Theorem 7.1. Similarly, $0x = 0$. Since

$$x(-y) + xy = x[(-y) + y] = x0 = 0,$$

$$xy + x(-y) = x[y + (-y)] = x0 = 0,$$

we conclude that $x(-y) = -(xy)$. Similarly, $(-x)y = -(xy)$. Replacing x by $-x$ in (9), we obtain

$$(-x)(-y) = -[(-x)y] = -[-(xy)] = xy,$$

from which (10) and the final assertion follow.

Identities (9) and (10) tell us how we must define multiplication on Z if we are to have any hope that it will be an extension of multiplication on N that is distributive over addition.

Definition. We extend multiplication on N to a composition on Z, again called multiplication, as follows:

If $m \in N$, $n \in N^*$, $m(-n) = -(mn)$.

If $m \in N^*$, $n \in N$, $(-m)n = -(mn)$.

If $m \in N^*$, $n \in N^*$, $(-m)(-n) = mn$.

If $m \in N$, $n \in N^*$, then

$$m(-n) = -(mn) = -(nm) = (-n)m$$

by definition and by the commutativity of multiplication on N, and similarly,

$$(-m)(-n) = mn = nm = (-n)(-m)$$

if $m, n \in N^*$. Thus, multiplication on Z is commutative. We note also that

(11) $(-m)a = -ma$

for all $m \in N^*$, $a \in Z$, for if $a \in N$, (11) holds by definition, and if $a = -n$ where $n \in N^*$,

$$(-m)a = (-m)(-n) = mn,$$

$$-(ma) = -[m(-n)] = -[-(mn)] = mn.$$

By (11), $(-m)1 = -(m \cdot 1) = -m$ for all $m \in N^*$; thus, 1 is the identity element for multiplication.

To show that multiplication is distributive over addition, let m, $n \in N$, $p \in N^*$. We shall first show that

$$(12) \qquad m(n - p) = mn - mp.$$

If $p \leq n$, then $n - p \in N$, so

$$m(n - p) + mp = m((n - p) + p) = mn$$

by the distributivity of multiplication over addition on N, from which (12) follows. If $n > p$, then $p - n \in N^*$, and a similar argument establishes that

$$m(p - n) = mp - mn,$$

whence

$$m(n - p) = m(-(p - n)) = -m(p - n)$$
$$= -(mp - mn) = mn - mp$$

by Theorem 14.3, the definition of multiplication, and what we have just proved. Moreover,

$$(13) \qquad m(-p + n) = -mp + mn$$

by (12), since $m(-p + n) = m(n - p) = mn - mp = -mp + mn$. Finally, if n, $p \in N^*$,

$$(14) \qquad m(-p - n) = -mp - mn,$$

since

$$m(-p - n) = m(-(n + p)) = -[m(n + p)]$$
$$= -[mn + mp] = -mp - mn$$

by the distributivity of multiplication over addition on N and by Theorem 4.4. Since $-mp = m(-p)$, $-mn = m(-n)$, we conclude from (12)–(14) that

$$(15) \qquad m(a + b) = ma + mb$$

for all $m \in N$, a, $b \in Z$. If $m \in N^*$, a, $b \in Z$, then

$$(-m)(a + b) = -[m(a + b)] = -[ma + mb]$$
$$= -ma - mb = (-m)a + (-m)b$$

by (11) applied twice, (15), and Theorem 4.4. Thus, multiplication is distributive over addition on Z.

To prove that multiplication on Z is associative is easy. By Theorem 14.5, it suffices to consider the case where none of the three terms is zero. If $m, n, p \in N^*$,

$$(mn)(-p) = -[(mn)p],$$
$$m[n(-p)] = m[-(np)] = -[m(np)]$$

by definition, since $np \in N^*$. Similar identities establish the remaining cases. We have thus proved the first sentence of the following theorem.

Theorem 14.6. Multiplication on Z is an associative, commutative composition that is distributive over addition, and 1 is the identity element for multiplication. Every nonzero integer is cancellable for multiplication. The only integers invertible for multiplication are 1 and -1. If a, b, and c are integers and if $c > 0$, then

1° $a < b$ if and only if $ca < cb$,
2° $a \le b$ if and only if $ca \le cb$.

Proof. To prove 1°, it suffices to prove that $b - a \in N^*$ if and only if $cb - ca \in N^*$; however, $b - a \in N^*$ if and only if $c(b - a) \in N^*$ by 1° of Theorem 11.13, and by distributivity and Theorem 14.5,

$$c(b - a) = cb + c(-a) = cb - ca.$$

Statement 2° and the cancellability for multiplication on Z of each $c \in N^*$ are immediate consequences of 1°. If $n \in N^*$ and if $(-n)x = (-n)y$, then $-nx = -ny$ by Theorem 14.5, so $nx = ny$, whence $x = y$ by what we have just proved; therefore, $-n$ is cancellable for multiplication

Let a be invertible for multiplication. If $a > 0$, then $a^{-1} > 0$, for otherwise $a^{-1} \le 0$, whence $1 = aa^{-1} \le a0 = 0$ by 2°, a contradiction; hence, if $a > 0$, then a is invertible for multiplication on N, so $a = 1$ by the Corollary of Theorem 11.13. If $a < 0$, then $-a$ is invertible by Theorem 14.5 and $-a > 0$, so $-a = 1$, whence $a = -1$.

We next wish to define $\triangle^n a$ for arbitrary integers n, where $\triangle$ is an associative composition and where a is an element invertible for $\triangle$.

Definition. Let $\triangle$ be an associative composition on E, and let a be an element of E invertible for $\triangle$. For each $n \in N^*$, we define $\triangle^{-n}a$ by

$$\triangle^{-n}a = \triangle^n a^*,$$

where a^* is the inverse of a.

Theorem 14.7. Let $\triangle$ be an associative composition on E, and let a and b be invertible elements of E for $\triangle$. For every $n \in Z$, $\triangle^n a$ is invertible, and

(16) $$(\triangle^n a)^* = \triangle^{-n} a = \triangle^n a^*.$$

For all $m, n \in Z$,

(17) $$\triangle^{n+m} a = (\triangle^n a)\triangle(\triangle^m a),$$

(18) $$\triangle^{nm} a = \triangle^n(\triangle^m a) = \triangle^m(\triangle^n a).$$

If a and b commute, then for all $n, m \in Z$,

(19) $$\triangle^n(a\triangle b) = (\triangle^n a)\triangle(\triangle^n b),$$

(20) $$(\triangle^m a)\triangle(\triangle^n b) = (\triangle^n b)\triangle(\triangle^m a).$$

Proof. Let e be the neutral element for $\triangle$. Since a and a^* commute, by (3) and (4) of Theorem 11.8,

$$(\triangle^n a)\triangle(\triangle^n a^*) = \triangle^n(a\triangle a^*) = e$$
$$= \triangle^n(a^*\triangle a) = (\triangle^n a^*)\triangle(\triangle^n a)$$

for all $n \in N$. Thus $\triangle^n a$ is invertible and (16) holds for all $n \geq 0$. To show (16) in general, it remains for us to show that for every $n \in N^*$, $\triangle^{-n} a$ is invertible and

$$(\triangle^{-n} a)^* = \triangle^{-(-n)} a = \triangle^{-n} a^*.$$

But $\triangle^{-n} a = \triangle^n a^*$ by definition, and

$$\triangle^{-(-n)} a = \triangle^n a = \triangle^n a^{**} = \triangle^{-n} a^*$$

by the definition of $\triangle^{-n} a^*$ and Theorem 4.3, so the preceding equalities also show that $\triangle^{-n} a$ is invertible and

$$(\triangle^{-n} a)^* = \triangle^{-(-n)} a = \triangle^{-n} a^*.$$

Thus, (16) holds for all $n \in Z$. It is now easy to verify (20) by (2) of Theorem 11.8, (16), and Theorem 4.5.

Clearly, (17) holds if either $n = 0$ or $m = 0$. Let $p, q \in N^*$. If $q \leq p$,

$$(\triangle^{p-q} a)\triangle(\triangle^q a) = \triangle^p a$$

by (1) of Theorem 11.8, so

$$\triangle^{p-q}a = (\triangle^p a)\triangle(\triangle^q a)^* = (\triangle^p a)\triangle(\triangle^{-q}a)$$

by (16). If $p < q$,

$$\triangle^{q-p}a = (\triangle^q a)\triangle(\triangle^{-p}a)$$

by what we have just proved, so

$$(\triangle^{q-p}a)^* = (\triangle^{-p}a)^*\triangle(\triangle^q a)^*$$

by Theorem 4.4, whence by (16),

$$\triangle^{p-q}a = (\triangle^p a)\triangle(\triangle^{-q}a).$$

Furthermore, by what we have just proved and (20),

$$\triangle^{-p+q}a = \triangle^{q-p}a = (\triangle^q a)\triangle(\triangle^{-p}a) = (\triangle^{-p}a)\triangle(\triangle^q a).$$

Finally, by (16),

$$\triangle^{-p-q}a = \triangle^{p+q}a^* = (\triangle^p a^*)\triangle(\triangle^q a^*) = (\triangle^{-p}a)\triangle(\triangle^{-q}a).$$

Thus, (17) holds for all n, $m \in \mathbf{Z}$.

By (4) of Theorem 11.8, (18) holds if either $n = 0$ or $m = 0$. Let p, $q \in \mathbf{N}^*$. By (16) and Theorem 11.11,

$$\triangle^{p(-q)}a = \triangle^{-pq}a = \triangle^{pq}a^* = \triangle^p(\triangle^q a^*) = \triangle^p(\triangle^{-q}a).$$

Moreover, since $(-p)q = p(-q)$,

$$\triangle^{(-p)q}a = \triangle^p(\triangle^{-q}a) = \triangle^p(\triangle^q a)^* = \triangle^{-p}(\triangle^q a)$$

by (16). Finally, by (16) and Theorem 11.11,

$$\triangle^{-p}(\triangle^{-q}a) = \triangle^p(\triangle^{-q}a^*) = \triangle^p(\triangle^q a) = \triangle^{pq}a = \triangle^{(-p)(-q)}a.$$

Thus, (18) holds for all n, $m \in \mathbf{Z}$.

If a and b commute and if $n > 0$,

$$\triangle^{-n}(a\triangle b) = \triangle^{-n}(b\triangle a) = \triangle^n(b\triangle a)^* = \triangle^n(a^*\triangle b^*)$$

$$= (\triangle^n a^*)\triangle(\triangle^n b^*) = (\triangle^{-n}a)\triangle(\triangle^{-n}b)$$

by (16), Theorem 4.4, and (3) of Theorem 11.8. Thus, (19) holds for all $n \in Z$.

For a composition denoted by $+$, (16) becomes

(21) $$-(n.a) = (-n).a = n.(-a),$$

and for a composition denoted by $\cdot$, (16) becomes

(22) $$(a^n)^{-1} = a^{-n} = (a^{-1})^n.$$

If $\triangle$ is denoted by $+$, equalities (17), (18), (19), and (20) become respectively (5), (14), (7), and (6) of §11, and if $\triangle$ is denoted by $\cdot$, they become (9), (15), (11), and (10) of §11.

Theorem 14.8. If n and m are integers, then $n.m = nm$.

Proof. The assertion is clear if either $n = 0$ or $m = 0$. Let $p, q \in N^*$. By the definition of multiplication on N, $p.q = pq$, and by (21) and the definition of multiplication on Z,

$$p.(-q) = -(p.q) = -(pq) = p(-q),$$
$$(-p).q = -(p.q) = -(pq) = (-p)q.$$

By the definitions of $(-p).(-q)$ and $(-p)(-q)$,

$$(-p).(-q) = p.q = pq = (-p)(-q).$$

Thus, $n.m = nm$ for all $n, m \in Z$.

One of the most important results of §11 is that to within isomorphism, there is only one naturally ordered semigroup. An analogous theorem holds for the ordered group of integers: To within isomorphism, $(Z, +, \leq)$ is the only totally ordered commutative group containing more than one element such that the set of elements ≥ 0 is well-ordered.

Theorem 14.9. Let $(Z', +', \leq')$ be a totally ordered commutative group, let $0'$ be the neutral element of Z', and let $N' = \{x \in Z' : x \geq' 0'\}$. If Z' contains at least two elements and if N' is well-ordered (for the ordering induced on N' by $\leq'$), then the function g from Z into Z' defined by

$$g(n) = n.1'$$

for all $n \in Z$ is an isomorphism from $(Z, +, \leq)$ onto $(Z', +', \leq')$, where $1'$ is the smallest element of $N' - \{0'\}$.

Proof. First, $N' - \{0'\}$ is not empty, for if z is an element of Z' distinct from $0'$, then either $z >' 0'$ or $-z >' 0'$ by Theorem 10.2. Consequently, $N' - \{0'\}$ does have a smallest element $1'$. Clearly, N' is an ordered sub-semigroup of Z' satisfying conditions (NO 1), (NO 2), and (NO 4) of §11. Also N' satisfies (NO 3), for if $0' \leq' x \leq' y$, then $0' \leq' (-x) +' y$; i.e., $(-x) +' y \in N'$, and $x +' [(-x) +' y] = y$. Consequently, $(N', +', \leq')$ is a naturally ordered semigroup, so by Theorem 11.14 the range of g contains N', and $p.1' >' 0'$ for all $p \in N^*$. Therefore, g is strictly increasing, for if $n < m$, then $m - n \in N^*$, so

$$0' <' (m - n).1' = m.1' +' (-n).1' = m.1' +' (-(n . 1'))$$

by (17) and (21), whence $n.1' <' m.1'$. Moreover, g is surjective, for if $y <' 0'$, then $-y >' 0'$, so $-y = g(n)$ for some $n \in N$, whence

$$y = -g(n) = -(n.1') = (-n).1' = g(-n)$$

by (21). Therefore, by (17) and Theorem 10.4, g is an isomorphism from $(Z, +, \leq)$ onto $(Z', +', \leq')$.

EXERCISES

14.1. Addition on Z, as defined in the text, is the only associative composition $\oplus$ on Z extending addition on N such that for each $n \in N^*$, n^- is the inverse of n for $\oplus$.

14.2. Complete the proof of Theorem 14.2.

14.3. (a) Complete the proof of (1).
(b) Complete the proof of (4).

14.4. Prove Theorem 14.3 by appealing to theorems of §4.

14.5. If $a, b \in Z$ and if $b > 0$, there exist unique integers q, r such that $a = bq + r$ and $0 \leq r < b$. [If $a < 0$, apply the Division Algorithm to $-a$ and b.]

14.6. Prove that multiplication on Z is associative.

14.7. For each $r \in Z$ let $\langle r \rangle$ be the composition on Z defined by

$$x \langle r \rangle y = xry$$

for all $x, y \in Z$.

(a) The composition $\langle r \rangle$ is associative, commutative, and distributive over addition. For which integers r is there an identity element for $\langle r \rangle$?

(b) If $\triangledown$ is a composition on Z that is distributive over addition, then there is a unique integer r such that $\triangledown$ is $\langle r \rangle$. [First use induction to calculate $1 \triangledown n$ for all $n \in N^*$.]

14.8. (a) For every integer n, $n + 1$ is the smallest integer greater than n.

(b) If p is an integer and if S is a subset of Z containing p and containing $n + 1$ whenever it contains n, then S contains every integer $\geq p$.

(c) Is the total ordering on Z a well-ordering? If p is an integer, is the restriction of the ordering on Z to $\{m \in Z : m \geq p\}$ a well-ordering?

14.9. Show that the function g of Theorem 14.9 is the only isomorphism from $(Z, +, \leq)$ onto $(Z', +', \leq')$.

15. Rings, Integral Domains, and Fields

Definition. A **ring** is an algebraic structure $(A, +, \cdot)$ with two compositions (the first called addition; the second, multiplication) such that $(A, +)$ is an abelian group and multiplication is an associative composition distributive over addition. A **commutative ring** is a ring whose multiplicative composition is commutative.

The set of nonzero elements of a ring A is denoted by A^*.

Example 15.1. We have formally proved in §14 that $(Z, +, \cdot)$ is a commutative ring. It is not difficult to verify that multiplication modulo m on N_m is distributive over addition modulo m. Consequently, $(N_m, +_m, \cdot_m)$ is a commutative ring for all $m \in N^*$; it is called the **ring of integers modulo** m.

Example 15.2. If E is a set and if $\triangle$ is the symmetric difference composition on $\mathfrak{P}(E)$ (Example 7.4), it is easy to verify that $(\mathfrak{P}(E), \triangle, \cap)$ is a commutative ring.

Example 15.3. If A is a set having just one element and if $+$ and $\cdot$ both denote the only possible composition on A, then $(A, +, \cdot)$ is a commutative ring. Any ring having just one element is called a **zero ring,** for its element must be the neutral element for addition. Any ring having at least two elements is consequently called a **nonzero ring.**

Example 15.4. Let $(A, +)$ be an abelian group. If $\cdot$ is the composition on A defined by

$$x \cdot y = 0$$

for all $x, y \in A$, then $(A, +, \cdot)$ is a commutative ring. Thus, any abelian group may be turned into a ring by defining the product of any two elements to be zero. A ring A whose multiplication satisfies $xy = 0$ for all $x, y \in A$ is called a **trivial ring.** A zero ring is an especially trivial ring.

Example 15.5. Let A be a ring, E a set. We introduce compositions $+$ and $\cdot$ on A^E, the set of all functions from E into A, by defining for all $f, g \in A^E$ the functions $f + g$ and fg as follows:

$$(f + g)(x) = f(x) + g(x),$$

$$(fg)(x) = f(x)g(x)$$

for all $x \in E$. It is easy to verify that $(A^E, +, \cdot)$ is a ring. The constant function $\mathbf{0}: x \mapsto 0$ from E into A is the additive neutral element, and for each $f \in A^E$, the additive inverse of f is the function $-f: x \mapsto -f(x)$, $x \in E$.

By Theorem 14.5, if A is a ring, then for all $x, y \in A$,

$$x0 = 0x = 0,$$

$$x(-y) = -xy = (-x)y,$$

$$(-x)(-y) = xy,$$

and if x is invertible for multiplication, then so is $-x$, and

$$(-x)^{-1} = -x^{-1}.$$

If A is a nonzero ring possessing an identity element 1 for multiplication, then $1 \neq 0$, for if b is a nonzero element, then

$$1b = b \neq 0 = 0b.$$

When we discuss rings A having a multiplicative identity element 1, it is convenient to exclude the possibility that $1 = 0$, i.e., that A is a zero ring. Therefore, we make the following definition:

Definition. A **ring with identity** (or with **unity**) is a nonzero ring possessing an identity element for multiplication.

The ring of even integers is an example of a ring without an identity element. The natural number 1 is the multiplicative identity of the ring of integers modulo m for all $m > 1$, and E is the multiplicative identity of the ring $\mathfrak{P}(E)$ of Example 15.2. If A is a ring with identity element 1 and if E is a set, the constant function $\mathbf{1}: x \mapsto 1$ is the multiplicative identity of the ring A^E discussed in Example 15.5.

Elements of a ring A that are invertible for multiplication are called simply **invertible elements** of A, and if an element a of A has an inverse for multiplication, that inverse is called the **multiplicative inverse** of a. By Theorem 8.5, if A is a ring with identity, then the set of all invertible elements of A is a subgroup of the semigroup $(A, \cdot)$.

By the General Distributivity Theorem (Theorem A.8), if $x_1, \dots, x_n$, $y, x, y_1, \dots, y_n$ are elements of a ring A, then

$$(x_1 + \dots + x_n)y = x_1 y + \dots + x_n y,$$

$$x(y_1 + \dots + y_n) = xy_1 + \dots + xy_n.$$

In particular, for all $x, y \in A$ and for every integer n,

(1) $$(n.x)y = n.(xy) = x(n.y).$$

Definition. An **integral domain** is a commutative ring with identity every nonzero element of which is cancellable for multiplication.

By Theorem 14.6, the ring of integers is an integral domain. The ring $\mathfrak{P}(E)$ of Example 15.2 is not an integral domain if E contains more than one element (Exercise 15.20). If m is an integer > 1 that is not a prime, then there exist nonzero integers $r, s \in N_m$ such that $rs = m$, whence

$$r \cdot_m s = 0 = r \cdot_m 0,$$

so the ring of integers modulo m is not an integral domain.

If A is a nonzero ring, then zero is not cancellable for multiplication. As we shall shortly see, to show that a nonzero element a is cancellable for multiplication, it suffices to show that if either $ax = 0$ or $xa = 0$, then $x = 0$.

Definition. An element a of a ring A is a **zero-divisor** of A if there exists $x \in A^*$ such that either $ax = 0$ or $xa = 0$. A **proper zero-divisor** of A is a nonzero zero-divisor of A. The ring A is a ring **without proper zero-divisors** if there are no proper zero-divisors of A.

If it is clear in the context what ring A is under consideration, we shall call a zero-divisor of A simply a zero-divisor.

Theorem 15.1. An element a of a nonzero ring A is a zero-divisor if and only if a is not cancellable for multiplication.

Proof. If $ax = 0$ or if $xa = 0$ for some $x \in A^*$, then a is not cancellable for multiplication since $a0 = 0a = 0$. Conversely, if $ax = ay$ where $x \neq y$, then

$$a(x - y) = ax - ay = 0$$

and $x - y \neq 0$, so a is a zero-divisor. Similarly, if $xa = ya$ where $x \neq y$, then a is a zero-divisor.

Corollary. Let A be a commutative ring with identity. The following statements are equivalent:

1° A is an integral domain.
2° A has no proper zero-divisors.
3° A^* is a subsemigroup of $(A, \cdot)$.

Thus, a ring is an integral domain if and only if it is a commutative ring with identity and without proper zero-divisors.

In elementary algebra one learns to solve equations such as $x^2 - x - 6 = 0$ by first factoring to obtain $(x - 3)(x + 2) = 0$ and then by setting each of the factors equal to zero, to obtain the answer $x = 3$ or $x = -2$. The justification for passing from the equation $(x - 3)(x + 2) = 0$ to the conclusion $x = 3$ or $x = -2$ is, of course, the fact that the ring Z is an integral domain and hence contains no proper zero-divisors.

Theorem 15.2. If A is a ring with identity and without proper zero-divisors, then an element x of A satisfies

$$x^2 = x$$

if and only if either $x = 0$ or $x = 1$.

Proof. If

$$x^2 = x = x \cdot 1$$

and if $x \neq 0$, then $x = 1$ as x is cancellable for multiplication.

It is obvious how to define "subring" and "subdomain," the analogues for rings and integral domains of subsemigroups and subgroups.

Definition. Let $(A, +, \cdot)$ be an algebraic structure with two compositions. If B is a subset of A that is stable for both $+$ and $\cdot$, then $(B, +_B, \cdot_B)$ is a **subring (subdomain)** of $(A, +, \cdot)$ if $(B, +_B, \cdot_B)$ is a ring (integral domain).

A stable subset B of $(A, +, \cdot)$ is itself also called a subring (subdomain) of $(A, +, \cdot)$ if $(B, +_B, \cdot_B)$ is a subring (subdomain).

If B is a subring of a ring A, then a zero-divisor of the ring B is clearly a zero-divisor of A, but an element of B may not be a zero-divisor of B and yet be a zero-divisor of A (Exercise 15.20(d)).

Theorem 15.3. A nonempty subset B of a ring A is a subring if and only if for all $x, y \in A$, if x and y belong to B, then $x + y$, $-y$, and xy also belong to B. A subset B of an integral domain A is a subdomain if and only if B is a subring of A containing the multiplicative identity of A.

Proof. The first assertion follows from Theorem 8.3. By Theorem 15.2, the multiplicative identity of a subdomain of an integral domain A is the multiplicative identity of A.

If A is a commutative ring with identity, for every element x of A and for every invertible element z of A it is customary to denote xz^{-1} by $\dfrac{x}{z}$ (or by x/z). (This notation is generally avoided in noncommutative rings, since it would be easy to forget whether x/z stood for xz^{-1} or for $z^{-1}x$.) The familiar rules learned in grammar school for combining fractions remain valid in any commutative ring with identity:

Theorem 15.4. Let A be a commutative ring with identity. If x and y are elements of A and if z and w are invertible elements of A, then zw, z/w and $-z$ are invertible, and

(2) $$(-z)^{-1} = -z^{-1} \text{ and } -\frac{x}{z} = \frac{-x}{z} = \frac{x}{-z},$$

(3) $$\frac{x}{z} + \frac{y}{w} = \frac{xw + yz}{zw},$$

(4) $$\frac{x}{z} = \frac{y}{w} \text{ if and only if } xw = yz,$$

(5) $$\frac{x}{z} \cdot \frac{y}{w} = \frac{xy}{zw},$$

(6) $$\left(\frac{z}{w}\right)^{-1} = \frac{w}{z}.$$

Proof. By Theorem 14.5, $-z^{-1}$ is the inverse of $-z$, that is, $-z^{-1} = (-z)^{-1}$. Therefore,
$$-xz^{-1} = (-x)z^{-1} = x(-z^{-1}) = x(-z)^{-1},$$

so (2) holds. The invertibility of zw and z/w and the remaining assertions follow from distributivity, Theorems 4.3 and 4.4, and the commutativity of multiplication. For example,

$$xz^{-1} + yw^{-1} = xww^{-1}z^{-1} + yzz^{-1}w^{-1} = (xw + yz)(zw)^{-1},$$

so (3) holds.

Definition. A **field** is a commutative ring with identity every nonzero element of which has a multiplicative inverse. A **subfield** of an algebraic structure with two compositions is a subring that is a field.

The rational numbers, real numbers, and complex numbers with the compositions of ordinary addition and multiplication are examples of fields. We shall formally construct the field of rational numbers in §17.

For many purposes, the commutativity of multiplication in a field is inessential, and rings that fail to be fields only because multiplication is not commutative are sufficiently important to warrant a special name:

Definition. A **division ring** is a ring with identity every nonzero element of which has a multiplicative inverse. A **division subring** of an algebraic structure with two compositions is a subring that is a division ring.

Thus, a field is simply a commutative division ring. Other words used in place of "division ring" are **skew field** and **sfield.** The simplest example of a division ring that is not a field is the division ring of quaternions [Exercise 33.22(g)].

Since an element invertible for an associative composition is cancellable for it, every nonzero element of a division ring is cancellable for multiplication and, therefore, by Theorem 15.1 is not a zero-divisor. Thus, *a division ring contains no proper zero-divisors*, and consequently, *every field is an integral domain.* There are no finite integral domains that are not fields by virtue of the following theorem.

Theorem 15.5. A finite integral domain A is a field.

Proof. Let $a \in A^*$. By hypothesis, the function L_a from A into A defined by

$$L_a(x) = ax$$

for all $x \in A$ is injective and hence is a permutation of A by Theorem 12.7. In particular, there exists $x \in A$ such that $ax = 1$. Every element of A^* is, therefore, invertible, so A is a field.

Definition. Let A be a ring. The **center** of A is the subset C defined by

$$C = \{a \in A : ax = xa \text{ for all } x \in A\}.$$

Theorem 15.6. The center C of a ring A is a subring of A. If an invertible element a of A belongs to C, then $a^{-1} \in C$.

Proof. Clearly $0 \in C$, so $C \neq \emptyset$. If $a, b \in C$, then $a + b$, $-a$, and $ab \in C$ since

$$(a + b)x = ax + bx = xa + xb = x(a + b),$$

$$(-a)x = -(ax) = -(xa) = x(-a),$$

$$abx = axb = xab$$

for all $x \in A$. The final assertion is a consequence of Theorem 4.5.

Corollary. The center of a division ring is a field.

The familiar Binomial Theorem of elementary algebra is valid in any commutative ring. We define, of course,

$$0! = 1,$$

$$n! = \prod_{k=1}^{n} k$$

for all $n \in N^*$; thus, $(n + 1)! = n! \, (n + 1)$ for all $n \in N$. We first need to establish that the "binomial coefficients" are indeed natural numbers.

Theorem 15.7. If $0 \leq m \leq n$, then $m! \, (n - m)! \mid n!$.

Proof. Let $S = \{n \in N : m! \, (n - m)! \mid n! \text{ for all } m \in [0, n]\}$. Clearly, $0 \in S$. Assume that $n \in S$. Obviously, $0! \, (n + 1)! \mid (n + 1)!$; if $0 < m \leq n$, then $m!(n - m)! \mid n!$ and $(m - 1)! \, (n - m + 1)! \mid n!$ since $n \in S$, so by an easy calculation,

$$(n + 1)! = n! \, (n + 1 - m) + n! \, m$$

$$= \left[\frac{n!}{m! \, (n - m)!} + \frac{n!}{(m - 1)! \, (n - m + 1)!} \right] m! \, (n + 1 - m)!.$$

Thus, $n + 1 \in S$. By induction, the proof is complete.

Hence, we are justified in making the following definition.

Definition. If $0 \leq m \leq n$, the natural number $\binom{n}{m}$ is defined by

$$\binom{n}{m} = \frac{n!}{m!\,(n-m)!}.$$

The natural numbers $\binom{n}{m}$ are called **binomial coefficients.**

The following identities are easily established (the last was implicitly used in the proof of Theorem 15.7).

Theorem 15.8. If n, $m \in N$ and if $0 < m \leq n$,

$$\binom{n}{0} = \binom{n}{n} = 1, \qquad \binom{n}{n-m} = \binom{n}{m},$$

$$\binom{n+1}{m} = \binom{n}{m} + \binom{n}{m-1}.$$

Theorem 15.9. (Binomial Theorem) Let $+$ be an associative, commutative composition on A, and let $\cdot$ be an associative composition on A that is distributive over $+$. If a and b are elements of A such that $ab = ba$, then for all integers $n \geq 2$,

(7) $$(a+b)^n = a^n + \sum_{m=1}^{n-1} \binom{n}{m} a^{n-m} b^m + b^n.$$

Proof. Let S be the set of all natural numbers $n \geq 2$ such that (7) holds. Clearly, $2 \in S$, for

$$(a+b)^2 = a(a+b) + b(a+b) = a^2 + 2ab + b^2.$$

Let $n \in S$. Then

$$(a+b)^{n+1} = (a+b)^n(a+b) = (a+b)^n a + (a+b)^n b$$
$$= a(a+b)^n + (a+b)^n b$$

by Theorem 11.8, as $(a+b)a = a^2 + ba = a^2 + ab = a(a+b)$. Since $n \in S$,

$$a(a+b)^n = a^{n+1} + \sum_{m=1}^{n-1} \binom{n}{m} a^{n-m+1} b^m + ab^n$$

$$= a^{n+1} + \sum_{m=1}^{n} \binom{n}{m} a^{n-m+1} b^m,$$

and

$$(a + b)^n b = a^n b + \sum_{m=1}^{n-1} \binom{n}{m} a^{n-m} b^{m+1} + b^{n+1}$$

$$= \sum_{m=0}^{n-1} \binom{n}{m} a^{n-m} b^{m+1} + b^{n+1}$$

$$= \sum_{m=1}^{n} \binom{n}{m-1} a^{n-m+1} b^{m} + b^{n+1}.$$

Adding, we obtain

$$(a + b)^{n+1} = a^{n+1} + \sum_{m=1}^{n} \binom{n+1}{m} a^{n+1-m} b^{m} + b^{n+1}$$

by Theorem 15.8, so $n + 1 \in S$. By induction, the proof is complete.

If there is an identity element for multiplication, (7) may be expressed more succinctly by

$$(a + b)^n = \sum_{m=0}^{n} \binom{n}{m} a^{n-m} b^{m}.$$

EXERCISES

15.1. Prove that $(N_m, +_m, \cdot_m)$ is a commutative ring by showing that multiplication modulo m is distributive over addition modulo m.

15.2. Determine whether A is a subring of R, where:
   (a) $A = \{n \in Z: \text{either } n = 0 \text{ or } n \text{ is odd}\}$.
   (b) $A = \{n \in Z: \text{either } n = 0 \text{ or } |n| \geq 17\}$.
   (c) $A = \{n + m\sqrt{3}: n, m \in Z\}$.
   (d) $A = \{n + m\sqrt[3]{3}: n, m \in Z\}$.
   (e) $A = \{n + m\sqrt[3]{3} + p\sqrt[3]{9}: n, m, p \in Z\}$.
   (f) $A = \{n + m\pi: n, m \in Z\}$. (Assume that π is not the root of a polynomial with integral coefficients.)
   (g) $A = \{p/q: p, q \in Z, \text{ and } q \text{ is not an integral multiple of 5}\}$.
   (h) $A = \{p/q: p, q \in Z, \text{ and } q \text{ is not an integral multiple of 4}\}$.
   (i) $A = \{p/q: p, q \in Z, \text{ and } q \text{ is not an integral multiple of 2, 5, or 7}\}$.
   (j) $A = \{5^{-n}k: k \in Z \text{ and } n \in N\}$.

15.3. If x, y, z, and w are elements of a ring, then

$$(x + y)(z + w) = xz + yz + xw + yw,$$

$$(x - y)(z - w) = (xz + yw) - (xw + yz).$$

*15.4. A **pseudo-ring** is an algebraic structure $(A, +, \cdot)$ such that $(A, +)$ is an abelian group, $(A, \cdot)$ is a semigroup, and

$$(x + y)(z + w) = xz + yz + xw + yw$$

for all $x, y, z, w \in A$.
(a) If $\cdot$ is the composition on N_3 defined by

$$x \cdot y = 1$$

for all $x, y \in N_3$, then $(N_3, +_3, \cdot)$ is a pseudo-ring but not a ring.
(b) Let $(A, +, \cdot)$ be a pseudo-ring, and let

$$z = 0 \cdot 0.$$

Show that $3.z = 0$ and that

$$a0 = 0a = az = za = z$$

for all $a \in A$. [Expand $(a + 0)(0 + 0)$, and first let $a = 0$.] Infer that if there is an element in A cancellable for multiplication, or if there exists $a \in A$ such that $a0 = 0$, then A is a ring.
(c) Let $(A, +, \cdot)$ be a pseudo-ring, and let $z = 0 \cdot 0$. Prove that if $\circ$ is the composition on A defined by

$$x \circ y = xy - z$$

for all $x, y \in A$, then $(A, +, \circ)$ is a ring and

$$x \circ z = 0 = z \circ x$$

for all $x \in A$.
(d) If $(A, +, \circ)$ is a ring and if z is an element of A satisfying $3.z = 0$ and $x \circ z = 0 = z \circ x$ for all $x \in A$, then $(A, +, \cdot)$ is a pseudo-ring where $\cdot$ is the composition defined by

$$xy = x \circ y + z$$

for all $x, y \in A$.

15.5. If $(A, +)$ is a group and if $\cdot$ is an associative composition on A that is distributive over $+$ and admits an identity element 1, then $+$ is commutative and hence $(A, +, \cdot)$ is a ring. [Expand $(x + y)(1 + 1)$ in two ways.]

*15.6. If $(A, +)$ is a group and if $\cdot$ is a composition on A admitting a left neutral element e (Exercise 4.5) such that $+$ is left distributive over $\cdot$ (Exercise 11.13), then $\cdot$ is the composition $\rightarrow$ on A. [First prove that $0z = z$ by considering $(e + 0)(e + z)$.]

15.7. If A has more than one element, are there compositions $+$ on A such that $(A, +, \to)$ is a ring?

15.8. Let $(A, +, \cdot)$ be a ring.
(a) Show that $(A, +, \circ)$ is a ring where $\circ$ is the composition defined by

$$x \circ y = yx$$

for all $x, y \in A$. (The ring $(A, +, \circ)$ is called the **reciprocal ring** of $(A, +, \cdot)$.)
(b) An **anti-isomorphism** from a ring A onto a ring B is a bijection f from A onto B such that

$$f(x + y) = f(x) + f(y),$$
$$f(xy) = f(y)f(x)$$

for all $x, y \in A$. Show that a bijection f from A onto B is an anti-isomorphism from A onto B if and only if f is an isomorphism from the reciprocal ring of A onto B.

15.9. (a) A ring A is an integral domain if and only if A^* is a commutative subsemigroup of $(A, \cdot)$ having an identity element.
(b) A ring A is a division ring if and only if A^* is a subgroup of the semigroup $(A, \cdot)$.

*15.10. If A is a ring with identity and if an element a of A has exactly one right inverse a' for multiplication, then a is invertible. [Show that a is left cancellable for multiplication, and then consider $aa'a$.]

15.11. A finite nonzero ring without proper zero-divisors is a division ring. [Use Exercise 12.1.]

15.12. What are the transplants under f of addition and multiplication on Q if f is the permutation of Q defined by
(a) $f(x) = -x$?
(b) $f(x) = 2 - x$?
(c) $f(x) = x^{-1}$ if $x \neq 0, f(0) = 0$?

15.13. What are the transplants of addition and multiplication on R under the hyperbolic tangent function from R onto $\{y \in R: -1 < y < 1\}$? Give an expression for the transplant of addition not involving tanh.

15.14. If $\triangle$ is the composition on R_+^* defined by

$$x \triangle y = x^{\log_2 y}$$

for all $x, y \in R_+^*$, then $(R_+^*, \cdot, \triangle)$ is a field isomorphic to the field of real numbers.

*15.15. (a) If K is a division ring, for what elements a of K does there exist an element $x \in K$ satisfying $a + x - ax = 0$?

(b) If K is a ring such that for every element a of K with exactly one exception there exists $x \in K$ satisfying $a + x - ax = 0$, then K is a division ring. [If e is the exceptional element, first show by contradiction that e is a left multiplicative identity, then show that e is a right identity, and consider finally $(e - a)(e - x)$.]

*15.16. Let C be the center of a ring A. If $x^2 - x \in C$ for all $x \in A$, then A is commutative. [Show first that $xy + yx \in C$ by considering $x + y$ and then show that $x^2 \in C$.]

*15.17. If A is a commutative nonzero ring such that for all $x, y \in A$ there exists $u \in A$ satisfying either $xu = y$ or $yu = x$, then A has an identity element. [Show first that there is an element u that is not a zero-divisor.]

15.18. Let A be a ring. An element b of A is **nilpotent** if $b^n = 0$ for some $n \in N^*$. If b is nilpotent, the smallest strictly positive integer m such that $b^m = 0$ is called the **index of nilpotency** of b.
(a) A nonzero nilpotent element of A is a proper zero-divisor.
(b) If A is a ring with identity and if b is a nilpotent element of A, then $1 - b$ is invertible. [Factor $1 - b^n$.]

15.19. Let b and c be elements of a ring A.
(a) If b is nilpotent, then $-b$ is nilpotent.
(b) If b is nilpotent and if $bc = cb$, then bc is nilpotent.
(c) If b and c are nilpotent and if $bc = cb$, then $b + c$ is nilpotent. [If $b^n = c^m = 0$, consider $(b + c)^{n+m}$.]

*15.20. Let E be a set.
(a) Show that $(\mathfrak{P}(E), \triangle, \cap)$ (Example 15.2) is a commutative ring. [Use Exercise 7.1.]
(b) A nonempty subset $\mathfrak{S}$ of $\mathfrak{P}(E)$ is a subring if and only if $X \cup Y$ and $X - Y$ belong to $\mathfrak{S}$ whenever X and Y belong to $\mathfrak{S}$.
(c) Every nonempty proper subset of E is a proper zero-divisor of the ring $\mathfrak{P}(E)$.
(d) If F is a proper subset of E, then $\mathfrak{P}(F)$ is a subring of $\mathfrak{P}(E)$, and F is a zero-divisor of $\mathfrak{P}(E)$ but not of the subring $\mathfrak{P}(F)$.

15.21. A **boolean ring** is a ring every element of which is idempotent for multiplication (Exercise 2.17).
(a) The ring $\mathfrak{P}(E)$ of Example 15.2 is a boolean ring.
(b) If A is a boolean ring, then $x = -x$ for all $x \in A$. [$x + x$ is idempotent.]
(c) A boolean ring is commutative. [$x + y$ is idempotent.]
(d) If a boolean ring contains at least three elements, then it contains a proper zero-divisor. [Consider $(x + y)xy$.]

*15.22. A ring A is a boolean ring if and only if A contains no nonzero nilpotent elements and $xy(x + y) = 0$ for all $x, y \in A$. [Show that $x^4 - x^5 = 0$, then consider $(x^2 - x^3)^2$.]

*15.23. If $(A, +, \cdot)$ is an algebraic structure such that $+$ is associative, every element of A is cancellable for $+$, $\cdot$ is distributive over $+$, and $xx = x$, $(xy)y = x(yy)$ for all $x, y \in A$, then $(A, +, \cdot)$ is a boolean ring. [To show that $x + x$ is the neutral element for $+$, consider $(x + x)^2$; to show that $\cdot$ is associative, consider $(xy)(z + x) + (yz)(z + x) + xy + yz$.]

*15.24. If $(A, +, \cdot)$ is an algebraic structure such that A has at least two elements, there is an identity element for multiplication, and

$$x + (y + y) = x,$$

$$[x(yy)]z = (zy)x,$$

$$x[(y + z) + w] = x(w + z) + xy$$

for all $x, y, z, w \in A$, then $(A, +, \cdot)$ is a boolean ring with identity.

16. Equivalence Relations

The concept of an equivalence relation is fundamental to all of mathematics. Our immediate need for it is in construction the field of rational numbers in §17.

Definition. A relation R on E is an **equivalence** relation if R is reflexive, symmetric, and transitive.

For example, the following are equivalence relations on the set of all people now alive:

(a) ____ and ... have the same parents.
(b) ... and ____ have the same given names.

On N, the following are equivalence relations:

(c) If ____ $\neq$ ..., then ____ ≥ 3, ... ≥ 3, and ____ $-$ ... is an integral multiple of 4.
(d) Upon division by 5, ____ and ... have the same quotient.
(e) Upon division by 5, ____ and ... have the same remainder.

Definition. A **partition** of a set E is a class $\mathscr{P}$ of nonempty subsets of E such that every element of E belongs to one and only one member of $\mathscr{P}$.

Thus, $\mathscr{P}$ is a partition of E if and only if $\emptyset \notin \mathscr{P}$, $\cup \mathscr{P} = E$, and any two distinct subsets of E belonging to $\mathscr{P}$ are disjoint. The principal facts about partitions are that every equivalence relation determines a partition and, conversely, every partition determines an equivalence relation (Theorems 16.1–16.3).

Definition. Let R be an equivalence relation on E. For each $x \in E$, the **equivalence class** of x determined by R is the set $\lfloor x \rfloor_R$ defined by

$$\lfloor x \rfloor_R = \{y \in E: x \, R \, y\}.$$

The set E/R of all equivalence classes determined by R [a subset of $\mathfrak{P}(E)$] is called the **quotient set** determined by R. The **canonical** or **natural surjection** from E onto E/R is the function φ_R defined by

$$\varphi_R(x) = \lfloor x \rfloor_R$$

for all $x \in E$.

If no confusion results, we shall drop the subscript "R" and denote the equivalence class of x determined by R simply by $\lfloor x \rfloor$.

Let "x" denote the reader. If R is relation (a), then $\lfloor x \rfloor_R$ is the set consisting of the reader and all his living brothers and sisters. If R is relation (b), then $\lfloor x \rfloor_R$ is the set consisting of all living persons whose given name is the same as the reader's.

The equivalence class $\lfloor 14 \rfloor$ consists of 6, 10, 14, 18, 22, etc. if R is (c), 10, 11, 12, 13, 14 if R is (d), and 4, 9, 14, 19, etc. if R is (e).

The equivalence class $\lfloor x \rfloor_R$ of x may be thought of pictorially as a box containing all those elements of E to which x bears R. Thus, E/R is the set of all such boxes. The next theorem implies that every element of E goes into exactly one box and that no box is empty, that is, that the set E/R of boxes is a partition of E.

Theorem 16.1. If R is an equivalence relation on E, then E/R is a partition of E.

Proof. As R is reflexive, $x \in \lfloor x \rfloor_R$ for every $x \in E$. Hence, the empty set does not belong to E/R, and every element of E belongs to at least one member of E/R. We shall show that if $\lfloor x \rfloor_R \cap \lfloor y \rfloor_R \neq \emptyset$, then $\lfloor x \rfloor_R = \lfloor y \rfloor_R$. Let $z \in \lfloor x \rfloor \cap \lfloor y \rfloor$. Since $x \, R \, z$ and since R is symmetric, we have $z \, R \, x$. If $u \in \lfloor x \rfloor$, then $x \, R \, u$, so as $y \, R \, z$, $z \, R \, x$, and $x \, R \, u$, we conclude by the transitivity of R that $y \, R \, u$ and hence that $u \in \lfloor y \rfloor$. Therefore, $\lfloor x \rfloor \subseteq \lfloor y \rfloor$, and similarly, $\lfloor y \rfloor \subseteq \lfloor x \rfloor$. Consequently, any two distinct members of E/R are disjoint, so E/R is a partition of E.

Definition. If $\mathscr{P}$ is a partition of E, we shall call the relation S on E satisfying $x \, S \, y$ if and only if x and y belong to the same member of $\mathscr{P}$ the **relation defined by** $\mathscr{P}$.

Theorem 16.2. The relation S on E defined by a partition $\mathscr{P}$ of E is an equivalence relation on E.

Proof. For each $x \in E$, there exists $P \in \mathscr{P}$ such that $x \in P$, so $x\,S\,x$ since $x \in P$ and $x \in P$. Thus, S is reflexive. If $x\,S\,y$, then there exists $P \in \mathscr{P}$ such that $x \in P$ and $y \in P$, whence $y \in P$ and $x \in P$, and therefore $y\,S\,x$. Thus, S is symmetric. If $x\,S\,y$ and if $y\,S\,z$, then there exist P and Q in $\mathscr{P}$ such that x and y belong to P and y and z belong to Q; consequently, $y \in P \cap Q$, so $P = Q$ as $\mathscr{P}$ is a partition, and therefore $x\,S\,z$. Thus, S is transitive. Hence, S is an equivalence relation.

Theorem 16.3. If R is an equivalence relation on E, then the relation S defined by the partition E/R of E is R itself. If $\mathscr{P}$ is a partition of E and if S is the relation defined by $\mathscr{P}$, then the partition E/S of E is $\mathscr{P}$ itself.

Proof. To prove the first statement, let R be an equivalence relation on E. If $x\,R\,y$, then $y \in \lfloor x \rfloor_R$ and $x \in \lfloor x \rfloor_R$, so as x and y belong to the same member $\lfloor x \rfloor_R$ of E/R, we conclude that $x\,S\,y$. Conversely, if $x\,S\,y$, then y belongs to the same member of E/R that x does, namely, $\lfloor x \rfloor_R$, so $x\,R\,y$. Therefore, $S = R$.

To prove the second statement, we shall first prove that if $P \in \mathscr{P}$, then $P \in E/S$. Let $x \in P$. Then $y \in \lfloor x \rfloor_S$ if and only if $x\,S\,y$, that is, if and only if y belongs to the same member of $\mathscr{P}$ that x does, namely, P. Therefore, $P = \lfloor x \rfloor_S$, and consequently, $P \in E/S$. Conversely, we shall prove that for every $x \in E$, $\lfloor x \rfloor_S \in \mathscr{P}$. Indeed, let P be the member of $\mathscr{P}$ to which x belongs. Then $y \in P$ if and only if $x\,S\,y$, that is, if and only if $y \in \lfloor x \rfloor_S$. Therefore, $\lfloor x \rfloor_S = P$, and consequently, $\lfloor x \rfloor_S \in \mathscr{P}$. Thus, $E/S = \mathscr{P}$.

Theorem 16.4. Let R be an equivalence relation on E. The following statements are equivalent:

    $1°$ $x\,R\,y$.
    $2°$ $x \in \lfloor y \rfloor_R$.
    $3°$ $y \in \lfloor x \rfloor_R$.
    $4°$ $\lfloor x \rfloor_R \cap \lfloor y \rfloor_R \neq \emptyset$.
    $5°$ $\lfloor x \rfloor_R = \lfloor y \rfloor_R$.

Proof. If $x\,R\,y$, then $y\,R\,x$ as R is symmetric, and so by definition $x \in \lfloor y \rfloor_R$ and $y \in \lfloor x \rfloor_R$. Thus, $1°$ implies both $2°$ and $3°$. Since $z \in \lfloor z \rfloor_R$ for all $z \in E$, either $2°$ or $3°$ implies $4°$. As E/R is a partition of E, $4°$ implies $5°$. Finally, if $\lfloor x \rfloor_R = \lfloor y \rfloor_R$, then $y \in \lfloor x \rfloor_R$ and so $x\,R\,y$; thus, $5°$ implies $1°$.

By virtue of Theorems 16.1–16.3, "equivalence relation" and "partition" are very similar concepts.

EXERCISES

16.1. Let E be the set of all people now alive. Verify that the following are equivalence relations on E:

(a) ____ and . . . have the same birthday anniversary.

(b) ____ and . . . have the same height.

(c) ____ is either married to or the same person as . . . (assume that no polygamous or polyandrous marriages exist).

(d) The given names, written in Latin letters, of ____ and . . . have the same initial letter.

(e) ____ and . . . were born in what is now the same country.

If "x" denotes the reader, what is $\lfloor x \rfloor$ for each of these relations?

16.2. Verify that the following are equivalence relations on the set of all straight lines in the plane of analytic geometry:

(a) ____ is parallel to . . . .

(b) Either both ____ and . . . are parallel to the Y-axis, or neither ____ nor . . . is parallel to the Y-axis, but either are identical or intersect on the Y-axis.

(c) ____ and . . . are equally distant from the origin.

If L is the line whose equation is $y = x + 1$, what is $\lfloor L \rfloor$ for each of these relations?

16.3. Complete the proof of Theorem 16.1 by showing that $\lfloor y \rfloor \subseteq \lfloor x \rfloor$.

16.4. If R is a symmetric and transitive relation on E and if for each $x \in E$ there exists $y \in E$ such that $x \, R \, y$, then R is an equivalence relation on E.

16.5. Let R be a relation on E. With the terminology of Exercise 5.8, prove that R is an equivalence relation if and only if R contains the diagonal subset of $E \times E$ and $R = R \circ R^{\leftarrow}$.

16.6. (a) If f is a function from E into F, the relation R_f on E satisfying $x \, R_f \, y$ if and only if $f(x) = f(y)$ is an equivalence relation.

(b) If f is the function from $N \times N$ into Z defined by $f(a, b) = a - b$, then $(a, b) \, R_f \, (c, d)$ if and only if $a + d = b + c$.

(c) If f is the function from $Z \times Z^*$ into Q defined by $f(a, b) = ab^{-1}$, then $(a, b) \, R_f \, (c, d)$ if and only if $ad = bc$.

16.7. A set of 10 elements is partitioned into three subsets, one each of five elements, three elements, and two elements. How many ordered couples belong to the equivalence relation defined by this partition?

17. The Rational Field

Here we wish formally to construct the field of rational numbers, that is, a field containing Z, every element of which is a quotient of integers. With no extra trouble we shall do much more. For an arbitrary integral domain A, we shall construct a field of which A is a subdomain and every element a quotient of elements of A.

Definition. A **quotient field** of an integral domain A is a field K of which A is a subdomain such that for each $z \in K$ there exist $x \in A$, $y \in A^*$ such that $z = xy^{-1}$.

If an integral domain A appears as a subdomain of a field L, it is easy to find a quotient field of A in L:

Theorem 17.1. If A is a subdomain of a field L and if

$$K = \left\{ \frac{x}{y} : x \in A, \, y \in A^* \right\},$$

then K is a quotient field of A.

Proof. The multiplicative identity of A is also that of L by Theorem 15.3. The sum and product of two elements of K belong to K by (3) and (5) of Theorem 15.4. The additive and multiplicative inverses of a nonzero element of K belong to K by (2) and (6) of that theorem. Thus, K is a subfield of L that clearly contains A and hence is a quotient field of A.

To gain some insight in how to proceed to construct the rationals, let us use our knowledge of fractions obtained in grammar school. We might begin by regarding a fraction as an ordered couple of integers, the first term corresponding to the numerator of the fraction, the second term to the denominator. Recalling the rules for adding and multiplying fractions, we would define addition and multiplication of ordered pairs of integers so as to mirror the addition and multiplication of fractions; for example, thinking of $(2, 3)$ as the fraction $\frac{2}{3}$ and $(1, 5)$ as the fraction $\frac{1}{5}$, we would have

$$(2, 3) + (1, 5) = (13, 15),$$

$$(2, 3) \cdot (1, 5) = (2, 15).$$

This would not literally make Z a subdomain of a field, but if we "identify" each integer n with the ordered couple $(n, 1)$ (corresponding to the fraction $n/1$), we might hope that our definitions of addition and multiplication on $Z \times Z^*$ make it a quotient field of Z.

This procedure does not work, however, and a moment's reflection reveals why. The ordered couples $(2, 3)$ and $(4, 6)$, for example, are distinct, but the fractions $\frac{2}{3}$ and $\frac{4}{6}$ are identical, that is, the expressions "$\frac{2}{3}$" and "$\frac{4}{6}$"

denote the same rational number. In short, a rational number is the quotient of (infinitely) many pairs of integers. A modification of the procedure outlined above to take into account this observation is the following: Instead of starting with the set of all ordered pairs (m, n) of integers (where $n \neq 0$), that is, with the set $Z \times Z^*$, we should "identify" ordered couples corresponding to the "same" fraction. In other words, we should start with the quotient set $(Z \times Z^*)/R$ where R is the equivalence relation which (m, n) and (p, q) bear each other if and only if they "represent" the same fraction. How may we make precise the notion that (m, n) and (p, q) represent the same fraction, that is, that the fraction whose numerator is m and whose denominator is n is identical with the faction whose numerator is p and whose denominator is q? The technique of "cross multiplication" learned in grammar school provides the answer; the fractions m/n and p/q are identical if and only if $mq = pn$. Seizing upon this fact, we may now formally define the relation R, which ordered couples are to bear each other if and only if, intuitively, they represent the same fraction, by declaring that $(m, n) \, R \, (p, q)$ if and only if $mq = pn$. Then we may proceed to define addition and multiplication on $(Z \times Z^*)/R$ as before by using the rules for combining fractions learned in grammar school, so that

$$\lfloor (m, n) \rfloor + \lfloor (p, q) \rfloor = \lfloor (mq + pn, nq) \rfloor$$

$$\lfloor (m, n) \rfloor \cdot \lfloor (p, q) \rfloor = \lfloor (mp, nq) \rfloor.$$

Before plunging ahead to verify that, with these compositions, $(Z \times Z^*)/R$ is a quotient field of a subdomain corresponding to Z, namely, the subdomain of all the equivalence classes $\lfloor (n, 1) \rfloor$ where $n \in Z$, we must verify that the above definitions are really "well defined." Indeed, we are defining addition and multiplication of equivalence classes of ordered couples of integers, but our definitions appear to depend on particular ordered couples; for example, if $a = \lfloor (1, 2) \rfloor$ and $b = \lfloor (4, 6) \rfloor$, then using $(1, 2)$ and $(4, 6)$ in the definitions, we obtain

$$a + b = \lfloor (14, 12) \rfloor, \qquad ab = \lfloor (4, 12) \rfloor;$$

but also, as $(1, 2) \, R \, (3, 6)$ and $(4, 6) \, R \, (2, 3)$, $a = \lfloor (3, 6) \rfloor$ and $b = \lfloor (2, 3) \rfloor$, and using $(3, 6)$ and $(2, 3)$ in the definitions, we obtain

$$a + b = \lfloor (21, 18) \rfloor, \qquad ab = \lfloor (6, 18) \rfloor;$$

thus, to make sure that we do not have conflicting interpretations of what $a + b$ and ab are to mean, we must verify that $\lfloor (14, 12) \rfloor = \lfloor (21, 18) \rfloor$ and $\lfloor (4, 12) \rfloor = \lfloor (6, 18) \rfloor$, i.e., that $(14, 12) \, R \, (21, 18)$ and $(4, 12) \, R \, (6, 18)$. All the needed verifications can be made, and our entire procedure works equally well for any integral domain. We present these considerations in a formal manner in the proof of the following theorem.

Theorem 17.2. Let A be an integral domain, and let R be the relation on $A \times A^*$ satisfying $(a, b) R (c, d)$ if and only if $ad = cb$. Then R is an equivalence relation; let K' be the quotient set $(A \times A^*)/R$. Addition and multiplication on K' are well defined by

$$\lfloor (a, b) \rfloor + \lfloor (c, d) \rfloor = \lfloor (ad + bc, bd) \rfloor,$$

$$\lfloor (a, b) \rfloor \cdot \lfloor (c, d) \rfloor = \lfloor (ac, bd) \rfloor$$

for all $\lfloor (a, b) \rfloor$, $\lfloor (c, d) \rfloor \in K'$. With these compositions, K' is a field and is, moreover, a quotient field of the subdomain $A' = \{\lfloor (a, 1) \rfloor : a \in A\}$; the function $f: a \mapsto \lfloor (a, 1) \rfloor$ is an isomorphism from the integral domain A onto A'.

Proof. Clearly R is reflexive; R is also symmetric, for if $ad = cb$, then $cb = ad$. To show that R is transitive, let $(a, b) R (c, d)$ and $(c, d) R (e, f)$. Then $ad = cb$ and $cf = ed$; hence,

$$afd = adf = cbf = cfb = edb = ebd,$$

so $af = eb$ and thus $(a, b) R(e, f)$ as d is cancellable. Therefore, R is an equivalence relation. To show that addition and multiplication are well defined, let $\lfloor (a, b) \rfloor = \lfloor (a', b') \rfloor$ and $\lfloor (c, d) \rfloor = \lfloor (c', d') \rfloor$; we must show that $\lfloor (ad + bc, bd) \rfloor = \lfloor (a'd' + b'c', b'd') \rfloor$ and $\lfloor (ac, bd) \rfloor = \lfloor (a'c', b'd') \rfloor$, i.e., that $(ad + bc, bd) R (a'd' + b'c', b'd')$ and $(ac, bd) R (a'c', b'd')$. But since $ab' = a'b$ and $cd' = c'd$,

$$(ad + bc)b'd' = adb'd' + bcb'd' = ab'dd' + cd'bb'$$
$$= a'bdd' + c'dbb' = a'd'bd + c'b'db$$
$$= (a'd' + c'b')bd,$$

so $(ad + bc, bd) R (a'd' + b'c', b'd')$, and

$$acb'd' = ab'cd' = a'bc'd = a'c'bd,$$

so $(ac, bd) R (a'c', b'd')$.

Addition on K' is associative, for

$$(\lfloor(a, b)\rfloor + \lfloor(c, d)\rfloor) + \lfloor(e, f)\rfloor = \lfloor(ad + bc, bd)\rfloor + \lfloor(e, f)\rfloor$$
$$= \lfloor(adf + bcf + bde, bdf)\rfloor,$$
$$\lfloor(a, b)\rfloor + (\lfloor(c, d)\rfloor + \lfloor(e, f)\rfloor) = \lfloor(a, b)\rfloor + \lfloor(cf + de, df)\rfloor$$
$$= \lfloor(adf + bcf + bde, bdf)\rfloor.$$

Similar calculations establish that addition is commutative and that multiplication is associative, commutative, and distributive over addition. It is easy also to establish that $\lfloor(0, 1)\rfloor$ is the neutral element for addition and that $\lfloor(1, 1)\rfloor$ is the identity element for multiplication. For each $c \in A^*$, $\lfloor(0, c)\rfloor = \lfloor(0, 1)\rfloor$, as $0 \cdot 1 = 0 \cdot c$; conversely, if $\lfloor(a, b)\rfloor = \lfloor(0, 1)\rfloor$, then $a = 0$, as $a = a \cdot 1 = 0 \cdot b = 0$. Consequently, for any $(a, b) \in A \times A^*$, $\lfloor(-a, b)\rfloor$ is the additive inverse of $\lfloor(a, b)\rfloor$, since

$$\lfloor(a, b)\rfloor + \lfloor(-a, b)\rfloor = \lfloor(ab + b(-a), b^2)\rfloor = \lfloor(0, b^2)\rfloor = \lfloor(0, 1)\rfloor.$$

For each $c \in A^*$, $\lfloor(c, c)\rfloor = \lfloor(1, 1)\rfloor$, as $c \cdot 1 = 1 \cdot c$. Hence, if $\lfloor(a, b)\rfloor \neq \lfloor(0, 1)\rfloor$, or equivalently, if $a \neq 0$, then $(b, a) \in A \times A^*$, and $\lfloor(a, b)\rfloor^{-1} = \lfloor(b, a)\rfloor$, since

$$\lfloor(a, b)\rfloor \cdot \lfloor(b, a)\rfloor = \lfloor(ab, ba)\rfloor = \lfloor(1, 1)\rfloor.$$

Therefore, K' is a field.

It is easy to verify that A' is a subdomain of K'. If $(a, b) \in A \times A^*$, then $\lfloor(1, b)\rfloor^{-1} = \lfloor(b, 1)\rfloor \in A'$ by the preceding, and

$$\lfloor(a, b)\rfloor = \lfloor(a, 1)\rfloor\lfloor(1, b)\rfloor = \lfloor(a, 1)\rfloor\lfloor(b, 1)\rfloor^{-1}.$$

Thus, K' is a quotient field of A'.

If $f(a) = f(b)$, then $(a, 1) R (b, 1)$, so $a = a \cdot 1 = b \cdot 1 = b$; thus, f is injective. By the definition of A' and f, f is surjective. Therefore, as

$$f(a) + f(b) = \lfloor(a, 1)\rfloor + \lfloor(b, 1)\rfloor = \lfloor(a + b, 1)\rfloor = f(a + b),$$
$$f(a)f(b) = \lfloor(a, 1)\rfloor\lfloor(b, 1)\rfloor = \lfloor(ab, 1)\rfloor = f(ab),$$

f is an isomorphism from A onto A'.

We have not literally constructed a quotient field K of A; rather, we have constructed a quotient field K' of an isomorphic copy A' of A. To make A

a subdomain of a field, we need somehow to "pull back" the quotient field K' of A' to a field K containing A. More precisely, we need to construct a field K that contains A algebraically in the sense of the following definition and an isomorphism g from K onto K' that is an extension of the constructed isomorphism f from A onto A'.

Definition. If $(E, \triangle)$ and $(A, \wedge)$ are algebraic structures, we shall say that $(E, \triangle)$ **contains** $(A, \wedge)$ **algebraically,** or that $(E, \triangle)$ is an **extension** of $(A, \wedge)$, if A is a subset of E stable for $\triangle$ and if $\wedge$ is the composition $\triangle_A$ induced on A by $\triangle$. Similarly, if $(E, \triangle, \triangledown)$ and $(A, \vee, \wedge)$ are algebraic structures, we shall say that $(E, \triangle, \triangledown)$ **contains** $(A, \wedge, \vee)$ **algebraically,** or that $(E, \triangle, \triangledown)$ is an **extension** of $(A, \wedge, \vee)$, if A is a subset of E stable for $\triangle$ and $\triangledown$ and if $\wedge$ and $\vee$ are respectively the compositions $\triangle_A$ and $\triangledown_A$ induced on A by $\triangle$ and $\triangledown$.

If one algebraic structure contains another algebraically, we shall also say that the second is **contained algebraically** or **embedded** in the first.

Theorem 17.3. (Pullback Theorem) If f is an isomorphism from an algebraic structure $(A, \wedge)$ onto an algebraic structure $(A', \wedge')$ embedded in $(E', \triangle')$, then there exist an algebraic structure $(E, \triangle)$ containing $(A, \wedge)$ algebraically and an isomorphism g from $(E, \triangle)$ onto $(E', \triangle')$ extending f.

Proof. First, suppose that A and E' are disjoint sets. Let E be the set $A \cup (E' - A')$, and let h be the function from E' into E defined by

$$h(z) = \begin{cases} z \text{ if } z \in E' - A', \\ f^{\leftarrow}(z) \text{ if } z \in A'. \end{cases}$$

As $E' - A'$ and A are disjoint, h is a bijection from E' onto E. Let $\triangle$ be the transplant of $\triangle'$ under h. Thus, for all $x, y \in E$,

$$x \triangle y = h(h^{\leftarrow}(x) \triangle' h^{\leftarrow}(y)).$$

If $x, y \in A$, then

$$h^{\leftarrow}(x) = f^{\leftarrow\leftarrow}(x) = f(x)$$

and similarly, $h^{\leftarrow}(y) = f(y)$, so

$$x \triangle y = h(f(x) \triangle' f(y)) = h(f(x) \wedge' f(y))$$
$$= h(f(x \wedge y)) = f^{\leftarrow}(f(x \wedge y))$$
$$= x \wedge y$$

as $f(x \wedge y) \in A'$. Thus, $(E, \triangle)$ contains $(A, \wedge)$ algebraically. Let $g = h^{\leftarrow}$. Then by the definition of $\triangle$, g is an isomorphism from $(E, \triangle)$ onto $(E', \triangle')$. If $x \in A$, then

$$h(f(x)) = f^{\leftarrow}(f(x)) = x$$

as $f(x) \in A'$, whence

$$g(x) = g(h(f(x))) = h^{\leftarrow}(h(f(x))) = f(x).$$

Thus, g is an extension of f.

Next, suppose that A and E' are not disjoint. Using the axiom we adopted on page 110 when constructing the integers, we shall prove that there is a bijection k from E' onto a set E'' disjoint from A. Indeed, by that axiom applied to the set $A \cup E'$, there is a bijection j from $A \cup E'$ onto a set B disjoint from $A \cup E'$; let $E'' = \{j(z): z \in E'\}$, and let k be the function obtained by restricting the domain of j to E' and the codomain of j to E''; then E'' is disjoint from A, and k is a bijection from E' onto E''. Let $\triangle''$ be the transplant of $\triangle'$ under k, and let

$$A'' = \{k(z): z \in A'\}.$$

Then A'' is stable for $\triangle''$, for if $x, y \in A''$, then $k^{\leftarrow}(x)$ and $k^{\leftarrow}(y)$ belong to A', whence $k^{\leftarrow}(x) \triangle' k^{\leftarrow}(y) \in A'$, and thus,

$$x \triangle'' y = k(k^{\leftarrow}(x) \triangle' k^{\leftarrow}(y)) \in A''.$$

Let $\wedge''$ be the composition on A'' induced by $\triangle''$. Then $(A'', \wedge'')$ is embedded in $(E'', \triangle'')$, and clearly the function $f_1: A \mapsto A''$ defined by $f_1(x) = k(f(x))$ for all $x \in A$ is an isomorphism from $(A, \wedge)$ onto $(A'', \wedge'')$. By what we have just proved, there exist an algebraic structure $(E, \triangle)$ containing $(A, \wedge)$ algebraically and an isomorphism g_1 from $(E, \triangle)$ onto $(E'', \triangle'')$ extending f_1. Let $g = k^{\leftarrow} \circ g_1$. Then as k is an isomorphism from $(E', \triangle')$ onto $(E'', \triangle'')$, g is an isomorphism from $(E, \triangle)$ onto $(E', \triangle')$, and g extends f since if $x \in A$, then

$$g(x) = k^{\leftarrow}(g_1(x)) = k^{\leftarrow}(f_1(x)) = k^{\leftarrow}(k(f(x))) = f(x).$$

Corollary. If f is an isomorphism from an algebraic structure $(A, \wedge, \vee)$ onto an algebraic structure $(A', \wedge', \vee')$ embedded in $(E', \triangle', \triangledown')$, then there exist an algebraic structure $(E, \triangle, \triangledown)$ containing $(A, \wedge, \vee)$ algebraically and an isomorphism g from $(E, \triangle, \triangledown)$ onto $(E', \triangle', \triangledown')$ extending f.

Proof. By Theorem 17.3, there exist an algebraic structure $(E, \triangle)$ containing $(A, \wedge)$ algebraically and an isomorphism g from $(E, \triangle)$ onto $(E', \triangle')$ extending f. Let $\triangledown$ be the transplant of $\triangledown'$ under $g^{\leftarrow}$. Then as $g^{\leftarrow}$ is an isomorphism from (E', ∇') onto (E, ∇), g is an isomorphism from (E, ∇) onto (E', ∇') and hence is an isomorphism from $(E, \triangle, \nabla)$ onto $(E', \triangle', \nabla')$. It remains for us to show that A is stable for ∇ and that the composition induced on A by ∇ is $\vee$. Let $x, y \in A$. Then $g(x) = f(x)$, $g(y) = f(y)$, $g(x \vee y) = f(x \vee y) = f(x) \vee' f(y)$, and all the elements $f(x), f(y), f(x \vee y)$ belong to A', so by the definition of ∇,

$$x \nabla y = g^{\leftarrow}(g(x) \nabla' g(y))$$
$$= g^{\leftarrow}(f(x) \vee' f(y))$$
$$= g^{\leftarrow}(f(x \vee y)) = g^{\leftarrow}(g(x \vee y))$$
$$= x \vee y.$$

Thus, $(A, \wedge, \vee)$ is embedded in $(E, \triangle, \nabla)$.

We shall refer also to this corollary as the "Pullback Theorem."

Theorem 17.4. If A is an integral domain, there is a quotient field of A.

Proof. By Theorem 17.2, there is an isomorphism f from A onto an integral domain A' that possesses a quotient field K'. By the Pullback Theorem, there exist an algebraic structure $(K, +, \cdot)$ containing the integral domain A algebraically and an isomorphism g from K onto the field K' that extends f. As K is isomorphic to a field, K is itself a field. To show that K is a quotient field of A, let $z \in K$. As K' is a quotient field of A' and as f is an isomorphism from A onto A', there exist $x \in A$, $y \in A^{*}$ such that $g(z) = f(x)f(y)^{-1}$. As g is an isomorphism extending f, therefore,

$$g(z) = g(x)g(y)^{-1} = g(x)g(y^{-1}) = g(xy^{-1}),$$

so $z = xy^{-1}$.

We have thus established that every integral domain has at least one quotient field. From our next result we may infer that a quotient field of an integral domain is essentially unique.

Theorem 17.5. If K and L are quotient fields of integral domains A and B respectively and if f is an isomorphism from A onto B, then there is one and only one isomorphism g from K onto L extending f, and moreover,

(1) $$g\left(\frac{x}{y}\right) = \frac{f(x)}{f(y)}$$

for all $x \in A$ and all $y \in A^{*}$.

Proof. Let us first show that we may use (1) as the definition of a function g from K into L. To do so, we must show that if $x, x' \in A$, $y, y' \in A^*$ are such that $x/y = x'/y'$, then $f(x)/f(y) = f(x')/f(y')$. If $x/y = x'/y'$, then $xy' = x'y$ by (4) of Theorem 15.4, whence

$$f(x)f(y') = f(xy') = f(x'y) = f(x')f(y),$$

and consequently, $f(x)/f(y) = f(x')/f(y')$ again by (4) of Theorem 15.4. Thus, g, defined by (1), is a well-defined function from K into L. As f is an isomorphism from A onto B and as L is a quotient field of B, g is surjective. If $g(x/z) = g(y/w)$, then $f(x)/f(z) = f(y)/f(w)$, whence

$$f(xw) = f(x)f(w) = f(y)f(z) = f(yz)$$

by (4) of Theorem 15.4, so $xw = yz$, and therefore, $x/z = y/w$ again by (4) of Theorem 15.4. Thus, g is injective. Let $x, y \in A$, $z, w \in A^*$. By two applications of (3) of Theorem 15.4,

$$g\left(\frac{x}{z} + \frac{y}{w}\right) = g\left(\frac{xw + yz}{zw}\right) = \frac{f(xw + yz)}{f(zw)} = \frac{f(x)f(w) + f(y)f(z)}{f(z)f(w)}$$

$$= \frac{f(x)}{f(z)} + \frac{f(y)}{f(w)} = g\left(\frac{x}{z}\right) + g\left(\frac{y}{w}\right),$$

and by two applications of (5) of Theorem 15.4,

$$g\left(\frac{x}{z} \cdot \frac{y}{w}\right) = g\left(\frac{xy}{zw}\right) = \frac{f(xy)}{f(zw)} = \frac{f(x)f(y)}{f(z)f(w)}$$

$$= \frac{f(x)}{f(z)} \cdot \frac{f(y)}{f(w)} = g\left(\frac{x}{z}\right)g\left(\frac{y}{w}\right).$$

Therefore, g is an isomorphism from K onto L that clearly extends f.

To show that g is the only isomorphism from K onto L extending f, let h be such an extension. For all $x \in A$, $y \in A^*$,

$$h(xy^{-1}) = h(x)h(y^{-1}) = h(x)h(y)^{-1} = f(x)f(y)^{-1} = g(xy^{-1}),$$

so $h = g$.

Corollary. If K and L are quotient fields of an integral domain A, then there is one and only one isomorphism g from K onto L satisfying $g(x) = x$ for all $x \in A$.

In view of the corollary, it is customary in a discussion concerning a given integral domain A to select a certain quotient field K of A and to call K *the* quotient field of A. If A is already a subdomain of a specified field L, the quotient field selected is usually the subfield of L consisting of all the elements x/y where $x \in A$ and $y \in A^*$ (Theorem 17.1).

There is, therefore, nothing arbitrary in our choice if we simply select some quotient field Q of Z and call its members "rational numbers." Every rational number is thus the quotient of an integer and a nonzero integer.

Our method of constructing the rationals from the integers may be generalized and, in fact, by replacing multiplication with addition, could have been applied to construct the additive group of integers from the additive semigroup of natural numbers. Indeed, every integer is the difference of two natural numbers, and since $m - n = p - q$ if and only if $m + q = p + n$, we could have constructed the integers by starting with the set $(N \times N)/R$, where $(m, n) \; R \; (p, q)$ if and only if $m + q = p + n$, by defining

$$\lfloor (m, n) \rfloor + \lfloor (p, q) \rfloor = \lfloor (m + p, n + q) \rfloor,$$

and by observing that $f : n \mapsto \lfloor (n, 0) \rfloor$ was an isomorphism from $(N, +)$ onto a stable subset of $(N \times N)/R$. A discussion and construction of the "inverse-completion" of a commutative semigroup having cancellable elements are presented in the exercises.

We next wish to extend the total ordering on Z to a total ordering on Q compatible with addition so that the product of any two positive numbers is positive. As we shall see, this can be done in only one way.

Definition. Let $(A, +, \cdot)$ be a ring. An ordering $\leq$ on A is **compatible with the ring structure** of A if $\leq$ is compatible with addition and if for all $x, y \in A$,

(OR)   if $x \geq 0$ and if $y \geq 0$, then $xy \geq 0$.

If $\leq$ is an ordering on A compatible with its ring structure, we shall say that $(A, +, \cdot, \leq)$ is an **ordered ring.** An element x of an ordered ring A is **positive** if $x \geq 0$, and x is **strictly positive** if $x > 0$.

The set of all positive elements of an ordered ring A is denoted by A_+, and the set of all strictly positive elements of A is denoted by A_+^*.

If $(A, +, \cdot, \leq)$ is an ordered ring and if $\leq$ is a total ordering, we shall, of course, call $(A, +, \cdot, \leq)$ a **totally ordered ring;** if $(A, +, \cdot)$ is a field, we shall call $(A, +, \cdot, \leq)$ an **ordered field,** and if, moreover, $\leq$ is a total ordering, we shall call $(A, +, \cdot, \leq)$ a **totally ordered field.**

It follows at once from Theorem 14.6 that $(Z, +, \cdot, \leq)$ is a totally ordered ring. If $(A, +, \cdot)$ is any ring, the equality relation on A is an ordering compatible with the ring structure of A, and for certain rings this is the only compatible ordering (Exercise 17.8).

The following theorem is a collection of important and easily proved facts about ordered and totally ordered rings.

Theorem 17.6. Let x, y, and z be elements of an ordered ring A.

1° $x < y$ if and only if $x + z < y + z$, and hence $x \leq y$ if and only if $x + z \leq y + z$.
2° $x < y$ if and only if $y - x > 0$, and hence $x \leq y$ if and only if $y - x \geq 0$.
3° $x > 0$ if and only if $-x < 0$, and hence $x \geq 0$ if and only if $-x \leq 0$.
4° $x < 0$ if and only if $-x > 0$, and hence $x \leq 0$ if and only if $-x \geq 0$.
5° If $x > 0$, then $n.x > 0$ for all $n \in N^*$.
6° If $x \leq y$ and if $z \geq 0$, then $xz \leq yz$ and $zx \leq zy$.
7° If $x \leq y$ and if $z \leq 0$, then $xz \geq yz$ and $zx \geq zy$.

If A is a totally ordered ring, the following also hold:

8° If $xy > 0$, then either $x > 0$ and $y > 0$, or $x < 0$ and $y < 0$.
9° If $xy < 0$, then either $x > 0$ and $y < 0$, or $x < 0$ and $y > 0$.
10° $x^2 \geq 0$; in particular, if A is a ring with identity element 1, then $1 > 0$.
11° If x is invertible, then $x > 0$ if and only if $x^{-1} > 0$, and $x < 0$ if and only if $x^{-1} < 0$.

Our next theorem shows that an ordering $\leq$ compatible with the ring structure of a ring A is completely determined by the set of its positive elements.

Theorem 17.7. If A is an ordered ring and if $P = A_+$, $-P = \{-x : x \in A_+\}$, then

(P 1) P is stable for addition,

(P 2) $P \cap (-P) = \{0\}$,

(P 3) P is stable for multiplication.

Furthermore, if A is a totally ordered ring, then

(P 4) $P \cup (-P) = A$.

Conversely, if A is a ring and if P is a subset of A satisfying (P 1), (P 2), and (P 3), then there is one and only one ordering $\leq$ on A compatible with the ring structure of A such that $P = A_+$. Furthermore, if (P 4) is also satisfied, then $\leq$ is a total ordering.

Proof. If $x \geq 0$ and if $y \geq 0$, then

$$x + y \geq x \geq 0,$$

so P is stable for addition. By $4°$ of Theorem 17.6,

$$-P = \{x \in A : x \leq 0\}.$$

Hence, (P 2) holds, for if $x \in P \cap (-P)$, then $x \geq 0$ and $x \leq 0$, so $x = 0$. Clearly, (P 3) is equivalent to (OR). Also, if $\leq$ is a total ordering, then (P 4) holds since for every $x \in A$, either $x \geq 0$ or $x \leq 0$.

Conversely, let P be a subset of a ring A satisfying (P 1), (P 2), and (P 3). By $2°$ of Theorem 17.6, there is at most one ordering on A compatible with the ring structure of A such that $P = A_+$, namely, that satisfying

$$x \leq y \text{ if and only if } y - x \in P.$$

It remains for us to show that the relation $\leq$ so defined has the requisite properties. For all $x \in A$, $x \leq x$ since $x - x \in P$ by (P 2). If $x \leq y$ and if $y \leq x$, then $y - x \in P$ and $-(y - x) = x - y \in P$, so by (P 2), $y - x = 0$ and thus $y = x$. If $x \leq y$ and if $y \leq z$, then $y - x$ and $z - y$ belong to P, so as $z - x = (z - y) + (y - x)$, $z - x$ also belongs to P by (P 1), whence $x \leq z$. If $x \leq y$, then $z + x \leq z + y$ since $(z + y) - (z + x) = y - x \in P$. Finally, (OR) holds by (P 3). Furthermore, if (P 4) holds, then for all x, $y \in A$, either $y - x \in P$ or $x - y = -(y - x) \in P$, that is, either $x \leq y$ or $y \leq x$.

If P is a subset of A satisfying (P 1)–(P 3), we shall say that the ordering $\leq$ on A satisfying $x \leq y$ if and only if $y - x \in P$ is the **ordering defined by** P.

Theorem 17.8. Let K be a quotient field of a totally ordered integral domain A. There is one and only one total ordering $\leq'$ on K that is compatible with its ring structure and induces on A its given total ordering $\leq$, namely, that defined by

$$P = \left\{ \frac{x}{y} \in K : x \in A_+ \text{ and } y \in A_+^* \right\}.$$

Proof. First, we observe that for every $z \in K$ there exist $x, y \in A$ such that $z = x/y$ and $y \in A_+^*$; indeed, if $z = x'/y'$ and if $y' \notin A_+^*$, then $y' < 0$ as A is totally ordered, so we need only let $x = -x'$, $y = -y'$.

Next, we shall show that P satisfies conditions (P 1)–(P 4). Clearly, P satisfies (P 1) and (P 3) by (3) and (5) of Theorem 15.4. To establish (P 2), let $z \in P \cap (-P)$. Then $z \in P$ and $-z \in P$, so there exist $x, u \in A_+$ and $y, v \in A_+^*$ such that $z = x/y$ and $-z = u/v$. Therefore, $x/y = -u/v$, so $xv = -uy$. But

$xv \geq 0$ and $-uy \leq 0$, so $xv = 0$. Hence, as $v > 0$, we conclude that $x = 0$ and therefore that $z = 0$. To show (P 4), let $z = x/y$ where $x \in A$ and $y \in A_+^*$. If $x \geq 0$, then $z \in P$, but if $x < 0$, then $-x > 0$, so $-z = (-x)/y \in P$, and hence $z = -(-z) \in -P$. Thus, $P \cup (-P) = K$. Consequently, by Theorem 17.7, the relation $\leq'$ on K defined by P is a total ordering on K compatible with its ring structure. To show that the ordering induced on A by $\leq'$ is $\leq$, it suffices by 2° of Theorem 17.6 to show that for all $z \in A$, $z \geq 0$ if and only if $z \in P$, that is, that $A_+ = A \cap P$. If $z \in A_+$, then $z = z/1 \in P$ since $1 \in A_+^*$. Conversely, if $z \in A \cap P$, then there exist $x \in A_+$ and $y \in A_+^*$ such that $z = x/y$; if $x = 0$, then $z = 0$, and if $x > 0$, then as $zy = x$ and as $y > 0$, we conclude that $z > 0$ by 8° of Theorem 17.6.

To show uniqueness, let $\leqslant$ be a total ordering on K that is compatible with its ring structure and induces on A the ordering $\leq$, and let $Q = \{z \in K: z \geqslant 0\}$. To show that $\leqslant$ is $\leq'$, it suffices by 2° of Theorem 17.6 to show that $Q = P$. If $x \in A_+$ and if $y \in A_+^*$, then $x \geqslant 0$, and $1/y \geqslant 0$ by 11° of Theorem 17.6, so $x/y \geqslant 0$ by (OR); hence, $P \subseteq Q$. Conversely, if $z \in Q$ and if $z = x/y$ where $x \in A$ and $y \in A_+^*$, then $x = zy \geqslant 0$ by (OR), so $x \geq 0$, and hence, $z \in P$; thus, $Q \subseteq P$.

We shall denote again by $\leq$ the unique total ordering on the field Q of rational numbers that is compatible with its ring structure and induces on Z its total ordering Thus, $(Q, +, \cdot, \leq)$ is a totally ordered field.

Definition. Let $(A, +, \cdot, \leq)$ and $(A', +', \cdot', \leq')$ be ordered rings. An **isomorphism** from A onto A' is a bijection f from A onto A' that is also an isomorphism from the ordered group $(A, +, \leq)$ onto the ordered group $(A', +', \leq')$ and an isomorphism from the semigroup $(A, \cdot)$ onto the semigroup $(A', \cdot')$. An **automorphism** of an ordered ring is an isomorphism from itself onto itself.

As before, the identity function is an automorphism of an ordered ring, a bijection f is an isomorphism from an ordered ring A onto an ordered ring A' if and only if $f^{\leftarrow}$ is an isomorphism from the ordered ring A' onto A, and the composite of two isomorphisms is again one.

Theorem 17.9. If K and L are totally ordered quotient fields of totally ordered integral domains A and B respectively and if f is an isomorphism from the ordered ring A onto the ordered ring B, then there is one and only one isomorphism g from the ordered field K onto the ordered field L extending f, and moreover,

$$g\left(\frac{x}{y}\right) = \frac{f(x)}{f(y)}$$

for all $x \in A$, $y \in A^*$.

Proof. By Theorem 17.5 it suffices to prove that if $x, u \in A$ and if $y, v \in A_+^*$, then

$$\frac{x}{y} \le \frac{u}{v} \qquad \text{if and only if} \qquad \frac{f(x)}{f(y)} \le \frac{f(u)}{f(v)}.$$

If $x/y \le u/v$, then

$$xv = \frac{x}{y}(yv) \le \frac{u}{v}(yv) = uy$$

as $yv > 0$, and conversely if $xv \le uy$, then

$$\frac{x}{y} = xv\left(\frac{1}{yv}\right) \le uy\left(\frac{1}{yv}\right) = \frac{u}{v}$$

as $1/yv > 0$. As $f(y)f(v) = f(yv) > 0$, similarly $f(x)/f(y) \le f(u)/f(v)$ if and only if $f(x)f(v) \le f(u)f(y)$. But as f is an isomorphism from the ordered ring A onto B, $xv \le uy$ if and only if $f(xv) \le f(uy)$, that is, if and only if $f(x)f(v) \le f(u)f(y)$.

EXERCISES

17.1. (a) If $A = \{m + 2n\sqrt{2}: m, n \in Z\}$, then A is a subdomain of R, and the quotient field of A is $\{r + s\sqrt{2}: r, s \in Q\}$.
(b) If $B = \{m + 5n\sqrt{3}: m, n \in Z\}$, then B is a subdomain of R. What subfield of R is its quotient field?
(c) Prove that the integral domains A and B are not isomorphic.

17.2. Let $C = \{m + in: m, n \in Z\}$, and let $D = \{m + in\sqrt{2}: m, n \in Z\}$.
(a) Show that C and D are subdomains of C. What subfields of C are their quotient fields?
(b) Prove that C and D are not isomorphic integral domains.
(c) Prove that a subdomain of C containing i cannot be isomorphic with a subdomain of R.

17.3. If K is a quotient field of an integral domain A, then K is also a quotient field of every subdomain of K containing A.

17.4. Let $\oplus$ be the composition on $Z \times Z^*$ defined by $(m, n) \oplus (p, q) = (mq + np, nq)$. Is $(Z \times Z^*, \oplus)$ a commutative group?

17.5. Complete the proof of Theorem 17.2:
(a) Show that addition is commutative and that $|(0, 1)|$ is the neutral element for addition.

(b) Show that multiplication is associative, commutative, and distributive over addition and that $\lfloor (1, 1) \rfloor$ is the identity element for multiplication.

(c) Show that A' is a subdomain of K'.

17.6. Prove Theorem 17.6.

17.7. (a) Let K be a quotient field of a totally ordered integral domain A that is not a field, and let $P = A_+$. If $\leq'$ is the ordering on K defined by P, then $(K, +, \cdot, \leq')$ is an ordered field. Which of $8°–11°$ of Theorem 17.6 hold for $\leq'$?

(b) What part of the assertion obtained by deleting the words "totally" and "total" from the statement of Theorem 17.7 is true?

17.8. If A is a boolean ring (Exercise 15.21), then the equality relation is the only ordering on A compatible with its ring structure. Show by a specific example that if E is a nonempty set, the ordering $\subseteq$ on $\mathfrak{P}(E)$ is not compatible with the ring structure of $\mathfrak{P}(E)$ (Example 15.2).

In the remaining exercises, $(E, \cdot)$ is a commutative semigroup that has at least one cancellable element, and C is the set of all cancellable elements of E. An **inverse-completion** of E is a semigroup $(G, \cdot)$ that contains $(E, \cdot)$ algebraically such that

1° Every element of C is invertible in $(G, \cdot)$,

2° For each $z \in G$ there exist $x \in E$, $y \in C$ such that $z = xy^{-1}$.

(We note that we could not hope to find a semigroup $(G, \cdot)$ containing $(E, \cdot)$ algebraically in which a noncancellable element of E had an inverse, for if an element of E has an inverse in G, it is a cancellable element of G by Theorem 7.1 and *a fortiori* is a cancellable element of the given semigroup $(E, \cdot)$.)

17.9. Rewrite the definition of an inverse-completion if the composition of E is denoted by $+$ rather than $\cdot$.

17.10. Prove that C is a stable subset of $(E, \cdot)$.

*17.11. Let $(G, \cdot)$ be an inverse-completion of $(E, \cdot)$.

(a) G is a commutative semigroup.

(b) Every cancellable element of G is invertible; i.e., G is an inverse-completion of itself.

(c) If E possesses a neutral element e, then e is also the neutral element of G.

(d) The inverse of an invertible element of $(E, \cdot)$ is also its inverse in G.

*17.12. Let R be the relation on $E \times C$ satisfying $(a, b) \, R \, (c, d)$ if and only if $ad = cb$.

(a) Show that R is an equivalence relation on $E \times C$.

(b) Let G' be the quotient set $(E \times C)/R$, and show that the composition on G' defined by

$$\lfloor (a, b) \rfloor \, \lfloor (c, d) \rfloor = \lfloor (ac, bd) \rfloor$$

for all $\lfloor (a, b) \rfloor$, $\lfloor (c, d) \rfloor \in G'$ is well defined.

(c) Let $s \in C$, and let $E' = \{|(as, s)| : a \in E\}$. Show that E' is a stable subset of G', and that $f : a \mapsto |(as, s)|$ is an isomorphism from E onto E'.
(d) If t is any other element of C, then $|(at, t)| = |(as, s)|$ for all $a \in E$.
(e) Show that G' is an inverse-completion of E'.
(f) Conclude that there is an inverse-completion G of E. [Modify the proofs of Theorems 17.2 and 17.4.]

*17.13. Let f be an isomorphism from $(E, \cdot)$ onto a semigroup $(F, \cdot)$, and let G and H be respectively inverse-completions of E and F. Show that the function $g : G \to H$ defined by

$$g(ab^{-1}) = f(a)f(b)^{-1}$$

for all $a \in E$, $b \in C$ is well-defined, and that g is an isomorphism from G onto H extending f. [Modify the proof of Theorem 17.5.] Conclude that if G and G' are inverse-completions of E, there is an isomorphism g from G onto G' satisfying $g(a) = a$ for all $a \in E$.

17.14. What subsemigroup of $(R, \cdot)$ is the inverse-completion of $(E, \cdot)$ if
(a) $E = Z$?
(b) E is the set of all positive even integers?
(c) E is the set of all positive odd integers?
(d) E is the set of all integers ≥ 5?
(e) $E = \{a + b\sqrt{2} : a, b \in Z\}$?
(f) $E = \{a + b\sqrt{2} : a, b \in N^*\}$? [First show that every strictly positive number of the form $a - b\sqrt{2}$ where $a, b \in N^*$ belongs to the inverse-completion.]

17.15. What subsemigroup of $(R, +)$ is the inverse completion of $(E, +)$ if
(a) $E = Z$?
(b) E is the set of negative integers?
(c) E is the set of all integers ≥ 5?
(d) E is the set of all positive even integers?

*17.16. Let $(E, +, \leq)$ be a totally ordered commutative semigroup all of whose elements are cancellable, and let $(G, +)$ be an inverse-completion of $(E, +)$. The relation $\leq'$ on G satisfying

$$x - y \leq' z - w \text{ if and only if } x + w \leq y + z$$

for all $x, y, z \; w \in E$ is a well-defined relation, and $\leq'$ is the only total ordering on G that is compatible with addition and induces the given ordering $\leq$ on E. [First show that for all $a, b, c \in E$, if $a + c \leq b + c$, then $a \leq b$.]

*17.17. Let $(G, +)$ be an inverse-completion of a commutative semigroup $(E, +)$ all of whose elements are cancellable, and let ∇ be a composition on E distributive over $+$.

(a) There is a unique composition ∇' on G that is distributive over addition on G and induces on E the given composition ∇. [Use the cancellability of suitably added terms in showing that ∇' is well defined by

$$(x - y)\nabla'(z - w) = (x\nabla z + y\nabla w) - (y\nabla z + x\nabla w),$$

for all $x, y, z, w \in E$.]
(b) If ∇ is associative, so is ∇'.
(c) If e is a neutral element in E for ∇, then e is also the neutral element for ∇'.
(d) If ∇ is commutative, so is ∇'.
(e) Every element of E cancellable for ∇ is also cancellable for ∇'.

CHAPTER III

GROUPS AND RINGS

Constructing a "quotient group" from a group and a special kind of subgroup called a "normal subgroup" and, correspondingly, constructing a "quotient ring" from a ring and a special kind of subring called an "ideal" are two of the most important techniques in algebra. The latter, for example, is the basic tool we shall use in Chapter VII to construct the real numbers from the rationals and the complex numbers from the real numbers. The general formation of quotient groups and quotient rings is the subject of §§18–19, and in §§20–21 we shall investigate certain functions, called homomorphisms, that differ from isomorphisms only in that they need not be either injective or surjective.

Readers are familiar with the fact that every nonzero integer is the product of primes in a unique way or the negative of such a product. The final section of this chapter is devoted to a discussion of those integral domains that share with the integers the property that each nonzero element is the product of an invertible element and prime elements in an essentially unique way.

18. Normal Subgroups and Quotient Groups

If $(E, \triangle)$ is an algebraic structure and if X and Y are subsets of E, the subset $X \triangle Y$ of E is defined by

$$X \triangle Y = \{x \triangle y \colon x \in X, y \in Y\}.$$

Thus, the given composition $\triangle$ on E induces in a natural way a composition on the set $\mathfrak{P}(E)$ of all subsets of E. A useful though obvious fact is that if $X \subseteq Y$, then $Z \triangle X \subseteq Z \triangle Y$ and $X \triangle Z \subseteq Y \triangle Z$ for every subset Z of E. If

$\triangle$ is associative, so is the composition it induces on $\mathfrak{P}(E)$, since for any subsets X, Y, Z of E,

$$(X \triangle Y)\triangle Z = \{(x\triangle y)\triangle z: x \in X, y \in Y, z \in Z\},$$
$$X\triangle(Y\triangle Z) = \{x\triangle(y\triangle z): x \in X, y \in Y, z \in Z\}.$$

Similarly, if $\triangle$ is commutative, then $X \triangle Y = Y \triangle X$ for any subsets X, Y of E. Clearly, a subset A of E is stable for $\triangle$ if and only if $A \triangle A \subseteq A$.

If $a \in E, X \subseteq E$, we usually write $a\triangle X$ for $\{a\}\triangle X$ and $X\triangle a$ for $X\triangle\{a\}$. Thus, $a\triangle X = \{a\triangle x: x \in X\}$, and $X\triangle a = \{x\triangle a: x \in X\}$.

Here we shall see how a certain kind of subgroup of a group gives rise to a new group, called a "quotient group."

Definition. Let H be a subgroup of a group $(G, \cdot)$. A subset X of G is a **left coset** of H in G if there exists $a \in G$ such that $X = aH$, and X is a **right coset** of H in G if there exists $a \in G$ such that $X = Ha$.

Note that if a belongs to the subgroup H, then $aH = H = Ha$. If the composition of G is denoted by $+$ rather than $\cdot$, then, of course, X is a left (right) coset of H in G if and only if there exists $a \in G$ such that $X = a + H$ $(X = H + a)$.

Theorem 18.1. If H is a subgroup of a group $(G, \cdot)$, the set of all left cosets of H is a partition of G, and the set of all right cosets of H is a partition of G.

Proof. For each $a \in G$, $a = ae \in aH$, where e is the neutral element; consequently, no left coset is the empty set, and each element of G belongs to at least one left coset. To show that the left cosets form a partition, therefore, it suffices to show that if aH and bH have an element in common, then $aH = bH$. Let $c \in aH \cap bH$. Then there exist $h, k \in H$ such that $c = ah$, $c = bk$, whence $a = ch^{-1}, b = ck^{-1}$. If $x \in H$,

$$ax = ch^{-1}x = b(kh^{-1}x) \in bH,$$

and

$$bx = ck^{-1}x = a(hk^{-1}x) \in aH.$$

Thus, $aH \subseteq bH$ and $bH \subseteq aH$, so $aH = bH$. Similarly, the right cosets of H form a partition of G.

Definition. Let H be a subgroup of a group G. We denote by G/H the set of all left cosets of H in G.

The choice of left cosets over right cosets in the definition is arbitrary, of course.

In proving certain theorems (such as Lagrange's Theorem below) here and in Chapter IV, we shall use without comment certain basic principles of counting familiar to everyone from childhood. The most basic of these is that if $B_1, \ldots, B_n$ are distinct subsets of a finite set A that form a partition of A and if each B_k has p_k elements, then A has $\sum_{k=1}^{n} p_k$ elements. In Appendix B we shall give formal proofs of these principles.

Theorem 18.2. Let H be a subgroup of a group $(G, \cdot)$. Each left coset of H is equipotent to H, and each right coset of H is equipotent to H.

Proof. Let $a \in G$. As a is cancellable, the function $x \mapsto ax$ from H into aH is a bijection, and similarly $x \mapsto xa$ from H into Ha is a bijection.

Theorem 18.3. (Lagrange) Let $(G, \cdot)$ be a finite group having n elements, and let H be a subgroup of G having m elements. Then $m \mid n$, and both the set G/H of left cosets of H and the set of right cosets of H have n/m members.

Proof. Let k be the number of left (right) cosets of H in G. As the left (right) cosets form a partition of G and as each left (right) coset has m elements by Theorem 18.2, the total number n of elements of G is km. Thus, $m \mid n$ and $k = n/m$.

The number of elements in a finite group is called the **order** of the group; an infinite group is said to have **infinite order**. If H is a subgroup of a finite group G, the number of left cosets of H in G is called the **index** of H in G. By Lagrange's Theorem,

$$\text{order } G = (\text{order } H)(\text{index } H).$$

Simple though it is, Lagrange's Theorem is fundamental to the study of finite groups. For example, in determining the subgroups of a finite group G, by virtue of Lagrange's Theorem we need only seek subgroups whose order divides that of G.

Here is a simple criterion for cosets to be identical.

Theorem 18.4. If H is a subgroup of a group $(G, \cdot)$ and if $a, b \in G$, then $aH = bH$ if and only if $b^{-1}a \in H$, and $Ha = Hb$ if and only if $ab^{-1} \in H$.

Proof. If $aH = bH$, then $a \in bH$, so $a = bh$ for some $h \in H$, whence $b^{-1}a = h \in H$. Conversely, if $b^{-1}a \in H$, then $a = b(b^{-1}a) \in bH$, so $aH \cap bH \neq \emptyset$, whence $aH = bH$ by Theorem 18.1. A similar argument establishes the assertion concerning right cosets.

Definition. A subgroup H of a group $(G, \cdot)$ is a **normal subgroup** of G if $aH = Ha$ for every $a \in G$.

If H is a normal subgroup of G, we shall call a left coset of H in G simply a coset of H in G, since it is also a right coset.

The importance of normal subgroups arises from the fact that if H is a normal subgroup of G, we may introduce a composition on G/H making it a group in a fairly natural way. Before doing this, we list some conditions equivalent to normality.

Theorem 18.5. Let H be a subgroup of a group $(G, \cdot)$. The following statements are equivalent:

1° H is a normal subgroup of G.
2° $aHa^{-1} \subseteq H$ for all $a \in G$.
3° $a^{-1}Ha \subseteq H$ for all $a \in G$.
4° $H \subseteq aHa^{-1}$ for all $a \in G$.
5° $H \subseteq a^{-1}Ha$ for all $a \in G$.

Proof. If H is a normal subgroup and if $a \in G$, then $aH = Ha$, so clearly

$$(aH)a^{-1} = (Ha)a^{-1} = H(aa^{-1}) = H.$$

Thus, 1° implies 2° and 4°. If 2° holds, then in particular for each $a \in G$, $a^{-1}H(a^{-1})^{-1} \subseteq H$, that is, $a^{-1}Ha \subseteq H$. Therefore, 2° implies 3°. If 3° holds, then for each $a \in G$, $a(a^{-1}Ha)a^{-1} \subseteq aHa^{-1}$, whence $H \subseteq aHa^{-1}$. Thus, 3° implies 4°. Similarly, 4° implies 5°, and 5° implies 2°. Therefore, statements 2°–5° are all equivalent. If they hold, then $aHa^{-1} = H$ for each $a \in G$, whence $(aHa^{-1})a = Ha$, that is, $aH = Ha$. Consequently, all five statements are equivalent.

Clearly, $\{e\}$ and G are normal subgroups of G. If G is an abelian group, every subgroup of G is normal. More generally, every subgroup of G contained in the center of G is clearly a normal subgroup of G. However, a normal subgroup of G need not be contained in its center, as the following example shows.

Example 18.1. Let $(G, \circ)$ be the group of symmetries of the square (Example 2.5). It is easy to verify that the center of G is $\{r_0, r_2\}$. The index of the subgroup $H = \{r_0, r_1, r_2, r_3\}$ is 2 by Lagrange's Theorem, and consequently, H is a normal subgroup of G not contained in its center by the following theorem:

Theorem 18.6. If H is a subgroup of a finite group $(G, \cdot)$ whose index is 2, then H is a normal subgroup of G.

Proof. If $a \in H$, then $aH = H = Ha$. Let $a \notin H$. Since there are exactly two left (right) cosets of H in G, one of which is H itself, the remaining left coset is aH and the remaining right coset is Ha; since the left (right) cosets of H form a partition of G,

$$aH = H^c = Ha.$$

Example 18.2. The subgroup $L = \{r_0, h\}$ of the group G of symmetries of the square is *not* a normal subgroup of G. Indeed, the left cosets of L are $\{r_0, h\}$, $\{r_1, d_2\}$, $\{r_2, v\}$, $\{r_3, d_1\}$, and the right cosets of L are $\{r_0, h\}$, $\{r_1, d_1\}$, $\{r_2, v\}$, $\{r_3, d_2\}$.

Let H be a subgroup of a group $(G, \cdot)$. We should like to define a composition on G/H by

(1) $$(aH) \cdot (bH) = abH$$

for all aH, $bH \in G/H$. We cannot simply declare that (1) defines a composition on G/H, for the right side of (1) appears to depend on the particular choice of elements in the two given left cosets of H. For (1) to define a composition, we must show that if $aH = a'H$ and $bH = b'H$, then $abH = a'b'H$. That this is not always the case is shown by Example 18.2, for there $r_1 \circ L = d_2 \circ L$ and $r_3 \circ L = r_3 \circ L$, but $(r_1 \circ r_3) \circ L = L \neq (d_2 \circ r_3) \circ L = v \circ L$. To see what conditions H must satisfy for (1) to define a composition, suppose that for all a, a', b, $b' \in G$, if $aH = a'H$ and $bH = b'H$, then $abH = a'b'H$. Letting $a = xh$, $a' = x$, $b = b' = x^{-1}$ where $x \in G$, $h \in H$, we must have $xhx^{-1}H = H$, that is, $xhx^{-1} \in H$. Thus, for (1) to define a composition on G/H, we must have $xHx^{-1} \subseteq H$ for all $x \in G$, so H must be a normal subgroup of G. Actually, the normality of H is sufficient for (1) to define a composition, and G/H, equipped with this composition, is a group.

Theorem 18.7. If H is a normal subgroup of a group $(G, \cdot)$, the composition on G/H defined by

$$(aH) \cdot (bH) = abH$$

for all aH, $bH \in G/H$ is well-defined, and $(G/H, \cdot)$ is a group whose neutral element is H; moreover, the inverse of aH is $a^{-1}H$ for all $a \in G$.

Proof. Let a, a', b, $b' \in G$ be such that $aH = a'H$ and $bH = b'H$. By the normality of H,

$$abH = a(bH) = a(b'H) = a(Hb') = (aH)b'$$
$$= (a'H)b' = a'(Hb') = a'(b'H) = a'b'H.$$

Thus, the composition is well-defined. Multiplication is associative, for if $a, b, c \in G$,

$$[(aH)(bH)](cH) = (abH)(cH) = (ab)cH,$$
$$(aH)[(bH)(cH)] = (aH)(bcH) = a(bc)H.$$

As $H = eH$, where e is the neutral element of G, clearly, H is the neutral element for multiplication on G/H. Consequently, $a^{-1}H$ is the inverse of aH for each $a \in G$, since

$$(aH)(a^{-1}H) = aa^{-1}H = H,$$
$$(a^{-1}H)(aH) = a^{-1}aH = H.$$

The group G/H is called the **quotient group** of G defined by H, and its composition is called the **composition induced on** G/H by that of G.

Example 18.3. In Example 18.1 we saw that $H = \{r_0, r_1, r_2, r_3\}$ is a normal subgroup of the group G of all symmetries of the square. There is exactly one other coset of H besides H itself, and it may be variously denoted by hH, vH, d_1H, d_2H (or Hh, etc.), since those cosets must all be identical. If that coset is denoted by hH, for example, the table for the induced composition on G/H is

	H	hH
H	H	hH
hH	hH	H

.

If the composition of G is denoted by $+$ rather than $\cdot$, then, of course, the composition induced on G/H is also denoted by $+$ and is defined by

$$(a + H) + (b + H) = (a + b) + H$$

for all $a + H$, $b + H \in G/H$; also the inverse of $a + H$ is $-a + H$.

Example 18.4. Let $H = \{0, 4, 8\}$. Clearly, H is a normal subgroup of the group $(N_{12}, +_{12})$ of integers modulo 12. The cosets of H are

$$H = \{0, 4, 8\} = 4 + H = 8 + H,$$
$$1 + H = \{1, 5, 9\} = 5 + H = 9 + H,$$
$$2 + H = \{2, 6, 10\} = 6 + H = 10 + H,$$
$$3 + H = \{3, 7, 11\} = 7 + H = 11 + H.$$

In constructing the addition table for the composition induced on N_{12}/H by $+_{12}$, it makes no difference, of course, how we choose to denote the four elements of N_{12}/H. We might choose to denote them by H, $1 + H$, $2 + H$, and $3 + H$, for example, or by $4 + H$, $5 + H$, $10 + H$, and $3 + H$. In the former case, the table for the induced composition is

	H	$1 + H$	$2 + H$	$3 + H$
H	H	$1 + H$	$2 + H$	$3 + H$
$1 + H$	$1 + H$	$2 + H$	$3 + H$	H
$2 + H$	$2 + H$	$3 + H$	H	$1 + H$
$3 + H$	$3 + H$	H	$1 + H$	$2 + H$

,

and in the latter case, the table for the induced composition is

	$4 + H$	$5 + H$	$10 + H$	$3 + H$
$4 + H$	$4 + H$	$5 + H$	$10 + H$	$3 + H$
$5 + H$	$5 + H$	$10 + H$	$3 + H$	$4 + H$
$10 + H$	$10 + H$	$3 + H$	$4 + H$	$5 + H$
$3 + H$	$3 + H$	$4 + H$	$5 + H$	$10 + H$

EXERCISES

In these exercises, the composition induced on $\mathfrak{P}(E)$ by a composition $\triangle$ on E is denoted by $\triangle_{\mathfrak{P}}$.

18.1. (a) Let $\triangle$ be a composition on E. If X, Y, and Z are subsets of E, then
$$X \triangle (Y \cup Z) = (X \triangle Y) \cup (X \triangle Z), \qquad X \triangle (Y \cap Z) \subseteq (X \triangle Y) \cap (X \triangle Z),$$
$$(Y \cup Z) \triangle X = (Y \triangle X) \cup (Z \triangle X), \qquad (Y \cap Z) \triangle X \subseteq (Y \triangle X) \cap (Z \triangle X).$$
(b) If A is a ring and if X, Y, and Z are subsets of A, then $X(Y + Z) \subseteq XY + XZ$. Show by an example where A is the ring of integers that $X(Y + Z)$ may be a proper subset of $XY + XZ$.

18.2. If $\triangle$ is a composition on E, the set E' of all subsets of E that contain exactly one element is stable for $\triangle_{\mathfrak{P}}$, and the function $G: x \mapsto \{x\}$ is an isomorphism from E onto E'. Infer that $\triangle_{\mathfrak{P}}$ is associative if and only if $\triangle$ is associative, and that $\triangle_{\mathfrak{P}}$ is commutative if and only if $\triangle$ is commutative.

18.3. If $\triangle$ is a composition on a nonempty set E, then a subset J of E is a neutral element for $\triangle_\mathfrak{P}$ if and only if there is a neutral element e for $\triangle$ and $J = \{e\}$.

18.4. Let $\triangle$ be a composition on E for which there is a neutral element e.
(a) If a subset X of E contains e and is invertible for $\triangle_\mathfrak{P}$, then $X = \{e\}$.
(b) If $E = N_3$ and if $\triangle$ is the composition defined by the table on page 28, then every nonempty subset of E not containing the neutral element is invertible for $\triangle_\mathfrak{P}$.
(c) If $\triangle$ is associative or if every element of E is cancellable for $\triangle$, then a subset X of E is invertible for $\triangle_\mathfrak{P}$ if and only if there exists an element $x \in E$ invertible for $\triangle$ such that $X = \{x\}$.

18.5. Let $\triangle$ be a composition on E. The set of all finite subsets of E is stable for $\triangle_\mathfrak{P}$. If $F \subseteq E$, then $\mathfrak{P}(F)$ is stable for $\triangle_\mathfrak{P}$ if and only if F is stable for $\triangle$.

18.6. If $\triangle$ is a commutative associative composition on E, then the set of all stable subsets of E is a stable subset of $(\mathfrak{P}(E), \triangle_\mathfrak{P})$. If $(E, \triangle)$ is a commutative group then the set of all subgroups of E is a subsemigroup of $(\mathfrak{P}(E), \triangle_\mathfrak{P})$.

18.7. Let $H = \{0, 6, 12, 18\}$.
(a) Show that H is a normal subgroup of $(N_{24}, +_{24})$.
(b) List the elements of each coset of H.
(c) Construct a table for the composition induced on N_{24}/H by addition modulo 24.

18.8. (a) Under multiplication, the set R_n of complex nth roots of unit is a subgroup of the multiplicative group of all nonzero complex numbers.
(b) Let $H = \{1, i, -1, -i\}$. Show that H is a normal subgroup of $(R_{12}, \cdot)$.
(c) List the elements of each coset of H.
(d) Construct a table for the composition induced on R_{12}/H by multiplication. [Let $\theta = \cos \pi/6 + i \sin \pi/6$.]

18.9. (a) Which subgroups of $\mathfrak{S}_3$ are normal subgroups?
(b) Let $H = \{1_3, (1, 2, 3), (1, 3, 2)\}$ where 1_3 denotes the identity permutation. List the elements of each coset of H. Construct a table for the composition on $\mathfrak{S}_3/H$ induced by the composition of $\mathfrak{S}_3$.

18.10. Let G be the group of symmetries of the square, let $H = \{r_0, r_2, d_1, d_2\}$, and let $K = \{r_0, d_2\}$. Then H is a normal subgroup of G, K is a normal subgroup of H, but K is not a normal subgroup of G.

18.11. If $(G, \triangle)$ is a group, then the set of all normal subgroups of G is a subsemigroup both of $(\mathfrak{P}(G), \triangle_\mathfrak{P})$ and of $(\mathfrak{P}(G), \cap)$.

*18.12. Let $(G, \triangle)$ be a group, and let $\mathfrak{L}$ be a subgroup of the semigroup $(\mathfrak{P}(G), \triangle_\mathfrak{P})$. There exist a subgroup H of G and a normal subgroup K of H such that $\mathfrak{L}$ is the group H/K if and only if the neutral element of $\mathfrak{L}$ is a subgroup of G.

18.13. If H is a normal subgroup of a group $(G, \cdot)$, then G/H is an abelian group if and only if $xyx^{-1}y^{-1} \in H$ for all $x, y \in G$.

The remaining exercises are devoted to congruence relations on semigroups. If $(E, \triangle)$ is an algebraic structure, an equivalence relation R on E is **compatible** with $\triangle$, or is a **congruence relation** for $\triangle$, if for all $x, x', y, y' \in E$, the following condition holds:

(C)   If $x\, R\, x'$ and $y\, R\, y'$, then $x \triangle y\, R\, x' \triangle y'$.

If R is compatible with $\triangle$, the **composition induced** on the quotient set E/R by $\triangle$ is the composition $\triangle_R$ defined by

$$\lfloor x \rfloor \triangle_R \lfloor y \rfloor = \lfloor x \triangle y \rfloor.$$

It is easy to verify that $\triangle_R$ is well defined by virtue of (C).

18.14. (a) Let R be the relation on C satisfying $z\, R\, w$ if and only if $z^4 = w^4$. Show that R is an equivalence relation. Is R compatible with multiplication? with addition? What numbers belong to $\lfloor 1 + i\sqrt{3} \rfloor_R$?

(b) Let R be the relation on Z satisfying $x\, R\, y$ if and only if $\sin(\pi x)/6 = \sin(\pi y)/6$. Show that R is an equivalence relation. Is R compatible with multiplication? with addition? What integers belong to $\lfloor 1 \rfloor_R$?

18.15. Let $(G, \cdot)$ be a group.
(a) If H is a normal subgroup of G, then the equivalence relation determined by the partition of G consisting of the cosets of H is a congruence relation.
(b) If R is a congruence relation on G and if $H = \lfloor e \rfloor_R$, where e is the neutral element of G, then H is a normal subgroup of G, and $G/R = G/H$.

18.16. An equivalence relation R on E is compatible with a composition $\triangle$ on E if and only if for all $x, y, z \in E$, if $x\, R\, y$, then $x \triangle z\, R\, y \triangle z$ and $z \triangle x\, R\, z \triangle y$.

18.17. Let R be an equivalence relation on E, and let $\triangle$ be a composition on E. If E/R is a stable subset of $\mathfrak{P}(E)$ for the composition $\triangle_\mathfrak{P}$, then R is compatible with $\triangle$, and the composition $\triangle_R$ induced by $\triangle$ on E/R is the composition induced on the subset E/R of $\mathfrak{P}(E)$ by $\triangle_\mathfrak{P}$.

18.18. Every equivalence relation on E is compatible with the compositions $\leftarrow$, $\rightarrow$, and $[c]$ for all $c \in E$ (Exercise 11.14).

*18.19. If E is a set having at least three elements and if $\triangle$ is a composition on E with which every equivalence relation on E is compatible, then $\triangle$ is either $\leftarrow$, $\rightarrow$, or $[c]$ for some $c \in E$. [First show that if $z \triangle z \neq z$ for some $z \in E$, then $\triangle$ is $[c]$ for a suitable element $c \in E$. In the contrary case, show that $x \triangle y$ is either x or y for all $x, y \in E$, and then consider whether or not there exist $u, v \in E$ such that $u \neq v$ and $u \triangle v = u$.]

*18.20. A subset of N is **convex** if for all x, y, $z \in N$, if x, $z \in A$ and if $x \leq y \leq z$, then $y \in A$. Let $\vee$ be the composition on N defined by

$$x \vee y = \max\{x, y\}.$$

(a) An equivalence relation R on N is compatible with $\vee$ if and only if each equivalence class defined by R is a convex subset.

(b) If $\triangle$ is a composition on N for which there is a neutral element e and with which every equivalence relation on N whose equivalence classes are all convex subsets is compatible, then $\triangle$ is $\vee$. [To show first that $e = 0$, use two different partitions of N into convex subsets in evaluating $0 \triangle (e + 1)$.]

*18.21. Let $m \in N$, $n \in N^*$, and let $R_{m,n}$ be the relation on N satisfying $x \, R_{m,n} \, y$ if and only if either $x = y$, or $x \geq m$, $y \geq m$, and $x - y \mid n$.

(a) $R_{m,n}$ is an equivalence relation compatible with addition and multiplication.

(b) If R is an equivalence relation on N compatible with addition and distinct from the equality relation on N, then there exist $m \in N$ and $n \in N^*$ such that $R = R_{m,n}$. [Let m be the smallest natural number such that $m \, R \, x$ for some $x > m$, and let n be the smallest integer > 0 such that $m \, R \, m + n$.]

*18.22. Let m, $n \in N^*$, and let $R_{m,n}^*$ be the restriction of $R_{m,n}$ (Exercise 18.21) to N^*.

(a) $R_{m,n}^*$ is compatible with addition and multiplication on N^*.

(b) If R is an equivalence relation on N^* compatible with addition and distinct from the equality relation on N^*, then there exist m, $n \in N^*$ such that $R = R_{m,n}^*$. [Either argue as in Exercise 18.21(b), or apply that exercise to $R' = R \cup \{(0, 0)\}$.]

19. Ideals and Quotient Rings

Let $\mathfrak{a}$ be an additive subgroup of a ring A. Since $(A, +)$ is a commutative group, $\mathfrak{a}$ is a normal subgroup of $(A, +)$, and by Theorem 18.4, cosets $x + \mathfrak{a}$ and $y + \mathfrak{a}$ are identical if and only if $y - x \in \mathfrak{a}$, or equivalently, if and only if $x - y \in \mathfrak{a}$. What further properties must $\mathfrak{a}$ possess so that we may define a composition on $A/\mathfrak{a}$ by

(1) $$(x + \mathfrak{a})(y + \mathfrak{a}) = xy + \mathfrak{a}$$

for all $x + \mathfrak{a}$, $y + \mathfrak{a} \in A/\mathfrak{a}$? We cannot simply declare that (1) defines a composition on $A/\mathfrak{a}$, for the right side of (1) appears to depend on the particular choice of elements in the two cosets of $\mathfrak{a}$. For (1) to define a composition, we must show that if $x + \mathfrak{a} = x' + \mathfrak{a}$ and $y + \mathfrak{a} = y' + \mathfrak{a}$, then $xy + \mathfrak{a} = x'y' + \mathfrak{a}$. In particular, if $z \in A$ and if $a \in \mathfrak{a}$, then choosing $x' = x = z$, $y = a$, $y' = 0$, we must have $za + \mathfrak{a} = \mathfrak{a}$, or equivalently, $za \in \mathfrak{a}$; similarly, choosing $x = a$, $x' = 0$, $y = y' = z$, we must have $az + \mathfrak{a} = \mathfrak{a}$, or equivalently, $az \in \mathfrak{a}$. Thus, for (1) to define a composition on $A/\mathfrak{a}$, $\mathfrak{a}$ must be an ideal of A in the following sense:

Definition. An **ideal** of a ring A is a subgroup $\mathfrak{a}$ of $(A, +)$ such that for all $x, y \in A$, if $x \in \mathfrak{a}$, then $xy \in \mathfrak{a}$ and $yx \in \mathfrak{a}$. A **proper ideal** of A is an ideal of A that is a proper subset of A.

An ideal of a ring A is necessarily a subring of A, but a subring of A need not be an ideal of A. For example, $\mathbf{Z}$ is a subring but not an ideal of the ring $\mathbf{R}$ of real numbers.

If $\mathfrak{a}$ is an ideal of A, then we may use (1) as the definition of a composition on A:

Theorem 19.1. If $\mathfrak{a}$ is an ideal of a ring A, then multiplication on $A/\mathfrak{a}$ is well-defined by

$$(x + \mathfrak{a})(y + \mathfrak{a}) = xy + \mathfrak{a}$$

for all $x + \mathfrak{a}, y + \mathfrak{a} \in A/\mathfrak{a}$, and the algebraic structure $(A/\mathfrak{a}, +, \cdot)$ is a ring. If A is commutative, so is $A/\mathfrak{a}$.

Proof. To show that multiplication is well-defined, let $x, x', y, y' \in A$ be such that $x + \mathfrak{a} = x' + \mathfrak{a}$, $y + \mathfrak{a} = y' + \mathfrak{a}$; we must show that $xy + \mathfrak{a} = x'y' + \mathfrak{a}$. By our assumption, $x - x' \in \mathfrak{a}$ and $y - y' \in \mathfrak{a}$, and we are to prove that $xy - x'y' \in \mathfrak{a}$. But

$$xy - x'\mathfrak{a}' = x(y - y') + (x - x')y' \in \mathfrak{a}$$

since $y - y'$, $x - x' \in \mathfrak{a}$. Thus, multiplication is well-defined. It is easy to verify that multiplication on $A/\mathfrak{a}$ is associative and distributive over addition and is, furthermore, commutative if A is a commutative ring.

If A is a ring, A itself and $\{0\}$ are clearly ideals of A. Any ideal of A other than $\{0\}$ is called, of course, a **nonzero ideal**. If A is a field, or more generally a division ring, these are the only ideals of A (Corollary of Theorem 19.4). We recall from §18 that if $a \in A$, by definition Aa is the set $\{xa: x \in A\}$.

Theorem 19.2. If A is a commutative ring with identity and if $a \in A$, then Aa is an ideal of A and is, moreover, the smallest ideal of A that contains a.

The proof is easy.

Definition. Let A be a commutative ring with identity. An ideal $\mathfrak{a}$ of A is a **principal ideal** if $\mathfrak{a} = Aa$ for some $a \in A$; if $\mathfrak{a} = Aa$, the element a is called a **generator** of the ideal $\mathfrak{a}$, and $\mathfrak{a}$ is called the **principal ideal generated** by a.

The principal ideal generated by a is often denoted by (a). For example, if A is a commutative ring with identity, then A itself is a principal ideal since $A = A \cdot 1 = (1)$, and $\{0\}$ is also a principal ideal since $\{0\} = A \cdot 0 = (0)$.

If $\mathfrak{a}$ and $\mathfrak{b}$ are subsets of a ring A, then $\mathfrak{a} + \mathfrak{b}$ is the set of all the elements $x + y$ where $x \in \mathfrak{a}$ and $y \in \mathfrak{b}$, by our definition in §18. By our discussion at the end of §11, applied to the composition induced on $\mathfrak{P}(A)$ by addition on A, we conclude that if $\mathfrak{a}_1, \ldots, \mathfrak{a}_n$ are subsets of A, then $\mathfrak{a}_1 + \ldots + \mathfrak{a}_n$ is by definition the set of all the elements $x_1 + \ldots + x_n$, where $x_i \in \mathfrak{a}_i$ for each $i \in [1, n]$.

Theorem 19.3. Let $\mathfrak{a}_1, \ldots, \mathfrak{a}_n$ be ideals of a ring A. Then $\mathfrak{a}_1 + \ldots + \mathfrak{a}_n$ and $\mathfrak{a}_1 \cap \ldots \cap \mathfrak{a}_n$ are ideals of A. Moreover, $\mathfrak{a}_1 + \ldots + \mathfrak{a}_n$ is the smallest ideal of A that contains each of $\mathfrak{a}_1, \ldots, \mathfrak{a}_n$, and $\mathfrak{a}_1 \cap \ldots \cap \mathfrak{a}_n$ is the largest ideal of A that is contained in each of $\mathfrak{a}_1, \ldots, \mathfrak{a}_n$.

Proof. Let $x, y \in \mathfrak{a}_1 + \ldots + \mathfrak{a}_n$. Then there exist $x_1, y_1 \in \mathfrak{a}_1, \ldots, x_n, y_n \in \mathfrak{a}_n$ such that $x = \sum_{k=1}^{n} x_k$, $y = \sum_{k=1}^{n} y_k$. Consequently,

$$x - y = \sum_{k=1}^{n} x_k - \sum_{k=1}^{n} y_k = \sum_{k=1}^{n} (x_k - y_k) \in \mathfrak{a}_1 + \ldots + \mathfrak{a}_n,$$

and for each $z \in A$,

$$zx = z\left(\sum_{k=1}^{n} x_k\right) = \sum_{k=1}^{n} zx_k \in \mathfrak{a}_1 + \ldots + \mathfrak{a}_n,$$

$$xz = \left(\sum_{k=1}^{n} x_k\right)z = \sum_{k=1}^{n} x_k z \in \mathfrak{a}_1 + \ldots + \mathfrak{a}_n.$$

Thus, $\mathfrak{a}_1 + \ldots + \mathfrak{a}_n$ is an ideal. If $\mathfrak{b}$ is an ideal containing each of $\mathfrak{a}_1, \ldots, \mathfrak{a}_n$, then $\mathfrak{b}$ contains $x_1, \ldots, x_n$ and, therefore, also $x_1 + \ldots + x_n = x$, so $\mathfrak{b}$ contains $\mathfrak{a}_1 + \ldots + \mathfrak{a}_n$. Thus, $\mathfrak{a}_1 + \ldots + \mathfrak{a}_n$ is the smallest ideal of A that contains each of $\mathfrak{a}_1, \ldots, \mathfrak{a}_n$.

The proof of the assertion about $\mathfrak{a}_1 \cap \ldots \cap \mathfrak{a}_n$ is easy.

Theorem 19.4. If an ideal $\mathfrak{a}$ of a ring A contains an invertible element, then $\mathfrak{a} = A$.

Proof. If x is an invertible element of $\mathfrak{a}$, then for every $y \in A$,

$$y = (yx^{-1})x,$$

an element of $\mathfrak{a}$, so $\mathfrak{a} = A$.

Corollary. The only ideals of a division ring K are K and $\{0\}$.

Theorem 19.5. If A is a commutative ring with identity, then A is a field if and only if A and $\{0\}$ are the only ideals of A.

Proof. The condition is necessary by the corollary of Theorem 19.4. Sufficiency: Let $a \in A^*$. Then (a) is a nonzero ideal and hence is A. Thus, $1 \in (a)$, so there exists $x \in A$ such that $xa = 1$, and therefore, a is invertible.

If A is a commutative ring with identity, what property characterizes those ideals $\mathfrak{a}$ of A such that $A/\mathfrak{a}$ is a field? This is an important question since, for example, our construction of the real numbers depends on the answer. An answer may be formulated by use of the notion of a maximal element of an ordered structure.

Definition. If $\leq$ is an ordering on E, an element a of E is **maximal** for the ordering $\leq$ if no other element of E is greater than a; that is, if for all $x \in E$, $x \geq a$ implies that $x = a$.

If an ordered structure has a greatest element, that element is certainly a maximal element and, moreover, is the only maximal element. However, a maximal element need not be a greatest element. In Figure 11 there are three maximal elements but no greatest element.

Figure 11

Definition. An ideal $\mathfrak{a}$ of a ring A is a **maximal ideal** of A if $\mathfrak{a}$ is a maximal element of the set of all proper ideals of A for the ordering $\subseteq$.

Thus, an ideal is maximal if and only if it is a proper ideal contained in no other proper ideal. The only ideal of a ring A properly containing a maximal ideal is thus A itself. The only maximal ideal of a division ring is $\{0\}$ by the corollary of Theorem 19.4.

Theorem 19.6. If $\mathfrak{a}$ is an ideal of a commutative ring A with identity, then the ring $A/\mathfrak{a}$ is a field if and only if $\mathfrak{a}$ is a maximal ideal of A.

Proof. Necessity: Let $\mathfrak{b}$ be an ideal of A properly containing $\mathfrak{a}$. Then there exists $x \in \mathfrak{b}$ such that $x \notin \mathfrak{a}$. Hence, $x + \mathfrak{a}$ is not the zero element of the field $A/\mathfrak{a}$, so $x + \mathfrak{a}$ is invertible in $A/\mathfrak{a}$. Let $y + \mathfrak{a}$ be the inverse of $x + \mathfrak{a}$. Then $xy + \mathfrak{a} = 1 + \mathfrak{a}$, so $xy - 1 \in \mathfrak{a}$. Since $x \in \mathfrak{b}$, however, $xy \in \mathfrak{b}$, and therefore, since $\mathfrak{a} \subset \mathfrak{b}$,

$$1 = xy - (xy - 1) \in \mathfrak{b},$$

whence $\mathfrak{b} = A$ by Theorem 19.4. Thus, $\mathfrak{a}$ is a maximal ideal.

Sufficiency: A maximal ideal is a proper ideal, and therefore, $1 \notin \mathfrak{a}$. Thus, $1 + \mathfrak{a} \neq \mathfrak{a}$, and it is easy to verify that $1 + \mathfrak{a}$ is the multiplicative identity of $A/\mathfrak{a}$. Let $\mathfrak{B}$ be a nonzero ideal of $A/\mathfrak{a}$, and let

$$\mathfrak{b} = \{x \in A : x + \mathfrak{a} \in \mathfrak{B}\}.$$

If $x, y \in \mathfrak{b}$, then $(x - y) + \mathfrak{a} = (x + \mathfrak{a}) - (y + \mathfrak{a}) \in \mathfrak{B}$, so $x - y \in \mathfrak{b}$. If $x \in \mathfrak{b}$ and $z \in A$, then $zx + \mathfrak{a} = (z + \mathfrak{a})(x + \mathfrak{a}) \in \mathfrak{B}$, so $zx \in \mathfrak{b}$. Hence, $\mathfrak{b}$ is an ideal. As $\mathfrak{a}$ is the zero element of $A/\mathfrak{a}$, $\mathfrak{b}$ clearly contains $\mathfrak{a}$. If $b + \mathfrak{a}$ is a nonzero element of $\mathfrak{B}$, then b is an element of $\mathfrak{b}$ not belonging to $\mathfrak{a}$. Thus, $\mathfrak{b}$ is an ideal of A strictly containing $\mathfrak{a}$, so $\mathfrak{b} = A$ by hypothesis. Therefore, for every $x \in A$, $x + \mathfrak{a} \in \mathfrak{B}$; that is, $\mathfrak{B} = A/\mathfrak{a}$. By Theorem 19.5, $A/\mathfrak{a}$ is a field.

We next wish to find all the ideals of the ring Z of integers. To do so we shall use the following extension of the Division Algorithm:

Theorem 19.7. If a and b are integers and if $b > 0$, there exist unique integers q and r such that

(2) $$a = bq + r,$$

(3) $$0 \leq r < b.$$

Proof. Existence: By the Division Algorithm, we may assume that $a < 0$. Then $-a \in N$, so there exist $q', r' \in N$ such that $-a = bq' + r'$ and $r' < b$ by the Division Algorithm. If $r' = 0$, let $q = -q'$, $r = 0$; if $r' > 0$, let $q = -q' - 1$, $r = b - r'$. Uniqueness: If $a = bq + r = bp + s$ where $0 \leq r \leq s < b$, then $s - r = b(q - p)$, so $b \mid s - r$ but $0 \leq s - r \leq s < b$, and consequently, $s - r = 0$; therefore, $s = r$ and also $b(q - p) = s - r = 0$, whence $q = p$.

We shall also refer to Theorem 19.7 as the "Division Algorithm."

Theorem 19.8. Every ideal of the ring Z of integers is a principal ideal. In fact, if $\mathfrak{a}$ is a nonzero ideal of Z, then $\mathfrak{a} = (b)$ where b is the smallest strictly positive integer belonging to $\mathfrak{a}$. Hence,

$$\psi : b \mapsto (b)$$

is a bijection from N onto the set of all ideals of Z.

Proof. If c is a nonzero member of $\mathfrak{a}$, both c and $-c$ belong to $\mathfrak{a}$ and one of them is strictly positive, so $\mathfrak{a}$ does contain strictly positive elements. Let b be the smallest strictly positive element of $\mathfrak{a}$. Clearly, $(b) \subseteq \mathfrak{a}$. To show that $\mathfrak{a} \subseteq (b)$, let $a \in \mathfrak{a}$. By the Division Algorithm, there exist integers q and r satisfying $a = bq + r$ and $0 \leq r < b$. Then, since a and b belong to $\mathfrak{a}$, so does $r = a - bq$, whence $r = 0$ by the definition of b. Hence, $a = bq \in (b)$. Thus, $\mathfrak{a} = (b)$. The function ψ from N into the set of all ideals of Z is injective, for, if $0 < b < c$, then $b \in (b)$ but $b \notin (c)$, so $(b) \neq (c)$, and certainly also $(b) \neq (0)$ if $b > 0$. Hence, ψ is indeed a bijection from N onto the set of all ideals of Z.

The following theorem, in conjunction with Theorem 19.8, shows that a quotient ring of Z determined by a nonzero ideal is isomorphic to the ring of integers modulo m for some $m > 0$ (Examples 2.3, 15.1).

Theorem 19.9. Let m be a strictly positive integer. The function φ_m from N_m into $Z/(m)$ defined by

$$\varphi_m(x) = x + (m)$$

is an isomorphism from the ring $(N_m, +_m, \cdot_m)$ of integers modulo m onto the quotient ring $Z/(m)$. In particular, $Z/(m)$ has m members.

Proof. To show that φ_m is injective, let x and y be such that $0 \leq x \leq y < m$ and $x + (m) = y + (m)$. Then $y - x \in (m)$ and $0 \leq y - x < m$, whence $y = x$, because m is the smallest strictly positive integer belonging to (m) by Theorem 19.8. To show that φ_m is surjective, let $z \in Z$. By the Division Algorithm, there exists $x \in N_m$ such that $z = mq + x$ for some integer q. As $mq \in (m)$, $z + (m) = x + (m) = \varphi_m(x)$.

We have only to show now that $(x +_m y) + (m) = [x + (m)] + [y + (m)]$ and $(x \cdot_m y) + (m) = [x + (m)][y + (m)]$ for all $x, y \in N_m$. By the Division Algorithm, there exist integers p, q, r, and s satisfying

$$x + y = mq + r, \qquad 0 \leq r < m,$$

$$xy = mp + s, \qquad 0 \leq s < m.$$

Then $x +_m y = r$ and $x \cdot_m y = s$, so as $mq, mp \in (m)$,

$$(x +_m y) + (m) = r + (m) = r + mq + (m) = x + y + (m)$$

$$= [x + (m)] + [y + (m)],$$

$$(x \cdot_m y) + (m) = s + (m) = s + mp + (m) = xy + (m)$$

$$= [x + (m)][y + (m)].$$

We shall denote the quotient ring $Z/(m)$ by Z_m for every $m \in N$. If $m > 0$, Z_m is frequently called the **ring of integers modulo** m since, as we have just seen, it is isomorphic in a natural way with the ring N_m.

If $\mathfrak{m}$ is an ideal of a ring A, the notation

$$a \equiv b \ (\text{mod } \mathfrak{m})$$

(read "a is congruent to b modulo $\mathfrak{m}$") is frequently used to mean $a - b \in \mathfrak{m}$, or equivalently, $a + \mathfrak{m} = b + \mathfrak{m}$. Clearly,

$$\underline{\quad\quad} \equiv \ldots \ (\text{mod } \mathfrak{m})$$

is an equivalence relation on A such that for all $x, x', y, y' \in A$, if

$$x \equiv x' \ (\text{mod } \mathfrak{m}),$$
$$y \equiv y' \ (\text{mod } \mathfrak{m}),$$

then

$$x + y \equiv x' + y' \ (\text{mod } \mathfrak{m}),$$
$$xy \equiv x'y' \ (\text{mod } \mathfrak{m}).$$

If $\mathfrak{m}$ is the ideal (m) of Z, it is customary to write "$x \equiv y \ (\text{mod } m)$" for "$x \equiv y \ (\text{mod } (m))$." To find, for example, all integers n satisfying

(4) $$2n + 1 \equiv 5 \ (\text{mod } 6),$$

it suffices to find all such integers in N_6, for an integer satisfies (4) if and only if it differs by an integral multiple of 6 from one in N_6 satisfying (4). The set of integers satisfying (4) is thus $(2 + (6)) \cup (5 + (6))$.

Extending our definition of division given in §11, we shall say that a non-zero integer m **divides** an integer n, or that m is a **divisor** of n, or that n is a **multiple** of m, and write $m \mid n$, if $mp = n$ for some integer p. An **even** integer is, of course, one divisible by 2, and an **odd** integer is an integer that is not even. By the Division Algorithm, if m is an odd integer, there exists one and only one integer q such that $m = 2q + 1$.

Definition. An integer p is a **prime** integer if $p > 1$ and if the only strictly positive divisors of p are 1 and p.

Theorem 19.10. If $m \in Z^*$ and if $n \in Z$, then $m \mid n$ if and only if $(n) \subseteq (m)$. A positive integer p is prime if and only if (p) is a maximal ideal of Z.

Proof. By definition, $m \mid n$ if and only if $n \in (m)$, but $n \in (m)$ if and only if $(n) \subseteq (m)$. Hence, by Theorem 19.8, if p is prime, then (p) is a maximal ideal, and conversely, if $p \in N$ and if (p) is a maximal ideal, then p is a prime.

Theorem 19.11. Let p be an integer greater than 1. The following statements are equivalent:

1° p is a prime.
2° The ring Z_p is an integral domain.
3° The ring Z_p is a field.

Proof. By Theorems 19.10 and 19.6, 1° implies 3°, and we have seen already that 3° implies 2°. To show that 2° implies 1°, it suffices by the corollary of Theorem 15.1 to show that if p is not a prime, then the ring Z_p contains proper zero-divisors. But if $p = mn$ where $1 < m < p$ and $1 < n < p$, then in the ring Z_p we have $m + (p) \neq (p)$, $n + (p) \neq (p)$, but

$$(m + (p))(n + (p)) = p + (p) = (p),$$

the zero element of Z_p.

To complete our discussion of quotient rings of Z, we note that Z_0 is isomorphic to Z and that Z_1 is a zero ring.

EXERCISES

19.1. Complete the proof of Theorem 19.1.

19.2. Prove Theorem 19.2.

19.3. Complete the proof of Theorem 19.3.

19.4. Let $\mathfrak{P}(E)$ be the ring of Example 15.2.
  (a) A nonempty subset $\mathfrak{S}$ of $\mathfrak{P}(E)$ is an ideal of $\mathfrak{P}(E)$ if and only if $\mathfrak{P}(X \cup Y) \subseteq \mathfrak{S}$ for all X, $Y \in \mathfrak{S}$.
  (b) The set of all finite subsets of E is an ideal of $\mathfrak{P}(E)$, and for every subset F of E, $\mathfrak{P}(F)$ is an ideal of $\mathfrak{P}(E)$.
  (c) If F is a subset of E, what is the principal ideal of the ring $\mathfrak{P}(E)$ generated by F?

19.5. An ideal $\mathfrak{a}$ of a ring A is **modular** (or **regular**) if $A/\mathfrak{a}$ is a ring with identity.
  (a) Every proper ideal of a ring with identity is modular.
  (b) Every proper ideal containing a modular ideal is modular.
  (c) The intersection of two modular ideals is modular. [If $e + \mathfrak{a}$ and $f + \mathfrak{b}$ are identities in $A/\mathfrak{a}$ and $A/\mathfrak{b}$, consider $e + f - ef$.]

19.6. Let A be a commutative ring with identity. An ideal $\mathfrak{p}$ of A is a **prime** ideal if the ring $A/\mathfrak{p}$ is an integral domain.
  (a) A proper ideal $\mathfrak{p}$ of A is a prime ideal if and only if for all $a, b \in A$, if $ab \in \mathfrak{p}$, then either $a \in \mathfrak{p}$ or $b \in \mathfrak{p}$.
  (b) A maximal ideal of A is a prime ideal.

19.7. Let $\mathfrak{p}$ be a proper ideal of a boolean ring A (Exercise 15.21). The following statements are equivalent:

$1°$ $\mathfrak{p}$ is a prime ideal.

$2°$ For all $x, y \in A$, either $x \in \mathfrak{p}$ or $y \in \mathfrak{p}$ or $x + y \in \mathfrak{p}$.

$3°$ $\mathfrak{p}$ is a maximal ideal.

$4°$ $A/\mathfrak{p}$ is isomorphic to the field Z_2.

[Consider $xy(x + y)$.]

19.8. (a) Find all the subgroups of the multiplicative groups Z_7^* and Z_{11}^*.

(b) Determine the invertible elements of the ring Z_m for each $m \in [2, 20]$.

19.9. Find the set of all integers m satisfying

(a) $4m \equiv 8 \pmod{12}$.

(b) $7m \equiv 10 \pmod{14}$.

(c) $6m + 5 \equiv 2 \pmod 3$.

(d) $2m^2 + 3m \equiv 5 \pmod 7$.

(e) $m^3 + 3m^2 + 5m \equiv 1 \pmod 6$.

*(f) $m^m \equiv m \pmod 4$.

19.10. Find the smallest positive integer m satisfying

(a) $m \equiv 4 \pmod 5$ and $3m \equiv 1 \pmod 8$.

(b) $2m \equiv 5 \pmod 7$ and $3m \equiv 2 \pmod 5$.

19.11. Find all the integers $n \in [0, 9]$ such that for no integer m does $m^2 \equiv n \pmod{10}$.

19.12. (a) If A is an integral domain, then for every $x \in A$, $x^2 = 1$ if and only if either $x = 1$ or $x = -1$.

(b) Let K be a field, and let L be the complement of $\{0, 1, -1\}$. For each $x \in L$, let $A_x = \{x, x^{-1}\}$. Then $\{A_x : x \in L\}$ is a partition of L, each member of which has two elements.

19.13. If K is a finite field, then

$$\prod_{x \in K^*} x = -1.$$

[Use Exercise 19.12(b).]

19.14. (Wilson's Theorem) If p is a prime, then

$$(p - 2)! \equiv 1 \pmod p.$$

[Use Exercise 19.13.]

19.15. Let A be a ring. A subset $\mathfrak{a}$ of A is a **left ideal** of the ring A if $\mathfrak{a}$ is a subgroup of $(A, +)$ such that for all $x, y \in A$, if $y \in \mathfrak{a}$, then $xy \in \mathfrak{a}$. A subset $\mathfrak{a}$ of A is a **right ideal** of the ring A if $\mathfrak{a}$ is a subgroup of $(A, +)$ such that for all $x, y \in A$, if $x \in \mathfrak{a}$, then $xy \in \mathfrak{a}$.

(a) A left (right) ideal of A is a subring.

(b) If $\mathfrak{a}$ is a left (right) ideal of A and if $c \in A$, then $\mathfrak{a}c$ (respectively, $c\mathfrak{a}$) is a left (right) ideal of A. In particular, for each $c \in A$, Ac is a left ideal and cA is a right ideal of A.

19.16. (a) If A is a ring with identity and if a left (right) ideal $\mathfrak{a}$ of A contains a left (right) invertible element (Exercise 4.11), then $\mathfrak{a} = A$.

(b) The only left (right) ideals of a division ring A are $\{0\}$ and A.

*19.17. Let A be a ring with identity, and let $\mathfrak{a}$, $\mathfrak{a}_L$, $\mathfrak{a}_R$ be respectively the set of all elements of A having no inverse, no left inverse, no right inverse (Exercise 4.11).

(a) $\mathfrak{a}_L$ is an additive subgroup if and only if for every $x \in A$, either x is left invertible or $1 - x$ is left invertible.

(b) If $\mathfrak{a}_L$ is an additive subgroup and if $yx = 1$, then either y or $1 - y$ is invertible. [If $z(1 - y) = 1$, apply (a) to z.]

(c) If $\mathfrak{a}_L$ is an additive subgroup and if x is left invertible, then x is invertible. [Apply (a) to xy, and use (b).]

(d) If $\mathfrak{a}_R$ is an additive subgroup, then every right invertible element of A is invertible. [Use (c) and Exercise 15.8.]

(e) If $\mathfrak{a}$ is an additive subgroup, then $\mathfrak{a}_L$ is an additive subgroup.

(f) If either $\mathfrak{a}_L$, $\mathfrak{a}_R$, or $\mathfrak{a}$ is an additive subgroup, then every left or right invertible element is invertible, and $\mathfrak{a}$ is an ideal containing every proper left or right ideal (Exercise 19.15).

(g) Conversely, if there is a largest proper left (right) ideal $\mathfrak{m}$, i.e., if $\mathfrak{m}$ is a proper left (right) ideal containing every proper left (right) ideal, then $\mathfrak{m} = \mathfrak{a}$, and in particular, $\mathfrak{m}$ is an ideal.

(h) If $A = \{m/(2n + 1): m \in Z \text{ and } n \in N\}$, then A is a subdomain of Q, and the set of noninvertible elements of A is a nonzero ideal of A.

20. Homomorphisms of Groups

Definition. Let f be a function from E into F. We define f_* to be the function from $\mathfrak{P}(E)$ into $\mathfrak{P}(F)$ given by

$$f_*(X) = \{f(x): x \in X\}$$

for all subsets X of E, and we define f^* to be the function from $\mathfrak{P}(F)$ into $\mathfrak{P}(E)$ given by

$$f^*(Y) = \{x \in E: f(x) \in Y\}$$

for all subsets Y of F.

Thus, the range of a function f from E into F is simply the set $f_*(E)$. Other customary notations are $f(X)$ for $f_*(X)$ and $f^{-1}(Y)$ for $f^*(Y)$.

Theorem 20.1. If $f: E \to F$ and if $g: F \to G$, then $(g \circ f)_* = g_* \circ f_*$ and $(g \circ f)^* = f^* \circ g^*$.

Proof. If $X \subseteq E$, then

$$(g \circ f)_*(X) = \{(g \circ f)(x): x \in X\} = \{g(f(x)): x \in X\},$$
$$g_*(f_*(X)) = g_*(\{f(x): x \in X\}) = \{g(f(x)): x \in X\}.$$

If $Z \subseteq F$, then

$$(g \circ f)^*(Z) = \{x \in E: (g \circ f)(x) \in Z\} = \{x \in E: g(f(x)) \in Z\},$$
$$f^*(g^*(Z)) = \{x \in E: f(x) \in g^*(Z)\} = \{x \in E: g(f(x)) \in Z\}.$$

Theorem 20.2. If f is a bijection from E onto F, then f_* and f^* are bijective, and $(f_*)^{\leftarrow} = (f^{\leftarrow})_* = f^*$, $(f^*)^{\leftarrow} = (f^{\leftarrow})^* = f_*$.

Proof. Clearly, $(1_E)_* = 1_{\mathfrak{P}(E)} = (1_E)^*$, and similarly $(1_F)_* = 1_{\mathfrak{P}(F)} = (1_F)^*$. By Theorem 20.1,

$$(f^{\leftarrow})_* \circ f_* = (f^{\leftarrow} \circ f)_* = (1_E)_* = 1_{\mathfrak{P}(E)},$$
$$f_* \circ (f^{\leftarrow})_* = (f \circ f^{\leftarrow})_* = (1_F)_* = 1_{\mathfrak{P}(F)}.$$

Therefore, f_* is a bijection, and $(f_*)^{\leftarrow} = (f^{\leftarrow})_*$ by Theorem 5.3. If Y is a subset of F, $x \in (f^{\leftarrow})_*(Y)$ if and only if $x = f^{\leftarrow}(y)$, that is, $f(x) = y$ for some $y \in Y$. But by definition, $f(x) \in Y$ if and only if $x \in f^*(Y)$. Thus, $(f^{\leftarrow})_*(Y) = f^*(Y)$. Similarly, $(f^*)^{\leftarrow} = (f^{\leftarrow})^* = f_*$.

If H is a normal subgroup of a group $(G, \cdot)$, we shall denote by φ_H the canonical surjection $x \mapsto xH$ from G onto G/H. By the definition of the induced composition on G/H, φ_H "preserves products"; that is,

$$\varphi_H(xy) = \varphi_H(x)\varphi_H(y)$$

for all $x, y \in G$. The surjection φ_H is an isomorphism only in the trivial case where $H = \{e\}$, for otherwise, φ_H is not injective. However, φ_H is an example of a homomorphism:

Definition. Let $(E, \triangle)$ and $(F, \triangledown)$ be algebraic structures with one composition. A **homomorphism** from $(E, \triangle)$ into $(F, \triangledown)$ is a function f from E into F satisfying

$$f(x \triangle y) = f(x) \triangledown f(y)$$

for all $x, y \in E$. Similarly, if $(E, \triangle, \wedge)$ and $(F, \triangledown, \vee)$ are algebraic structures with two compositions, a **homomorphism** from $(E, \triangle, \wedge)$ into $(F, \triangledown, \vee)$ is a function f from E into F that is both a homomorphism from $(E, \triangle)$ into

$(F, \triangledown)$ and a homomorphism from $(E, \wedge)$ into $(F, \vee)$. An **endomorphism** of an algebraic structure is a homomorphism from itself into itself.

An easy inductive argument establishes that if f is a homomorphism from $(E, \triangle)$ into $(F, \triangledown)$, then $f(\triangle^n a) = \triangledown^n f(a)$ for each $a \in E$, $n \in N^*$, and more generally, $f(a_1 \triangle \ldots \triangle a_n) = f(a_1) \triangledown \ldots \triangledown f(a_n)$ for any sequence $(a_k)_{1 \le k \le n}$ of elements of E.

Theorem 20.3. Let f be a homomorphism from $(E, \triangle)$ into $(F, \triangledown)$. If A is a stable subset of E, then $f_*(A)$ is a stable subset of F. In particular, the range of f is a stable subset of F. If B is a stable subset of F, then $f^*(B)$ is a stable subset of E.

Proof. If $z, w \in f_*(A)$, then there exist $x, y \in A$ such that $z = f(x)$ and $w = f(y)$, whence

$$z \triangledown w = f(x) \triangledown f(y) = f(x \triangle y),$$

an element of $f_*(A)$, since A is stable for $\triangle$. If $x, y \in f^*(B)$, then both $f(x)$ and $f(y)$ belong to B, so as $f(x \triangle y) = f(x) \triangledown f(y)$, which belongs to B since B is stable for $\triangledown$, we conclude that $x \triangle y \in f^*(B)$.

Definition. An **epimorphism** from an algebraic structure E onto an algebraic structure F is a surjective homomorphism from E onto F. A **monomorphism** from E into F is an injective homomorphism from E into F.

For example, if H is a normal subgroup of a group G, then $\varphi_H : x \mapsto xH$ is an epimorphism (called the **canonical epimorphism**) from the group G onto the group G/H, and $\iota_H : x \to x$ is a monomorphism from the group H into the group G.

Let f be a homomorphism from an algebraic structure E into an algebraic structure F, and let E' be the range of f, which is a stable subset of F by Theorem 20.3. The function f', obtained by restricting the codomain of f to E', is clearly an epimorphism from the algebraic structure E onto the algebraic structure E'. Moreover, if f is a monomorphism, f' is an isomorphism.

The following list summarizes the definitions just made:

$$\text{Homomorphism} + \begin{cases} \text{Surjection} \\ \text{Injection} \\ \text{Bijection} \\ \text{Permutation} \\ \text{Function with same} \\ \quad \text{domain and codomain} \end{cases} = \begin{cases} \text{Epimorphism} \\ \text{Monomorphism} \\ \text{Isomorphism} \\ \text{Automorphism} \\ \text{Endomorphism} \end{cases}$$

Epimorphisms preserve many algebraic properties, as the following theorem shows.

Theorem 20.4. Let f be an epimorphism from $(E, \triangle)$ onto $(F, \triangledown)$. If $\triangle$ is associative (commutative), then $\triangledown$ is also associative (commutative). If e is a neutral element for $\triangle$, then $f(e)$ is a neutral element for $\triangledown$. If x^* is an inverse of x for $\triangle$, then $f(x^*)$ is an inverse of $f(x)$ for $\triangledown$.

The proof is similar to that of Theorem 6.2.

Corollary. If f is an epimorphism from a semigroup (group) $(E, \cdot)$ onto an algebraic structure $(F, \cdot)$, then $(F, \cdot)$ is a semigroup (group); moreover, if E is commutative, so is F.

Theorem 20.5. Let $(E, \triangle)$ and $(F, \triangledown)$ be algebraic structures possessing neutral elements e and e' respectively, and let f be a homomorphism from E into F.

1° If every element of F is cancellable, then $f(e) = e'$.
2° If $f(e) = e'$ and if x^* is an inverse of x for $\triangle$, then $f(x^*)$ is an inverse of $f(x)$ for $\triangledown$.
3° If $(F, \triangledown)$ is a group, then $f(e) = e'$, and if x^* is an inverse of an element x of E for $\triangle$, then $f(x^*)$ is the inverse of $f(x)$.

Proof. If every element of F is cancellable, then as

$$f(e)\triangledown f(e) = f(e\triangle e) = f(e),$$

we conclude that $f(e) = e'$ by Theorem 8.1. If $f(e) = e'$ and if x^* is an inverse of x for $\triangle$, then

$$f(x)\triangledown f(x^*) = f(x\triangle x^*) = e' = f(x^*\triangle x) = f(x^*)\triangledown f(x),$$

so $f(x^*)$ is an inverse of $f(x)$ for $\triangledown$. Finally, 3° follows from 1°, 2°, and the corollary of Theorem 7.1.

Corollary. If f is a homomorphism from a semigroup $(E, \cdot)$ into a group $(G, \cdot)$ and if a is an invertible element of E, then

$$f(a^n) = f(a)^n$$

for all $n \in \mathbf{Z}$.

Proof. We observed earlier that the equality holds if $n > 0$, and it holds also if $n = 0$ by 3° of Theorem 20.5. If $m > 0$,

$$f(a^{-m}) = f((a^{-1})^m) = f(a^{-1})^m$$
$$= (f(a)^{-1})^m = f(a)^{-m}$$

by 3° of Theorem 20.5 and Theorem 14.7.

Of particular importance is the case where f is a homomorphism from a group E into a group F. If the compositions of both groups are denoted additively and if the neutral elements of both are denoted by "0," then

$$f(0) = 0,$$
$$f(-x) = -f(x),$$
$$f(n.x) = n.f(x)$$

for all $x \in E$, $n \in Z$; if the compositions of both groups are denoted multiplicatively and if the neutral elements of both are denoted by "1," then

$$f(1) = 1,$$
$$f(x^{-1}) = f(x)^{-1},$$
$$f(x^n) = f(x)^n$$

for all $x \in E$, $n \in Z$.

Not all algebraic properties are preserved by epimorphisms. Indeed, the image of a cancellable element under an epimorphism need not be cancellable. For example, 2 is cancellable for multiplication on Z, but $2 + (4)$ is not cancellable for multiplication on Z_4.

Theorem 20.6. If f is a homomorphism from $(E, \triangle)$ into $(F, \triangledown)$ and if g is a homomorphism from $(F, \triangledown)$ into $(G, \vee)$, then $g \circ f$ is a homomorphism from $(E, \triangle)$ into $(G, \vee)$.

The proof is similar to that of 3° of Theorem 6.1.

Definition. Let f be a homomorphism from a group $(G, \cdot)$ into a group $(G', \cdot)$, and let e' be the neutral element of G'. The **kernel** of f is the subset $\ker f$ of G defined by

$$\ker f = f^*(\{e'\}).$$

Thus, $x \in \ker f$ if and only if $f(x) = e'$.

Theorem 20.7. The kernel K of a homomorphism f from a group $(G, \cdot)$ into a group $(G', \cdot)$ is a normal subgroup of $(G, \cdot)$.

Proof. Let e and e' be the neutral elements of G and G' respectively. By $3°$ of Theorem 20.5, $f(e) = e'$, so $e \in K$, and thus $K \neq \emptyset$. If $x, y \in K$ and if $a \in G$, then

$$f(xy^{-1}) = f(x)f(y^{-1}) = f(x)f(y)^{-1} = e'e'^{-1} = e',$$
$$f(axa^{-1}) = f(a)f(x)f(a^{-1}) = f(a)e'f(a)^{-1} = e'$$

by $2°$ of Theorem 20.5, so $xy^{-1} \in K$, $axa^{-1} \in K$. Thus, K is a normal subgroup by Theorems 8.3 and 18.5.

Theorem 20.8. Let f be a homomorphism from a group $(G, \cdot)$ into a group $(G', \cdot)$. If H is a normal subgroup of G that is contained in the kernel K of f, then

$$g: xH \mapsto f(x)$$

is a well-defined homomorphism from G/H into G' satisfying $g \circ \varphi_H = f$.

Proof. If $xH = yH$, then $y^{-1}x \in H \subseteq K$, so $f(y^{-1}x) = e'$, the neutral element of G', whence

$$f(x) = f(y(y^{-1}x)) = f(y)f(y^{-1}x) = f(y)e' = f(y).$$

Thus, g is well-defined, and clearly, $g \circ \varphi_H = f$. If $x, y \in G$,

$$g((xH)(yH)) = g(xyH) = f(xy)$$
$$= f(x)f(y) = g(xH)g(yH).$$

Hence, g is a homomorphism.

Theorem 20.9. (Factor Theorem for Groups) Let f be an epimorphism from a group $(G, \cdot)$ onto a group $(G', \cdot)$, and let K be the kernel of f. Then

$$g: xK \mapsto f(x)$$

is a well-defined isomorphism from G/K onto G' satisfying $g \circ \varphi_K = f$. Moreover, f is an isomorphism if and only if $K = \{e\}$.

Proof. By Theorem 20.8, g is a well-defined homomorphism, and clearly g is surjective, since f is. If $f(x) = f(y)$, then $f(y^{-1}x) = f(y^{-1})f(x) = f(y)^{-1}f(x) = e'$, the neutral element of G', whence $y^{-1}x \in K$, and thus $xK = yK$. Therefore, g is injective and hence is an isomorphism. If f is an isomorphism, then φ_K is injective as $g \circ \varphi_K = f$, whence $K = \{e\}$. Conversely, if $K = \{e\}$, then φ_K is clearly an isomorphism, whence f is also.

The Factor Theorem shows that an epimorphism from one group onto another can be "factored" into an isomorphism and the canonical epimorphism from a group onto a quotient group.

Example 20.1. The **circle group** is the group $(T, \cdot)$ where

$$T = \{z \in C : |z| = 1\}.$$

Let f be the function from R into T defined by

$$f(x) = \cos x + i \sin x.$$

By a theorem of analysis, f is surjective. Consequently, f is an epimorphism from $(R, +)$ onto $(T, \cdot)$, for

$$\begin{aligned} f(x)f(y) &= (\cos x + i \sin x)(\cos y + i \sin y) \\ &= (\cos x \cos y - \sin x \sin y) + i(\sin x \cos y + \cos x \sin y) \\ &= \cos(x + y) + i \sin(x + y) \\ &= f(x + y) \end{aligned}$$

for all $x, y \in R$. The kernel of f is the subgroup $2\pi Z$ of all integral multiples of 2π. Hence, by the Factor Theorem,

$$g \colon x + 2\pi Z \mapsto \cos x + i \sin x$$

is an isomorphism from the quotient group $(R/2\pi Z, +)$ onto $(T, \cdot)$.

Example 20.2. Let n be a strictly positive integer, and let C^* be the set of all nonzero complex numbers. The function f defined by

$$f(z) = z^n$$

is an epimorphism from $(C^*, \cdot)$ onto itself. Indeed,

$$f(wz) = (wz)^n = w^n z^n = f(w)f(z)$$

for all $w, z \in C^*$, and f is surjective, for if

$$w = r(\cos \alpha + i \sin \alpha),$$

then $w = f(z)$ where

$$z = \sqrt[n]{r}\left(\cos \frac{\alpha}{n} + i \sin \frac{\alpha}{n}\right).$$

The kernel of f is the set R_n of all complex nth roots of unity, so by the Factor Theorem,

$$g: zR_n \mapsto z^n$$

is an isomorphism from $(C^*/R_n, \cdot)$ onto $(C^*, \cdot)$. Thus, if $n \neq 1$, f is an example of a surjective endomorphism that is not an automorphism. For a more concrete picture of the quotient group and the associated isomorphism, let us consider the case where $n = 3$. The set R_3 of all complex cube roots of unity has three members, namely, 1,

$$\omega = \cos \frac{2\pi}{3} + i \sin \frac{2\pi}{3},$$

and

$$\omega^2 = \cos \frac{4\pi}{3} + i \sin \frac{4\pi}{3}.$$

For each $z \in C^*$, the coset zR_3 is thus the set $\{z, z\omega, z\omega^2\}$, and multiplication on the set C^*/R_3 of all such cosets satisfies

$$\{z_1, z_1\omega, z_1\omega^2\} \cdot \{z_2, z_2\omega, z_2\omega^2\} = \{z_1z_2, z_1z_2\omega, z_1z_2\omega^2\}.$$

The associated isomorphism g from C^*/R_3 onto C^* takes an equivalence class $\{z, z\omega, z\omega^2\}$ into the cube $z^3 = (z\omega)^3 = (z\omega^2)^3$ of any one of its members.

Example 20.3. For each complex number z, we denote the real part of z by $\mathscr{R}z$ and the imaginary part of z by $\mathscr{I}z$. Thus, if $z = x + iy$ where x and y are real numbers, then $\mathscr{R}z = x$ and $\mathscr{I}z = y$. Clearly, $\mathscr{R}$ is an endomorphism of $(C, +)$ whose range is R. The kernel of $\mathscr{R}$ is the subgroup iR of purely imaginary complex numbers, so by the Factor Theorem,

$$g: z + iR \mapsto \mathscr{R}z$$

is an isomorphism from $(C/iR, +)$ onto $(R, +)$. The set iR of purely imaginary complex numbers may be described geometrically as the Y-axis of the plane of analytic geometry. For each $z \in C$, $z + iR$ is simply the set of all the complex numbers having the same real part as z, i.e., the set of all points of the plane whose abscissa is the real part of z; consequently, $z + iR$ may be described geometrically as the line through the point z parallel to the Y-axis. Thus, C/iR is the set of all lines parallel to the Y-axis. The sum of two such lines L_1 and L_2 is the line all of whose points have for abscissa the sum of the number that is the abscissa of all the points of L_1 and the number that is the abscissa of all the points of L_2. The associated isomorphism g takes a given line parallel to the Y-axis into the number that is the abscissa of all its points.

Similarly, $\mathscr{I}$ is an endomorphism of $(C, +)$ whose range is R. The kernel of $\mathscr{I}$ is the subgroup R of real numbers, so

$$h\colon z + R \to \mathscr{I}z$$

is an isomorphism from $(C/R, +)$ onto $(R, +)$. Thus, $\mathscr{I}$ is an example of an endomorphism whose kernel is the same as its range. However, $\mathscr{R}$ and $\mathscr{I}$ are not endomorphisms of $(C, \cdot)$, since

$$\mathscr{R}i^2 = -1 \neq 0 = (\mathscr{R}i)^2,$$
$$\mathscr{I}i^2 = 0 \neq 1 = (\mathscr{I}i)^2.$$

Example 20.4. To illustrate the use of Theorem 20.9, we shall prove the following theorem: *If H is a normal subgroup of a group G, if K is a normal subgroup of G/H, and if $L = \varphi_H^*(K)$, then L is a normal subgroup of G, and there is an isomorphism f from $(G/H)/K$ onto G/L satisfying*

$$f \circ \varphi_K \circ \varphi_H = \varphi_L.$$

Indeed, $\varphi_K \circ \varphi_H$ is an epimorphism from G onto $(G/H)/K$. For every $x \in G$, $x \in \ker(\varphi_K \circ \varphi_H)$ if and only if $\varphi_K(\varphi_H(x)) = K$, the neutral element of $(G/H)/K$, or equivalently, if and only if $\varphi_H(x) \in \ker \varphi_K$, which is the set K; but $\varphi_H(x) \in K$ if and only if $x \in \varphi_H^*(K) = L$. Thus, L is the kernel of $\varphi_K \circ \varphi_H$. By Theorem 20.9, L is a normal subgroup of G, and there is an isomorphism g from G/L onto $(G/H)/K$ satisfying $g \circ \varphi_L = \varphi_K \circ \varphi_H$. Let $f = g^{\leftarrow}$. Then f is an isomorphism from $(G/H)/K$ onto G/L, and

$$f \circ \varphi_K \circ \varphi_H = f \circ g \circ \varphi_L = \varphi_L.$$

As a final illustration, we shall construct an important epimorphism from the group $\mathfrak{S}_n$ of all permutations on $\{1, 2, \ldots, n\}$ onto the multiplicative group $\{1, -1\}$; this epimorphism will enable us to identify an important subgroup of $\mathfrak{S}_n$.

We recall that if $a_1, \ldots, a_n$ are distinct elements of a set E, then $(a_1, \ldots, a_n)$ denotes the permutation τ of E defined by

$$\tau(a_k) = \begin{cases} a_{k+1} \text{ if } k \in [1, n-1], \\ a_1 \text{ if } k = n, \end{cases}$$

$$\tau(x) = x \text{ if } x \notin \{a_1, \ldots, a_n\}.$$

(The expression $(a_1, \ldots, a_n)$ has also been used to denote an n-tuple. Custom dictates this multiple use of the expression $(a_1, \ldots, a_n)$, but in any given context it will be clear which meaning is intended.) The permutation $(a_1, \ldots, a_n)$ is called an **n-cycle** or a **cycle of length n**. A **transposition** is a 2-cycle.

Definition. Let $\sigma \in \mathfrak{S}_n$. The **number of inversions** of σ is the number of ordered couples (i, j) such that $1 \le i < j \le n$ and $\sigma(i) > \sigma(j)$. We shall say that σ is an **even permutation** or an **odd permutation** according as the number of inversions of σ is an even integer or an odd integer. The **signature** of σ, which we shall denote by sgn σ, is defined to be 1 or -1 according as σ is an even or an odd permutation.

If J_σ is the number of inversions of σ, it follows at once from the definition that

$$\text{sgn } \sigma = (-1)^{J_\sigma}.$$

The following two theorems show that if $n > 1$, sgn is an epimorphism from $\mathfrak{S}_n$ into the multiplicative group $\{1, -1\}$.

Theorem 20.10. For all permutations σ and τ of $[1, n]$,

$$\text{sgn } \sigma\tau = (\text{sgn } \sigma)(\text{sgn } \tau),$$

$$\text{sgn } \sigma^{\leftarrow} = \text{sgn } \sigma.$$

Proof. Let T be the set of all ordered couples (i, j) such that $1 \le i < j \le n$. For each $\rho \in \mathfrak{S}_n$ we shall denote by ρ^* the function from T into T defined by

$$\rho^*(i, j) = \begin{cases} (\rho(i), \rho(j)) \text{ if } \rho(i) < \rho(j), \\ (\rho(j), \rho(i)) \text{ if } \rho(i) > \rho(j). \end{cases}$$

It is easy to verify that ρ^* is a permutation of T. Also, for each $(i, j) \in T$, we shall define $m_\rho(i,j)$ by

$$m_\rho(i,j) = \begin{cases} 0 \text{ if } \rho(i) < \rho(j), \\ 1 \text{ if } \rho(i) > \rho(j). \end{cases}$$

Then the number J_σ of inversions of ρ is given by

$$J_\rho = \sum_{(i,j)\in T} m_\rho(i,j).$$

A consideration of the four possible cases shows that

$$m_{\sigma\tau}(i,j) \equiv m_\sigma(\tau^*(i,j)) + m_\tau(i,j) \qquad \text{(mod 2)}$$

for all $(i,j) \in T$, and hence that

$$\sum_{(i,j)\in T} m_{\sigma\tau}(i,j) \equiv \sum_{(i,j)\in T} m_\sigma(\tau^*(i,j)) + \sum_{(i,j)\in T} m_\tau(i,j). \qquad \text{(mod 2)}$$

But since τ^* is a permutation of T,

$$\sum_{(i,j)\in T} m_\sigma(\tau^*(i,j)) = \sum_{(i,j)\in T} m_\sigma(i,j).$$

Consequently,

$$J_{\sigma\tau} \equiv J_\sigma + J_\tau, \qquad \text{(mod 2)}$$

whence

$$\text{sgn } \sigma\tau = (-1)^{J_{\sigma\tau}} = (-1)^{J_\sigma + J_\tau}$$
$$= (-1)^{J_\sigma}(-1)^{J_\tau}$$
$$= (\text{sgn } \sigma)(\text{sgn } \tau).$$

In particular,

$$1 = \text{sgn } \sigma\sigma^\leftarrow = (\text{sgn } \sigma)(\text{sgn } \sigma^\leftarrow),$$

so $\text{sgn } \sigma^\leftarrow = \text{sgn } \sigma$.

From Theorem 20.10, we conclude that a product of two even permutations or of two odd permutations is an even permutation, and that the product of an even permutation and an odd permutation is an odd permutation.

Theorem 20.11. A transposition is an odd permutation; more generally, an m-cycle is an odd or even permutation according as m is even or odd.

Proof. An m-cycle is a product of $m - 1$ transpositions, since clearly,

$$(a_1, \ldots, a_m) = (a_1, a_m)(a_1, a_{m-1}) \ldots (a_1, a_2).$$

To show that an m-cycle is odd or even according as m is even or odd, it suffices by Theorem 20.10 and what we have just proved to show that a transposition is an odd permutation.

Let $\tau = (r, s)$, where $1 \leq r < s \leq n$. With the notation of Theorem 20.10,

$$m_\tau(r, s) = 1,$$

$$m_\tau(i, r) = m_\tau(i, s) = 0 \text{ if } i < r,$$

$$m_\tau(i, s) = 1 \text{ if } r < i < s,$$

$$m_\tau(r, j) = m_\tau(s, j) = 0 \text{ if } j > s,$$

$$m_\tau(r, j) = 1 \text{ if } r < j < s,$$

$$m_\tau(i, j) = 0 \text{ if } i < j \text{ and } \{i, j\} \cap \{r, s\} = \emptyset.$$

Consequently, $J_\tau = 1 + 2(s - r - 1)$, so τ is an odd permutation.

Theorem 20.12. The set $\mathfrak{A}_n$ of all even permutations of $[1, n]$ is a normal subgroup of $\mathfrak{S}_n$. If $n \geq 2$, then $\mathfrak{A}_n$ is a subgroup of $\mathfrak{S}_n$ of index 2 and order $\frac{1}{2}n!$.

Proof. Let $n \geq 2$. Then there exist transpositions and hence odd permutations of $[1, n]$, so sgn is an epimorphism from $\mathfrak{S}_n$ onto the multiplicative group $\{-1, 1\}$ by Theorem 20.10. By definition, the kernel of sgn is the set of even permutations. Consequently, $\mathfrak{A}_n$ is a normal subgroup of $\mathfrak{S}_n$ of index 2 and hence of order $\frac{1}{2}n!$ by Theorems 20.7, 20.9, and 18.3.

Definition. The group $\mathfrak{A}_n$ of all even permutations of $[1, n]$ is called the **alternating group** on n objects.

EXERCISES

20.1. If a and b are real numbers such that $a \leq b$, we shall denote by $[a, b]$ the set $\{x \in \mathbf{R}: a \leq x \leq b\}$. Determine $f_*([-1,0])$, $f^*([-1,0])$, $f_*([-1, 1])$, $f^*([-1, 1])$, $f_*([2, 3])$, $f^*(\{0\})$, $f^*(\{1\})$, and $f^*(\{2\})$ if f is the function from R into R defined by

(a) $f(x) = x^2$.   (c) $f(x) = \sin \pi x$.

(b) $f(x) = x^3$.   (d) $f(x) = 2^x$.

20.2. Let f be a function from E into F.

(a) If A and B are subsets of E, then

$$f_*(A \cup B) = f_*(A) \cup f_*(B),$$
$$f_*(A \cap B) \subseteq f_*(A) \cap f_*(B).$$

Show by an example where E has two elements and F one element that $f_*(A \cap B)$ need not be $f_*(A) \cap f_*(B)$.

(b) If C and D are subsets of F, then

$$f^*(C \cup D) = f^*(C) \cup f^*(D),$$
$$f^*(C \cap D) = f^*(C) \cap f^*(D).$$

(c) If A is a subset of E, then

$$f^*(f_*(A)) \supseteq A.$$

Show by an example that $f^*(f_*(A))$ need not be A.

(d) If D is a subset of F, then

$$f^*(D^c) = f^*(D)^c,$$
$$f_*(f^*(D)) = D \cap f_*(E).$$

20.3. Prove Theorem 20.4 and its corollary.

20.4. State the analogues of Theorems 20.3 and 20.6 for algebraic structures with two compositions.

20.5. (a) Describe the elements of the set C^*/R_4 of Example 20.2, how multiplication is defined on C^*/R_4, and the associated isomorphism g.
(b) Describe geometrically the elements of the set C/R of Example 20.3, how addition is defined on C/R, and the associated isomorphism h.

20.6. Construct an endomorphism of $(N_4, +_4)$ whose kernel and range are both $\{0, 2\}$.

20.7. Determine whether f is an endomorphism of the group R^* of nonzero real numbers under multiplication and, if so, what the kernel and range of f are, where $f(x) =$

(a) $|x|$.   (c) x^2.   (e) 2^x.   (g) $3x$.

(b) $-x$.   (d) x^3.   (f) $\dfrac{1}{x}$.   (h) $\sqrt{|x|}$.

20.8. Let F be a subset of E. Determine whether f is an endomorphism of $(\mathfrak{P}(E),\ \cup)$ or of $(\mathfrak{P}(E),\ \cap)$, where $f(X) =$

(a) $X \cap F$.   (c) X^c.

(b) $X \cup F$.   (d) $X \triangle F$ (Example 7.4).

Determine also whether f is a homomorphism from $(\mathfrak{P}(E),\ \cup)$ into $(\mathfrak{P}(E),\ \cap)$ or from $(\mathfrak{P}(E),\ \cap)$ into $(\mathfrak{P}(E),\ \cup)$. For each case where f is a homomorphism, determine its range and kernel.

20.9. Let $\vee$ and $\wedge$ be the compositions on Z defined by

$$x \vee y = \max\{x, y\},$$
$$x \wedge y = \min\{x, y\}.$$

Prove that f is an endomorphism of $(Z, \vee)$ or of $(Z, \wedge)$ if and only if f is increasing, and that f is a homomorphism from $(Z, \vee)$ into $(Z, \wedge)$ or from $(Z, \wedge)$ into $(Z, \vee)$ if and only if f is decreasing.

20.10. If H and K are normal subgroups of a group G such that $H \subseteq K$, then H is a normal subgroup of K, K/H is a normal subgroup of G/H, and there is an isomorphism g from $(G/H)/(K/H)$ onto G/K satisfying

$$g \circ \varphi_{K/H} \circ \varphi_H = \varphi_K.$$

[First use Theorem 20.8 to show that there is an epimorphism g_1 from G/H onto G/K satisfying $g_1 \circ \varphi_H = \varphi_K$.]

20.11. Let $(G, \cdot)$ be a group, let H be a normal subgroup of G, and let L be a subgroup of G.
(a) LH is a subgroup of G, and H is a normal subgroup of LH.
(b) The function f from L into LH/H defined by $f(x) = xH$ for all $x \in L$ is an epimorphism whose kernel is $L \cap H$.
(c) The function

$$g: x(L \cap H) \mapsto xH$$

is an isomorphism from $L/(L \cap H)$ onto $(LH)/H$.

20.12. Let f be an epimorphism from a group $(G, \cdot)$ onto a group $(G', \cdot)$, and let K be the kernel of f.

(a) If H is a subgroup (normal subgroup) of G, then $f_*(H)$ is a subgroup (normal subgroup) of G', and

$$f^*(f_*(H)) = HK.$$

(b) If H' is a subgroup (normal subgroup) of G', then $f^*(H')$ is a subgroup (normal subgroup) of G.

(c) If H is a normal subgroup of G and if $H' = f_*(H)$, then there is one and only one epimorphism g from G/H onto G'/H' satisfying

$$g \circ \varphi_H = \varphi_{H'} \circ f,$$

and moreover, the kernel of g is $(HK)/H$.

20.13. Prove the congruence concerning $m_{\sigma\tau}(i, j)$ used in the proof of Theorem 20.10.

20.14. If f is a homomorphism from $(E, \triangle)$ into $(F, \triangledown)$, then the relation R_f (Exercise 16.6) defined by f is compatible with $\triangle$ (Exercises, §18).

The remaining exercises classify to within isomorphism all "inductive" and "strictly inductive" semigroups. If $m \in N$, $n \in N^*$, we denote by $D_{m,n}$ the quotient set $N/R_{m,n}$ (Exercise 18.21) and by $+_{m,n}$ the composition induced on $D_{m,n}$ by addition on N. If $m, n \in N^*$, we denote by $D^*_{m,n}$ the quotient set $N^*/R^*_{m,n}$ (Exercise 18.22) and by $+^*_{m,n}$ the composition induced on $D^*_{m,n}$ by addition on N^*.

20.15. A semigroup $(E, +)$ is **strictly inductive** if there exists $\beta \in E$ such that the only subset of E containing β and containing $x + \beta$ whenever it contains x is E itself.

(a) The following assertions are equivalent:

1° $(E, +)$ is a strictly inductive semigroup.
2° There is an epimorphism from $(N^*, +)$ onto $(E, +)$.
3° Either $(E, +)$ is isomorphic to $(N^*, +)$, or there exist $m, n \in N^*$ such that $(E, +)$ is isomorphic to $(D^*(m, n), +^*_{m,n})$.

[Consider $f: n \mapsto n.\beta$, and use Exercises 20.14 and 18.22.]

(b) Every strictly inductive semigroup is an inductive semigroup (Exercise 11.6).

(c) The following semigroups are inductive but not strictly inductive: $(N, +)$, $(D(m, n), +_{m,n})$ for all $m, n \in N^*$, $(N_2, \to)$, and $(N_2, \leftarrow)$.

*20.16. If $(E, +)$ is a semigroup containing a subgroup G such that $E = G \cup \{\alpha\}$ where $\alpha \notin G$, then one and only one of the following possibilities occurs:

1° For all $x \in E$, $\alpha + x = \alpha = x + \alpha$.
2° For all $x \in E$, $\alpha + x = x = x + \alpha$.
3° There exists $b \in G$ such that $\alpha + x = b + x$ and $x + \alpha = x + b$ for all $x \in G$, and $\alpha + \alpha = b + b$.

$4°$ There is only one element β belonging to G, and one of the following
three conditions holds:

$4_1°$ $(E, +)$ is isomorphic to $(N_2, +_2)$.

$4_2°$ The composition $+$ is $\leftarrow$.

$4_3°$ The composition $+$ is $\rightarrow$.

[Suggested outline of proof:

(a) Either $\alpha + x = \alpha$ for all $x \in G$, or $\alpha + x \in G$ for all $x \in G$, and similarly
for $x + \alpha$.

(b) If G has more than one element, either $\alpha + x = x + \alpha = \alpha$ for all
$x \in G$, or $\alpha + x$ and $x + \alpha$ belong to G for all $x \in G$.

(c) Let β be the neutral element of G. If G has more than one element,
consider the four cases determined by whether or not $\alpha + \alpha = \alpha$ and
whether or not $\alpha + \beta = \alpha = \beta + \alpha$.] Conversely, if $(G, +)$ is a group
and if $\alpha \notin G$, then the composition $+$ may be extended to an associative
composition on $G \cup \{\alpha\}$ by use of $1°, 2°, 3°$, or, if G has only one element
β, by use of $4_1°, 4_2°$, or $4_3°$.

20.17. Let $K_n = N_n \cup \{\alpha\}$ where α is an object not in N_n, and let $m \in N_n$. We
extend the composition addition modulo n on N_n to a composition $\oplus_{m,n}$
on K_n by making the following definitions:

$$\alpha \oplus_{m,n} \alpha = m +_n m,$$

$$\alpha \oplus_{m,n} x = x \oplus_{m,n} \alpha = m +_n x$$

for all $x \in N_n$. Then $(K_n, \oplus_{m,n})$ is an inductive semigroup. [Use Exercise
20.16.] Show how to label the points $(1, 0)$, $(0, 1)$, $(-1, 0)$, $(0, -1)$, and
$(0, -3)$ of the plane of analytic geometry by elements of K_4 so that the
configuration obtained by drawing the broken line joining successively
$0, 1, 2, 3, 0$ and, for each $z \in K_4$, the broken line joining successively z,
$z \oplus_{2,4} \alpha$, $z \oplus_{2,4} 2.\alpha$, $z \oplus_{2,4} 3.\alpha$, $z \oplus_{2,4} 4.\alpha$ looks like a kite. Similarly, if
$m \geq 2$, the vertices of a regular polygon of n sides together with one more
point outside the polygon may be used as a model for $(K_n, \oplus_{m,n})$ in such
a way that the broken line joining successively $0, 1, \ldots, n-1, 0$ is the
edge of an n-sided kite and the broken lines joining successively $z, z \oplus_{m,n} \alpha$,
$z \oplus_{m,n} 2.\alpha, \ldots, z \oplus_{m,n} n.\alpha$ for all $z \in K_n$ form a tail and frame for the
kite. Draw the models for $(K_6, \oplus_{2,6})$, $(K_6, \oplus_{3,6})$, $(K_8, \oplus_{2,8})$, $(K_8, \oplus_{3,8})$,
and $(K_8, \oplus_{4,8})$.

20.18. Let α and β be objects not in N, and for each $n \in N^*$ let $E_n = N_n \cup \{\alpha, \beta\}$.
If $n > 1$, we extend the composition addition modulo n on N_n to a com-
position $+$ on E_n by making the following definitions:

$$\alpha + \alpha = \alpha, \qquad\qquad \alpha + \beta = \beta,$$

$$\beta + \beta = 1 +_n 1, \qquad \beta + \alpha = 1,$$

$$\alpha + x = x + \alpha = x,$$

$$\beta + x = x + \beta = 1 +_n x$$

for all $x \in N_n$. We also define a composition $+$ on E_1 by the following table:

$+$	0	α	β
0	0	0	0
α	0	α	β
β	0	0	0

Prove that for each $n \in N^*$, $(E_n, +)$ is an inductive semigroup. [Use Exercise 20.16.]

*20.19. A **dipper** is a semigroup isomorphic either to $(D(m, n), +_{m,n})$ for some $m \in N$, $n \in N^*$ or to $(D^*(m, n), +^*_{m,n})$ for some $m, n \in N^*$. A **kite** is a semigroup isomorphic to $(K_n, \oplus_{m,n})$ for some $n \in N^*$, $m \in N$.

(a) A commutative inductive semigroup either is a dipper or a kite, or is isomorphic either to $(N, +)$ or to $(N^*, +)$.
(b) An inductive semigroup that is not commutative is isomorphic either to $(N_2, \rightarrow)$ or to $(E_n, +)$ for some $n \in N^*$ (Exercise 20.18). [Suggested outline of proof: Let

$$D' = \{\alpha + k.\beta : k \in N^*\},$$

$$D = \{k.\beta : k \in N^*\}.$$

Show that $E = D' \cup \{\alpha\}$, and eliminate the case where $\alpha = \beta$. Show that if $\alpha + \alpha \neq \alpha$, then $E = D \cup \{\alpha\}$, and D is isomorphic to $(N_n, +_n)$ for some $n \in N^*$; for this, show that $\alpha + \beta \in D$ by expressing the fact that $\alpha + \alpha$ and β belong to D'. Show that if $\alpha + \alpha = \alpha$, then $\alpha + x = x$ for all $x \in E$ and consequently, $E = D \cup \{\alpha\}$. Consider separately the possibilities $\beta + \alpha = \alpha$, $\beta + \alpha = \beta$, $\beta + \alpha = r.\beta$ where $r > 1$.]

21. Homomorphisms of Rings

Theorem 21.1. If f is an epimorphism from a ring $(A, +, \cdot)$ onto an algebraic structure $(A', +', \cdot')$, then $(A', +', \cdot')$ is a ring. Moreover, if A is a commutative ring, so is A'.

Proof. By the Corollary of Theorem 20.4, $(A', +')$ is a commutative group. By Theorem 20.4, $\cdot'$ is associative and is, moreover, commutative if A is a commutative ring. It is easy to make the remaining verification that multiplication is distributive over addition.

Corollary. If f is a homomorphism from a ring $(A, +, \cdot)$ into an algebraic structure $(A', +', \cdot')$, then the range $f_*(A)$ is a subring of $(A', +', \cdot')$. Moreover, if A is a commutative ring, so is $f_*(A)$.

Proof. By Theorem 20.3, $f_*(A)$ is a stable subset of $(A', +', \cdot')$. The assertion, therefore, follows from Theorem 21.1, since the function obtained by restricting the codomain of f to $f_*(A)$ is an epimorphism from A onto $f_*(A)$.

If $\mathfrak{a}$ is an ideal of a ring A, the canonical surjection $\varphi_{\mathfrak{a}}$ from A onto $A/\mathfrak{a}$ is an epimorphism from the ring A onto the ring $A/\mathfrak{a}$ by the definition of multiplication on $A/\mathfrak{a}$.

Theorem 21.2. The kernel $\mathfrak{a}$ of a homomorphism f from a ring A into a ring A' is an ideal of A.

Proof. By Theorem 20.7, $\mathfrak{a}$ is a subgroup under addition. If $x \in \mathfrak{a}$, $a \in A$, then $f(x) = 0$, so $f(ax) = f(a)f(x) = 0$ and $f(xa) = f(x)f(a) = 0$, whence $ax, xa \in \mathfrak{a}$.

Theorem 21.3. Let f be a homomorphism from a ring A into a ring A'. If $\mathfrak{b}$ is an ideal of A that is contained in the kernel $\mathfrak{a}$ of f, then

$$g: x + \mathfrak{b} \mapsto f(x)$$

is a well-defined homomorphism from $A/\mathfrak{b}$ into A' satisfying $g \circ \varphi_{\mathfrak{b}} = f$.

Proof. By Theorem 20.8, we need only show that $g((x + \mathfrak{b})(y + \mathfrak{b})) = g(x + \mathfrak{b})g(y + \mathfrak{b})$. But

$$g((x + \mathfrak{b})(y + \mathfrak{b})) = g(xy + \mathfrak{b}) = f(xy)$$
$$= f(x)f(y) = g(x + \mathfrak{b})g(y + \mathfrak{b}).$$

Theorem 21.4. (Factor Theorem for Rings) Let f be an epimorphism from a ring A onto a ring A', and let $\mathfrak{a}$ be the kernel of f. Then

$$g: x + \mathfrak{a} \mapsto f(x)$$

is a well-defined isomorphism from $A/\mathfrak{a}$ onto A' satisfying $g \circ \varphi_{\mathfrak{a}} = f$. Moreover, f is an isomorphism if and only if $\mathfrak{a} = (0)$.

The assertion is an immediate consequence of Theorems 21.3 and 20.9.

Theorem 21.5. If f is an epimorphism from a field K onto a ring A, then either A is a zero ring, or A is a field and f is an isomorphism from K onto A.

Proof. The kernel of f is an ideal and hence is either K or (0) by the Corollary of Theorem 19.4. In the former case, $f(x) = 0$ for all $x \in K$, so A is a zero ring; in the latter case, f is an isomorphism by Theorem 21.4.

We consider next homomorphisms from the ring Z of integers into a ring with identity. We recall that if $m > 0$, the ring $Z/(m)$ is denoted by Z_m.

Theorem 21.6. Let A be a ring with identity 1. The function g from Z into A defined by

$$g(n) = n.1$$

for all $n \in Z$ is a homomorphism from the ring Z into A. The range B of g is the smallest subring of A that contains 1. If A has no proper zero-divisors, then g is the only nonzero homomorphism from Z into A, and the kernel of g is either (0), in which case g is an isomorphism from Z onto B, or (p) for some prime p, in which case B is isomorphic to the field Z_p.

Proof. By (17) of Theorem 14.7,

$$(n + m).1 = n.1 + m.1,$$

and by (1) of §15 and (18) of Theorem 14.7,

$$(n.1)(m.1) = n.(m.1) = (nm).1$$

for all $n, m \in Z$. Consequently, g is a homomorphism from Z into A, and its range B is, therefore, a subring by the Corollary of Theorem 21.1. An inductive argument establishes that every subring of A containing 1 also contains B, so B is the smallest subring of A that contains 1. Let us assume further that A has no proper zero-divisors. By Theorem 19.8, the kernel of g is (p) for some $p \in N$. By Theorem 21.4, applied to the function obtained by restricting the codomain of g to B, B is isomorphic to Z_p and also has no proper zero-divisors, so either $p = 0$ or p is a prime by Theorem 19.11. To show the uniqueness of g, let h be a nonzero homomorphism from Z into A. Since

$$h(1) = h(1^2) = h(1)^2,$$

either $h(1) = 1$ or $h(1) = 0$ by Theorem 15.2. But

$$h(n) = h(n.1) = n.h(1)$$

for all $n \in Z$ by the Corollary of Theorem 20.5. Therefore, if $h(1)$ were 0, then $h(n)$ would be $n.0 = 0$ for all $n \in Z$, and hence, h would be the zero homomorphism, a contradiction. Consequently, $h(1) = 1$, so

$$h(n) = n.1 = g(n)$$

for all $n \in Z$.

Definition. The **characteristic** of a ring A with identity 1 is the natural number p such that (p) is the kernel of the homomorphism $g: n \mapsto n.1$ from Z into A.

By Theorem 19.8, if $p \in N^*$, then p is the smallest strictly positive integer belonging to the principal ideal (p) generated by p. Thus *a ring A with identity 1 has characteristic $p > 0$ if and only if p is the smallest strictly positive integer such that $p.1 = 0$; and A has characteristic zero if and only if $n.1 \neq 0$ for all $n > 0$.* For example, for each $p \in N^*$, the ring Z_p has characteristic p, for clearly p is the smallest of the strictly positive integers m such that $m.(1 + (p)) = (p)$, since $m.(1 + (p)) = m + (p)$.

By Theorem 21.6, *the characteristic of a ring with identity and without proper zero-divisors is either zero or a prime. In particular, the characteristic of a division ring or integral domain is either zero or a prime.*

Theorem 21.7. Let A be a ring with identity 1, let p be the characteristic of A, let $a \in A$, and let g_a be the function from Z into A defined by

$$g_a(n) = n.a$$

for all $n \in Z$.

1° The function g_a is a homomorphism from the additive group $(Z, +)$ into $(A, +)$, and the kernel of g_a contains (p). Consequently, if $p > 0$, then $n.a = 0$ whenever $p \mid n$.

2° If a is not a zero-divisor of A, then the kernel of g_a is (p), and consequently if $p = 0$, then $n.a = 0$ only if $n = 0$, but if $p > 0$, then $n.a = 0$ if and only if $p \mid n$.

Proof. By Theorem 14.7, g_a is a homomorphism. For every $n \in Z$,

$$n.a = (n.a)1 = a(n.1)$$

by (1) of §15. Therefore, $n.a = 0$ whenever $n.1 = 0$, and if a is not a zero-divisor, $n.a = 0$ if and only if $n.1 = 0$, or equivalently, if and only if $n \in (p)$.

For example, if A is a ring with identity whose characteristic is 2, then for all $a \in A$, we have $2.a = 0$, or equivalently, $a = -a$.

Definition. A **prime field** is a field containing no proper subfields.

Theorem 21.8. A division ring K contains one and only one prime subfield P, and P is the smallest division subring of K. If the characteristic of K is a prime p, then Z_p is isomorphic to P; if the characteristic of K is zero, then Q is isomorphic to P under an isomorphism g satisfying $g(n) = n.1$ for all $n \in Z$.

Proof. By Theorem 21.6, there is a smallest subring B of K that contains 1. The center C of K is a subfield by the Corollary of Theorem 15.6, so $B \subseteq C$. Let P be the quotient field of B in C. Any division subring of K contains 1, and hence B, and therefore also P, since each element of P is the product of an element of B and the inverse of a nonzero element of B. Hence, P is contained in every division subring of K, and in particular, P is the smallest subfield of K. As every subfield of P is a subfield of K, P has no proper subfields and hence is a prime field. Since P is contained in every subfield of K, P is the only subfield of K that is a prime field. If the characteristic of K is a prime p, then Z_p is isomorphic to B by Theorem 21.6, and hence, $P = B$ as B is a field. If the characteristic of K is zero, then $f: n \mapsto n.1$ is an isomorphism from Z onto B by Theorem 21.6, so there is an extension g of f that is an isomorphism from Q onto P by Theorem 17.5.

Corollary. To within isomorphism, the only prime field whose characteristic is a prime p is Z_p, and the only prime field of characteristic zero is Q.

EXERCISES

21.1. Let $(G, +)$ be a commutative group. We denote by $\text{End}(G)$ the set of all endomorphisms of $(G, +)$. Prove that $(\text{End}(G), +, \circ)$ is a ring, where for all $u, v \in \text{End}(G)$, $u + v$ is defined to be the endomorphism $x \mapsto u(x) + v(x)$ of G.

*21.2. For each integer p, let L_p be the endomorphism of $(Z, +)$ defined by

$$L_p(x) = px$$

for all $x \in Z$, and let L be the function from Z into $\text{End}(Z)$ defined by

$$L(p) = L_p$$

for all $p \in Z$.
(a) The function L is an isomorphism from the ring Z of integers onto the ring $\text{End}(Z)$ of endomorphisms of the abelian group $(Z, +)$.

(b) What are the automorphisms of $(Z, +)$?

(c) The rings $(Z, +, \langle r \rangle)$ and $(Z, +, \langle s \rangle)$ (Exercise 14.7) are isomorphic if and only if either $r = s$ or $r = -s$.

(d) If $(A, +, \cdot)$ is a ring such that $(A, +)$ is isomorphic to $(Z, +)$, then there is one and only one natural number r such that $(A, +, \cdot)$ is isomorphic to $(Z, +, \langle r \rangle)$. Thus, to within isomorphism, there are denumerably many rings whose additive group is isomorphic to the additive group of integers.

21.3. If $\mathfrak{a}$ and $\mathfrak{b}$ are ideals of a ring A and if $\mathfrak{a} \subseteq \mathfrak{b}$, then $\mathfrak{a}$ is an ideal of the ring $\mathfrak{b}$, $\mathfrak{b}/\mathfrak{a}$ is an ideal of the ring $A/\mathfrak{a}$, and there is an isomorphism g from $(A/\mathfrak{a})/(\mathfrak{b}/\mathfrak{a})$ onto $A/\mathfrak{b}$ satisfying

$$g \circ \varphi_{\mathfrak{b}/\mathfrak{a}} \circ \varphi_{\mathfrak{a}} = \varphi_{\mathfrak{b}}.$$

[Use Theorem 21.3.]

21.4. Let A be a ring, let $\mathfrak{a}$ be an ideal of A, and let $\mathfrak{b}$ be a subring of A.

(a) $\mathfrak{b} + \mathfrak{a}$ is a subring of A, and $\mathfrak{a}$ is an ideal of $\mathfrak{b} + \mathfrak{a}$.

(b) The function f from $\mathfrak{b}$ onto $(\mathfrak{b} + \mathfrak{a})/\mathfrak{a}$ defined by $f(x) = x + \mathfrak{a}$ for all $x \in \mathfrak{b}$ is an epimorphism whose kernel is $\mathfrak{b} \cap \mathfrak{a}$.

(c) The function $g: x + (\mathfrak{b} \cap \mathfrak{a}) \mapsto x + \mathfrak{a}$ is an isomorphism from $\mathfrak{b}/(\mathfrak{b} \cap \mathfrak{a})$ onto $(\mathfrak{b} + \mathfrak{a})/\mathfrak{a}$.

21.5. Let f be an epimorphism from a ring A onto a ring A', and let $\mathfrak{a}$ be the kernel of f.

(a) If $\mathfrak{b}$ is a subring (an ideal) of A, then $f_*(\mathfrak{b})$ is a subring (an ideal) of A', and $f^*(f_*(\mathfrak{b})) = \mathfrak{b} + \mathfrak{a}$.

(b) If $\mathfrak{b}'$ is a subring (an ideal) of A', then $f^*(\mathfrak{b}')$ is a subring (an ideal) of A and $f_*(f^*(\mathfrak{b}')) = \mathfrak{b}'$.

(c) If $\mathfrak{b}$ is an ideal of A and if $\mathfrak{b}' = f(\mathfrak{b})$, then there is one and only one epimorphism g from $A/\mathfrak{b}$ onto $A'/\mathfrak{b}'$ satisfying

$$g \circ \varphi_{\mathfrak{b}} = \varphi_{\mathfrak{b}'} \circ f,$$

and moreover, the kernel of g is $(\mathfrak{b} + \mathfrak{a})/\mathfrak{b}$.

21.6. Let A be a ring, and for each $a \in A$, let L_a be the endomorphism of $(A, +)$ defined by

$$L_a(x) = ax$$

for all $x \in A$. The **left regular representation** of the ring A is the function $L: a \mapsto L_a$ from A into $\text{End}(A)$.

(a) The function L is a homomorphism from the ring A into the ring $\text{End}(A)$.

(b) If A is a ring with identity and if $u \in \text{End}(A)$, then $u = L_a$ for some $a \in A$ if and only if $u(xy) = u(x)y$ for all $x, y \in A$.

21.7. (a) If A is a commutative ring containing an element that is not a zero-divisor, then L_a is an endomorphism of the ring A if and only if a is a multiplicative idempotent (Exercise 2.17).

(b) If A is a ring whose center contains an element that is not a zero-divisor, then L_a is an endomorphism of the ring A for all $a \in A$ if and only if A is a boolean ring (Exercise 15.21).

21.8. If f is an epimorphism from a ring onto a division ring, then the kernel of f is a maximal ideal.

*21.9. If $\mathfrak{a}$ is an ideal of a ring A and if f is an epimorphism from the ring $\mathfrak{a}$ onto a ring with identity B, then there is one and only one epimorphism g from A onto B extending f. [Show first that if $e \in f^*(\{1\})$, then $f(ex) = f(exe) = f(xe)$ for all $x \in A$.]

*21.10. Let D be a subdomain of Q.
(a) Prove that $D \supseteq Z$.
(b) If f is a homomorphism from the group $(D, +)$ into the group $(Q, +)$, then there exists $a \in f_*(D)$ such that $f(x) = ax$ for all $x \in D$.
(c) The left regular representation (Exercise 21.6) of D is an isomorphism from the ring D onto the ring $\text{End}(D)$.
(d) The only nonzero homomorphism from the ring D into the ring Q is the function $\iota_D : x \mapsto x$.
(e) If E is a subdomain of Q isomorphic to D, then $E = D$.

*21.11. If K is a field, then the groups $(K, +)$ and $(K^*, \cdot)$ are not isomorphic. [Consider the final example of §6.]

*21.12. If K is a field whose characteristic is not 2 and if f is a function from K into K satisfying

$$f(x + y) = f(x) + f(y)$$

for all $x, y \in K$ and

$$f(x)f(x^{-1}) = 1$$

for all $x \in K^*$, then either f or $-f$ is a monomorphism from K into K. [Assume that $f(1) = 1$. If x is neither 0 nor -1, show that $f(x^2) = f(x)^2$ by expanding $f((1 + x^{-1})^2)$ in two ways.]

21.13. If A is a totally ordered integral domain whose set of positive elements is well-ordered, then $g : n \mapsto n.1$ is an isomorphism from the ordered ring of integers onto the ordered ring A. [Use Theorem 14.9.]

22. Principal Ideal Domains

We next wish to extend the definition of divisibility given for the integral domain Z in §19. Although we shall formulate divisibility concepts in general for commutative rings with identity, we shall investigate them only for integral domains.

Definition. If a and b are elements of a commutative ring with identity A and if $b \neq 0$, we shall say that b **divides** a in A, or that b is a **divisor** or a **factor** of a in A, or that a is **divisible** by b in A or a **multiple** of b in A if there exists $c \in A$ such that $bc = a$.

The symbol $|$ means "divides," so that

$$b \mid a$$

means that b divides a.

If A is a subring of a commutative ring with identity K containing the identity element of K and if a and b are nonzero elements of A, it is entirely possible that b divides a in K but not in A. For example, 2 divides 3 in Q but not in Z. In discussing divisibility, therefore, we must always keep in mind what ring is being considered; if it is clear from the context what that ring is, we shall usually omit explicit reference to it.

We recall that if A is a commutative ring with identity and if $b \in A$, the ideal Ab of A is denoted by (b). Consequently, if $b \in A^*$,

(1) $b \mid a$ if and only if $(a) \subseteq (b)$.

Definition. Let a and b be nonzero elements of a commutative ring with identity A. We shall say that a is an **associate** of b in A if $b \mid a$ and $a \mid b$ in A. A **unit** of A is an associate of 1 in A.

Theorem 22.1. Let a and b be nonzero elements of an integral domain A. The following statements are equivalent:

1° a is an associate of b.
2° $(a) = (b)$.
3° There is an invertible element u of A such that $a = bu$.

Proof. The equivalence of 1° and 2° follows from (1). Condition 3° implies 1°, for if $a = bu$ where u is invertible, then also $b = au^{-1}$, so $b \mid a$ and $a \mid b$. Condition 1° implies 3°, for if $a = bu$ and if $b = av$, then $a = avu$, so $1 = vu$ and hence u is invertible.

Corollary 22.1.1. Let u be a nonzero element of an integral domain A. The following statements are equivalent:

$1°$ u is a unit of A.

$2°$ $(u) = A$.

$3°$ u is an invertible element of A.

Corollary 22.1.2. Let A be an integral domain, and let G be the set of invertible elements of A. Then the relation "____ is an associate of ..." on A^* is an equivalence relation, and for every $a \in A^*$ the equivalence class of a for this relation is the set Ga.

Definition. Let $a_1, \ldots, a_n$ be nonzero elements of a commutative ring with identity A. A **common divisor** of $a_1, \ldots, a_n$ is any element of A^* dividing each of $a_1, \ldots, a_n$. An element d of A^* is a **greatest common divisor** of $a_1, \ldots, a_n$ if d is a common divisor of $a_1, \ldots, a_n$ and if every common divisor of $a_1, \ldots, a_n$ is also a divisor of d.

Since $(a_k) \subseteq (d)$ for all $k \in [1, n]$ if and only if $(a_1) + \ldots + (a_n) \subseteq (d)$ by Theorem 19.3, from (1) we conclude that

(2) d is a common divisor of $a_1, \ldots, a_n$ if and only if

$$(a_1) + \ldots + (a_n) \subseteq (d).$$

Definition. Let $a_1, \ldots, a_n$ be nonzero elements of a commutative ring with identity A. A **common multiple** of $a_1, \ldots, a_n$ is any element of A^* that is a multiple of each of $a_1, \ldots, a_n$. An element m of A^* is a **least common multiple** of $a_1, \ldots, a_n$ if m is a common multiple of $a_1, \ldots, a_n$ and if every common multiple of $a_1, \ldots, a_n$ is also a multiple of m.

Since $(m) \subseteq (a_k)$ for all $k \in [1, n]$ if and only if $(m) \subseteq (a_1) \cap \ldots \cap (a_n)$, from (1) we also conclude that

(3) m is a common multiple of $a_1, \ldots, a_n$ if and only if

$$(m) \subseteq (a_1) \cap \ldots \cap (a_n).$$

The following theorem follows easily from (1) and Theorem 22.1.

Theorem 22.2. Let $a_1, \ldots, a_n$ be nonzero elements of an integral domain A. If d is a greatest common divisor of $a_1, \ldots, a_n$, then the set of all greatest common divisors of $a_1, \ldots, a_n$ is the set of all associates of d. If m is a least common multiple of $a_1, \ldots, a_n$, then the set of all least common multiples of $a_1, \ldots, a_n$ is the set of all associates of m.

A paraphrase of Theorem 22.2 is that in an integral domain, a greatest common divisor or a least common multiple of $a_1, \ldots, a_n$ is "unique to within an associate." The use of the superlative adjectives "greatest" and "least" in the preceding definition thus does not imply that there is at most one greatest common divisor or at most one least common multiple, even though in ordinary English a singular noun modified by a superlative adjective names at most one member of the class considered.

Definition. If $a_1, \ldots, a_n$ are nonzero elements of a commutative ring with identity, we shall say that $a_1, \ldots, a_n$ are **relatively prime** if 1 is a greatest common divisor of $a_1, \ldots, a_n$.

Theorem 22.3. If m is a least common multiple of nonzero elements a and b of an integral domain A, then there is a greatest common divisor d of a and b satisfying $md = ab$.

Proof. Since ab is a common multiple of a and b, $m \mid ab$; it remains for us to show that $d = ab/m$ is a greatest common divisor of a and b. Since $b \mid m, ab \mid am$, and consequently, $d \mid a$; since $a \mid m, ab \mid mb$, and consequently $d \mid b$; hence, d is a common divisor of a and b. Let c be a common divisor of a and b, and let $x, y \in A^*$ be such that $cx = a$, $cy = b$. Then $a \mid cxy$ and $b \mid cxy$, so $m \mid cxy$ and

$$ c\left(\frac{cxy}{m}\right) = \frac{cxcy}{m} = \frac{ab}{m} = d, $$

whence $c \mid d$. Thus, d is a greatest common divisor of a and b.

Corollary. If m is a least common multiple of nonzero elements a and b of an integral domain A, then d' is a greatest common divisor of a and b if and only if md' is an associate of ab.

Proof. Clearly md' is an associate of ab if and only if d' is an associate of $ab/m = d$, so the assertion follows from Theorems 22.3 and 22.2.

For a deeper investigation of divisibility, we shall limit our discussion to a special class of integral domains.

Definition. A **principal ideal domain** is an integral domain A satisfying the following two conditions:

(PID)　　　Every ideal of A is a principal ideal.

(N)　　　Every nonempty set of ideals of A, ordered by $\subseteq$, possesses a maximal element.

Theorem 22.4. The ring Z of integers is a principal ideal domain.

Proof. By Theorem 19.8, Z satisfies (PID). To show that Z satisfies (N), let $\mathfrak{A}$ be a nonempty set of ideals of Z. If $\mathfrak{A}$ contains only the zero ideal, then that ideal is a maximal element of $\mathfrak{A}$. In the contrary case, let n be the smallest of those strictly positive integers m for which $(m) \in \mathfrak{A}$. If s is a strictly positive integer such that $(s) \in \mathfrak{A}$ and $(s) \supseteq (n)$, then $s \mid n$, whence $s \leq n$ and, therefore, $s = n$ by the definition of n. Consequently, (n) is a maximal element of $\mathfrak{A}$ by Theorem 19.8.

Theorem 22.5. Let $a_1, \ldots, a_n$ be nonzero elements of a principal ideal domain A.

    1° There exists a greatest common divisor of $a_1, \ldots, a_n$.
    2° If d is a greatest common divisor of $a_1, \ldots, a_n$, then there exist $x_1, \ldots, x_n \in A$ such that

$$d = a_1 x_1 + \ldots + a_n x_n.$$

    3° An element d of A is a greatest common divisor of $a_1, \ldots, a_n$ if and only if $(d) = (a_1) + \ldots + (a_n)$.

Proof. Since $(a_1) + \ldots + (a_n)$ is an ideal of A by Theorem 19.3, there exists $d_1 \in A$ such that $(d_1) = (a_1) + \ldots + (a_n)$. Consequently, d_1 is a common divisor of $a_1, \ldots, a_n$ by (2). If s is a common divisor of $a_1, \ldots, a_n$, then $(d_1) = (a_1) + \ldots + (a_n) \subseteq (s)$ by (2), so s is also a divisor of d_1 by (1). Thus, d_1 is a greatest common divisor of $a_1, \ldots, a_n$. Let d be any greatest common divisor of $a_1, \ldots, a_n$. By Theorem 22.2, d is an associate of d_1. By Theorem 22.1, therefore, $(d) = (d_1) = (a_1) + \ldots + (a_n)$, so there exist $x_1, \ldots, x_n \in A$ such that

$$d = a_1 x_1 + \ldots + a_n x_n.$$

Conversely, if $(d) = (a_1) + \ldots + (a_n)$, then $(d) = (d_1)$, so d is an associate of d_1 and hence is a greatest common divisor of $a_1, \ldots, a_n$ by Theorem 22.2.

    Since there exist $x_1, \ldots, x_n \in A$ such that $a_1 x_1 + \ldots + a_n x_n = 1$ if and only if $(1) = (a_1) + \ldots + (a_n)$, we obtain from 2° and 3° the following corollary.

Corollary. (Bezout's Identity) If $a_1, \ldots, a_n$ are nonzero elements of a principal ideal domain A, then $a_1, \ldots, a_n$ are relatively prime if and only if there exist $x_1, \ldots, x_n \in A$ such that

$$a_1 x_1 + \ldots + a_n x_n = 1.$$

Theorem 22.6. Let $a_1, \ldots, a_n$ be nonzero elements of a principal ideal domain A.

1° There exists a least common multiple of $a_1, \ldots, a_n$.
2° An element m of A is a least common multiple of $a_1, \ldots, a_n$ if and only if $(m) = (a_1) \cap \ldots \cap (a_n)$.

Proof. As $(a_1) \cap \ldots \cap (a_n)$ is an ideal, there exists $m_1 \in A$ such that $(m_1) = (a_1) \cap \ldots \cap (a_n)$. By (3), m_1 is a common multiple of $a_1, \ldots, a_n$. If s is a common multiple of $a_1, \ldots, a_n$, then $(s) \subseteq (a_1) \cap \ldots \cap (a_n) = (m_1)$ by (3), so s is also a multiple of m_1 by (1). Therefore, m_1 is a least common multiple of $a_1, \ldots, a_n$.

By Theorems 22.1 and 22.2, m is a least common multiple of $a_1, \ldots, a_n$ if and only if $(m) = (m_1) = (a_1) \cap \ldots \cap (a_n)$.

If A is a commutative ring with identity, a nonzero element of A is divisible by all of its associates and also by every unit of A.

Definition. A nonzero element p of a commutative ring with identity A is an **irreducible** element of A if p is not a unit of A and if the only divisors of p in A are its associates and the units of A.

For example, 2 is an irreducible element of Z, but 2 is not an irreducible element of Q. Indeed, a field contains no irreducible elements, since every nonzero element of a field is a unit.

An associate of an irreducible element is irreducible, for associates have exactly the same divisors.

Theorem 22.7. Let p be a nonzero element of a principal ideal domain A. The following statements are equivalent:

1° p is irreducible.
2° (p) is a maximal ideal.
3° p is not a unit, and for all nonzero elements a and b of A, if $p \mid ab$, then either $p \mid a$ or $p \mid b$.

Proof. Condition 1° implies 2°: As p is not a unit, (p) is a proper ideal of A. If $(a) \supseteq (p)$, then $a \mid p$, so a is either a unit of A or an associate of p, and consequently, (a) is either A or (p). Therefore, as every ideal of A is principal, (p) is a maximal ideal.

Condition 2° implies 3°: Since (p) is a proper ideal of A by 2°, p is not a unit of A. Suppose that $p \mid ab$ but that $p \nmid a$. Then (a) is not contained in (p), so $(p) + (a)$ is an ideal properly containing (p), and consequently, $(p) + (a) = A$ by 2°. Therefore, there exist $s, t \in A$ such that $sp + ta = 1$.

Let $c \in A$ be such that $pc = ab$. Then

$$b = spb + tab = spb + tpc = p(sb + tc),$$

so $p \mid b$.

Condition 3° implies 1°: Let a be a divisor of p, and let b be such that $ab = p$. By 3°, either $p \mid a$ or $p \mid b$. If $p \mid a$, then a and p are associates. If $p \mid b$ and if c is such that $pc = b$, then

$$abc = pc = b,$$

so $ac = 1$, and therefore, a is a unit.

Definition. A subset P of an integral domain A is a **representative set of irreducible elements** of A if each element of P is irreducible and if each irreducible element of A is an associate of one and only one member of P.

The word "prime" often occurs in discussions of divisibility. Sometimes "prime" is used as a synonym for "irreducible." More often, however, there is a "natural" representative set of irreducible elements, and its members are called primes to distinguish them from irreducible elements not belonging to the representative set. For example, the set P of all positive irreducible integers is a representative set of irreducible elements of Z, for the only units of Z are 1 and -1; elements of P are called **prime numbers** in accordance with definition of §19.

The "Fundamental Theorem of Arithmetic," which we shall prove shortly, is the assertion that every nonzero integer other than 1 and -1 is either the product of a sequence of prime numbers in an essentially unique way or the additive inverse of such a product. A natural question is: What other integral domains satisfy a similar statement? This question may be separated into two for an arbitrary integral domain A:

1° Is every element of A^* either a unit or a product of a sequence of irreducible elements?

2° Is the factorization of an element of A^* into a product of irreducible elements unique?

In answering the second question, we need a reasonable interpretation of the word "unique"; for example,

$$15 = 3 \cdot 5 = 5 \cdot 3 = (-3)(-5) = (-5)(-3),$$

so $(3, 5)$, $(5, 3)$, $(-3, -5)$, and $(-5, -3)$ are all sequences of irreducible integers whose product is 15. However, these four sequences are very similar, for they all have two terms, and if (q_1, q_2) is any one of them, there is a permutation φ of $\{1, 2\}$ such that $q_{\varphi(1)}$ is an associate of 3 and $q_{\varphi(2)}$ is an associate of 5. We are led, therefore, to the following definition:

Definition. An integral domain A is a **unique factorization domain** (or **gaussian domain**) if the following two conditions hold:

(UFD 1)   Every nonzero element of A is either a unit or a product of irreducible elements.

(UFD 2)   If $(p_i)_{1 \leq i \leq n}$ and $(q_j)_{1 \leq j \leq m}$ are any sequences of irreducible elements of A such that

$$\prod_{i=1}^{n} p_i = \prod_{j=1}^{m} q_j, \tag{4}$$

then $n = m$ and there is a permutation φ of $[1, n]$ such that p_i and $q_{\varphi(i)}$ are associates for all $i \in [1, n]$.

Theorem 22.8. If A is an integral domain satisfying (UFD 1), then A satisfies (UFD 2) if and only if A satisfies the following condition:

(UFD 3)   If p is an irreducible element of A, then for all elements a, b of A^*, if $p \mid ab$, then either $p \mid a$ or $p \mid b$.

Proof. We shall first show that (UFD 1) and (UFD 2) imply (UFD 3). Let $c \in A^*$ be such that $pc = ab$. If a is a unit, then $pca^{-1} = b$, so $p \mid b$, and similarly, if b is a unit, then $p \mid a$. Therefore, by (UFD 1) we may assume that there exist sequences $(a_k)_{1 \leq k \leq m}$ and $(b_k)_{1 \leq k \leq n}$ of irreducible elements whose products are respectively a and b. Consequently, $a_1, \ldots, a_m, b_1, \ldots, b_n$ is a sequence of irreducible elements whose product is pc. If c is a unit, then pc is irreducible, so by (UFD 2) pc, and hence also p, are associates of one of $a_1, \ldots, a_m, b_1, \ldots, b_n$, and consequently, p divides either a or b; in the contrary case, there is a sequence $(c_k)_{1 \leq k \leq r}$ of irreducible elements whose product is c by (UFD 1), so $p, c_1, \ldots, c_r$ is a sequence of irreducible elements whose product is pc, and consequently, by (UFD 2) p is an associate of one of $a_1, \ldots, a_m, b_1, \ldots, b_n$ and hence divides either a or b.

Next we shall show that (UFD 1) and (UFD 3) imply (UFD 2). First, we note that by an easy inductive argument, we obtain from (UFD 3) the following assertion:

(UFD 3')   If p is an irreducible element of A, then for every sequence $(a_k)_{1 \leq k \leq m}$ of elements of A^*, if $p \mid a_1 \ldots a_m$, then there exists $i \in [1, m]$ such that $p \mid a_i$.

Let $(p_i)_{1 \le i \le n}$ and $(q_j)_{1 \le j \le m}$ be sequences of irreducible elements such that (4) holds, and let S be the set of all $k \in [1, n]$ for which there is an injection ρ from $[1, k]$ into $[1, m]$ such that p_i and $q_{\rho(i)}$ are associates for each $i \in [1, k]$. Since p_1 divides $\prod_{j=1}^{m} q_j$, by (UFD 3′) there exists $r \in [1, m]$ such that p_1 divides and hence is an associate of q_r. Consequently, $1 \in S$. Suppose that $s \in S$ and that $s < n$. Then there exist an injection σ from $[1, s]$ into $[1, m]$ and, for each $i \in [1, s]$, a unit u_i such that $p_i = u_i q_{\sigma(i)}$. Consequently, $u = \prod_{i=1}^{n} u_i$ is a unit, and

$$\prod_{i=1}^{s} p_i = u \left(\prod_{i=1}^{s} q_{\sigma(i)} \right). \tag{5}$$

If σ were surjective, then from (4) and (5) we would obtain

$$u \left(\prod_{i=1}^{s} p_i \right) \left(\prod_{i=s+1}^{n} p_i \right) = u \left(\prod_{j=1}^{m} q_j \right) = u \left(\prod_{i=1}^{s} q_{\sigma(i)} \right) = \prod_{i=1}^{s} p_i,$$

whence

$$u \left(\prod_{i=s+1}^{n} p_i \right) = 1,$$

and therefore, p_{s+1} would be a unit, a contradiction. Hence, the complement J of the range of σ in $[1, m]$ is not empty. From (4) and (5) we obtain

$$u \left(\prod_{i=1}^{s} p_i \right) \left(\prod_{i=s+1}^{n} p_i \right) = u \left(\prod_{j=1}^{m} q_j \right) = u \left(\prod_{i=1}^{s} q_{\sigma(i)} \right) \left(\prod_{j \in J} q_j \right)$$

$$= \left(\prod_{i=1}^{s} p_i \right) \left(\prod_{j \in J} q_j \right),$$

whence

$$u \left(\prod_{i=s+1}^{n} p_i \right) = \prod_{j \in J} q_j. \tag{6}$$

Therefore, by (6), p_{s+1} divides $\prod_{j \in J} q_j$, so by (UFD 3′) there exists $t \in J$ such that p_{s+1} divides and hence is an associate of q_t. The function τ from $[1, s + 1]$ into $[1, m]$ defined by

$$\tau(i) = \begin{cases} \sigma(i) \text{ for all } i \in [1, s], \\ t \text{ if } i = s + 1 \end{cases}$$

is then an injection, and p_i is an associate of $q_{\tau(i)}$ for each $i \in [1, s + 1]$. Therefore, $s + 1 \in S$. By induction, $n \in S$, so there is an injection φ from $[1, n]$ into $[1, m]$ such that p_i and $q_{\varphi(i)}$ are associates for each $i \in [1, n]$. Let v_i be the unit such that $q_{\varphi(i)} = v_i p_i$ for each $i \in [1, n]$, and let $v = \prod\limits_{i=1}^{n} v_i$. If the complement L in $[1, m]$ of the range of φ were not empty, we would have

$$\left(\prod_{i=1}^{n} q_{\varphi(i)} \right) = v \left(\prod_{i=1}^{n} p_i \right) = v \left(\prod_{j=1}^{m} q_j \right) = v \left(\prod_{i=1}^{n} q_{\varphi(i)} \right) \left(\prod_{k \in L} q_k \right),$$

whence upon cancelling $\prod\limits_{i=1}^{n} q_{\varphi(i)}$ we would obtain

$$1 = v \left(\prod_{k \in L} q_k \right) ;$$

consequently, q_k would be a unit for each $k \in L$, a contradiction. Thus, $L = \emptyset$, so φ is surjective and hence is a permutation of $[1, n]$.

We may now identify an important class of unique factorization domains:

Theorem 22.9. Every principal ideal domain is a unique factorization domain.

Proof. By Theorems 22.7 and 22.8, we need only show that if A is a principal ideal domain, then A satisfies (UFD 1). For this, we shall first show that a nonzero element a of A that is not a unit is divisible by an irreducible element. Let $\mathfrak{A}$ be the set of all proper ideals of A containing (a). As a is not a unit, $(a) \in \mathfrak{A}$, and therefore, $\mathfrak{A}$ is not empty. By (N) there exists $p \in A$ such that (p) is a maximal element of $\mathfrak{A}$, ordered by $\subseteq$. But then (p) is a maximal ideal of A, for a proper ideal of A containing (p) belongs to $\mathfrak{A}$ and hence is (p). Consequently, p is irreducible by Theorem 22.7, and $p \mid a$ by (1).

Finally, we shall show that if a nonzero element b is not a unit, then b is a product of irreducible elements. Let Y be the set of all $y \in A^*$ such that $sy = b$ for some product s of irreducible elements, and let

$$\mathfrak{B} = \{(y) \colon y \in Y\}.$$

Since b is divisible by an irreducible element, $\mathfrak{B}$ is not empty and hence possesses a maximal element (u). Let $p_1, \ldots, p_n$ be irreducible elements such that $p_1 \ldots p_n u = b$. If u were not a unit, there would exist an irreducible element q such that $q \mid u$ by what we have just proved. If $qv = u$, then $p_1 \ldots p_n q v = b$, so $(v) \in \mathfrak{B}$ and $(u) \subset (v)$ as v divides but is not an associate of

u, a contradiction of the maximality of (u). Hence, u is a unit, so b is the product of the irreducible elements $p_1u, p_2, \ldots, p_n$.

The following lemmas lead to the theorem that any finite sequence of nonzero elements of a unique factorization domain has a greatest common divisor and a least common multiple.

Lemma 22.1. Let A be a unique factorization domain, and let $a \in A^*$. If $(p_k)_{1 \leq k \leq n}$ is a sequence of irreducible elements of A no two members of which are associates such that every irreducible divisor of a is an associate of some p_k, then there exist a unit u and a sequence $(r_k)_{1 \leq k \leq n}$ of natural numbers such that

$$a = u \prod_{k=1}^{n} p_k^{r_k}. \tag{7}$$

Proof. If a is a unit, we may let $u = a$ and $r_k = 0$ for all $k \in [1, n]$. Therefore, we may assume that a is not a unit. By (UFD 1), there is a sequence $(q_j)_{1 \leq j \leq m}$ of irreducible elements such that $a = q_1 \ldots q_m$. By hypothesis, for each $j \in [1, m]$, there is a unique $k \in [1, n]$ such that $q_j = u_j p_k$ for some unit u_j. Consequently, (7) holds where $u = u_1 \ldots u_m$ and where r_k is the number of integers $j \in [1, m]$ such that q_j is an associate of p_k.

Lemma 22.2. Let a and b be nonzero elements of a unique factorization domain A. If

$$a = u \prod_{k=1}^{n} p_k^{r_k} \tag{7}$$

where u is a unit, $(p_k)_{1 \leq k \leq n}$ is a sequence of irreducible elements no two of which are associates, and $(r_k)_{1 \leq k \leq n}$ is a sequence of natural numbers, then $b \mid a$ if and only if there exist a unit v and a sequence $(s_k)_{1 \leq k \leq n}$ of natural numbers such that

$$b = v \prod_{k=1}^{n} p_k^{s_k}, \tag{8}$$

$$s_k \leq r_k \text{ for all } k \in [1, n].$$

Proof. The condition is clearly sufficient. Necessity: Every irreducible divisor of a is an associate of some p_k by (UFD 3'). Therefore, as every irreducible divisor of b is also an irreducible divisor of a, by Lemma 22.1 there exist a unit v and a sequence $(s_k)_{1 \leq k \leq n}$ of natural numbers satisfying (8). If $s_j > r_j$ for some $j \in [1, n]$, then

$$p_j^{s_j} \, \bigg| \, \prod_{k=1}^{n} p_k^{r_k},$$

so

$$p_j^{s_j-r_j} \left| \left(\prod_{k=1}^{j-1} p_k^{r_k} \right) \left(\prod_{k=j+1}^{n} p_k^{r_k} \right), \right.$$

whence by (UFD 3′) p_j would divide and hence be an associate of one of $p_1, \ldots, p_{j-1}, p_{j+1}, \ldots, p_n$, in contradiction to our hypothesis. Therefore, $s_k \le r_k$ for all $k \in [1, n]$.

Lemma 22.3. If A is a unique factorization domain and if

$$a = u \prod_{k=1}^{n} p_k^{r_k}, \qquad b = v \prod_{k=1}^{n} p_k^{s_k}$$

where $(p_k)_{1 \le k \le n}$ is a sequence of irreducible elements, no two of which are associates, where u and v are units, and where $(r_k)_{1 \le k \le n}$ and $(s_k)_{1 \le k \le n}$ are sequences of natural numbers, then the elements m and d defined by

$$m = \prod_{k=1}^{n} p_k^{\max\{r_k, s_k\}}, \qquad d = \prod_{k=1}^{n} p_k^{\min\{r_k, s_k\}}$$

are respectively a least common multiple and a greatest common divisor of a and b.

The assertion follows easily from Lemma 22.2.

Theorem 22.10. If $a_1, \ldots, a_h$ is a sequence of nonzero elements of a unique factorization domain A, then there exist a sequence $(p_k)_{1 \le k \le n}$ of irreducible elements, no two of which are associates, and for each $j \in [1, h]$ a unit u_j and a sequence $(r_{jk})_{1 \le k \le n}$ of natural numbers such that

$$a_j = u_j \prod_{k=1}^{n} p_k^{r_{jk}} \tag{9}$$

for each $j \in [1, h]$, and the elements m and d defined by

$$m = \prod_{k=1}^{n} p_k^{\max\{r_{1k}, \ldots, r_{hk}\}}, \qquad d = \prod_{k=1}^{n} p_k^{\min\{r_{1k}, \ldots, r_{hk}\}}$$

are respectively a least common multiple and a greatest common divisor of $a_1, \ldots, a_h$.

Proof. By (UFD 1),

$$\prod_{i=1}^{h} a_i = \prod_{j=1}^{t} q_j$$

where $q_1, \ldots, q_t$ are irreducible elements. Let $p_1, \ldots, p_n$ be chosen from $q_1, \ldots, q_t$ so that p_i and p_j are not associates if $i \neq j$, but each q_i is an associate of some p_k. By Lemma 22.1, there exist for each $j \in [1, h]$ a unit u_j and a sequence $(r_{jk})_{1 \leq k \leq n}$ of natural numbers such that (9) holds. An inductive argument based on Lemma 22.3 then establishes the final assertion.

Theorem 22.11. Let a and b be relatively prime elements of a unique factorization domain A, and let $c \in A$.

1° If $a \mid bc$, then $a \mid c$.
2° If $a \mid c$ and $b \mid c$, then $ab \mid c$.

Proof. By Theorem 22.10, a and b have a least common multiple, and hence ab is a least common multiple by Theorem 22.3. If $a \mid bc$, then bc is a common multiple of a and b, so $ab \mid bc$, whence $a \mid c$. If $a \mid c$ and $b \mid c$, then c is a common multiple of a and b, so $ab \mid c$.

The following notational convention is frequently convenient: If $\triangle$ is an associative, commutative composition on E for which there is a neutral element e and if $(x_\alpha)_{\alpha \in A}$ is a family of elements of E indexed by a (possibly infinite) set A, where $x_\alpha = e$ for all but finitely many indices α, then $\underset{\alpha \in A}{\triangle} x_\alpha$ is defined as being $\underset{\alpha \in B}{\triangle} x_\alpha$ where $B = \{\alpha \in A : x_\alpha \neq e\}$ if $B \neq \emptyset$, and $\underset{\alpha \in A}{\triangle} x_\alpha$ is defined as being e if B is empty.

If P is a representative system of irreducible elements of a unique factorization domain A, then for each $a \in A^*$ there exist a unique unit u and a unique family $(n_p)_{p \in P}$ of natural numbers such that $n_p = 0$ for all but finitely many $p \in P$ and

$$a = u \prod_{p \in P} p^{n_p}.$$

Indeed, the existence is an immediate consequence of Lemma 22.1, and the uniqueness of u and $(n_p)_{p \in P}$ follows from (UFD 2). From Theorems 19.8 and 22.9, therefore, we obtain the following theorem, since 1 and -1 are the only units of Z:

Theorem 22.12. (Fundamental Theorem of Arithmetic) Let P be the set of prime numbers. For each $a \in N^*$ there exists a unique family $(n_p)_{p \in P}$ of natural numbers such that $n_p = 0$ for all but finitely many $p \in P$ and

$$a = \prod_{p \in P} p^{n_p}.$$

EXERCISES

22.1. (a) If a and b are nonzero elements of a commutative ring with identity A, then a and b are associates if and only if $(a) = (b)$.
(b) Show the correctness of the statement obtained from Corollary 22.1.1 by replacing "an integral domain" with "a commutative ring with identity."

22.2. If A is a principal ideal domain and if (m) is a proper ideal of A, then every ideal of $A/(m)$ is principal. [Use Exercise 21.5(b).]

22.3. Let (m) be a proper nonzero ideal of a principal ideal domain A, and for each $x \in A$, let $\bar{x}$ denote the coset $x + (m)$.
(a) If $a \in A^*$ and if d is a greatest common divisor of a and m, then $(\bar{d}) = (\bar{a})$.
(b) If c and d are divisors of m such that $(\bar{c}) = (\bar{d})$, then c and d are associates.
(c) Let D be a set of divisors of m such that each divisor of m is an associate of one and only one element of D. Then $d \mapsto (\bar{d})$ is a bijection from D onto the set of all ideals of $A/(m)$. [Use Exercise 22.2.]
(d) If $a, b \notin (m)$, then $\bar{a}$ and $\bar{b}$ are associates in $A/(m)$ if and only if there is a unit $\bar{u}$ of $A/(m)$ such that $\overline{au} = \bar{b}$.

22.4. Let A be a ring. For each $a \in A$, let $\langle a \rangle$ be the composition on A defined by

$$x \langle a \rangle y = xay$$

for all $x, y \in A$.
(a) For each $a \in A$, $(A, +, \langle a \rangle)$ is a ring.
(b) If A is a commutative ring with identity and if a and b are associates, then $(A, +, \langle a \rangle)$ and $(A, +, \langle b \rangle)$ are isomorphic.

22.5. Let $m > 1$, and for each $x \in Z$, let $\bar{x}$ denote the coset $x + (m)$.
(a) The left regular representation of the ring Z_m (Exercise 21.6) is an isomorphism from Z_m onto $\text{End}(Z_m)$.
(b) If $\circ$ is a composition on Z_m distributive over addition modulo m, then relative to ordinary multiplication modulo m on Z_m, $\circ$ is $\langle \bar{r} \rangle$ (Exercise 22.4) for some $r \in Z$.
(c) For all $\bar{r}, \bar{s} \in Z_m$, the rings $(Z_m, +, \langle \bar{r} \rangle)$ and $(Z_m, +, \langle \bar{s} \rangle)$ are isomorphic if and only if $\bar{r}$ and $\bar{s}$ are associates in the ring Z_m.
(d) Let D_m be the set of all positive divisors of m. If A is a ring whose additive group is isomorphic to the additive group of integers modulo m, then there is one and only one $r \in D_m$ such that A is isomorphic to $(Z_m, +, \langle \bar{r} \rangle)$. [Use Exercise 22.3.]

22.6. If $\mathfrak{p}$ is a nonzero ideal of a principal ideal domain A, then $\mathfrak{p}$ is a prime ideal of A (Exercise 19.6) if and only if $\mathfrak{p}$ is a maximal ideal. [Use Theorem 22.7.]

22.7. Let a and b be nonzero elements of a commutative ring with identity A.
(a) The element a is irreducible if and only if (a) is a maximal element of the set of all proper principal ideals of A, ordered by $\subseteq$.
(b) There is a greatest common divisor of a and b if and only if the set of all principal ideals of A containing $(a) + (b)$, ordered by $\subseteq$, possesses a smallest member (i.e., a member contained in every other member).
(c) There is a least common multiple of a and b if and only if $(a) \cap (b)$ is a principal ideal.

22.8. Let A be an integral domain, and let $a_1, \ldots, a_n \in A^*$.
(a) If b divides each of $a_1, \ldots, a_n$, then b divides $a_1 + \ldots + a_n$.
(b) If d is a greatest common divisor of $a_1, \ldots, a_n$ and if $dc_i = a_i$ for each $i \in [1, n]$, then $c_1, \ldots, c_n$ are relatively prime.
(c) Let K be a quotient field of A. If every pair of elements of A^* admits a greatest common divisor, then for every $x \in K^*$ there exist relatively prime elements a and b of A^* such that $x = a/b$.

22.9. Let a and b be positive integers.
(a) If $a = bq + r$ where $0 \le r < b$, then a greatest common divisor of a and b is also a greatest common divisor of b and r.
(b) If $(r_k)_{0 \le k \le n+1}$ and $(q_k)_{1 \le k \le n}$ are sequences of positive integers such that

$$r_0 = a, \qquad r_1 = b,$$

$$r_n \ne 0, \qquad r_{n+1} = 0,$$

$$r_{k-1} = r_k q_k + r_{k+1},$$

$$0 \le r_{k+1} < r_k$$

for all $k \in [1, n]$, then r_n is a greatest common divisor of a and b.
(c) Use (b) and Theorem 22.3 to calculate the positive greatest common divisor and least common multiple of the following pairs of integers:

$$143, 117;$$

$$1241, 2336;$$

$$4224, 10692.$$

*22.10. Let A be an integral domain such that every pair of nonzero elements of A admits a greatest common divisor. We shall denote a greatest common divisor of a and b by (a, b), and we shall write $a \sim b$ if a and b are associates. For all $a, b, c \in A^*$, the following three conditions hold:

1° $(a, (b, c)) \sim ((a, b), c)$.
2° $(ca, cb) \sim c(a, b)$.
3° If $(a, b) \sim 1$ and $(a, c) \sim 1$, then $(a, bc) \sim 1$.

[For 3°, use 2° to make a substitution for c in $(a, c) \sim 1$, and apply 1°.] Infer that A satisfies (UFD 3).

22.11. If A is an integral domain satisfying (UFD 3), then for every $p \in A$, (p) is a prime ideal of A (Exercise 19.6) if and only if p is either zero or irreducible.

22.12. If A is a unique factorization domain, for each $a \in A^*$ the **length** $\lambda(a)$ of a is defined as follows: if a is a unit, then $\lambda(a) = 0$; if $a = p_1 \ldots p_n$ where $(p_k)_{1 \le k \le n}$ is a sequence of (not necessarily distinct) irreducible elements, then $\lambda(a) = n$. (By (UFD 2), $\lambda(a)$ is well-defined.)
(a) If λ is the length function of a principal ideal domain A, then for all $a, b \in A^*$ the following three conditions hold:

$1°$ If $b \mid a$, then $\lambda(b) \le \lambda(a)$.
$2°$ If $b \mid a$ and if $\lambda(b) = \lambda(a)$, then $a \mid b$.
$3°$ If $b \nmid a$ and $a \nmid b$, then there exist $p, q \in A^*$ such that $pa + qb \ne 0$ and $\lambda(pa + qb) < \min\{\lambda(a), \lambda(b)\}$.

(b) If A is an integral domain and if λ is a function from A^* into N satisfying $1°$, $2°$, and $3°$ of (a) for all $a, b \in A^*$, then A is a principal ideal domain. [Generalize the proofs of Theorems 22.4 and 19.8.]

22.13. Let A be an integral domain. A **euclidean stathm** on A is a function λ from A^* into N satisfying the following two conditions for all $a, b \in A^*$:

$1°$ If $b \mid a$, then $\lambda(b) \le \lambda(a)$.
$2°$ There exist $q, r \in A$ such that $a = bq + r$ and either $r = 0$ or $\lambda(r) < \lambda(b)$.

The integral domain A is a **euclidean domain** if there exists a euclidean stathm on A.
(a) If λ is a euclidean stathm on A, then λ satisfies conditions $1°$, $2°$ and $3°$ of Exercise 22.12(a), and hence, A is a principal ideal domain.
(b) If λ is a euclidean stathm on A, then a nonzero element x of A is a unit if and only if $\lambda(x) = \lambda(1)$.
(c) The function $\lambda: n \mapsto |n|$, $n \in Z^*$, is a euclidean stathm on Z.
(d) If λ is a euclidean stathm on A and if f is a strictly increasing function from N into N, then $f \circ \lambda$ is also a euclidean stathm on A.

22.14. Let A be an integral domain such that every nonempty set of principal ideals of A, ordered by $\subseteq$, possesses a maximal element, and let a be a nonzero, noninvertible element of A.
(a) There exists an irreducible element of A dividing a. [Modify the first paragraph of the proof of Theorem 22.9.]
(b) The element a is a product of irreducible elements. [Modify the second paragraph of the proof of Theorem 22.9.]

22.15. If A is an integral domain and if N is a function from A^* into N^* satisfying

$1°$ $N(ab) = N(a)N(b)$,
$2°$ $N(a) = 1$ if and only if a is a unit

for all $a, b \in A^*$, then A satisfies (UFD 1). [Use Exercise 22.14.]

*22.16. Let m be either the integer 1 or a positive integer that is not a square, and let $Z[i\sqrt{m}] = \{a + bi\sqrt{m}: a, b \in Z\}$. Let N be the function defined by

$$N(a + bi\sqrt{m}) = a^2 + mb^2$$

for all nonzero elements $a + bi\sqrt{m}$ of $Z[i\sqrt{m}]$.

(a) The function N satisfies $1°$ and $2°$ of Exercise 22.15, and hence, $Z[i\sqrt{m}]$ satisfies (UFD 1).

(b) If $m = 1$, then N is a euclidean stathm. (Elements of $Z[i]$ are called **gaussian integers**.) [If z and w are nonzero elements of $Z[i]$, let $z/w = x + iy$, and consider $b = u + iv$ where u and v are integers satisfying $|u - x| \le \frac{1}{2}$, $|v - y| \le \frac{1}{2}$.]

(c) The numbers 3, $2 + i\sqrt{5}$, $2 - i\sqrt{5}$ are irreducible elements of $Z[i\sqrt{5}]$. [If z is a divisor, then $N(z)$ divides 9.] Infer that $Z[i\sqrt{5}]$ satisfies (UFD 1) but not (UFD 2).

*22.17. If $\mathfrak{p}$ is an ideal of a unique factorization domain A, then $\mathfrak{p} = (p)$ for some irreducible element p of A if and only if $\mathfrak{p}$ is a minimal member of the set of all nonzero prime ideals of A (Exercise 19.6), ordered by $\subseteq$. [Let p be an element of $\mathfrak{p}$ of smallest possible length, and use Exercise 22.11.]

22.18. Let A be a unique factorization domain.

(a) If a is a nonzero element of A of length n, there are at most 2^n principal ideals of A containing a.

(b) If $\mathfrak{a}$ is a nonzero ideal of A, there are only a finite number of principal ideals containing $\mathfrak{a}$.

(c) Every nonempty set of principal ideals of A, ordered by $\subseteq$, contains a maximal element.

*22.19. An integral domain A is a principal ideal domain if and only if the following three conditions hold:

$1°$ A is a unique factorization domain.

$2°$ Every nonzero prime ideal of A (Exercise 19.6) is a maximal ideal.

$3°$ Every proper ideal of A is contained in a maximal ideal.

[To prove (PID), first show that every prime ideal of A is principal by use of $2°$ and Exercise 22.17. If $\mathfrak{a}$ is an arbitrary nonzero ideal of A, use Exercise 22.18(b) to show that there is a minimal member (c) of the set of principal ideals of A containing $\mathfrak{a}$, ordered by $\subseteq$; if $\mathfrak{b}$ is the set of all $x \in A$ such that $xc \in \mathfrak{a}$, then $\mathfrak{a} = \mathfrak{b}c$; if $\mathfrak{b} \ne A$, apply $3°$ to show that $\mathfrak{b} \subseteq (d)$ where d is not a unit.]

22.20. (a) If P is a finite representative set of irreducible elements of a unique factorization domain A, then $(\prod_{p \in P} p) + 1$ is a unit.

(b) There are infinitely many prime numbers in Z.

*22.21. A **noetherian** ring is a commutative ring with identity satisfying condition (N) (page 198).

(a) Every ideal of a noetherian ring is a sum of principal ideals.

(b) If A is an integral domain, then A is a principal ideal domain if and only if A is noetherian and the sum $\mathfrak{a} + \mathfrak{b}$ of any two principal ideals $\mathfrak{a}$ and $\mathfrak{b}$ of A is a principal ideal.

(c) If A is an integral domain, then A is a principal ideal domain if and only if A is noetherian and for all $a, b \in A^*$ there exist $s, t \in A$ such that $sa + tb$ is a greatest common divisor of a and b.

*22.22. Let A be a principal ideal domain, let P be a representative set of irreducible elements of A, and let K be a quotient field of A. For each subset S of P, let A_S be the set of all $m/n \in K$ such that $m \in A$ and n is either 1 or a product of elements of S.

(a) A_S is a subdomain of K containing A.

(b) If m and n are relatively prime elements of A such that $m/n \in A_S$, then $1/n \in A_S$.

(c) If $\mathfrak{a}$ is an ideal of A_S and if $\mathfrak{a} \cap A$ is the principal ideal Ab of A, then $\mathfrak{a}$ is the principal ideal $A_S b$ of A_S.

(d) The integral domain A_S is a principal ideal domain, and the complement $P - S$ of S in P is a representative set of irreducible elements of A_S.

(e) If D is a subdomain of K containing A, then $D = A_S$ for some subset S of P.

(f) The function $S \mapsto A_S$ is a bijection from $\mathfrak{P}(P)$ onto the set of all subdomains of K containing A. In particular, if P has n elements, there are 2^n subdomains of K containing A.

(g) There is a bijection from the set of all subdomains of Q onto $\mathfrak{P}(N)$, and no two subdomains of Q are isomorphic (Exercise 21.10).

CHAPTER IV

FINITE GROUPS

23. Cyclic Groups

Let A be a subset of a group $(G, \cdot)$, and let H be the intersection of all the subgroups of G that contain A. Since the neutral element e of G belongs to each subgroup of G, $e \in H$. If $x, y \in H$, then xy^{-1} belongs to each subgroup of G containing A by Theorem 8.3, whence $xy^{-1} \in H$. Thus, H is a subgroup of G by Theorem 8.3. By its very definition, H is the smallest subgroup of G that contains A. Hence, we have proved the following theorem:

Theorem 23.1. If A is a subset of a group G, then of all the subgroups of G that contain A there is a smallest one, namely, the intersection of all the subgroups of G that contain A.

Definition. Let A be a subset of a group G. The smallest subgroup H of G that contains A is called the **subgroup generated** by A, and A is called a **set of generators** or a **generating set** for H. If A is a finite set $\{a_1, \ldots, a_n\}$, the subgroup H generated by A is also said to be generated by $a_1, \ldots, a_n$, and the elements $a_1, \ldots, a_n$ are called **generators** of H.

The central problem of group theory is to describe all groups in some concrete fashion. The general problem is best attacked, however, by considering only a part of it at a time, that is, by seeking to describe in some way all groups belonging to some prescribed class. One might seek to describe, for example, all finite abelian groups, or all finite groups having a certain prescribed number of elements. In this section we shall see how the properties of the *ring* Z lead to a complete description of one very important class of groups—those generated by a single element.

Definition. A **cyclic** group is a group that is generated by one of its elements.

If a is an element of a group G, the subgroup of G generated by a will be denoted by $[a]$ in this chapter.

Theorem 23.2. If a is an element of a group $(G, \cdot)$, then

$$[a] = \{a^n : n \in \mathbf{Z}\},$$

and the function g from $\mathbf{Z}$ into $[a]$ defined by

$$g(n) = a^n$$

for all $n \in \mathbf{Z}$ is an epimorphism from $(\mathbf{Z}, +)$ onto the subgroup $[a]$.

Proof. An easy inductive argument establishes that for each $m \in \mathbf{N}$, a^m belongs to any subgroup containing a, and in particular, $a^m \in [a]$. Since for each $m \in \mathbf{N}^*$, $a^{-m} = (a^m)^{-1}$ by Theorem 14.8, a^{-m} also belongs to any subgroup of G containing a. Thus, $[a] \supseteq \{a^m : m \in \mathbf{Z}\}$. But since $a^n a^m = a^{n+m}$ and $(a^n)^{-1} = a^{-n}$ for all $n, m \in \mathbf{Z}$ by Theorem 14.8, $\{a^m : m \in \mathbf{Z}\}$ is a subgroup by Theorem 8.3, and it clearly contains a. Hence, $[a] \subseteq \{a^m : m \in \mathbf{Z}\}$, and also g is an epimorphism.

If the composition of G is denoted additively, of course,

$$[a] = \{n.a : n \in \mathbf{Z}\}.$$

Theorem 23.3. The group $(\mathbf{Z}, +)$ is cyclic. Every subgroup of $(\mathbf{Z}, +)$ is an ideal of the ring $\mathbf{Z}$.

Proof. For every $n \in \mathbf{Z}$,

$$n = n.1 \in [1]$$

by Theorems 23.2 and 14.8, so 1 is a generator of $(\mathbf{Z}, +)$. Let H be a subgroup of $(\mathbf{Z}, +)$. If $n \in \mathbf{Z}$ and if $h \in H$, then

$$nh = n.h \in [h] \subseteq H$$

by Theorems 23.2 and 14.8. Hence, H is an ideal of the ring $\mathbf{Z}$.

It is easy to see that -1 is also a generator of $(\mathbf{Z}, +)$ and that 1 and -1 are the only generators of $(\mathbf{Z}, +)$.

Theorem 23.4. A cyclic group is abelian. If a is a generator of a cyclic group $(G, \cdot)$ and if H is a subgroup of G, then G/H is a cyclic group, and aH is a generator of G/H.

Proof. The first assertion is a consequence of Theorem 23.2 and the corollary of Theorem 20.4. Since $G = \{a^n : n \in Z\}$ and since $\varphi_H : x \mapsto xH$ is an epimorphism from G onto G/H,

$$G/H = \{\varphi_H(a^n) : n \in Z\} = \{\varphi_H(a)^n : n \in Z\} = \{(aH)^n : n \in Z\}$$

by the Corollary of Theorem 20.5, whence G/H is cyclic with generator aH by Theorem 23.2.

Theorem 23.5. For every $m \in N$, the group $(Z_m, +)$ is a cyclic group, and every subgroup of $(Z_m, +)$ is an ideal of the ring Z_m.

Proof. The first assertion follows from Theorems 23.3 and 23.4. Let H be a subgroup of $(Z_m, +)$, and let φ_m be the canonical epimorphism from Z onto Z_m. If $h + (m) \in H$ and if $n \in Z$, then by Theorem 14.8 and the Corollary of Theorem 20.5,

$$(n + (m))(h + (m)) = nh + (m) = \varphi_m(nh)$$
$$= \varphi_m(n.h) = n.\varphi_m(h),$$

an element of $[\varphi_m(h)]$ and thus of H by Theorem 23.2. Hence, H is an ideal of the ring Z_m.

We recall that the order of a finite group is the number of elements in it, and that an infinite group is said to have infinite order. If a is an element of a group and if $[a]$ is finite, the **order** of a is the order of the cyclic group $[a]$, but if $[a]$ is infinite, a is said to have **infinite order.**

Theorem 23.6. Let a be an element of a group $(G, \cdot)$, and let g be the function from Z into $[a]$ defined by

$$g(n) = a^n$$

for all $n \in Z$. If a has infinite order, then g is an isomorphism from $(Z, +)$ onto $[a]$. If a has finite order m, then g is an epimorphism from $(Z, +)$ onto $[a]$ whose kernel is (m), $[a]$ is, therefore, isomorphic to the additive group $(Z_m, +)$, and m is the smallest strictly positive integer such that a^m is the neutral element of G.

Proof. By Theorem 23.2, g is an epimorphism from $(Z, +)$ onto $[a]$. The kernel K of g is a subgroup of $(Z, +)$, and therefore, by Theorems 23.3 and

19.8, $K = (m)$ for some $m \in N$, whence $[a]$ is isomorphic to the additive group $(Z_m, +)$. By Theorem 19.9, for every $m \in N^*$, Z_m is finite and has m elements. Hence, if the order of a is finite and if $[a]$ is isomorphic to $(Z_m, +)$, then m is the order of a, and furthermore, m is the smallest strictly positive integer such that a^m is the neutral element of G since m is the smallest strictly positive integer in (m); if the order of a is infinite, then $m = 0$, so g is an isomorphism from $(Z, +)$ onto $[a]$.

Theorem 23.7. If a is an element of a group $(G, \cdot)$ and if $n \in Z^*$, then $a^n = e$, the neutral element of G, if and only if the order of a is finite and divides n.

Proof. If $a^n = e$, then a has finite order by Theorem 23.6. Let m be the order of a. By Theorem 23.6, $a^n = e$ if and only if $n \in (m)$, or equivalently, if and only if $m \mid n$.

Theorem 23.8. If $(G, \cdot)$ is a finite group of order n and if $a \in G$, then the order of a divides n, and $a^n = e$, the neutral element of G.

Proof. The first assertion is a consequence of Lagrange's theorem, and the second follows from Theorem 23.7.

Corollary. A finite group whose order is a prime number is cyclic, and each of its elements other than the neutral element is a generator.

We investigate next subgroups of cyclic groups.

Theorem 23.9. Let a be a generator of a finite cyclic group $(G, \cdot)$ of order n. If m is a positive divisor of n, then $[a^{n/m}]$ is a subgroup of G of order m, and moreover, $[a^{n/m}]$ is the only subgroup of order m. Thus,

$$F: m \mapsto [a^{n/m}]$$

is a bijection from the set of all positive divisors of n onto the set of all subgroups of G.

Proof. By Theorem 23.6, n is the smallest strictly positive integer such that $a^n = 1$, and the order of $a^{n/m}$ is the smallest strictly positive integer s such that $(a^{n/m})^s = 1$. Consequently, if $1 \le r < m$, then $(n/m) \cdot r < n$, so $1 \ne a^{(n/m)r} = (a^{n/m})^r$. Therefore, since $(a^{n/m})^m = a^n = 1$, the order of $a^{n/m}$ is m. Let H be a subgroup of G of order m. If $a^t \in H$, then $a^{tm} = (a^t)^m = 1$ by Theorem 23.8, so $tm \in (n)$ by Theorem 23.6, whence $n \mid tm$. Consequently, $(n/m) \mid t$, so $a^t \in [a^{n/m}]$. Therefore, $H \subseteq [a^{n/m}]$, but as both H and $[a^{n/m}]$ have m elements, we conclude that $H = [a^{n/m}]$. By Lagrange's

Theorem, therefore, F is a bijection from the set of all positive divisors of n onto the set of all subgroups of G.

Corollary. Every subgroup of a cyclic group is cyclic.

Proof. The assertion for finite cyclic groups follows at once from Theorem 23.9. An infinite cyclic group is isomorphic to $(Z, +)$ by Theorem 23.6, so we need only prove that every subgroup of $(Z, +)$ is cyclic. If H is a subgroup of $(Z, +)$, then $H = (m)$ for some $m \in N$ by Theorems 23.3 and 19.8. But m is a generator of the subgroup (m) of $(Z, +)$, for if $n \in Z$, then

$$nm = n.m \in [m]$$

by Theorems 23.2 and 14.8.

We have seen that the only generators of the additive group of integers are 1 and -1; these are precisely the invertible elements of the ring of integers. Similarly, for the group of integers modulo m we have the following result:

Theorem 23.10. Let $m > 1$, and let $k \in Z^*$. The following statements are equivalent:

1° $k + (m)$ is a generator of the cyclic group $(Z_m, +)$.
2° $k + (m)$ is an invertible element of the ring Z_m.
3° k and m are relatively prime integers.

Proof. Let φ_m be the canonical epimorphism from Z onto Z_m, and let $s \in Z$. The assertions

(1) $s.[k + (m)] = 1 + (m)$,

(2) $[s + (m)][k + (m)] = 1 + (m)$,

(3) $sk + tm = 1$ for some $t \in Z$

are all equivalent. Indeed, if (1) holds, then

$$[s + (m)][k + (m)] = sk + (m) = \varphi_m(sk) = \varphi_m(s.k)$$
$$= s.\varphi_m(k) = s.[k + (m)] = 1 + (m)$$

by Theorem 14.8 and the Corollary of Theorem 20.5, so (2) holds; if (2) holds, then $sk + (m) = 1 + (m)$, so $1 - sk \in (m)$, whence $1 - sk = tm$ for some $t \in Z$, and thus (3) holds; if (3) holds, then $sk + (m) = sk +$

$tm + (m) = 1 + (m)$, so as before,

$$s.[k + (m)] = s.\varphi_m(k) = \varphi_m(s.k) = \varphi_m(sk)$$
$$= sk + (m) = 1 + (m),$$

whence (1) holds. Now $2°$ is equivalent to the assertion that (2) holds for some $s \in \mathbf{Z}$, and by Bezout's Identity, $3°$ is equivalent to the assertion that (3) holds for some $s \in \mathbf{Z}$. Condition $1°$ implies that (1) holds for some $s \in \mathbf{Z}$; conversely, if (1) holds for some $s \in \mathbf{Z}$, $1 + (m)$ belongs to the cyclic subgroup generated by $k + (m)$, so the smallest subgroup containing $1 + (m)$, namely, $\mathbf{Z}_m$, is contained in the cyclic subgroup generated by $k + (m)$, and thus, $k + (m)$ is a generator of $\mathbf{Z}_m$. Therefore, all three assertions are equivalent.

Definition. The **Euler function** is the function φ from N^* into N^* defined by

$$\varphi(m) = \text{the number of generators of a cyclic group of order } m.$$

Clearly, $\varphi(1) = 1$. If $m > 1$, $\varphi(m)$ is the number of invertible elements of the ring $\mathbf{Z}_m$, and $\varphi(m)$ is also the number of natural numbers $< m$ that are relatively prime to m by Theorems 23.10 and 19.9.

Theorem 23.11. If p is a prime and if $n > 0$, then $\varphi(p^n) = p^n - p^{n-1}$.

Proof. By Theorem 23.9, a cyclic subgroup G of order p^n has exactly one subgroup H of order p^{n-1}. Let $a \in G$; the order of a is p^m for some $m \in [0, n]$, and $[a]$ is the only subgroup of G of order p^m. If $m < n$, then H contains a unique subgroup of order p^m by Theorem 23.9, so that subgroup must be $[a]$, whence $a \in H$; conversely, if $a \in H$, then $p^m \mid p^{n-1}$ and hence $m < n$. Thus, H^c is the set of all generators of G, so $\varphi(p^n) = p^n - p^{n-1}$.

EXERCISES

23.1. If G and G' are finite cyclic groups and if a and a' are generators of G and G' respectively, then there exists an isomorphism f from G onto G' such that $f(a) = a'$ if and only if G and G' have the same order.

23.2. Let $m > 1$. With the notation of Exercise 21.6, $\text{Aut}(\mathbf{Z}_m) = \{L_a : a \text{ is an invertible element of the ring } \mathbf{Z}_m\}$, and hence, $\text{Aut}(\mathbf{Z}_m)$ is a group of order $\varphi(m)$. [Use Exercise 22.5(a).]

23.3. If A is an integral domain such that every subgroup of the group $(A, +)$ is a subring of the ring A, then A is isomorphic either to the ring Z or to the field Z_p for some prime p.

23.4. If A is a ring, the set of all elements of A having finite additive order is an ideal of A.

23.5. (a) (Fermat's Theorem) If a is an integer and if p is a prime not dividing a, then
$$a^{p-1} \equiv 1 \pmod{p},$$
and if b is any integer, then
$$b^p \equiv b \pmod{p}.$$

[Consider the field Z_p.]
(b) More generally, if $m > 1$ and if a is an integer relatively prime to m, then
$$a^{\varphi(m)} \equiv 1 \pmod{m}.$$

[Consider the ring Z_m.]

23.6. Let a and b be elements of finite order of a group $(G, \cdot)$, and let m and n be the orders of a and b respectively. If $ab = ba$ and if m and n are relatively prime, then the order of ab is mn. Is it true more generally that if $ab = ba$, then the positive least common multiple of m and n is the order of ab?

23.7. Let H be a subgroup of a finite cyclic group G.
(a) If α is an automorphism of G, then $\alpha_*(H) = H$.
(b) If for each $\alpha \in \text{Aut}(G)$ the function obtained by restricting the domain and codomain of α to H is denoted by α_H, then $\alpha \mapsto \alpha_H$ is a homomorphism from $\text{Aut}(G)$ into $\text{Aut}(H)$.

23.8. Let $(G, \cdot)$ be a group having more than one element.
(a) If G and $\{e\}$ are the only subgroups of G, then G is a cyclic group of prime order.
(b) If there are only a finite number of subgroups of G, then G is finite.

*23.9. If A is a nonzero ring whose only left ideals (Exercise 19.15) are A and $\{0\}$, then either A is a division ring, or A is a trivial ring whose additive group is cyclic of prime order. [Use Exercise 23.8. Suppose that a is an element such that $Aa \neq \{0\}$. What is Aa? Show that $\{x \in A : xa = 0\}$ is a left ideal. If $ea = a$, show successively that $e^2 = e$, that e is a right identity, that
$$\{x \in A : ex = 0\}$$
an ideal, and that e is the identity.]

*23.10. Construct the addition and multiplication tables for a field having four elements. [Show that the nonzero elements may be denoted by 1, a, and

a^2; to construct the addition table, determine first the characteristic of the field.]

*23.11. (a) A finite group of even order possesses an element of order 2. [See Exercise 19.12(b).]

(b) If $(G, \cdot)$ is a finite group, then for every $a \in G$ there exists $x \in G$ such that $x^2 = a$ if and only if the order of G is odd. [Use (a) and Theorem 23.8.]

*23.12. Let $(G, +)$ be a cyclic group of order $n = rq$, let H be the subgroup of G of order q, and let σ be an automorphism of H. We shall denote the positive greatest common divisor of integers a and b by (a, b).

(a) There is an automorphism α of G such that $\alpha_H = \sigma$ (Exercise 23.7(b)). [Show that $r = r_1 r_2$ where $(r_1, q) = 1$ and every prime dividing r_2 also divides q. If $\sigma(x) = s.x$ for all $x \in H$, use 3° of Theorem 22.5 to show that there exists k such that $(s + kq, n) = 1$.]

(b) There are $\dfrac{\varphi(n)}{\varphi(q)}$ automorphisms α of G such that $\alpha_H = \sigma$. [Use (a) and Exercises 23.2, 23.7.]

24. Direct Products

Let $(G_1, \cdot_1)$ and $(G_2, \cdot_2)$ be groups. We shall form a new group from these groups as follows: We define a composition $\cdot$ on $G_1 \times G_2$ by

$$(x_1, x_2) \cdot (y_1, y_2) = (x_1 \cdot_1 y_1, x_2 \cdot_2 y_2).$$

For ease of notation, we denote the various multiplications simply by juxtaposition. Multiplication on $G_1 \times G_2$ is associative, for

$$[(x_1, x_2)(y_1, y_2)](z_1, z_2) = (x_1 y_1, x_2 y_2)(z_1, z_2) = ((x_1 y_1)z_1, (x_2 y_2)z_2),$$

$$(x_1, x_2)[(y_1, y_2)(z_1, z_2)] = (x_1, x_2)(y_1 z_1, y_2 z_2) = (x_1(y_1 z_1), x_2(y_2 z_2)).$$

If e_1 and e_2 are respectively the neutral elements of G_1 and G_2, then (e_1, e_2) is the neutral element of $G_1 \times G_2$ since

$$(x_1, x_2)(e_1, e_2) = (x_1 e_1, x_2 e_2) = (x_1, x_2),$$

$$(e_1, e_2)(x_1, x_2) = (e_1 x_1, e_2 x_2) = (x_1, x_2).$$

Moreover, (x_1^{-1}, x_2^{-1}) is the inverse of (x_1, x_2) in $G_1 \times G_2$ since

$$(x_1, x_2)(x_1^{-1}, x_2^{-1}) = (x_1 x_1^{-1}, x_2 x_2^{-1}) = (e_1, e_2),$$

$$(x_1^{-1}, x_2^{-1})(x_1, x_2) = (x_1^{-1} x_1, x_2^{-1} x_2) = (e_1, e_2).$$

Thus, $G_1 \times G_2$ is a group, called the **cartesian product** of the groups G_1 and G_2. Furthermore, if G_1 and G_2 are commutative, then so is $G_1 \times G_2$, since

$$(x_1, x_2)(y_1, y_2) = (x_1y_1, x_2y_2) = (y_1x_1, y_2x_2) = (y_1, y_2)(x_1, x_2).$$

Of course, if the compositions of G_1 and G_2 are denoted by symbols similar to $+$, the composition of $G_1 \times G_2$ is also customarily denoted by $+$.

This method of forming new groups may be generalized in an obvious way. Let $(G_1, \cdot_1), \ldots, (G_n, \cdot_n)$ be a sequence of groups. We define a composition $\cdot$ on $G_1 \times \ldots \times G_n$ by

$$(x_1, x_2, \ldots, x_n) \cdot (y_1, y_2, \ldots, y_n) = (x_1y_1, x_2y_2, \ldots, x_ny_n).$$

Equipped with this composition, $G_1 \times \ldots \times G_n$ is a group, called the **cartesian product** of the groups $G_1, \ldots, G_n$. Indeed, multiplication is associative as before, the neutral element of $G_1 \times \ldots \times G_n$ is $(e_1, \ldots, e_n)$, where e_k is the neutral element of G_k for each $k \in [1, n]$, and the inverse of $(x_1, \ldots, x_n)$ is $(x_1^{-1}, \ldots, x_n^{-1})$. Furthermore, it is easy to verify as before that if $G_1, \ldots, G_n$ are commutative groups, then $G_1 \times \ldots \times G_n$ is also commutative.

The following theorem is easy to prove.

Theorem 24.1. If $G_1, \ldots, G_n, G_1', \ldots, G_n'$ are groups and if for each $k \in [1, n], f_k$ is an isomorphism from G_k onto G_k', then

$$f: (x_1, \ldots, x_n) \mapsto (f_1(x_1), \ldots, f_n(x_n))$$

is an isomorphism from the group $G_1 \times \ldots \times G_n$ onto the group $G_1' \times \ldots \times G_n'$.

Sometimes a group is isomorphic in a "natural" way to the cartesian product of certain of its subgroups. Before we indicate precisely what is meant by "natural," let us consider an example. The cyclic group Z_6 contains a subgroup H of order 2, namely, $\{0, 3\}$, and a subgroup K of order 3, namely, $\{0, 2, 4\}$. A simple calculation shows that the restriction of addition on Z_6 to $H \times K$ is an isomorphism from the group $H \times K$ onto Z_6, that is, that the function $A: (x, y) \mapsto x + y$ from $H \times K$ into Z_6 is an isomorphism. For example,

$$A((3, 2) + (0, 4)) = A((3, 0)) = 3 + 0 = 3,$$
$$A((3, 2)) + A((0, 4)) = 5 + 4 = 3.$$

Thus, Z_6 is isomorphic to $H \times K$ under a "natural" isomorphism, the restriction of its composition, addition, to $H \times K$.

Definition. A group $(G, \cdot)$ is the **direct product** of subgroups H and K if the function M from $H \times K$ into G defined by

$$M(x, y) = xy$$

for all $(x, y) \in H \times K$ is an isomorphism from the group $H \times K$ onto the group G.

If the composition of G is denoted by a symbol similar to $+$, we say that G is the **direct sum** rather than the direct product of H and K.

Theorem 24.2. If H and K are subgroups of a group $(G, \cdot)$, then G is the direct product of H and K if and only if the following three conditions hold:

1° Every element of H commutes with every element of K.
2° $HK = G$.
3° $H \cap K = \{e\}$, where e is the neutral element of G.

Proof. Let M be the function from $H \times K$ into G defined by $M(x, y) = xy$. We shall show that M is a homomorphism if and only if 1° holds. Indeed, if M is a homomorphism and if $x \in H$, $y \in K$, then as

$$(x, y) = (e, y)(x, e),$$

$$xy = M(x, y) = M(e, y)M(x, e) = eyxe = yx.$$

Conversely, if every element of H commutes with every element of K, then for all $x, x' \in H$, $y, y' \in K$,

$$M((x, y)(x', y')) = M(xx', yy') = xx'yy'$$
$$= xyx'y' = M(x, y)M(x', y').$$

If M is a homomorphism, then M is a monomorphism if and only if 3° holds. Indeed, if M is a monomorphism and if $z \in H \cap K$, then $z \in H$ and $z^{-1} \in K$, so $M(z, z^{-1}) = e$, whence $(z, z^{-1}) = (e, e)$ and thus, $z = e$. Conversely, if $H \cap K = \{e\}$ and if $M(x, y) = e$, then $xy = e$, so $y = x^{-1} \in H$ and hence $y \in H \cap K$, and consequently, $y = e$, $x = e$. Clearly, M is surjective if and only if 2° holds.

Theorem 24.3. If H and K are subgroups of a group $(G, \cdot)$, then G is the direct product of H and K if and only if the following three conditions hold:

1° H and K are normal subgroups of G.

2° $HK = G$.

3° $H \cap K = \{e\}$, where e is the neutral element of G.

Proof. By Theorem 24.2, it suffices to show that if 2° and 3° hold, then 1° holds if and only if each element of H commutes with each element of K. Necessity: Let $x \in H$, $y \in K$. As K is normal, $xyx^{-1} \in K$, so $xyx^{-1}y^{-1} \in K$; since H is normal and since $x^{-1} \in H$, $yx^{-1}y^{-1} \in H$, so $xyx^{-1}y^{-1} \in H$; hence, $xyx^{-1}y^{-1} \in H \cap K = \{e\}$, so $xy = yx$. Sufficiency: Let $a \in G$. By 2°, $a = hk$ for some $h \in H$, $k \in K$. If $x \in H$, then $axa^{-1} = hkxk^{-1}h^{-1} = hxkk^{-1}h^{-1} = hxh^{-1} \in H$, as x commutes with k; thus, H is normal. If $y \in K$, then $aya^{-1} = h(kyk^{-1})h^{-1} = (kyk^{-1})hh^{-1} = kyk^{-1} \in K$ as kyk^{-1} commutes with h; thus, K is normal.

If H and K are subgroups of a group $(G, \cdot)$, it is sometimes desirable to know when the subgroup generated by the set $H \cup K$ is the direct product of H and K. To identify that subgroup the following theorem is helpful.

Theorem 24.4. If H and K are subgroups of a group $(G, \cdot)$ and if H is a normal subgroup of G, then HK is a subgroup of G and is, moreover, the smallest subgroup of G containing H and K; if H and K are both normal subgroups of G, then HK is a normal subgroup of G.

Proof. If $x_1, x_2 \in H$, $y_1, y_2 \in K$, then $y_1 y_2^{-1} x_2^{-1} = z y_1 y_2^{-1}$ for some $z \in H$ as H is normal, so

$$(x_1 y_1)(x_2 y_2)^{-1} = x_1 y_1 y_2^{-1} x_2^{-1} = (x_1 z)(y_1 y_2^{-1}) \in HK.$$

Thus, by Theorem 8.3, HK is a subgroup that contains H and K, since $H = He \subseteq HK$, $K = eK \subseteq HK$. Any subgroup containing H and K clearly contains HK. Thus, HK is the smallest subgroup of G containing H and K. If H and K are both normal subgroups and if $a \in G$, then

$$a(HK) = (aH)K = (Ha)K = H(aK) = H(Ka) = (HK)a,$$

so HK is itself a normal subgroup.

An inductive argument establishes the following corollary:

Corollary. If $H_1, H_2, \ldots, H_n$ are normal subgroups of a group $(G, \cdot)$, then $H_1 H_2 \ldots H_n$ is a normal subgroup of G and is, moreover, the smallest subgroup of G containing $H_1, H_2, \ldots, H_n$.

It is easy to extend the definition of direct product:

Definition. A group $(G, \cdot)$ is the **direct product** (**direct sum** if the composition of G is denoted by a symbol similar to $+$) of the sequence $H_1, H_2, \ldots, H_n$ of subgroups of G if the function M from $H_1 \times H_2 \times \ldots \times H_n$ into G defined by

$$M(x_1, x_2, \ldots, x_n) = x_1 x_2 \ldots x_n$$

is an isomorphism from the group $H_1 \times H_2 \times \ldots \times H_n$ onto the group G.

If $n = 1$ and $H_1 = G$, then M is simply the identity automorphism of G; hence, G is the direct product of the sequence whose only term is G itself.

Necessary and sufficient conditions for a group to be the direct product of a sequence of its subgroups analogous to those of Theorem 24.3 may be given (Exercise 24.3), but we shall actually use only the following theorem.

Theorem 24.5. If a group $(G, \cdot)$ is the direct product of subgroups $H_1, H_2, \ldots,$ H_n, then $H_1, H_2, \ldots, H_n$ are normal subgroups of G.

Proof. By hypothesis, the function $M: (x_1, x_2, \ldots, x_n) \mapsto x_1 x_2 \ldots x_n$ from $H_1 \times H_2 \times \ldots \times H_n$ into G is an isomorphism. To show that H_i is normal, let $h \in H_i$, $a \in G$. Then $a = M(a_1, a_2, \ldots, a_n)$ where $a_k \in H_k$ for each $k \in [1, n]$. Hence,

$$
\begin{aligned}
aha^{-1} &= M(a_1, \ldots, a_n)M(e, \ldots, e, h, e, \ldots, e)M(a_1, \ldots, a_n)^{-1} \\
&= M(a_1, \ldots, a_n)M(e, \ldots, e, h, e, \ldots, e)M(a_1^{-1}, \ldots, a_n^{-1}) \\
&= M((a_1, \ldots, a_n)(e, \ldots, e, h, e, \ldots, e)(a_1^{-1}, \ldots, a_n^{-1})) \\
&= M(e, \ldots, e, a_i h a_i^{-1}, e, \ldots, e) = a_i h a_i^{-1} \in H_i
\end{aligned}
$$

as $a_i \in H_i$. Thus, H_i is normal.

Theorem 24.6. Let H and K be subgroups of a group $(G, \cdot)$. If H is the direct product of subgroups $H_1, \ldots, H_m$ and if K is the direct product of subgroups $K_1, \ldots, K_n$, then G is the direct product of H and K if and only if G is the direct product of $H_1, \ldots, H_m, K_1, \ldots, K_n$.

Proof. The function

$$J: (x_1, \ldots, x_m, y_1, \ldots, y_n) \mapsto ((x_1, \ldots, x_m), (y_1, \ldots, y_n))$$

is easily seen to be an isomorphism from the group $H_1 \times \ldots \times H_m \times K_1 \times \ldots \times K_n$ onto the group $(H_1 \times \ldots \times H_m) \times (K_1 \times \ldots \times K_n)$. By Theorem 24.1,

$$P: ((x_1, \ldots, x_m), (y_1, \ldots, y_n)) \mapsto (x_1 \ldots x_m, y_1 \ldots y_n)$$

is an isomorphism from $(H_1 \times \ldots \times H_m) \times (K_1 \times \ldots \times K_n)$ onto $H \times K$. Let M be the function from $H \times K$ into G defined by $M(x, y) = xy$. Then

$$(M \circ P \circ J)(x_1, \ldots, x_m, y_1, \ldots, y_n) = x_1 \ldots x_m y_1 \ldots y_n.$$

As P and J are isomorphisms, M is an isomorphism if and only if $M \circ P \circ J$ is an isomorphism; i.e., G is the direct product of H and K if and only if G is the direct product of $H_1, \ldots, H_m, K_1, \ldots, K_n$.

Theorem 24.7. Let $r_1, \ldots, r_m$ be natural numbers such that r_i and r_j are relatively prime whenever $i, j \in [1, m]$, $i \neq j$. If G is a group of order $r_1 r_2 \ldots r_m$ that contains a normal subgroup H_k of order r_k for each $k \in [1, m]$, then

$1°$ for each $n \in [1, m]$, $G_n = H_1 H_2 \ldots H_n$ is a normal subgroup of G and is, moreover, the direct product of $H_1, H_2, \ldots, H_n$,

$2°$ G is the direct product of $H_1, H_2, \ldots, H_m$.

Proof. By the Corollary of Theorem 24.4, G_n is a normal subgroup of G. To prove $1°$, it suffices by induction to show that if G_n is the direct product of $H_1, \ldots, H_n$ where $n < m$, then G_{n+1} is the direct product of $H_1, \ldots, H_{n+1}$. Suppose, therefore, that G_n is the direct product of $H_1, \ldots, H_n$. Then G_n has $r_1 r_2 \ldots r_n$ elements. Since r_{n+1} is relatively prime to $r_1, \ldots, r_n$, we conclude that r_{n+1} is relatively prime to $r_1 r_2 \ldots r_n$ by (UFD 3′). If $a \in G_n \cap H_{n+1}$, then the order of a divides $r_1 r_2 \ldots r_n$ and r_{n+1} and hence is 1, so $a = e$. Thus, by Theorem 24.3, G_{n+1} is the direct product of G_n and H_{n+1}, so by Theorem 24.6, G_{n+1} is the direct product of $H_1, \ldots, H_n, H_{n+1}$. Thus, $1°$ holds. In particular, as G_m is the direct product of $H_1, \ldots, H_m$, we conclude that G_m has $r_1 r_2 \ldots r_m$ elements and hence is G; thus, $2°$ holds.

Corollary. Let $r_1, \ldots, r_m$ be natural numbers such that r_i and r_j are relatively prime whenever $i, j \in [1, m]$, $i \neq j$. A cyclic group of order $r_1 r_2 \ldots r_m$ is the direct product of its cyclic subgroups of orders $r_1, \ldots, r_m$.

Theorem 24.8. Let $r_1, \ldots, r_m$ be natural numbers such that r_i and r_j are relatively prime whenever $i, j \in [1, m]$, $i \neq j$. If G is a group of order $r_1 r_2 \ldots r_m$ that contains a normal subgroup H_k of order r_k for each $k \in [1, m]$, and if $a_k \in H_k$ for each $k \in [1, m]$, then $a_1 a_2 \ldots a_m$ is a generator of G if and only if a_k is a generator of H_k for each $k \in [1, m]$.

Proof. For each $n \in [1, m]$, $G_n = H_1 H_2 \ldots H_n$ is a normal subgroup of G and is the direct product of $H_1, \ldots, H_n$ by Theorem 24.7. For each $n \in [1, m]$ let S_n be the statement: $a_1 a_2 \ldots a_n$ is a generator of G_n if and only if a_k is a generator of H_k for each $k \in [1, n]$. Clearly, S_1 is true. Assume that S_n is true where $n < m$; we shall show that S_{n+1} is true. Since S_n is true, S_{n+1} is equivalent to the following statement: $a_1 a_2 \ldots a_{n+1}$ is a generator of G_{n+1} if and only if $a_1 a_2 \ldots a_n$ is a generator of G_n and a_{n+1} is a generator of H_{n+1}. The order of G_n is $r_1 r_2 \ldots r_n$ since G_n is the direct product of $H_1, \ldots, H_n$; as before, r_{n+1} and $r_1 r_2 \ldots r_n$ are relatively prime. Let $a = a_1 a_2 \ldots a_n$, $b = a_{n+1}$, $r = r_1 r_2 \ldots r_n$, $s = r_{n+1}$. We are, therefore, to prove that if $a \in G_n$, $b \in H_{n+1}$, then ab is a generator of the group G_{n+1} of order rs if and only if a is a generator of the subgroup G_n of order r and b is a generator of the subgroup H_{n+1} of order s.

Necessity: As ab is a generator of G_{n+1}, G_{n+1} is cyclic (and hence abelian), and $(ab)^s$ clearly has order r. But $(ab)^s = a^s b^s = a^s$ by Theorem 23.8, since H_{n+1} has order s; therefore, a^s has order r and hence is a generator of G_n, because the cyclic group G_{n+1} of order rs has only one subgroup of order r by Theorem 23.9; hence, $[a] \subseteq G_n = [a^s] \subseteq [a]$, so a is a generator of G_n. Similarly, b is a generator of H_{n+1}. Sufficiency: Since G_{n+1} is the direct product of $H_1, \ldots, H_{n+1}$ and since G_n is the direct product of $H_1, \ldots, H_n$, G_{n+1} is the direct product of G_n and H_{n+1} by Theorem 24.6. Thus, as a is a generator of G_n and as b is a generator of H_{n+1}, G_{n+1} is the direct product of two cyclic (and hence abelian) subgroups and therefore is abelian. By Bezout's Identity, there exist $x, y \in Z$ such that $xr + ys = 1$. Consequently, because the orders of a and b are r and s respectively,

$$(ab)^{ys} = a^{ys} b^{ys} = a^{1-xr} = a,$$

$$(ab)^{xr} = a^{xr} b^{xr} = b^{1-ys} = b.$$

Thus, $[ab] \supseteq [a] = G_n$, $[ab] \supseteq [b] = H_{n+1}$, so $[ab] \supseteq G_n H_{n+1} = G_{n+1}$. Therefore, ab is a generator of G_{n+1}.

By induction, therefore, S_n is true for each $n \in [1, m]$. In particular, S_m is true. As before, $G_m = G$, and consequently, the proof is complete.

Corollary 24.8.1. Let $r_1, \ldots, r_m$ be natural numbers such that r_i and r_j are relatively prime whenever $i, j \in [1, m]$, $i \neq j$. If G is a group of order $r_1 r_2 \ldots r_m$ that contains a normal cyclic subgroup of order r_k for each $k \in [1, m]$, then G is cyclic.

Corollary 24.8.2. Let $r_1, \ldots, r_m$ be natural numbers such that r_i and r_j are relatively prime whenever $i, j \in [1, m]$, $i \neq j$. If a group G is the direct product of cyclic subgroups of orders $r_1, \ldots, r_m$, then G is cyclic.

The assertion follows at once from Theorem 24.5 and Corollary 24.8.1.

Corollary 24.8.3. Let $r_1, \ldots, r_m$ be natural numbers such that r_i and r_j are relatively prime whenever $i, j \in [1, m]$, $i \neq j$. The Euler function φ then satisfies

$$\varphi(r_1 r_2 \ldots r_m) = \varphi(r_1)\varphi(r_2) \ldots \varphi(r_m).$$

The assertion follows from the Corollary of Theorem 24.7 and Theorem 24.8.

Every group $(G, \cdot)$ is the direct product of itself and the subgroup $\{e\}$ consisting only of the neutral element. Groups which are the direct product of no other pairs of subgroups are called indecomposable:

Definition. A group G is **decomposable** if it is the direct product of two proper subgroups. A group is **indecomposable** if it is not decomposable.

Theorem 24.9. If $(G, \cdot)$ is a cyclic group of order n and if $n = p_1^{r_1} \ldots p_m^{r_m}$ is the prime decomposition of n, then G is the direct product of $H_1, \ldots, H_m$ where H_k is the subgroup of G of order $p_k^{r_k}$ for each $k \in [1, m]$, and moreover, each H_k is indecomposable.

Proof. By the Corollary of Theorem 24.7, we need only show that if H is a cyclic group of order p^r where p is a prime, then H is indecomposable. Let H_1 and H_2 be subgroups of H; their orders are p^{r_1} and p^{r_2} respectively for some natural numbers r_1, r_2 by Lagrange's Theorem; we may assume that $r_2 \leq r_1$. Since H_1 is cyclic, it has a unique subgroup of order p^{r_2} by Theorem 23.9, and that subgroup must be the unique subgroup of H of order p^{r_2}, namely, H_2. Thus, $H_2 \subseteq H_1$. Consequently, by Theorem 24.3, H is the direct product of H_1 and H_2 only if $H_2 = \{e\}$, whence $H_1 = H_1\{e\} = H$. Thus, H is indecomposable.

To illustrate further conditions under which a group is the direct product of certain of its subgroups, we consider finite commutative groups whose elements other than the neutral element all have order p, where p is a prime. We note first that the order of each element of the additive group $Z_p \times \ldots \times Z_p$ other than the zero element is p.

Theorem 24.10. Let p be a prime, and let $(G, \cdot)$ be a finite group containing more than one element. If G is commutative and if all of its elements other than the neutral element e have order p, then the order of G is p^m for some $m \in N^*$, and G is the direct product of a sequence of m subgroups, each cyclic of order p.

Proof. There exist natural numbers $k > 0$ such that G contains a subgroup that is the direct product of k cyclic subgroups of order p; for example, if $a \neq e$, then $[a]$ is cyclic of order p by hypothesis, so 1 is such a natural number. Let m be the largest natural number such that G contains a subgroup of order p^m that is the direct product of m subgroups, all cyclic of order p, and let G' be such a subgroup. We wish to prove that $G' = G$. Suppose that G' is a proper subgroup of G; then there exists $b \notin G'$. Consequently, $G' \cap [b] = \{e\}$, for if x were an element of $G' \cap [b]$ other than e, then x would be a generator of $[b]$ by the Corollary of Theorem 23.8, so $b \in [x] \subseteq G'$, a contradiction. By Theorem 24.4, $G'[b]$ is a group and is the direct product of G' and $[b]$ by Theorem 24.2. Since $[b]$ is cyclic of order p, $G'[b]$ is the direct product of $m + 1$ subgroups, each cyclic of order p, by Theorem 24.6. This contradiction shows that $G' = G$.

The hypothesis of Theorem 24.10 that G be commutative may be omitted if $p = 2$:

Theorem 24.11. If $(G, \cdot)$ is a group all of whose elements other than the neutral element e have order 2, then G is commutative.

Proof. By hypothesis, $z^2 = e$ for all $z \in G$, whence $z^{-1} = z$. In particular, if $x, y \in G$,

$$xyx^{-1}y^{-1} = xyxy = e,$$

so $xy = yx$.

By the Corollary of Theorem 23.8, a group of prime order p is cyclic and hence, by Theorem 23.6, is isomorphic to the (additive) cyclic group Z_p. One consequence of Theorems 24.10 and 24.11 is the determination of all groups of order 4:

Theorem 24.12. A group G of order 4 is either cyclic or the direct product of two cyclic subgroups of order 2. Thus, to within isomorphism the only groups of order 4 are Z_4 and $Z_2 \times Z_2$.

Proof. Suppose that G is not cyclic. Then every element of G other than the neutral element has order 2 by Theorem 23.8. Consequently, by Theorems 24.10 and 24.11, G is the direct product of two cyclic subgroups of order 2.

.   The group $Z_2 \times Z_2$ is often called the **four group**; it is isomorphic to the group of all symmetries of a rectangle that is not a square.

EXERCISES

24.1. Verify that if $G_1, \ldots, G_n$ are groups, then $G_1 \times \ldots \times G_n$ with the indicated composition is a group, and that $G_1 \times \ldots \times G_n$ is commutative if each G_k is.

24.2. Prove Theorem 24.1.

24.3. A group $(G, \cdot)$ is the direct product of a sequence $H_1, H_2, \ldots, H_n$ of subgroups if and only if the following three conditions hold:

1° $H_1, H_2, \ldots, H_n$ are normal subgroups of G.
2° $H_1 H_2 \ldots H_n = G$.
3° For each $i \in [2, n]$, $(H_1 \ldots H_{i-1}) \cap H_i = \{e\}$.

[Show by induction that $H_1 H_2 \ldots H_i$ is the direct product of $H_1, H_2, \ldots, H_i$ for each $i \in [2, n]$.]

24.4. Prove the Corollary of Theorem 24.4.

24.5. Verify that the function J of Theorem 24.5 is an isomorphism.

24.6. If a group $(G, \cdot)$ is the direct product of subgroups H and K, then $f \colon x \mapsto xK$, $x \in H$, is an isomorphism from H onto G/K, and $g \colon y \mapsto yH$, $y \in K$, is an isomorphism from K onto G/H.

24.7. Let $(G, \cdot)$ be the group of symmetries of the square, let G' be the multiplicative group $\{1, -1\}$, and let E be the cartesian product group $G \times G'$. Let $H = G \times \{1\}$, $K = \{(r_0, 1), (h, -1)\}$. Show that H and K are subgroups of E, that $HK = E$ and $H \cap K = \{(r_0, 1)\}$, that E is isomorphic to the group $H \times K$, but that E is not the direct product of H and K.

24.8. If K_1 and K_2 are normal subgroups of groups G_1 and G_2 respectively, then $K_1 \times K_2$ is a normal subgroup of the group $G_1 \times G_2$, and the group $(G_1 \times G_2)/(K_1 \times K_2)$ is isomorphic to the group $(G_1/K_1) \times (G_2/K_2)$.

24.9. If f is a function from a group G_1 into a group G_2, then f is a homomorphism if and only if the graph of f is a subgroup of the group $G_1 \times G_2$.

24.10. Let $(G, \cdot)$ be a group, and let $\cdot$ be the composition on $G \times \mathrm{Aut}(G)$ defined by

$$(x, \alpha) \cdot (y, \beta) = (x\alpha(y), \alpha\beta).$$

Then $(G \times \mathrm{Aut}(G), \cdot)$ is a group, $x \mapsto (x, 1_G)$ is an isomorphism from G onto a normal subgroup of $G \times \mathrm{Aut}(G)$, and $\alpha \mapsto (e, \alpha)$ is an isomorphism from $\mathrm{Aut}(G)$ onto a subgroup of $G \times \mathrm{Aut}(G)$.

24.11. If $n > 1$ and if $n = p_1^{r_1} \ldots p_m^{r_m}$ is its prime decomposition (where $r_k \geq 1$ for all $k \in [1, m]$), then

$$\varphi(n) = n \prod_{k=1}^{m} \left(1 - \frac{1}{p_k}\right).$$

*24.12. Prove that $\lim_{n \to \infty} \varphi(n) = +\infty$. [Show that if p is the rth prime and if $n \geq 2^r p$, then $\varphi(n) \geq p$; for this proof use Exercise 24.11 and consider first the case where each prime factor of n is $\leq p$.]

25. Extensions of Cyclic Groups by Cyclic Groups*

Nonabelian groups were introduced in §2 by the example of the group of symmetries of a square. To gain greater familiarity with nonabelian groups, we shall first scrutinize the group of symmetries of a regular polygon P_m of m sides (where $m \geq 3$). As observed in §8, each symmetry is completely determined by its behavior on the vertices of P_m. Since we wish to employ addition modulo m, we shall number the vertices of P_m by $0, 1, \ldots, m-1$ in a counterclockwise fashion; for simplicity we shall denote addition modulo m on N_m simply by $+$. What permutations of N_m correspond to symmetries of P_m? For each $k \in N_m$, the rotation of P_m in its plane about its center through $2\pi k/m$ radians is clearly a symmetry that takes vertex j into vertex $j + k$ for each $j \in N_m$. Consequently, for each $k \in N_m$, we shall call the permutation $r_k : j \longmapsto j + k$ of N_m a **rotation.**

Other symmetries are determined by rotating P_m in space through $180°$ about an axis of symmetry (in Figure 12, the dashed lines indicate the axes

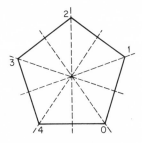 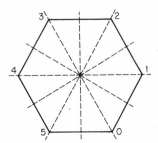

Figure 12

of symmetry for a regular pentagon and hexagon); such symmetries are called **reflections.** First, suppose that m is odd. Then the axis of symmetry through vertex n bisects the edge joining vertices $n + (m - 1)/2$ and $n + (m + 1)/2$. The reflection of P_m about the axis of symmetry through vertex n takes vertex $n - x$ into vertex $n + x$ for each $x \in N_m$; let vertex k be the vertex this reflection takes vertex 0 into, so that $k \equiv 2n \pmod{m}$; it is then easy to verify that this reflection takes vertex j into vertex $k - j$ for all $j \in N_m$. Since m is odd, 2 is invertible for multiplication modulo m, and hence, each $k \in N_m$ satisfies the congruence $k \equiv 2n \pmod{m}$ for one and only one number $n \in N_m$.

Next, suppose that m is even. Then the axis of symmetry through vertex n passes also through vertex $n + (m/2)$. Consequently, to consider those axes

---

* Subsequent chapters do not depend on this section.

of symmetry passing through vertices, it suffices to consider those vertices n where $0 \leq n < (m/2)$. As before, the reflection of P_m about the axis of symmetry through vertex n takes vertex $n - x$ into vertex $n + x$ for each $x \in N_m$; let vertex k be the vertex this reflection takes vertex 0 into, so that $k \equiv 2n \pmod{m}$; this reflection then takes vertex j into vertex $k - j$ for all $j \in N_m$. Since m is even and $k \equiv 2n \pmod{m}$, k is even; conversely, each even $k \in N_m$ satisfies the congruence $k \equiv 2n \pmod{m}$ for one and only one number $n \in N_{m/2}$. There are, however, additional axes of symmetry, namely, those which bisect an edge. The axis of symmetry bisecting the edge of P_m joining vertices n and $n + 1$ also bisects the edge joining vertices $n + (m/2)$ and $n + 1 + (m/2)$. Consequently, to consider those axes of symmetry bisecting edges, it suffices to consider those bisecting the edges joining vertices n and $n + 1$ where $0 \leq n < (m/2)$. The reflection of P_m about the axis of symmetry bisecting the edge joining vertices n and $n + 1$ takes vertex $n - x$ into vertex $n + 1 + x$ for each $x \in N_m$; let vertex k be the vertex this reflection takes vertex 0 into, so that $k \equiv 2n + 1 \pmod{m}$; it is then easy to verify that this reflection takes vertex j into vertex $k - j$ for all $j \in N_m$. Since m is even, if $k \equiv 2n + 1 \pmod{m}$, then k is odd; conversely, each odd $k \in N_m$ satisfies the congruence $k \equiv 2n + 1 \pmod{m}$ for one and only one number $n \in N_{m/2}$.

In view of these considerations, for each $k \in N_m$ we shall call the permutation $s_k: j \mapsto k - j$ of N_m a **reflection**. Is every permutation of N_m corresponding to a symmetry of P_m a rotation or reflection? To see that the answer to this question is yes, we observe that any motion of P_m that does not distort its shape must take adjacent vertices into adjacent vertices. In particular, if σ is a permutation of N_m corresponding to a symmetry of P_m, then for each $j \in N_m$, since vertex $j + 1$ is adjacent to vertex j, vertex $\sigma(j + 1)$ must be adjacent to vertex $\sigma(j)$; that is, $\sigma(j + 1)$ must be either $\sigma(j) + 1$ or $\sigma(j) - 1$ (if σ is a rotation, $\sigma(j + 1) = \sigma(j) + 1$ for all $j \in N_m$, and if σ is a reflection, $\sigma(j + 1) = \sigma(j) - 1$ for all $j \in N_m$). The following theorem, therefore, shows that the only permutations of N_m corresponding to symmetries of P_m are the rotations and reflections.

Theorem 25.1. If σ is a permutation of N_m such that for each $j \in N_m$, $\sigma(j + 1)$ is either $\sigma(j) + 1$ or $\sigma(j) - 1$ (where $+$ denotes addition modulo m), then there exists $k \in N_m$ such that either $\sigma(j) = k + j$ for all $j \in N_m$ or $\sigma(j) = k - j$ for all $j \in N_m$.

Proof. Let $k = \sigma(0)$. Then either $\sigma(1) = k + 1$ or $\sigma(1) = k - 1$. Case 1: $\sigma(1) = k + 1$. Let $S = \{j \in N_m: \sigma(x) = k + x \text{ for all } x \in [0, j]\}$. Certainly $0 \in S$, $1 \in S$. Suppose that $j \in S$, where $1 \leq j < m - 1$. Then $\sigma(j + 1)$ is either $\sigma(j) + 1$ or $\sigma(j) - 1$, and hence, since $j \in S$, $\sigma(j + 1)$ is either $k + j + 1$ or $k + j - 1$; but $\sigma(j - 1) = k + j - 1$, since $j \in S$, so

$\sigma(j+1) = k + j + 1$, because σ is a permutation. By induction, $S = N_m$, so $\sigma(j) = k + j$ for all $j \in N_m$. Case 2: $\sigma(1) = k - 1$. An argument similar to that of Case 1 establishes that $\sigma(j) = k - j$ for all $j \in N_m$.

Let r be the rotation r_1, s the reflection s_0. An easy inductive argument establishes that $r_k = r^k$ for all $k \in N_m$; in particular, the order of r is m. The reflection s_k is simply $r^k s$, since

$$(r^k s)(j) = r_k(-j) = k - j = s_k(j)$$

for all $j \in N_m$. Since $m \geq 3$, the order of s is clearly two, and s is not a rotation since $s(0) = 0$ but the only rotation taking 0 into 0 is the identity permutation. Consequently, the elements $r^k s^j$, where $k \in N_m$, $j \in N_2$ are mutually distinct, for if $r^k s$ were the rotation r^j, then s would be the rotation r^{j-k}, a contradiction. Therefore, the group G of permutations of N_m corresponding to symmetries of P_m is a group of order $2m$ that is generated by r and s; this group is called the **dihedral group** of order $2m$. Moreover,

(1) $$srs^{-1} = r^{-1},$$

for

$$(srs^{-1})(j) = (srs)(j) = sr(-j) = s(1-j)$$

$$= j - 1 = m - 1 + j = r_{m-1}(j) = r^{m-1}(j) = r^{-1}(j).$$

Since $G = \{r^k s^j : k \in N_m, j \in N_2\}$, it follows easily from (1) that $[r]$ is a normal subgroup of order m. However, G is not the direct product of $[r]$ and $[s]$, since s does not commute with r. In sum, for each $m \geq 3$, the dihedral group of order $2m$ is generated by elements r and s where $[r]$ is a normal cyclic subgroup of order m, s is an element of order 2 not belonging to $[r]$, and $srs^{-1} = r^{-1}$.

In §26 we shall prove fundamental theorems set forth by the Norwegian mathematician M. Ludwig Sylow (1832–1918) about finite groups. To illustrate the power of these theorems, we shall use them to classify all groups of order ≤ 15. If H and L are groups, a group G is said to be an **extension of H by L** if G contains a normal subgroup H' that is isomorphic to H such that the quotient group G/H' is isomorphic to L. In preparation for our study of groups of order ≤ 15, we shall next investigate extensions of a finite cyclic group by a finite cyclic group. The dihedral groups are examples of such extensions.

Suppose that G is a finite group having a normal cyclic subgroup H of order m such that G/H is cyclic of order n. If a is a generator of H and bH a generator of G/H, then $bab^{-1} = a^s$ for some $s \in Z^*$, as H is normal, and $b^n = a^r$ for some $r \in N_m$, as $b^n H = (bH)^n = H$.

Theorem 25.2. Let $(G, \cdot)$ be a group that contains a normal cyclic subgroup H of order m whose corresponding quotient group G/H is cyclic of order n. Let $a, b \in G$ be such that a is a generator of H and bH is a generator of G/H, and let $r \in N_m$, $s \in Z^*$ be such that

$$(2) \qquad b^n = a^r, \qquad bab^{-1} = a^s.$$

Then

$$(3) \qquad b^i a^k b^{-i} = a^{s^i k}$$

for all $i \in N$, $k \in Z$,

$$(4) \qquad s^n \equiv 1 \pmod{m}, \qquad sr \equiv r \pmod{m},$$

and for each $x \in G$, there exist unique integers $i \in N_m$, $j \in N_n$ such that $x = a^i b^j$. Moreover, G is commutative if and only if $s \equiv 1 \pmod{m}$. Finally, if $(G', \cdot)$ is a group that contains a normal cyclic subgroup H' of order m whose corresponding quotient group G'/H' is cyclic of order n, and if $c, d \in G'$ are such that c is a generator of H' and dH' a generator of G'/H' satisfying $d^n = c^r$, $dcd^{-1} = c^s$, then $f: a^i b^j \mapsto c^i d^j$, $i \in N_m$, $j \in N_n$, is an isomorphism from G onto G'.

Proof. An inductive argument establishes that $ba^k b^{-1} = a^{sk}$ for all $k \in N$, for if $ba^i b^{-1} = a^{si}$, then

$$ba^{i+1}b^{-1} = ba^i b^{-1} bab^{-1} = a^{si} a^s = a^{s(i+1)}.$$

Since $ba^{-i}b^{-1} = (ba^i b^{-1})^{-1} = (a^{si})^{-1} = a^{s(-i)}$, we conclude that $ba^k b^{-1} = a^{sk}$ for all $k \in Z$. In particular, upon setting $k = r$, we obtain $a^r = b^n = bb^n b^{-1} = ba^r b^{-1} = a^{sr}$, whence $sr \equiv r \pmod{m}$, since a has order m.

An inductive argument also establishes that $b^i a^k b^{-i} = a^{s^i k}$ for all $i \in N$, for if $b^j a^k b^{-j} = a^{s^j k}$, then

$$b^{j+1} a^k b^{-(j+1)} = b(b^j a^k b^{-j})b^{-1} = ba^{s^j k} b^{-1} = a^{s^{j+1} k}.$$

In particular, upon setting $i = n$, $k = 1$, we obtain

$$a = a^r a a^{-r} = b^n a b^{-n} = a^{s^n},$$

whence $s^n \equiv 1 \pmod{m}$ as a has order m. If $a^i b^j = a^u b^v$ where $i, u \in N_m$, $j, v \in N_n$, then $b^{j-v} = a^{u-i} \in H$, so $(bH)^{j-v} = b^{j-v} H = H$, whence $j \equiv v \pmod{n}$ and thus $j = v$, since both belong to N_n; consequently, $a^i = a^u$, so $i \equiv u \pmod{m}$, whence $i = u$, as both belong to N_m. By Lagrange's Theorem, G has nm elements. Thus, every element of G is uniquely of the form $a^i b^j$ where $i \in N_m$, $j \in N_n$. Consequently, if $s \equiv 1 \pmod{m}$, then G is commutative by (2) of Theorem 11.8; conversely, if G is commutative, then $bab^{-1} = a$, so $s \equiv 1 \pmod{m}$.

To prove the final assertion, we first note that if $i, u \in N_m$, $j, v \in N_n$, and if $x, y \in N_m$ are such that $x \equiv i + s^j u \pmod{m}$, $y \equiv i + s^j u + r \pmod{m}$, then

$$(a^i b^j)(a^u b^v) = \begin{cases} a^x b^{j+v} & \text{if } j + v < n \\ a^y b^{j+v-n} & \text{if } j + v \geq n, \end{cases}$$

since

$$(a^i b^j)(a^u b^v) = a^i(b^j a^u b^{-j})b^{j+v} = a^{i+s^j u}b^{j+v} = \begin{cases} a^{i+s^j u}b^{j+v} & \text{if } j + v < n \\ a^{i+s^j u+r}b^{j+v-n} & \text{if } j + v \geq n. \end{cases}$$

By the same token, each element of G' is uniquely of the form $c^i d^j$ where $i \in N_m$, $j \in N_n$, and

$$(c^i d^j)(c^u d^v) = \begin{cases} c^x d^{j+v} & \text{if } j + v < n \\ c^y d^{j+v-n} & \text{if } j + v \geq n. \end{cases}$$

Thus, $f: a^i b^j \mapsto c^i d^j$ is an isomorphism, and the proof is complete.

Conversely, if r and s satisfy (4), then there is an extension of a cyclic group of order m by a cyclic group of order n generated by elements a and b satisfying (2), and by Theorem 25.2, there is to within isomorphism only one such group:

Theorem 25.3. Let $m, n \in N^*$, and let $r \in N_m$, $s \in Z^*$ be such that

$$sr \equiv r \pmod{m}, \qquad s^n \equiv 1 \pmod{m}.$$

Let $G(m, n, r, s)$ be the set $N_m \times N_n$, and let $\cdot$ be the composition on $G(m, n, r, s)$ defined by

$$(x, y)(z, w) = \begin{cases} (x +_m s^y.z, y + w) & \text{if } y + w < n, \\ (x +_m s^y.z +_m r, y + w - n) & \text{if } y + w \geq n. \end{cases}$$

Let $a = (1, 0)$, $b = (0, 1)$. Then $(G(m, n, r, s), \cdot)$ is a group, the cyclic subgroup H generated by a is a normal subgroup of order m, $G(m, n, r, s)/H$ is a cyclic group of order n generated by bH, and

$$b^n = a^r, \qquad bab^{-1} = a^s.$$

Proof. Since $sr \equiv r \pmod{m}$, an inductive argument establishes that $s^y.r \equiv r$ $\pmod{m}$ for all $y \in N$. To show that $[(x, y)(z, w)](u, v) = (x, y)[(z, w)(u, v)]$, six cases must be considered: (1) $y + w + v < n$; (2) $y + w < n, w + v < n$, $y + w + v \geq n$; (3) $y + w \geq n$, $w + v < n$; (4) $y + w < n$, $w + v \geq n$; (5) $y + w \geq n$, $w + v \geq n$, $y + v + w < 2n$; (6) $y + w \geq n$, $w + v \geq n$, $y + v + w \geq 2n$. In Case (6), for example,

$$[(x, y)(z, w)](u, v) = (x +_m s^y.z +_m r, y + w - n)(u, v)$$

$$= (x +_m s^y.z +_m r +_m s^{y+w-n}.u +_m r, y + w + v - 2n),$$

$$(x, y)[(z, w)(u, v)] = (x, y)(z +_m s^w.u +_m r, w + v - n)$$

$$= (x +_m s^y.[z +_m s^w.u +_m r] +_m r, y + w + v - 2n).$$

Since $s^n \equiv 1 \pmod{m}$, $s^{y+w-n}.u \equiv s^{y+w}.u \pmod{m}$, and $s^y.r \equiv r \pmod{m}$ as noted earlier. A similar discussion establishes the identity in the remaining five cases. Clearly, $(0, 0)$ is the neutral element. If $0 < x < m$, the inverse of $(x, 0)$ is $(m - x, 0)$; if $0 < y < n$, the inverse of (x, y) is $(z, n - y)$ where $z \in N_m$ satisfies $z \equiv -(s^{n-y}.x +_m r) \pmod{m}$, for, since $s^n \equiv 1 \pmod{m}$ and $s^y.r \equiv r \pmod{m}$,

$$(z, n - y)(x, y) = (z +_m s^{n-y}.x +_m r, 0) = (0, 0),$$

$$(x, y)(z, n - y) = (x +_m s^y.z +_m r, 0) = (0, 0).$$

Thus, $G(m, n, r, s)$ is a group.

Since $(x, 0)(z, 0) = (x +_m z, 0)$, it follows that a has order m and that $a^i = (i, 0)$ for all $i \in N_m$. Let $s' \in N_m$ be such that $s' \equiv s \pmod{m}$; then $a^{s'} = a^s$. By our calculation above, $b^{-1} = (m - r, n - 1)$ if $r > 0$, and $b^{-1} = (0, n - 1)$ if $r = 0$. In the former case,

$$bab^{-1} = (0, 1)(1, 0)(m - r, n - 1) = (0, 1)(1 +_m (m - r), n - 1)$$

$$= (s.(1 +_m (m - r)) +_m r, 0) = (s.(1 +_m (m - r)) +_m s.r, 0)$$

$$= (s.(1 +_m (m - r) +_m r), 0) = (s', 0) = a^{s'} = a^s,$$

and in the latter,

$$bab^{-1} = (0, 1)(1, 0)(0, n - 1) = (0, 1)(1, n - 1)$$

$$= (s', 0) = a^{s'} = a^s.$$

It is easy to see that $b^j = (0, j) \notin [a]$ if $1 \leq j < n$ but that $b^n = (r, 0) = a^r$. Each element of $G(m, n, r, s)$ is of the form $a^i b^j$ where $i \in N_m, j \in N_n$, since $(i, j) = (i, 0)(0, j) = a^i b^j$. This fact, together with the identity $bab^{-1} = a^s$, shows that the subgroup H generated by a is a normal subgroup of order m. Moreover, bH is a generator of $G(m, n, r, s)/H$ of order n as $b^n = a^r \in H$ but $b^j \notin H$ if $0 < j < n$. This completes the proof.

The dihedral group of order $2m$ is isomorphic to the group $G(m, 2, 0, -1)$ by virtue of the last statement of Theorem 25.2. For a further example, let $q \in N^*$, and let $m = 2q$, $n = 2$, $r = q$, $s = -1$. Then (4) of Theorem 25.2 holds; the group $G(2q, 2, q, -1)$ is called the **dicyclic group** of order $4q$; it contains a normal cyclic subgroup $[a]$ of order $2q$ such that $G/[a]$ is a (cyclic) group of order 2 generated by the coset of an element b satisfying

$$b^2 = a^q, \qquad bab^{-1} = a^{-1}.$$

The dicyclic group of order 4 is simply the cyclic group of order 4, for as $b^2 = a$ and as a has order 2, b is a generator of the group. The dicyclic group of order 8 is called the **quaternionic group**; the quaternionic group contains a normal cyclic subgroup $[a]$ of order 4 such that $G/[a]$ is a (cyclic) group of order 2 generated by the coset of an element b satisfying

$$b^2 = a^2, \qquad bab^{-1} = a^{-1}.$$

The dihedral group of order 8 contains only two elements of order 4, whereas the quaternionic group contains six elements of order 4; thus, the dihedral group of order 8 is not isomorphic to the quaternionic group. To illustrate the use of Theorem 25.2, we classify all groups of order 8:

Theorem 25.4. A group $(G, \cdot)$ of order 8 is isomorphic to one and only one of the following five groups: Z_8, $Z_4 \times Z_2$, $Z_2 \times Z_2 \times Z_2$, the dihedral group of order 8, the quaternionic group.

Proof. By Theorem 23.8, the order of each element of G other than the neutral element e is either 2, 4, or 8. Case 1: There exists an element of order 8. Then G is isomorphic to Z_8 by Theorem 23.6. Case 2: Every element of G other than e has order 2. By Theorems 24.10, 24.11, and 24.1, G is isomorphic to $Z_2 \times Z_2 \times Z_2$.

Case 3: There exists an element a of order 4 but no element of order 8. By Theorem 18.6, $[a]$ is a normal cyclic subgroup. The quotient group $G/[a]$ has order 2 and thus is cyclic, so we may apply Theorem 25.2. Let $b \notin [a]$, and let $r \in N_4$, $s \in Z^*$ be such that $b^2 = a^r$, $bab^{-1} = a^s$. Then $s^2 \equiv 1 \pmod 4$, so either $s \equiv 1 \pmod 4$ or $s \equiv -1 \pmod 4$.

Suppose that $s \equiv 1 \pmod 4$. By Theorem 25.2, G is commutative. If r were 1 or 3, then, as $b^2 = a^r$, b would be a generator of G, since a and a^3 are generators of $[a]$, in contradiction to our assumption that G is not cyclic. If $r = 2$, let $b_1 = ab$; then $b_1 \notin [a]$, and $b_1{}^2 = a^2b^2 = a^4 = e$, so G is the direct product of $[a]$ and $[b_1]$ by Theorem 24.3, whence G is isomorphic to $Z_4 \times Z_2$ by Theorem 24.1. By the same reasoning, G is also isomorphic to $Z_4 \times Z_2$ if $r = 0$. Thus, if $s \equiv 1 \pmod 4$, then G is isomorphic to $Z_4 \times Z_2$.

Suppose finally that $s \equiv -1 \pmod 4$. Then $bab^{-1} = a^{-1}$, and $r \equiv -r$ (mod 4), so r is either 0 or 2. If $r = 0$, then $b^2 = e$, $bab^{-1} = a^{-1}$, so G is isomorphic to the dihedral group of order 8 by Theorem 25.2; if $r = 2$, then $b^2 = a^2$, $bab^{-1} = a^{-1}$, so G is isomorphic to the quaternionic group again by Theorem 25.2.

EXERCISES

25.1. Complete the proof of Theorem 25.1.

25.2. Complete the proof of Theorem 25.3.

25.3. If m is odd, the dihedral group D_{2m} of order $4m$ is isomorphic to the cartesian product of a cyclic group of order 2 and the dihedral group D_m of order $2m$. [If $a^{2m} = e$, show that D_{2m} is the direct product of $[a^m]$ and another group.]

25.4. If $m \geq 2$, the center of the dicyclic group of order $4m$ contains exactly two elements.

25.5. Let Q be the quaternionic group; let 1 be its neutral element, -1 the other element in its center (Exercise 25.4). For each $x \in Q$, denote $(-1)x$ by $-x$.
(a) If x is not in the center of Q, then the order of x is 4, $x^2 = -1$, and the cyclic subgroup generated by x is $\{x, -1, -x, 1\}$.
(b) Let i be an element of Q not in the center, let j be an element of Q not in the subgroup generated by i, and let $k = ij$. Then

$$Q = \{1, -1, i, -i, j, -j, k, -k\},$$

$$i^2 = j^2 = k^2 = -1,$$

$$ij = k, \qquad jk = i, \qquad ki = j,$$

$$ji = -k, \qquad kj = -i, \qquad ik = -j.$$

25.6. (a) If $m - r$ is odd, $G(m, 2, r, 1)$ is cyclic. [Find j so that $(a^jb)^2 = a$.]
(b) If $m - r$ is even, $G(m, 2, r, 1)$ is isomorphic to $Z_m \times Z_2$. [Find j so that the order of a^jb is 2.]

25.7. Let $m \geq 3$, let J_m be the set of all integers in $[1, m - 1]$ that are relatively prime to m, and let a and b be generators of the dihedral group D_m of order $2m$ satisfying $a^m = b^2 = e$, $bab^{-1} = a^{-1}$.
(a) For each $i \in N_m$ and each $s \in J_m$, there is one and only one automorphism $f_{i,s}$ of D_m satisfying $f_{i,s}(a) = a^s$ and $f_{i,s}(b) = a^ib$.
(b) For each $s \in J_m$, let L_s be the automorphism of the additive group Z_m defined by $L_s(x) = s.x$ (Exercise 23.2). The function $F: (i + (m), L_s) \mapsto f_{i,s}$

is a well-defined isomorphism from the group $(Z_m \times \text{Aut}(Z_m),\, .)$ (Exercise 24.10) onto the group $\text{Aut}(D_m)$.

25.8. Let $m \geq 3$, let J_{2m} be the set of all integers in $[1, 2m - 1]$ that are relatively prime to $2m$, and let a and b be generators of the dicyclic group G of order $4m$ satisfying $a^{2m} = e$, $b^2 = a^m$, $bab^{-1} = a^{-1}$.
(a) For each $i \in N_{2m}$ and each $s \in J_{2m}$, there is one and only one automorphism $f_{i,s}$ of G satisfying $f_{i,s}(a) = a^s$ and $f_{i,s}(b) = a^i b$.
(b) For each $s \in J_{2m}$, let L_s be the automorphism of the additive group Z_{2m} defined by $L_s(x) = s.x$ (Exercise 23.2). The function $F: (i + (2m), L_s) \mapsto f_{i,s}$ is a well-defined isomorphism from the group $(Z_{2m} \times \text{Aut}(Z_{2m}),\, .)$ onto the group $\text{Aut}(G)$.

26. Sylow's Theorems*

Let G be a finite group of order n. By Lagrange's Theorem, if H is a subgroup of G of order m, then $m \mid n$. If G is cyclic, the converse holds: if $m \mid n$, then a cyclic group of order n contains a subgroup of order m (Theorem 23.9). However, it is not true in general that if m is a divisor of n, then G has a subgroup of order m (later we shall see, for example, that the alternating group $\mathfrak{A}_4$ of order 12 contains no subgroup of order 6). For those divisors m of n that are powers of a prime, however, G necessarily contains a subgroup of order m; this is the content of Sylow's First Theorem. Further information concerning the nature and number of subgroups of G of order m, where m is a power of a prime, is given by Sylow's Second and Third Theorems.

In this section we shall denote by $n(X)$ the number of elements in a finite set X, and by e the neutral element of a group whose composition is denoted multiplicatively. The proofs we shall give of Sylow's Theorems depend on certain counting arguments, a general framework for which is contained in the following definition.

Definition. Let $(G, \cdot)$ be a group. A **G-space** is an ordered couple $(E, .)$ where E is a set and $.$ is a function from $G \times E$ into E (whose value at $(a, z) \in G \times E$ is denoted by $a.z$) satisfying the following two conditions for all $a, b \in G$ and all $z \in E$:

$1°$ $e.z = z$.
$2°$ $(ab).z = a.(b.z)$.

The function $.$ is called the **action** of G on E. If $(E, .)$ is a G-space, we also say that the **group G acts on E under** $.$, or that E is a **set with group of operators** G. If $(E, .)$ is a G-space, for each $z \in E$ the set $G.z$ defined by

$$G.z = \{a.z : a \in G\}$$

---

* Except for §47, subsequent sections do not depend upon this section.

is called the **orbit** of z, and the subset G_z of G defined by

$$G_z = \{a \in G: a.z = z\}$$

is called the **stabilizer** of z; an **orbit** of the G-space E is simply the orbit of z for some $z \in E$.

Theorem 26.1. Let $(E, .)$ be a G-space.

1° The orbits of E form a partition of E.

2° For each $z \in E$, the stabilizer G_z of z is a subgroup of G, and

$$aG_z \mapsto a.z$$

is a well-defined bijection from the set G/G_z onto the orbit of z.

Proof. 1° Since $e.z = z$, $z \in G.z$. Thus, no orbit is the empty set, and every element of E belongs to at least one orbit. It remains to show that if $x \in G.z \cap G.w$, then $G.z = G.w$. By our assumption, there exist $a, b \in G$ such that $x = a.z = b.w$. For any $c \in G$,

$$c.z = (ca^{-1}a).z = (ca^{-1}).(a.z) = (ca^{-1}).(b.w)$$
$$= (ca^{-1}b).w \in G.w,$$

and similarly, $c.w \in G.z$. Thus, $G.z = G.w$, so the orbits partition E.

2° Clearly, $e \in G_z$, so $G_z \neq \emptyset$. If $a, b \in G_z$, then $a.z = z = b.z$, so

$$ab^{-1}.z = ab^{-1}.(b.z) = (ab^{-1})b.z = a.z = z,$$

whence $ab^{-1} \in G_z$. Thus, by Theorem 8.3, G_z is a subgroup of G. We have only to show now that $aG_z = bG_z$ if and only if $a.z = b.z$. But $aG_z = bG_z$ if and only if $b^{-1}a \in G_z$, that is, if and only if $b^{-1}a.z = z$; but if $z = b^{-1}a.z$, then $b.z = b.(b^{-1}a.z) = (bb^{-1}a).z = a.z$, and if $b.z = a.z$, then

$$z = (b^{-1}b).z = b^{-1}.(b.z) = b^{-1}.(a.z) = b^{-1}a.z.$$

Theorem 26.2. If $(G, \cdot)$ is a group, for each $a \in G$ the function

$$\kappa_a: x \mapsto axa^{-1}$$

is an automorphism of G.

The proof is easy.

An automorphism g of $(G, \cdot)$ is called an **inner automorphism** of G if $g = \kappa_a$ for some $a \in G$. If $x \in G$, an element y of G is a **conjugate** of x if $y = axa^{-1}$ for some $a \in G$, that is, if y is the image of x under an inner automorphism of G. We shall denote by $\text{Conj}(x)$ the set of all conjugates of x; a subset C of G is called a **conjugate class** of G if $C = \text{Conj}(x)$ for some $x \in G$. If x belongs to the center $Z(G)$ of G, then $axa^{-1} = x$ for all $a \in G$, so x itself is the only conjugate of x; conversely, if $\text{Conj}(x) = \{x\}$, then $axa^{-1} = x$ for all $a \in G$, whence $ax = xa$ for all $a \in G$, so $x \in Z(G)$. Therefore, $n(\text{Conj}(x)) = 1$ if and only if $x \in Z(G)$.

Our first goal is to prove that if G is a group whose order is a power of a prime, then the center of G contains at least one element other than the neutral element. To prove this, we shall study a suitable action of the group G on the set G.

Let G be a group, and let . be the function from $G \times G$ into G defined by $a.x = axa^{-1}$. Clearly, $e.x = x$, and

$$ab.x = abx(ab)^{-1} = a(bxb^{-1})a^{-1} = a.bxb^{-1} = a.(b.x)$$

for all a, b, $x \in G$. Thus, $(G, .)$ is a G-space; . is called the **action of G on G by conjugation.** Clearly, the orbit of x for this action is simply $\text{Conj}(x)$. Thus, by 1° of Theorem 26.1, the sets $\text{Conj}(x)$, where $x \in G$, form a partition of G.

Suppose further that G is finite. Selecting exactly one element from each conjugate class of G, we obtain a subset F of G such that $\{\text{Conj}(x): x \in F\}$ is a partition of G, and $\text{Conj}(x) \neq \text{Conj}(y)$ if $x, y \in F$ and $x \neq y$. Thus,

$$n(G) = \sum_{x \in F} n(\text{Conj}(x)).$$

Let $x_1, \ldots, x_r$ be the elements of F not in the center Z of G. If $x \in Z$, then $\text{Conj}(x) = \{x\}$, so x must belong to F; thus, $F = Z \cup \{x_1, \ldots, x_r\}$, whence

$$n(G) = n(Z) + \sum_{i=1}^{r} n(\text{Conj}(x_i)).$$

Finally, assume that the order of G is p^m where p is a prime and $m \geq 1$. Since $\text{Conj}(x_i)$ is an orbit of G for the action of conjugation, by 2° of Theorem 26.1, $n(\text{Conj}(x_i)) = n(G/G_{x_i})$ where $G_{x_i} = \{a \in G: ax_ia^{-1} = x_i\}$. Since $p^m = n(G) = n(G/G_{x_i})n(G_{x_i})$ by Lagrange's Theorem, $n(G/G_{x_i})$ divides p^m; since $x_i \notin Z$, $n(G/G_{x_i}) = n(\text{Conj}(x_i)) > 1$; hence, $n(\text{Conj}(x_i)) = p^{m_i}$ for some $m_i > 0$. Thus,

$$n(Z) = n(G) - \sum_{i=1}^{r} n(\text{Conj}(x_i)) = p^m - \sum_{i=1}^{r} p^{m_i}.$$

Since $m_i > 0$ for all $i \in [1, r]$, therefore, $p \mid n(Z)$. Consequently, Z contains at least p elements, and in particular is not $\{e\}$. Let b be an element of Z distinct from e. The order of b is then p^k for some $k \in [1, m]$ by Theorem 23.8, so, as Z is a subgroup of G, $b^{p^{k-1}}$ is an element of Z of order p. We have completed the proof of the following theorem:

Theorem 26.3. If G is a group of order p^m, where p is a prime and $m \geq 1$, then the center of G contains an element of order p.

Using Theorem 26.3, we may generalize Theorem 24.12:

Theorem 26.4. If p is a prime, a group of order p^2 is either cyclic or the direct product of two cyclic subgroups of order p. Thus, to within isomorphism, the only groups of order p^2 are Z_{p^2} and $Z_p \times Z_p$.

Proof. Let G be a noncyclic group of order p^2. Each element of G other than the neutral element e then has order p. By Theorem 26.3, the center Z of G contains an element a of order p. Let b be an element of G not belonging to $[a]$. Then $[a] \cap [b]$ is a proper subgroup of $[a]$, a group of order p, so $[a] \cap [b] = \{e\}$. As $[a] \subseteq Z$, each element of $[a]$ commutes with every element of G; consequently, $[a]$ is a normal subgroup of G, so $[a][b]$ is a subgroup of G by Theorem 24.4 and is, moreover, the direct product of $[a]$ and $[b]$ by Theorem 24.2. In particular, $[a][b]$ has order p^2, so $[a][b] = G$. Thus, G is the direct product of two (cyclic) subgroups of order p.

As originally stated, Sylow's First Theorem is the assertion that if p^s is the highest power of a prime p that divides the order of a finite group G, then G contains a subgroup of order p^s. We shall prove more: if p^s is *any* power of a prime p that divides the order of G, then G contains a subgroup of order p^s. We shall call this statement "Sylow's First Theorem," although it is somewhat inaccurate, historically, to do so. The proof we shall give depends on a certain property of binomial coefficients:

Lemma. If p is a prime and if $s, m \in N^*$, then

$$p \nmid \binom{p^s m - 1}{p^s - 1}.$$

Proof. Since

$$\binom{p^s m - 1}{p^s - 1} = \frac{(p^s m - 1)(p^s m - 2) \ldots (p^s m - p^s + 1)}{1 \cdot 2 \cdot \ldots \cdot (p^s - 1)},$$

it suffices to show that for each $k \in [1, p^s - 1]$, if $p^t \mid (p^s m - k)$, then $p^t \mid k$.

Suppose that $p^t \mid (p^s m - k)$, where $k < p^s$. Then $p^s \nmid k$, so $p^s \nmid (p^s m - k)$, whence $t < s$; therefore, $p^t \mid p^s m$, and consequently, $p^t \mid [p^s m - (p^s m - k)]$, that is, $p^t \mid k$.

Theorem 26.5. (Sylow's First Theorem) If $(G, \cdot)$ is a finite group of order g and if $p^s \mid g$ where p is a prime, then G contains a subgroup of order p^s.

Proof. We may assume that $s \geq 1$. Let $\mathscr{E}$ be the set of all subsets of G having p^s elements. If $X \in \mathscr{E}$, then $aX \in \mathscr{E}$ for each $a \in G$ since $x \mapsto ax$ is a permutation of G. Thus, we may define a function . from $G \times \mathscr{E}$ into $\mathscr{E}$ by

$$a.X = aX$$

for all $a \in G$, $X \in \mathscr{E}$. Clearly, $e.X = X$ and

$$a.(b.X) = a.(bX) = a(bX) = (ab)X = (ab).X$$

for all $a, b \in G$, $X \in \mathscr{E}$. Thus, $(\mathscr{E}, .)$ is a G-set. Let $g = p^s m$, and let p^r be the highest power of p that divides m. By Theorem B.6, $\mathscr{E}$ has $\begin{pmatrix} p^s m \\ p^s \end{pmatrix}$ members. Since

$$\begin{pmatrix} p^s m \\ p^s \end{pmatrix} = \frac{p^s m}{p^s} \cdot \frac{(p^s m - 1)(p^s m - 2) \ldots (p^s m - p^s + 1)}{1 \cdot 2 \cdot \ldots \cdot (p^s - 1)} = m \begin{pmatrix} p^s m - 1 \\ p^s - 1 \end{pmatrix},$$

by the Lemma, p^r is also the highest power of p that divides $\begin{pmatrix} p^s m \\ p^s \end{pmatrix}$. Because the orbits of $\mathscr{E}$ form a partition of $\mathscr{E}$ by Theorem 26.1, there exists an orbit $G.Y$ of $\mathscr{E}$ such that $p^{r+1} \nmid n(G.Y)$, for otherwise p^{r+1} would divide $\begin{pmatrix} p^s m \\ p^s \end{pmatrix}$. By Theorem 26.1, the stabilizer G_Y of Y is a subgroup of G such that $n(G/G_Y) = n(G.Y)$. By Lagrange's Theorem,

$$p^s m = n(G) = n(G/G_Y)n(G_Y) = n(G.Y)n(G_Y).$$

As $p^{s+r} \mid n(G)$ and as $p^{r+1} \nmid n(G.Y)$, we conclude that $p^s \mid n(G_Y)$, and in particular, $p^s \leq n(G_Y)$. To show that $p^s \geq n(G_Y)$, let $y \in Y$; then for each $a \in G_Y$, $ay \in a.Y = Y$, so $Y \supseteq G_Y y$ and thus, $p^s = n(Y) \geq n(G_Y y) = n(G_Y)$, since $a \mapsto ay$ is a permutation of G. Thus, G_Y is a subgroup of G having p^s elements.

Corollary. If p is a prime and if G is a finite group the order of each element of which is a power of p, then the order of G is p^m for some $m \in N$.

Proof. Let $p^m t$ be the order of G, where $p \nmid t$. If $t \neq 1$, then there is a prime q such that $q \mid t$, so G has a subgroup of order q and hence an element of order q by Sylow's First Theorem, in contradiction to our hypothesis. Therefore, $t = 1$.

Definition. Let p be a prime. A finite group G is a **p-group** if the order of G is a power of p, or equivalently, if the order of each element of G is a power of p.

Let G be a group of order p^m, where p is a prime. By Sylow's First Theorem, G has a subgroup of order p^k for each $k \in [0, m]$. Another important fact, which follows from Theorem 26.3, is that every subgroup of G of order p^{m-1} is a normal subgroup (Exercise 26.2).

Our first application of Theorem 26.5 is the classification of groups of order $2p$, p an odd prime.

Theorem 26.6. *If p is an odd prime, a group of order $2p$ is either cyclic or isomorphic to the dihedral group of order $2p$.*

Proof. By Sylow's First Theorem, a group G of order $2p$ contains a subgroup of order p and hence an element a of order p. By Theorem 18.6, $[a]$ is a normal subgroup and $G/[a]$ is a (cyclic) group of order 2. Let $b \notin [a]$. By Theorem 25.2, there exist $r \in N_p$ and $s \in Z^*$ such that $b^2 = a^r$, $bab^{-1} = a^s$, $s^2 \equiv 1 \pmod p$, $sr \equiv r \pmod p$, and moreover, G is commutative if $s \equiv 1 \pmod p$. Since Z_p is a field, the only elements x of Z_p satisfying $x^2 = 1$ are 1 and -1; hence, either $s \equiv 1 \pmod p$ or $s \equiv -1 \pmod p$. In the former case, G is commutative and also contains an element of order 2 by Sylow's First Theorem, so G is cyclic by Corollary 24.8.1. Suppose, therefore, that $s \equiv -1 \pmod p$. Then $-r \equiv r \pmod p$, so $2r \equiv 0 \pmod p$, whence $r \equiv 0 \pmod p$, because p is odd, and consequently, $r = 0$. Thus, $b^2 = e$, $bab^{-1} = a^{-1}$, so G is isomorphic to the dihedral group of order $2p$ by the final assertion of Theorem 25.2.

Definition. Let G be a finite group of order g, and let p be a prime. If p^m is the highest power of p that divides g, any subgroup of G of order p^m is called a **Sylow p-subgroup** of G.

Our next application of Sylow's First Theorem is to classify all groups of order 12. We know two abelian groups of order 12, namely, Z_{12} (isomorphic to $Z_4 \times Z_3$ by the Corollary of Theorem 24.7) and $Z_6 \times Z_2$ (isomorphic to $Z_3 \times Z_2 \times Z_2$). These two groups are not isomorphic, as the latter is not cyclic. We also know two nonabelian groups of order 12, the dihedral group (the group of symmetries of a regular hexagon) and the dicyclic group $G(6, 2, 3, -1)$. These groups are not isomorphic, since the dihedral group has seven elements of order 2, whereas the dicyclic group has only one.

The alternating group $\mathfrak{A}_4$, consisting of all even permutations of $[1, 4]$, is also a group of order 12, since $4!/2 = 12$. By Theorem 20.11, its elements are the identity permutation, $(1, 2, 3)$, $(1, 3, 2)$, $(1, 2, 4)$, $(1, 4, 2)$, $(1, 3, 4)$, $(1, 4, 3)$, $(2, 3, 4)$, $(2, 4, 3)$, $(1, 2)(3, 4)$, $(1, 3)(2, 4)$, and $(1, 4)(2, 3)$. This group is also the subgroup of $\mathfrak{S}_4$ corresponding to the group of symmetries of a regular tetrahedron and hence is called the **tetrahedral group**. Indeed, if the vertices of a regular tetrahedron are labelled $1, 2, 3, 4$, then the rotations through $120°$ and $240°$ about each of the four altitudes correspond to the eight cycles of length 3, and the rotations through $180°$ about each of the three axes that join midpoints of opposite edges correspond to the permutations $(1, 2)(3, 4)$, $(1, 3)(2, 4)$, and $(1, 4)(2, 3)$. This group contains no element of order 6 and hence is not isomorphic to any of the four previously mentioned groups. To show that the five groups listed are, to within isomorphism, all the groups of order 12, we require a preliminary theorem.

Theorem 26.7. If G is a subgroup of order 12 of the symmetric group $\mathfrak{S}_4$, then $G = \mathfrak{A}_4$.

Proof. An easy calculation establishes that if $\tau \in \mathfrak{S}_n$ and if $(a_1, \ldots, a_m)$ is any cycle in $\mathfrak{S}_n$, then

$$\tau \circ (a_1, \ldots, a_m) \circ \tau^{\leftarrow} = (\tau(a_1), \ldots, \tau(a_m)).$$

Consequently, if a normal subgroup of $\mathfrak{S}_n$ contains a cycle of length m, then it contains every cycle of length m.

By Theorem 18.6, G is a normal subgroup of $\mathfrak{S}_4$. The elements of $\mathfrak{S}_4$ not in $\mathfrak{A}_4$ are the six cycles of length 4 and the six transpositions. Suppose first that G contained a cycle of length 4. Then G would contain all six cycles of length 4; hence, G would contain $(1, 2, 3, 4)(1, 2, 4, 3) = (1, 3, 2)$ and, therefore, all eight cycles of length 3; consequently, G would contain at least 14 elements, a contradiction. Suppose next that G contained a transposition. Then G would contain all six transpositions; hence, G would contain $(1, 2)(1, 3) = (1, 3, 2)$ and, therefore, all eight cycles of length 3; consequently, G would contain once again at least 14 elements, a contradiction. Therefore, G contains no cycle of length 4 or 2, so $G \subseteq \mathfrak{A}_4$, and thus $G = \mathfrak{A}_4$.

Theorem 26.8. If G is a group of order 12, then G is isomorphic to one and only one of the following five groups: $\mathbf{Z}_{12}$, $\mathbf{Z}_6 \times \mathbf{Z}_2$, $\mathfrak{A}_4$, the dihedral group of order 12, the dicyclic group of order 12.

Proof. By Sylow's First Theorem, G contains an element a of order 3 and a subgroup H of order 4. By Theorem 24.12, H is either cyclic or the direct product of two cyclic groups of order 2.

Case 1: $[a]$ is a normal subgroup. We shall first show that H contains an element of order 2 that commutes with a. Suppose first that H is cyclic. Let b be a generator of H. Then bab^{-1} is either a or a^2 as $[a]$ is normal; in the former case, b commutes with a and hence b^2 does also; in the latter case, $b^2ab^{-2} = b(bab^{-1})b^{-1} = ba^2b^{-1} = (bab^{-1})^2 = a^4 = a$; thus, in either case b^2 is an element of order two that commutes with a. Suppose next that H is the direct product of two cyclic subgroups of order 2. If b_1 and b_2 are elements of H of order 2 that do not commute with a, then $b_1ab_1^{-1} = a^2 = b_2ab_2^{-1}$ as $[a]$ is normal, so

$$(b_1b_2)a(b_1b_2)^{-1} = b_1(b_2ab_2^{-1})b_1^{-1} = b_1a^2b_1^{-1}$$
$$= (b_1ab_1^{-1})^2 = a^4 = a,$$

and moreover b_1b_2 is not e and hence is an element of order 2, since each element of H is its own inverse. Thus, in either case H contains an element c of order 2 that commutes with a and hence also with a^2. By Theorem 24.4, $[a][c]$ is a subgroup and is, moreover, the direct product of $[a]$ and $[c]$ by Theorem 24.2; thus, $K = [a][c]$ is a cyclic group of order 6 by Corollary 24.8.2.

By Theorem 18.6, K is a normal subgroup, and G/K is a (cyclic) group of order 2. Let x be a generator of $K, y \notin K$. By Theorem 25.2, there exist $r \in N_6, s \in Z^*$ such that $y^2 = x^r, yxy^{-1} = x^s, s^2 \equiv 1 \pmod 6$, $sr \equiv r \pmod 6$, and moreover, G is commutative if $s \equiv 1 \pmod 6$. As $s^2 \equiv 1 \pmod 6$, either $s \equiv 1 \pmod 6$ or $s \equiv 5 \equiv -1 \pmod 6$.

Case 1(a): $s \equiv 1 \pmod 6$ and r is even. Let $r = 2t$ and let $z = yx^{3-t}$. Then $z \notin K$ and $z^2 = y^2x^{-2t} = e$, so $K[z]$ is a subgroup by Theorem 24.4 that is the direct product of K and $[z]$ by Theorem 24.2; hence, $K[z]$ has 12 elements and thus is G. Therefore, G is the direct product of cyclic subgroups of orders 6 and 2 and hence is isomorphic to $Z_6 \times Z_2$. Case 1(b): $s \equiv 1 \pmod 6$ and r is odd. Let $r = 2t + 1$, and let $z = yx^{3-t}$. Then $z^2 = y^2x^{-2t} = x$, so the order of z divides 12, but since $z^6 = x^3 \neq e$ and $z^4 = x^2 \neq e$, the order of z is a divisor of neither 6 nor 4; hence, the order of z is 12, so G is cyclic and thus isomorphic to Z_{12}. Case 1(c): $s \equiv -1 \pmod 6$. Then $-r \equiv r \pmod 6$, so either $r = 0$ or $r = 3$. By Theorem 25.2, G is isomorphic to the dihedral group of order 12 in the former case and to the dicyclic group of order 12 in the latter.

Case 2: $[a]$ is not a normal subgroup. Let $S = [a]$, a Sylow 3-subgroup of G. We shall first show that for each $c \in G$, $c^\wedge: xS \mapsto cxS$ is a well-defined injection from G/S into G/S. Indeed, $xS = yS$ if and only if $y^{-1}x \in S$, and $cxS = cyS$ if and only if $(cy)^{-1}(cx) \in S$; thus, as $(cy)^{-1}(cx) = y^{-1}c^{-1}cx = y^{-1}x$, we conclude that $xS = yS$ if and only if $cxS = cyS$. Furthermore, $c^\wedge$ is a permutation of G/S, for if $z \in G$, $c^\wedge(c^{-1}zS) = zS$. Clearly,

$(cd)^\wedge = c^\wedge d^\wedge$ for all $c, d \in G$. Thus, $F: c \mapsto c^\wedge$ is a homomorphism from G into the group $\mathfrak{S}_{G/S}$ of all permutations of G/S. If $c^\wedge$ is the identity permutation, then $cxS = xS$ for all $x \in G$, so in particular $c \in ceS = eS = S$. The kernel of F is, therefore, a normal subgroup of G contained in S. Since S has prime order, $\{e\}$ and S are the only subgroups of S; consequently, because S is not normal, the kernel of F is $\{e\}$, so G is isomorphic to a group of permutations of the set G/S. As G/S has four elements, $\mathfrak{S}_{G/S}$ is clearly isomorphic to $\mathfrak{S}_4$ (Exercise 7.5); hence, G is isomorphic to a subgroup of $\mathfrak{S}_4$ of order 12. By Theorem 26.7, G is isomorphic to $\mathfrak{A}_4$.

Our further study of subgroups of a finite group requires the following general theorem.

Theorem 26.9. If K is a normal subgroup of a group $(G, \cdot)$ and if H is a subgroup of G, then $f: x \mapsto xK$ is an epimorphism from H onto HK/K with kernel $H \cap K$, and consequently, $g: x(H \cap K) \mapsto xK$ is an isomorphism from $H/(H \cap K)$ onto HK/K.

Proof. By Theorem 24.4, HK is indeed a subgroup of G. Clearly, f is a homomorphism. If $x \in H$, $y \in K$, then $xyK = xK = f(x)$; hence, f is surjective. The kernel of f is clearly $H \cap K$. The final assertion, therefore, follows from Theorem 20.9.

Let S be a subset of a group $(G, \cdot)$. A subset T of G is a **conjugate** of S if $T = aSa^{-1}$ for some $a \in G$. Thus, the conjugates of S are simply the images of S under the inner automorphisms of G. If S is a subgroup, then a conjugate of S is a subgroup and hence is called a **conjugate subgroup** of S. We shall denote by $\mathrm{Conj}(S)$ the set of all conjugates of a subset S of G.

Let H be a subgroup of G. A subset T of G is an H-**conjugate** of S if $T = aSa^{-1}$ for some $a \in H$. Thus, the H-conjugates of S are simply the images of S under the inner automorphisms of G arising from elements of H. We shall denote by $\mathrm{Conj}_H(S)$ the set of all H-conjugates of S.

Definition. If S is a subgroup of a group $(G, \cdot)$, the **normalizer** of S is the set $N(S)$ defined by

$$N(S) = \{a \in G: aSa^{-1} = S\}.$$

If H and S are subgroups of G, the H-**normalizer** of S is the set $N_H(S)$ defined by

$$N_H(S) = \{a \in H: aSa^{-1} = S\}.$$

Thus, $N_H(S) = N(S) \cap H$.

Theorem 26.10. If S is a subgroup of a group $(G, \cdot)$, then $N(S)$ is a subgroup of G that contains S, and S is a normal subgroup of $N(S)$.

The proof is easy.

Theorem 26.11. Let p be a prime. If H is a subgroup of order p^r of a finite group G and if T is a Sylow p-subgroup of G, then

$$N_H(T) = H \cap T.$$

Proof. Clearly $H \cap T \subseteq N_H(T) \subseteq H$, so we need only prove that $N_H(T) \subseteq T$. Let p^m be the highest power of p that divides the order g of G, and let $g = p^m q$, so that $p \nmid q$. By Theorem 26.10, T is a normal subgroup of $N(T)$, and $N_H(T)$ is also a subgroup of $N(T)$. Hence, by Theorem 26.9, the groups $N_H(T)/(T \cap N_H(T))$ and $N_H(T)T/T$ are isomorphic. The order of the former divides the order of $N_H(T)$, a subgroup of H, and hence is p^s for some $s \in [0, r]$. The order of $N_H(T)T$ is a multiple of the order p^m of T and a divisor of the order $p^m q$ of G; hence, the order of $N_H(T)T$ is $p^m t$ where $p \nmid t$, and consequently, the order of $N_H(T)T/T$ is t. Since $t = p^s$ and $p \nmid t$, we conclude that $t = 1$. Thus, $T \cap N_H(T) = N_H(T)$, so $N_H(T) \subseteq T$.

Theorem 26.12. Let p be a prime, let G be a finite group of order g, let p^m be the highest power of p that divides g, and let S be a Sylow p-subgroup of G.

1° (Sylow's Second Theorem) If H is a p-subgroup of G, then H is contained in a conjugate subgroup of S. In particular, every Sylow p-subgroup of G is a conjugate subgroup of S.

2° (Sylow's Third Theorem) The number n_p of Sylow p-subgroups of G is a divisor of g/p^m and satisfies the congruence

$$n_p \equiv 1 \pmod{p};$$

moreover,

$$n_p = 1 + k_1 p + k_2 p^2 + \ldots + k_m p^m$$

where for each $r \in [1, m]$, the number of Sylow p-subgroups T for which $S \cap T$ has order p^{m-r} is $k_r p^r$.

Proof. Let $g = p^m q$, whence $p \nmid q$. For any subgroup H of G, $(\text{Conj}(S), \cdot)$ is an H-space where . is defined by

$$a.T = aTa^{-1}$$

for all $a \in H$, $T \in \text{Conj}(S)$. Indeed, if $T \in \text{Conj}(S)$, clearly, $aTa^{-1} \in \text{Conj}(S)$, and it is easy to verify that $e.T = T$, $ab.T = a.(b.T)$ for all $T \in \mathrm{Conj}(S)$ and all $a, b \in H$. The orbit of T for this action is clearly $\mathrm{Conj}_H(T)$, and the stabilizer of T is $N_H(T)$.

Applying these remarks to the case where $H = G$, $T = S$, we conclude from 2° of Theorem 26.1 that

$$n(\mathrm{Conj}(S)) = n(G/N(S)).$$

Since S is a subgroup of $N(S)$, the order of $N(S)$ is a multiple of the order p^m of S. Therefore, $n(G/N(S))$ is a divisor of $q = g/p^m$, and in particular, $p \nmid n(\mathrm{Conj}(S))$.

Now let H be a subgroup of order p^r, and let $T \in \mathrm{Conj}(S)$. By 2° of Theorem 26.1 and by Theorem 26.11,

$$n(\mathrm{Conj}_H(T)) = n(H/N_H(T)) = n(H/(H \cap T)),$$

a divisor of p^r. Hence, if $H \subseteq T$, then $n(\mathrm{Conj}_H(T)) = 1$ as $H \cap T = H$; but if $H \nsubseteq T$, then $n(\mathrm{Conj}_H(T)) = p^s$ for some $s \geq 1$ as $H \cap T \subset H$. By 1° of Theorem 26.1, the sets $\mathrm{Conj}_H(T)$ where $T \in \mathrm{Conj}(S)$ form a partition of $\mathrm{Conj}(S)$. Hence, if H were not contained in any conjugate subgroup T of S, $\mathrm{Conj}(S)$ would be the union of disjoint sets, the number of elements in each of which being a strictly positive power of p, so p would divide $n(\mathrm{Conj}(S))$, in contradiction to the result of the preceding paragraph. Thus, $H \subseteq T$ for some $T \in \mathrm{Conj}(S)$, and 1° is proved. In particular, $\mathrm{Conj}(S)$ is the set of all Sylow p-subgroups of G.

To prove 2°, we apply our remarks of the first paragraph of the proof to the case where $H = S$. Let $T \in \mathrm{Conj}(S)$. Then

$$n(\mathrm{Conj}_S(T)) = n(S/N_S(T)) = n(S/(S \cap T))$$

by 2° of Theorem 26.1 and by Theorem 26.11. Hence, $S \cap T$ has p^{m-r} elements if and only if $n(\mathrm{Conj}_S(T)) = p^r$. For each $r \in [0, m]$, let k_r be the number of orbits of the S-space $\mathrm{Conj}(S)$ that have p^r members. Then

$$n(\mathrm{Conj}(S)) = k_0 + k_1 p + \ldots + k_m p^m.$$

However, $S \cap T$ has p^m elements if and only if $T = S$, so $n(\mathrm{Conj}_S(T)) = 1$ if and only if $T = S$, and consequently, $k_0 = 1$. As $\mathrm{Conj}(S)$ is the set of all Sylow p-subgroups of G, therefore,

$$n_p = 1 + k_1 p + k_2 p^2 + \ldots + k_m p^m,$$

and in particular,

$$n_p \equiv 1 \pmod{p}.$$

The set of all Sylow subgroups T such that $S \cap T$ has p^{m-r} members is thus the union of k_r mutually disjoint sets, each containing p^r members. Hence, there are $k_r p^r$ Sylow p-subgroups T such that the order of $S \cap T$ is p^{m-r}.

Our final application of Sylow's theorems is to classify groups of order pq, where p and q are primes such that $p < q$. If G is a commutative group of order pq, G contains (normal) subgroups of orders p and q by Sylow's First Theorem and hence is cyclic by Corollary 24.8.1.

Suppose, therefore, that there exists a noncommutative group G of order pq. Let n_q be the number of Sylow q-subgroups of G. Because each has prime order q, the intersection of any two is a proper subgroup of each and hence is $\{e\}$. Thus, each element of order q belongs to only one Sylow q-subgroup. Since each Sylow q-subgroup contains $q - 1$ elements of order q, G contains $n_q(q - 1)$ elements of order q. Suppose that $n_q > 1$. Then $n_q \geq q + 1$ by Sylow's Third Theorem, so G would have at least $(q + 1)(q - 1) = q^2 - 1$ elements of order q. The neutral element is not among them, so G would have at least q^2 elements, in contradiction to the fact that the order of G is pq where $p < q$. Hence, $n_q = 1$, so G contains only one Sylow q-subgroup S. The only conjugate subgroup of S is, therefore, S itself, so S is a normal subgroup of G, and G/S is a (cyclic) group of order p. Let a be a generator of S, and let $b \notin [a]$. By Theorem 25.2, there exist $s \in Z^*$ and $r \in N_q$ such that $b^p = a^r$, $bab^{-1} = a^s$, $s^p \equiv 1 \pmod{q}$, $sr \equiv r \pmod q$, and moreover, $s \not\equiv 1 \pmod q$, since G is not commutative. Denoting by $\bar{x}$ the coset $x + (q)$ in Z_q corresponding to an integer x, we conclude that $\bar{s}^p = 1$ but $\bar{s} \neq 1$, so $\bar{s}$ is an element of order p in the multiplicative group Z_q^*. Hence, $p \mid q - 1$. Thus, if there is a noncommutative group of order pq, then $q \equiv 1 \pmod p$.

Conversely, suppose that $q \equiv 1 \pmod p$. Then $p \mid q - 1$, so by Sylow's First Theorem, applied to the multiplicative group Z_q^*, there exists $s \in Z^*$ such that $s^p \equiv 1 \pmod q$, $s \not\equiv 1 \pmod q$. Consequently, $G(q, p, 0, s)$, for example, is a nonabelian group of order pq. Thus, we have proved the following theorem:

Theorem 26.13. Let p and q be primes such that $p < q$. If $q \not\equiv 1 \pmod p$, then a group of order pq is cyclic. If $q \equiv 1 \pmod p$, an abelian group of order pq is cyclic, but there exist nonabelian groups of order pq.

In §38 we shall prove that the multiplicative group F^* of a finite field F is cyclic. This information enables us to classify completely all groups of order pq. Once again, assume that $p \mid q - 1$. By Theorem 23.9, the group Z_q^* contains exactly one subgroup of order p. As we have seen, $G(q, p, 0, s)$ is a nonabelian group of order pq, where $s^p \equiv 1 \pmod q$, $s \not\equiv 1 \pmod q$. To show that, to within isomorphism, this is the only nonabelian group of order pq, it suffices to consider only the groups $G(q, p, r, t)$ where $t^p \equiv 1$ (mod q), $t \not\equiv 1$ (mod q), $tr \equiv r$ (mod q) by Theorem 25.2, for we have seen that a group of order pq contains a normal subgroup of order q. Since Z_q is a field, if $r \not\equiv 0$ (mod q), then from $tr \equiv r$ (mod q) we conclude that $t \equiv 1$ (mod q), a contradiction; hence, $r \equiv 0$ (mod q), whence $r = 0$, as $r \in N_q$. Since Z_q^* has only one subgroup of order p, both $\bar{s}$ and $\bar{t}$ are generators of that subgroup, so $\bar{s} = \bar{t}^i$ for some $i \in [1, q-1]$, whence $s \equiv t^i$ (mod q). Let a, b be generators of $G(q, p, 0, t)$ such that $a^q = b^p = e$, $bab^{-1} = a^t$, and let $b_1 = b^i$. Then b_1 is also a generator of $[b]$ as $[b]$ has prime order p, and $b_1 a b_1^{-1} = a^{t^i} = a^s$ by Theorem 25.2, so $G(q, p, 0, t)$ is isomorphic to $G(q, p, 0, s)$ by the final assertion of that theorem. In sum, using the fact that the multiplicative group of nonzero elements of a finite field is cyclic, we have proved the following theorem:

Theorem 26.14. If p and q are primes such that $p < q$ and $q \equiv 1$ (mod p), then a group of order pq is either cyclic or isomorphic to $G(q, p, 0, s)$ where $s \in Z^*$ is such that $s^p \equiv 1$ (mod q), $s \not\equiv 1$ (mod q).

EXERCISES

26.1. (a) Prove Theorem 26.2.
(b) Prove that $a \mapsto \kappa_a$ is a homomorphism from G into Aut(G) whose kernel is the center $Z(G)$ of G.

*26.2. Let p be a prime. For every $m \in N^*$, every subgroup of order p^{m-1} of a group of order p^m is a normal subgroup. [Proceed by induction on m by considering $G/[a]$, where a is an element of order p in $Z(G)$; use Exercise 20.12.]

26.3. Verify the assertion made at the beginning of the proof of Theorem 26.7.

26.4. Prove Theorem 26.10.

26.5. Show that the group of symmetries of a cube is isomorphic to $\mathfrak{S}_4$. [Number the diagonals of the cube.]

26.6. (a) If a group of order 104 contains no normal subgroup of order 8, how many subgroups of order 8 does it contain?
(b) If a group of order 80 contains no normal subgroup of order 5, how many elements of order 5 does it contain?

26.7. If the Sylow subgroups of a finite group G are all normal, then G is their direct product.

26.8. For each $n \in [1, 15]$, list to within isomorphism all the groups of order n.

*26.9. Let q be a prime > 3, and let G be a group of order $4q$.

(a) G contains a normal subgroup S of order q.

(b) If G contains a cyclic subgroup of order $2q$, then G is isomorphic to one and only one of the following four groups: Z_{4q}, $Z_{2q} \times Z_2$, the dihedral group of order $4q$, the dicyclic group of order $4q$. [Use Theorem 25.2.]

(c) If G contains no cyclic subgroup of order $2q$, then $q \equiv 1 \pmod 4$, and G is isomorphic to $G(q, 4, 0, s)$ where $s^4 \equiv 1 \pmod q$, $s^2 \not\equiv 1 \pmod q$. [First show that G/S is cyclic; argue as in the proof of Theorem 26.14.]

(d) If $q = 5$, which of the five groups mentioned in (b) and (c) is $\mathrm{Aut}(D_5)$ (Exercise 25.7) isomorphic to?

*26.10. (a) To within isomorphism there are two groups of order 847, both of which are abelian.

(b) State and prove a theorem concerning groups of order p^2q where p and q are distinct primes that generalizes (a).

*26.11. (a) To within isomorphism there are four groups of order 1225, all of which are abelian.

(b) State and prove a theorem concerning groups of order p^2q^2 where p and q are distinct primes that generalizes (a).

CHAPTER V

VECTOR SPACES

An algebraic structure is by definition a set together with one or two compositions on that set. In this chapter we shall consider vector spaces and modules, which are important examples of a new kind of algebraic object, consisting of an algebraic structure E, a ring K, and a function from $K \times E$ into E.

27. Vector Spaces and Modules

Addition on the set R of real numbers induces on the plane R^2 of analytic geometry a composition, also called addition, defined by

$$(\alpha_1, \alpha_2) + (\beta_1, \beta_2) = (\alpha_1 + \beta_1, \alpha_2 + \beta_2).$$

Thus, $(R^2, +)$ is simply the cartesian product of $(R, +)$ and $(R, +)$. Under addition, R^2 is an abelian group. If (α_1, α_2), (β_1, β_2), and the origin $(0, 0)$ do not lie on the same straight line, the line segments joining the origin $(0, 0)$ to (α_1, α_2) and to (β_1, β_2) respectively are two adjacent sides of a parallelogram, and $(\alpha_1, \alpha_2) + (\beta_1, \beta_2)$ and $(0, 0)$ are the endpoints of one of the diagonals of that parallelogram (Figure 13). For every $n \in N^*$, by definition $n.(\alpha_1, \alpha_2)$ is the sum of (α_1, α_2) with itself n times; it is easy to see by induction that $n.(\alpha_1, \alpha_2) = (n\alpha_1, n\alpha_2)$, an equality that holds also if $n \leq 0$. Consequently, if we define

$$\lambda.(\alpha_1, \alpha_2) = (\lambda\alpha_1, \lambda\alpha_2)$$

for every real number λ, we have generalized for this particular group the notion of adding an element to itself n times. If $(\alpha_1, \alpha_2) \neq (0, 0)$, then

252

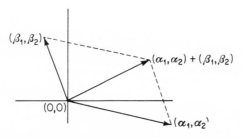

Figure 13

$\{\lambda.(\alpha_1, \alpha_2): \lambda \in \mathbf{R}\}$ is simply the set of all points on the line through (α_1, α_2) and the origin (Figure 14). The directed line segment from the origin to $\lambda.(\alpha_1, \alpha_2)$ is $|\lambda|$ times as long as that from the origin to (α_1, α_2) and has the same direction as the latter if $\lambda > 0$ but the opposite direction if $\lambda < 0$. The function $(\lambda, x) \to \lambda.x$ from $\mathbf{R} \times \mathbf{R}^2$ into $\mathbf{R}^2$ is denoted simply by . and is called "scalar multiplication." It is easy to verify that with addition and scalar multiplication so defined, $(\mathbf{R}^2, +, .)$ is a vector space over the field of real numbers:

Definition. Let K be a division ring. A **vector space over** K or a K**-vector space** is an ordered triple $(E, +, .)$ such that $(E, +)$ is an abelian group and . is a function from $K \times E$ into E satisfying

(VS 1) $\lambda.(x + y) = \lambda.x + \lambda.y$

(VS 2) $(\lambda + \mu).x = \lambda.x + \mu.x$

(VS 3) $(\lambda\mu).x = \lambda.(\mu.x)$

(VS 4) $1.x = x$

for all $x, y \in E$ and all $\lambda, \mu \in K$. Elements of E are called **vectors,** elements of K are called **scalars,** and . is called **scalar multiplication.**

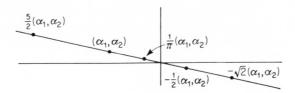

Figure 14

Similarly, $(R^n, +, .)$ is a vector space over R for each $n \in N^*$ where addition is defined by

$$(\alpha_1, \ldots, \alpha_n) + (\beta_1, \ldots, \beta_n) = (\alpha_1 + \beta_1, \ldots, \alpha_n + \beta_n)$$

and scalar multiplication by

$$\lambda.(\alpha_1, \ldots, \alpha_n) = (\lambda\alpha_1, \ldots, \lambda\alpha_n).$$

For the case $n = 3$, addition and scalar multiplication may be described geometrically exactly as before.

In physics a vector is sometimes described as an entity having magnitude and direction (such as a force acting on a point, a velocity, or an acceleration) and is represented by a directed line segment or an arrow lying either in the coordinate plane of analytic geometry or in space. In this description, two arrows represent the same vector if they are parallel and similarly directed and if they have the same length. Consequently, each vector is represented uniquely by an arrow emanating from the origin. Geometric definitions of the sum of two vectors and the product of a real number and a vector are given in such descriptions in terms of arrows and parallelograms, but it is apparent that what essentially is being described in geometric language is the vector space $(R^2, +, .)$ or $(R^3, +, .)$, depending on whether the arrows representing the vectors are considered to lie all in a plane or in space.

An important generalization of vector spaces is obtained by relaxing the requirement that K be a division ring.

Definition. Let K be a ring with identity. A **module over** K or a K-**module** is an ordered triple $(E, +, .)$ such that $(E, +)$ is an abelian group and . is a function from $K \times E$ into E satisfying conditions (VS 1), (VS 2), (VS 3), and (VS 4) for all $x, y \in E$ and all $\lambda, \mu \in K$. Elements of K are called **scalars,** and K itself is called the **scalar ring.**

Thus, a vector space is a module whose scalar ring is a division ring. Our primary interest is in vector spaces, but we shall prove theorems for modules when specializing to vector spaces would yield no simplifications.

Except in the unusual case where E and K are the same set, scalar multiplication is not a composition on a set, and hence a module is not an algebraic structure as we have defined the term in §6. To provide a suitable framework for the definition of "isomorphism" we make the following definition:

Definition. Let K be a ring. A K-**algebraic structure with one composition** is an ordered triple $(E, +, .)$ where $(E, +)$ is an algebraic structure with one composition and where . is a function (called **scalar multiplication**) from

$K \times E$ into E. A K-**algebraic structure with two compositions** is an ordered quadruple $(E, +, \cdot, .)$ where $(E, +, \cdot)$ is an algebraic structure with two compositions and where $.$ is a function from $K \times E$ into E. A K-**algebraic structure** is simply a K-algebraic structure with either one or two compositions.

Our definition is quite artificial in allowing only binary operations on E rather than n-ary operations, in restricting their number to one or two, and in demanding that K be a ring, but it is sufficiently general for our purposes.

The following definition formally expresses what is meant by saying that two K-algebraic structures are "just like" each other:

Definition. If $(E, +, .)$ and $(F, \oplus, .)$ are K-algebraic structures with one composition, a bijection f from E onto F is an **isomorphism** from $(E, +, .)$ onto $(F, \oplus, .)$ if

$$f(x + y) = f(x) \oplus f(y)$$
$$f(\lambda.x) = \lambda.f(x)$$

for all $x, y \in E$ and all $\lambda \in K$. If $(E, +, \cdot, .)$ and $(F, \oplus, \odot, .)$ are K-algebraic structures with two compositions, a bijection f from E onto F is an **isomorphism** from $(E, +, \cdot, .)$ onto $(F, \oplus, \odot, .)$ if the above two conditions hold and if, in addition,

$$f(x \cdot y) = f(x) \odot f(y)$$

for all $x, y \in E$. An **automorphism** of a K-algebraic structure is an isomorphism from itself onto itself. If there exists an isomorphism from one K-algebraic structure onto another, we shall say that they are **isomorphic**.

The analogue of Theorem 6.1 for K-algebraic structures is valid and easily proved:

Theorem 27.1. Let K be a ring, and let E, F, and G be K-algebraic structures with the same number of compositions.

1° The identity function 1_E is an automorphism of the K-algebraic structure E.
2° If f is a bijection from E onto F, then f is an isomorphism from the K-algebraic structure E onto the K-algebraic structure F if and only if $f^{\leftarrow}$ is an isomorphism from F onto E.
3° If f and g are isomorphisms from E onto F and from F onto G respectively, then $g \circ f$ is an isomorphism from E onto G.

It is easy to verify also that *a K-algebraic structure isomorphic to a K-module is itself a K-module*. In particular, if K is a division ring, *a K-algebraic structure isomorphic to a K-vector space is itself a K-vector space*.

If $(E, +, .)$ is a vector space over K, it is customary to speak of "the vector space E" when $(E, +, .)$ is meant, "the additive group E" when $(E, +)$ is meant. A similar convention applies to modules.

In any discussion concerning a K-module E, *the symbol "0" has two possible meanings:* It denotes the zero element of the scalar ring K and also the zero element of the additive group E. In any context it will be clear which is meant. In both (1) and (8) below, for example, the first and third occurrences of "0" denote the zero element of E, but the second occurrence denotes the zero element of K. In the notation for scalar multiplication, the dot in "$\lambda.x$" is usually omitted, and thus, "λx" often denotes the scalar product of λ and x.

Theorem 27.2. Let E be a K-module. If $x \in E$, $\lambda \in K$, and $n \in Z$, then

(1) $$\lambda 0 = 0x = 0$$

(2) $$\lambda(-x) = (-\lambda)x = -(\lambda x)$$

(3) $$\lambda(n.x) = n.(\lambda x) = (n.\lambda)x$$

(4) $$(-1)x = -x$$

(5) $$n.x = (n.1)x.$$

If $(x_k)_{1 \leq k \leq m}$ is a sequence of elements of E and if $(\lambda_k)_{1 \leq k \leq m}$ is a sequence of scalars, then

(6) $$\lambda \left(\sum_{k=1}^{m} x_k \right) = \sum_{k=1}^{m} \lambda x_k$$

(7) $$\left(\sum_{k=1}^{m} \lambda_k \right) x = \sum_{k=1}^{m} \lambda_k x$$

If K is a division ring, then for every vector x and every scalar λ,

(8)   if $\lambda x = 0$, then either $\lambda = 0$ or $x = 0$.

Proof. By (VS 1), $y \mapsto \lambda y$ is an endomorphism of the additive group E. Hence, $\lambda 0 = 0$ and $\lambda(-x) = -(\lambda x)$ by Theorem 20.5, (6) holds by induction, and $\lambda(n.x) = n.(\lambda x)$ by the Corollary of Theorem 20.5. By (VS 2), $\lambda \mapsto \lambda x$ is a homomorphism from the additive group K into the additive group E. Hence, $0x = 0$ and $(-\lambda)x = -(\lambda x)$ by Theorem 20.5 (in particular, $(-1)x = -x$ by (VS 4)), (7) holds by induction, and $(n.\lambda)x = n.(\lambda x)$ by the Corollary of Theorem 20.5 (in particular, $(n.1)x = n.x$ by (VS 4)). Finally, if K is a division ring and if $\lambda x = 0$ but $\lambda \neq 0$, then by (1),

$$0 = \lambda^{-1}.0 = \lambda^{-1}(\lambda x) = (\lambda^{-1}\lambda)x = 1.x = x.$$

Example 27.1. The definition of the R-vector space R^n given earlier may be generalized by replacing R with any ring with identity K: For every $n \in N^*$, addition on K^n and scalar multiplication are defined by

$$(\alpha_1, \ldots, \alpha_n) + (\beta_1, \ldots, \beta_n) = (\alpha_1 + \beta_1, \ldots, \alpha_n + \beta_n)$$

$$\lambda.(\alpha_1, \ldots, \alpha_n) = (\lambda\alpha_1, \ldots, \lambda\alpha_n).$$

It is easy to verify that $(K^n, +, .)$ is indeed a K-module, and when we refer to the K-module K^n, we shall always have the K-module $(K^n, +, .)$ just defined in mind.

Example 27.2. If L is a ring with identity and if K is a subring of L containing the identity element 1, then $(L, +, .)$ is a K-module where $+$ is the additive composition of L and where scalar multiplication is the restriction to $K \times L$ of the multiplicative composition of L. Whenever L is a ring with identity and K is a subring of L containing the identity element 1, by the K-module L we shall always have the K-module $(L, +, .)$ just defined in mind. In particular, if K is a division subring of a ring L and if the identity element of K is that of L, we may regard L as a K-vector space. With this definition of scalar multiplication, for example, both R and C are R-vector spaces, and any division ring is a vector space over its prime subfield.

Example 27.3. If L is a ring with identity, if K is a subring of L containing the identity element 1, and if $(E, +, .)$ is an L-module, then $(E, +, ._K)$ is a K-module where $._K$ is the restriction of . to $K \times E$. The K-module $(E, +, ._K)$ is called the *K-module obtained from $(E, +, .)$ by restricting scalar multiplication*. In particular, if K is a division subring of a division ring L and if E is an L-vector space, E may also be regarded in this way as a K-vector space. Example 27.2 is the special case of this example where E is the L-module L.

Example 27.4. Let F be a K-module, E a set. Then $(F^E, +, .)$ is a K-module where $+$ is defined by

$$(f + g)(x) = f(x) + g(x)$$

for all $x \in E$, and where . is defined by

$$(\lambda f)(x) = \lambda f(x)$$

for all $x \in E$. Indeed, as noted in the discussion of Example 15.5, $+$ on F^E is associative, the neutral element for $+$ is the function $x \mapsto 0$, and the additive inverse of $f \in F^E$ is the function $-f\colon x \mapsto -f(x)$. The verification of (VS 1)–(VS 4) is also easy. The most important case of this example is that where F is the K-module K. Whenever we refer to the K-module F^E of all functions from E into F, we shall have the K-module $(F^E, +, .)$ just defined in mind.

Example 27.5. If $E_1, \ldots, E_n$ are K-modules and if $(E, +)$ is the cartesian product of the additive groups $E_1, \ldots, E_n$, then $(E, +, .)$ is a K-module where scalar multiplication is defined by

$$\lambda.(x_1, \ldots, x_n) = (\lambda.x_1, \ldots, \lambda.x_n).$$

Example 27.1 is the special case of this example where each E_k is the K-module K.

Example 27.6. If $(E, +)$ is an abelian group, then $(E, +, .)$ is a Z-module where . is the function from $Z \times E$ into E defined in §11 and §14. The Z-module $(E, +, .)$ is called the Z-module *associated* with $(E, +)$.

EXERCISES

27.1. Draw a diagram illustrating (VS 1) for the R-vector space R^2 similar to that illustrating vector addition on R^2.

27.2. Prove Theorem 27.1 and the assertion following.

27.3. Verify that (VS 1)–(VS 4) hold in Examples 27.1–27.5.

27.4. Which of (VS 1)–(VS 4) are satisfied if scalar multiplication is the function from $C \times C$ into C defined by $z.w = |z| w$? by $z.w = \mathscr{R}(z)w$? by $z.w = 0$? Which are satisfied if scalar multiplication is the function from $C \times C^2$ into C^2 defined by

$$z.(u, v) = \begin{cases} (zu, zv) & \text{if } v \neq 0 \\ (\bar{z}u, 0) & \text{if } v = 0? \end{cases}$$

Infer that no one of (VS 1)–(VS 4) is implied by the other three.

27.5. If E is a K-vector space containing more than one element, then for each $\lambda \in K^*$, the function $x \mapsto \lambda x$ is an automorphism of E if and only if λ belongs to the center of K.

27.6. An isomorphism from an abelian group E onto an abelian group F is also an isomorphism from the associated Z-module E onto the associated Z-module F.

27.7. What are $\oplus$ and $.$ if f is an isomorphism from the R-vector space R^2 onto the R-vector space $(R^2, \oplus, .)$, where f is defined by

$$f(x, y) = (x + y - 1, x - y + 1)?$$

where f is defined by

$$f(x, y) = \begin{cases} (x^2 - y^2, x - y) \text{ if } x \neq y \\ (x, 0) \text{ if } x = y? \end{cases}$$

27.8. If E is a K-module, then $\{\lambda \in K : \lambda x = 0 \text{ for all } x \in E\}$ is an ideal of K, called the **annihilator** of E.

27.9. Let E be a module over an integral domain K. What are the possible orders of a nonzero element of the abelian group E if the characteristic of K is a prime p? if the characteristic of K is zero? What are the possible orders if K is a field whose characteristic is zero?

27.10. Let $(E, +)$ be an abelian group.
   (a) There is exactly one function $.$ from $Z \times E$ into E such that $(E, +, .)$ is a Z-module.
   (b) If the order of every element of the abelian group E is finite and divides n, there is exactly one function $.$ from $Z_n \times E$ into E such that $(E, +, .)$ is a Z_n-module.
   (c) In particular, if p is a prime and if every nonzero element of E has order p, there is a unique function $.$ from $Z_p \times E$ into E such that $(E, +, .)$ is a vector space over Z_p.

*27.11. If $(E, +, .)$ is a Q-algebraic structure satisfying (VS 2), (VS 3), and (VS 4) for all $\lambda \in Q$ and all $x \in E$ and if $(E, +)$ is an abelian group, then $(E, +, .)$ is a Q-vector space.

*27.12. An abelian group $(E, +)$ is **divisible** if $n.E = E$ for every $n \in Z^*$; that is, if for each $y \in E$ and each $n \in Z^*$ there exists $x \in E$ such that $n.x = y$. If $(E, +)$ is an abelian group, then there is a function $.$ from $Q \times E$ into E such that $(E, +, .)$ is a Q-vector space if and only if $(E, +)$ is a divisible group all of whose nonzero elements have infinite order.

*27.13. Let $(E, +, .)$ be a K-module. The module E is **torsion-free** if for all $\lambda \in K$ and all $x \in E$, if $\lambda x = 0$, then either $\lambda = 0$ or $x = 0$. The module E is a **divisible** K-module if $\lambda E = E$ for all $\lambda \in K^*$; that is, if for each $y \in E$ and each $\lambda \in K^*$ there exists $x \in E$ such that $\lambda x = y$. If K is an integral domain and if L is a quotient field of K, there is a function . from $L \times E$ into E such that $(E, +, .)$ is an L-vector space and $(E, +, .)$ is the K-module obtained from $(E, +, .)$ by restricting scalar multiplication if and only if $(E, +, .)$ is a divisible torsion-free module. Infer the statement of Exercise 27.12 as a special case.

28. Subspaces and Bases

If E is a K-algebraic structure, a subset M of E is said to be **stable for scalar multiplication** if $\lambda.x \in M$ for all $\lambda \in K$ and all $x \in M$, and M is called a **stable subset** of E if M is stable for scalar multiplication and the composition(s) of E. Analogous to the definition of subgroup and subsemigroup is the following definition of vector subspace and submodule:

Definition. Let K be a division ring (a ring with identity), let $(E, +, .)$ be a K-algebraic structure with one composition, and let M be a stable subset of E. If $(M, +_M, \cdot_M)$ is a K-vector space (a K-module) where $+_M$ is the restriction of addition to $M \times M$ and where $\cdot_M$ is the restriction of scalar multiplication to $K \times M$, we shall call $(M, +_M, \cdot_M)$ a **(vector) subspace** (a **submodule**) of E.

As for subgroups and subsemigroups, however, a stable subset M of E is called a subspace (a submodule) of E if $(M, +_M, \cdot_M)$ is a subspace (a submodule) according to our formal definition.

Theorem 28.1. If E is a K-module, then a nonempty subset M of E is a submodule if and only if for all $x, y \in M$ and all $\lambda \in K$, $x + y$ and λx belong to M, that is, if and only if M is a stable subset of E.

Proof. If the condition is fulfilled, then $-x = (-1)x \in M$ if $x \in M$, so M is a subgroup of $(E, +)$ by Theorem 8.3 and consequently is a submodule.

Example 28.1. If E is a K-module, then E and $\{0\}$ are submodules of E.

Example 28.2. Let us find all subspaces of the R-vector space R^2. Let M be a nonzero subspace of R^2. Then M contains a nonzero vector (α_1, α_2) and consequently also $\{\lambda(\alpha_1, \alpha_2): \lambda \in R\}$, a set that may be described geometrically as the line through (α_1, α_2) and the origin. Suppose that M contains a vector (β_1, β_2) not on that line. Then $\alpha_1\beta_2 - \alpha_2\beta_1 \neq 0$, for otherwise (β_1, β_2) would be $\zeta(\alpha_1, \alpha_2)$ where $\zeta = \beta_1/\alpha_1$ or $\zeta = \beta_2/\alpha_2$ according as $\alpha_1 \neq 0$ or $\alpha_2 \neq 0$. But then

$M = \mathbf{R}^2$, for if (γ_1, γ_2) is any vector whatever,

$$(\gamma_1, \gamma_2) = \lambda(\alpha_1, \alpha_2) + \mu(\beta_1, \beta_2)$$

where

$$\lambda = \frac{\gamma_1\beta_2 - \gamma_2\beta_1}{\alpha_1\beta_2 - \alpha_2\beta_1}, \qquad \mu = \frac{\alpha_1\gamma_2 - \alpha_2\gamma_1}{\alpha_1\beta_2 - \alpha_2\beta_i},$$

as we see by solving simultaneously the equations

$$\alpha_1\lambda + \beta_1\mu = \gamma_1$$
$$\alpha_2\lambda + \beta_2\mu = \gamma_2.$$

Thus, a subspace of $\mathbf{R}^2$ is either the whole plane $\mathbf{R}^2$, or a line through the origin, or $\{0\}$. Conversely, every line through the origin consists of all scalar multiples of any given nonzero vector on it and hence is easily seen to be a subspace.

Example 28.3. If F is a K-module and if E is a set, the set $F^{(E)}$ of all functions f from E into F such that $f(x) = 0$ for all but finitely many elements x of E is a submodule of the K-module F^E.

Example 28.4. Let K be a commutative ring with identity. For each $k \in N$, let p_k be the function from K into K defined by $p_k(x) = x^k$ for all $x \in K$. A function p from K into K is a **polynomial function** on K if there exists a sequence $(\alpha_k)_{0 \leq k \leq n}$ of elements of K such that

$$p = \sum_{k=0}^{n} \alpha_k p_k.$$

The set $P(K)$ of all polynomial functions on K is a submodule of the K-module K^K. If m is a given natural number, the set $P_m(K)$ of all the polynomial functions $\sum_{k=0}^{m-1} \alpha_k P_k$ where $(\alpha_k)_{0 \leq k \leq m-1}$ is any sequence of m terms of K is a submodule of $P(K)$.

Example 28.5. Let $J = \{x \in R: a \leq x \leq b\}$. Familiar subspaces of the R-vector space R^J encountered in calculus are the space $\mathscr{C}(J)$ of all real-valued continuous functions on J, the space $\mathscr{D}(J)$ of all real-valued differentiable functions on J, the space $\mathscr{C}^{(m)}(J)$ of all real-valued functions on J having continuous derivatives of order m, the space $\mathscr{C}^{(\infty)}(J)$ of all real-valued functions on J having derivatives of all orders, and the space $\mathscr{R}(J)$ of all Riemann-integrable functions on J.

Theorem 28.2. Let E be a K-module. If $M_1, \ldots, M_n$ are submodules of E, then $M_1 + \ldots + M_n$ and $M_1 \cap \ldots \cap M_n$ are also submodules. The intersection of any set of submodules of E is a submodule; consequently, if S is a subset of E, the intersection of the set of all submodules of E containing S is the smallest submodule of E containing S.

The proof is easy. Theorem 28.2 enables us to make the following definition.

Definition. If S is a subset of a K-module E, the submodule **generated** (or **spanned**) by S is the smallest submodule M of E containing S, and S is called a **set of generators** for M. The module E is **finitely generated** if there is a finite set of generators for E.

We shall next characterize the elements of the submodule generated by an arbitrary subset S of a K-module, but before doing so, we need two preliminary definitions.

Definition. Let $(a_k)_{1 \leq k \leq n}$ be a sequence of elements of a K-module E. An element b of E is a **linear combination** of $(a_k)_{1 \leq k \leq n}$ if there exists a sequence $(\lambda_k)_{1 \leq k \leq n}$ of scalars such that

$$b = \sum_{k=1}^{n} \lambda_k a_k.$$

Definition. Let S be a subset of a K-module E. If S is not empty, an element b of E is a **linear combination** of S if b is a linear combination of some sequence $(a_k)_{1 \leq k \leq n}$ of elements of S; the zero element is the only element of E called a **linear combination** of the empty set.

If $(a_k)_{1 \leq k \leq n}$ is a sequence of elements of a K-module E, then an element b of E is a linear combination of the sequence $(a_k)_{1 \leq k \leq n}$ if and only if b is a linear combination of the set $\{a_1, \ldots, a_n\}$. Indeed, every linear combination of $(a_k)_{1 \leq k \leq n}$ is clearly a linear combination of $\{a_1, \ldots, a_n\}$. Conversely, let b be a linear combination of $\{a_1, \ldots, a_n\}$. Then there exist a sequence $(c_j)_{1 \leq j \leq m}$ of elements of $\{a_1, \ldots, a_n\}$ and a sequence $(\mu_j)_{1 \leq j \leq m}$ of scalars such that $b = \sum_{j=1}^{m} \mu_j c_j$. For each $k \in [1, n]$, if $a_k \in \{c_1, \ldots, c_m\}$ and if $a_i \neq a_k$ for all indices i such that $1 \leq i < k$, let λ_k be the sum of all the scalars μ_j such that $c_j = a_k$, but if $a_k \notin \{c_1, \ldots, c_m\}$ or if $a_i = a_k$ for some index i

such that $1 \leq i < k$, let $\lambda_k = 0$. Then clearly,

$$b = \sum_{j=1}^{m} \mu_j c_j = \sum_{k=1}^{n} \lambda_k a_k.$$

Consequently, if $(a_k)_{1 \leq k \leq n}$ and $(b_j)_{1 \leq j \leq m}$ are sequences of elements of E such that the sets $\{a_1, \ldots, a_n\}$ and $\{b_1, \ldots, b_m\}$ are identical, then an element is a linear combination of $(a_k)_{1 \leq k \leq n}$ if and only if it is a linear combination of $(b_j)_{1 \leq j \leq m}$.

Theorem 28.3. If S is a subset of a K-module E, then the submodule M generated by S is the set of all linear combinations of S.

Proof. The smallest submodule of E containing the empty set is $\{0\}$, and by definition, $\{0\}$ is the set of all linear combinations of $\emptyset$. Let S be a non-empty subset of E, and let L be the set of all linear combinations of S. Since $x = 1.x$ for all $x \in S$, $S \subseteq L$. But L is stable for addition and scalar multiplication and so by Theorem 28.1 is a submodule. Consequently, $M \subseteq L$, but as every linear combination of S clearly belongs to any sub-module of E containing S, we also have $L \subseteq M$.

Definition. Let E be a K-module. A sequence $(a_k)_{1 \leq k \leq n}$ of elements of E is **linearly independent** if for every sequence $(\lambda_k)_{1 \leq k \leq n}$ of scalars, if $\sum_{k=1}^{n} \lambda_k a_k = 0$, then $\lambda_1 = \ldots = \lambda_n = 0$. A sequence $(a_k)_{1 \leq k \leq n}$ of elements of E is **linearly dependent** if it is not linearly independent.

Thus, a sequence $(a_k)_{1 \leq k \leq n}$ of elements of a K-module E is linearly independent if and only if zero is a linear combination of $(a_k)_{1 \leq k \leq n}$ in only one way, and $(a_k)_{1 \leq k \leq n}$ is linearly dependent if and only if there is a sequence $(\lambda_k)_{1 \leq k \leq n}$ of scalars, not all of which are zero, such that $\sum_{k=1}^{n} \lambda_k a_k = 0$.

Let $(a_k)_{1 \leq k \leq n}$ be a sequence of elements of a K-module E. *If $a_j = 0$ for some $j \in [1, n]$, then $(a_k)_{1 \leq k \leq n}$ is linearly dependent*; indeed, if $\lambda_j = 1$ and $\lambda_k = 0$ for all indices k other than j, then $\sum_{k=1}^{n} \lambda_k a_k = 0$. *If $a_i = a_j$ for distinct indices i and j, then $(a_k)_{1 \leq k \leq n}$ is linearly dependent*; indeed, if $\lambda_i = 1, \lambda_j = -1$, and $\lambda_k = 0$ for all indices k other than i and j, then $\sum_{k=1}^{n} \lambda_k a_k = 0$.

Definition. Let E be a K-module. A subset S of E is **linearly independent** if every sequence $(a_k)_{1 \leq k \leq n}$ of *distinct* terms of S is a linearly independent sequence. A subset S of E is **linearly dependent** if S is not linearly independent.

Thus, a subset S of E is linearly dependent if there exist a sequence $(a_k)_{1 \leq k \leq n}$ of distinct elements of S and a sequence $(\lambda_k)_{1 \leq k \leq n}$ of scalars, not all of which are zero, such that $\sum_{k=1}^{n} \lambda_k a_k = 0$. *The empty set is linearly independent*, for there are no sequences at all of n terms of the empty set for any $n > 0$. As we saw above, *any subset of E containing zero is linearly dependent*. Clearly also, *any subset of a linearly independent set is linearly independent*, and consequently, *any set containing a linearly dependent set is linearly dependent*. By (8) of Theorem 27.2, *if a is a nonzero vector of a vector space, then $\{a\}$ is a linearly independent set*.

If $(a_k)_{1 \leq k \leq n}$ is a sequence of distinct terms of a K-module E, then $(a_k)_{1 \leq k \leq n}$ is a linearly independent sequence if and only if $\{a_1, \ldots, a_n\}$ is a linearly independent set. Indeed, if $\{a_1, \ldots, a_n\}$ is linearly independent, clearly, $(a_k)_{1 \leq k \leq n}$ is linearly independent. Conversely, suppose that $(a_k)_{1 \leq k \leq n}$ is a linearly independent sequence, let $(b_j)_{1 \leq j \leq m}$ be a sequence of distinct terms of $\{a_1, \ldots, a_n\}$, and let $(\mu_j)_{1 \leq j \leq m}$ be a sequence of scalars such that $\sum_{j=1}^{m} \mu_j b_j = 0$. For each $k \in [1, n]$, if $a_k \in \{b_1, \ldots, b_m\}$, let $\lambda_k = \mu_j$ where j is the unique index such that $a_k = b_j$, and let $\lambda_k = 0$ if $a_k \notin \{b_1, \ldots, b_m\}$. Then clearly,

$$0 = \sum_{j=1}^{m} \mu_j b_j = \sum_{k=1}^{n} \lambda_k a_k,$$

so $\lambda_k = 0$ for all $k \in [1, n]$, whence $\mu_j = 0$ for all $j \in [1, m]$, since $\{\mu_1, \ldots, \mu_m\} \subseteq \{\lambda_1, \ldots, \lambda_n\}$.

Consequently, if $(a_k)_{1 \leq k \leq n}$ is a linearly independent sequence and if $(b_j)_{1 \leq j \leq m}$ is a sequence of distinct terms such that $\{b_1, \ldots, b_m\} \subseteq \{a_1, \ldots, a_n\}$, then $(b_j)_{1 \leq j \leq m}$ is also a linearly independent sequence.

Definition. Let E be a K-module. A **basis** of E is a linearly independent set of generators for E. The module E is **free** if there exists a basis of E.

As we shall shortly see, some modules have infinite bases, others have finite bases, and still others have no bases at all. If a module has a finite basis, it is often convenient to have the elements of that basis arranged in a definite order, and therefore we make the following definition.

Definition. An **ordered basis** of a K-module E is a linearly independent sequence $(a_k)_{1 \leq k \leq n}$ of elements of E such that $\{a_1, \ldots, a_n\}$ is a set of generators for E.

Thus, by Theorem B.8, each basis of n elements determines $n!$ ordered bases.

Example 28.6. Let K be a ring with identity, let n be a strictly positive integer, and for each $j \in [1, n]$ let e_j be the ordered n-tuple of elements of K whose jth entry is 1 and all of whose other entries are 0. Then $(e_k)_{1 \le k \le n}$ is an ordered basis of the K-module K^n, since

$$\sum_{k=1}^{n} \lambda_k e_k = (\lambda_1, 0, 0, \ldots, 0) + (0, \lambda_2, 0, \ldots, 0) + \ldots + (0, 0, 0, \ldots, \lambda_n)$$

$$= (\lambda_1, \lambda_2, \lambda_3, \ldots, \lambda_n).$$

This ordered basis is called the **standard ordered basis** of K^n, and the corresponding set $\{e_1, \ldots, e_n\}$ is called the **standard basis** of K^n.

Example 28.7. The set B of all the functions p_n where $n \in N$ is a basis of the R-vector space $P(R)$ of all polynomial functions on R (Example 28.4). By definition, every polynomial function is a linear combination of B. If $\sum_{k=0}^{m} \alpha_k p_k = 0$ where $\alpha_m \neq 0$, we would obtain by differentiating m times

$$m! \, \alpha_m = 0,$$

whence $\alpha_m = 0$, a contradiction. Hence, B is linearly independent and therefore is a basis of $P(R)$.

Example 28.8. Let K be a ring with identity, let A be a set, and for each $a \in A$ let f_a be the function from A into K defined by

$$f_a(x) = \begin{cases} 1 \text{ if } x = a \\ 0 \text{ if } x \neq a. \end{cases}$$

Then $B = \{f_a : a \in A\}$ is a basis of $K^{(A)}$ (Example 28.3); indeed, if $(a_k)_{1 \le k \le n}$ is a sequence of distinct terms of A and if $(\lambda_k)_{1 \le k \le n}$ is a sequence of scalars, then $\sum_{k=1}^{n} \lambda_k f_{a_k}$ is the function whose value at a_k is λ_k and whose value at any x not in $\{a_1, \ldots, a_n\}$ is 0; consequently, B is a linearly independent set of generators of $K^{(A)}$.

Theorem 28.4. A sequence $(a_k)_{1 \le k \le n}$ of elements of a K-module E is an ordered basis of E if and only if for every $x \in E$ there is one and only one sequence $(\lambda_k)_{1 \le k \le n}$ of scalars such that $x = \sum_{k=1}^{n} \lambda_k a_k$.

Proof. Necessity: Every element of E is a linear combination of the set $\{a_1, \ldots, a_n\}$ of generators for E by Theorem 28.3. If

$$\sum_{k=1}^{n} \lambda_k a_k = \sum_{k=1}^{n} \mu_k a_k,$$

then

$$\sum_{k=1}^{n} (\lambda_k - \mu_k) a_k = \sum_{k=1}^{n} (\lambda_k a_k - \mu_k a_k) = \sum_{k=1}^{n} \lambda_k a_k - \sum_{k=1}^{n} \mu_k a_k = 0,$$

so $\lambda_k = \mu_k$ for all $k \in [1, n]$, since $(a_k)_{1 \le k \le n}$ is a linearly independent sequence. Sufficiency: Clearly $\{a_1, \ldots, a_n\}$ generates E. If

$$\sum_{k=1}^{n} \lambda_k a_k = 0,$$

then since also

$$\sum_{k=1}^{n} 0 . a_k = 0,$$

by hypothesis, we have $\lambda_k = 0$ for all $k \in [1, n]$. Therefore, $(a_k)_{1 \le k \le n}$ is a linearly independent sequence.

If $(a_k)_{1 \le k \le n}$ is an ordered basis of a K-module E and if $x = \sum_{k=1}^{n} \lambda_k a_k$, the scalars $\lambda_1, \ldots, \lambda_n$ are sometimes called the **coordinates** of x relative to the ordered basis $(a_k)_{1 \le k \le n}$, and $(a_k)_{1 \le k \le n}$ itself is called a **coordinate system.** The geometric reason for this terminology is that if (a_1, a_2) is an ordered basis of the plane R^2, for example, and if the distance from the origin to a_1 (to a_2) is taken as the "unit" of length on the line L_1 (the line L_2) through the origin and a_1 (a_2), then the coordinates λ_1 and λ_2 of $x = \lambda_1 a_1 + \lambda_2 a_2$ locate the vector x as the intersection of the line parallel to L_2 and λ_1 units along L_1 from it and the line parallel to L_1 and λ_2 units along L_2 from it. Figure 15 illustrates this for the case where $a_1 = (2, -1)$, $a_2 = (-\frac{1}{2}, 1)$, and $x = (3, \frac{3}{2}) = \frac{5}{2} a_1 + 4 a_2$.

Theorem 28.5. If $(a_k)_{1 \le k \le n}$ is an ordered basis of a K-module E, then

$$\psi : (\lambda_k)_{1 \le k \le n} \mapsto \sum_{k=1}^{n} \lambda_k a_k$$

is an isomorphism from the K-module K^n onto the K-module E.

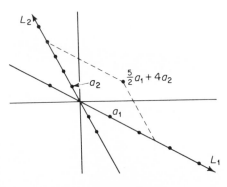

Figure 15

Proof. By Theorem 28.4, ψ is bijective. Since

$$\sum_{k=1}^{n} \lambda_k a_k + \sum_{k=1}^{n} \mu_k a_k = \sum_{k=1}^{n} (\lambda_k a_k + \mu_k a_k) = \sum_{k=1}^{n} (\lambda_k + \mu_k) a_k$$

and since

$$\beta \sum_{k=1}^{n} \lambda_k a_k = \sum_{k=1}^{n} \beta(\lambda_k a_k) = \sum_{k=1}^{n} (\beta \lambda_k) a_k,$$

ψ is an isomorphism.

Consequently, *any two K-modules having bases of n elements are isomorphic*, for they are both isomorphic to the K-module K^n.

The remaining theorems in this section concern only vector spaces, and their proofs depend heavily on the existence of multiplicative inverses of nonzero scalars.

Theorem 28.6. A sequence $(a_k)_{1 \le k \le n}$ of distinct nonzero vectors of a K-vector space E is linearly dependent if and only if a_p is a linear combination of $(a_k)_{1 \le k \le p-1}$ for some $p \in [2, n]$.

Proof. Necessity: By hypothesis, the set of all integers $r \in [1, n]$ such that $(a_k)_{1 \le k \le r}$ is linearly dependent is not empty; let p be its smallest member. Then $p \ge 2$ as $a_1 \ne 0$, and there exist scalars $\lambda_1, \ldots, \lambda_p$, not all of which are zero, such that

$$\sum_{k=1}^{p} \lambda_k a_k = 0.$$

If λ_p were zero, then $(a_k)_{1 \le k \le p-1}$ would be linearly dependent since $\lambda_1, \ldots, \lambda_{p-1}$ could not then all be zero, a contradiction of the definition of

p. Hence $\lambda_p \neq 0$, so as

$$\lambda_p a_p = -\sum_{k=1}^{p-1} \lambda_k a_k,$$

we have

$$a_p = \sum_{k=1}^{p-1} (-\lambda_p^{-1}\lambda_k)a_k.$$

Sufficiency: If

$$a_p = \sum_{k=1}^{p-1} \mu_k a_k,$$

then

$$\sum_{k=1}^{n} \lambda_k a_k = 0$$

where $\lambda_k = \mu_k$ for all $k \in [1, p-1]$, $\lambda_p = -1$, and $\lambda_k = 0$ for all $k \in [p+1, n]$.

Theorem 28.7. If L is a linearly independent subset of a finitely generated K-vector space E and if G is a finite set of generators for E that contains L, then there is a basis B of E such that $L \subseteq B \subseteq G$.

Proof. Let $\mathscr{S}$ be the set of all the subsets S of E such that S is a set of generators for E and $L \subseteq S \subseteq G$. Since $G \in \mathscr{S}$, $\mathscr{S}$ is not empty, and every member of $\mathscr{S}$ is finite as G is. Let n be the smallest of those integers r for which there is a member of $\mathscr{S}$ having r elements, and let B be a member of $\mathscr{S}$ having n elements. Then $0 \notin B$, for otherwise $B - \{0\}$ would be a set of generators for E having only $n - 1$ elements, and $B - \{0\}$ would, therefore, belong to $\mathscr{S}$ as L, being linearly independent, does not contain 0. Let m be the number of elements in L, and let $(a_k)_{1 \leq k \leq n}$ be a sequence of distinct vectors such that $L = \{a_1, \ldots, a_m\}$ and $B = \{a_1, \ldots, a_n\}$. Suppose that B were linearly dependent. By Theorem 28.6, there would then exist $p \in [2, n]$ and scalars $\mu_1, \ldots, \mu_{p-1}$ such that

$$a_p = \sum_{k=1}^{p-1} \mu_k a_k.$$

As L is linearly independent, $p > m$, and therefore, the set $B' = B - \{a_p\}$ would contain L. Also, B' would be a set of generators for E, for if $x = \sum_{k=1}^{n} \lambda_k a_k$, then

$$x = \sum_{k=1}^{p-1} (\lambda_k + \lambda_p \mu_k)a_k + \sum_{k=p+1}^{n} \lambda_k a_k.$$

Consequently, B' would be a member of $\mathscr{S}$ having only $n - 1$ members, a contradiction. Therefore, B is linearly independent and hence is a basis.

Theorem 28.8. Every finitely generated vector space has a finite basis.

Proof. We need only apply Theorem 28.7 to the case where G is a finite set of generators and L is the empty set.

There exist finitely generated modules over integral domains that have no basis at all (Exercise 28.10).

Theorem 28.9. If E is a K-vector space generated by a finite set of p vectors, then every linearly independent subset of E is finite and contains at most p vectors.

Proof. Let L be a finite linearly independent subset of E containing m vectors. We shall first prove that every finite set H of generators for E contains at least m vectors, whence in particular, $p \geq m$. For this proof we shall proceed by induction on the number of vectors in the relative complement $L - H$ of H in L.

Let S be the set of all natural numbers n such that for every finite set H of generators for E, if the relative complement $L - H$ has n vectors, then H contains at least m vectors. We shall prove that $S = N$. Clearly, $0 \in S$, for if $L - H = \emptyset$, then $L \subseteq H$, so H contains at least m vectors. Assume that $n \in S$, and let H be a finite set of generators for E such that $L - H$ has $n + 1$ vectors. Let a be a vector in $L - H$. Then $(L \cap H) \cup \{a\}$ is a subset of L and hence is linearly independent, so by Theorem 28.7 there is a basis B of E such that

$$(L \cap H) \cup \{a\} \subseteq B \subseteq H \cup \{a\}.$$

Then $L - B = (L - H) - \{a\}$, so $L - B$ has n vectors. Consequently, as $n \in S$, B has at least m vectors. Since a is a linear combination of H but $a \notin H$, $H \cup \{a\}$ is linearly dependent, and therefore, B is a proper subset of $H \cup \{a\}$. Thus, if b is the number of vectors in B and if h is the number of vectors in H, we have

$$m \leq b < h + 1,$$

so $m \leq h$. Hence, $n + 1 \in S$, so by induction $S = N$.

Thus, if L is a finite linearly independent subset of E, then L contains no more than p vectors. If there existed an infinite linearly independent subset L of E, L would contain a subset having $p + 1$ vectors, which would again be linearly independent, a contradiction. Thus, every linearly independent subset of E is finite and contains at most p vectors.

Since a basis is both linearly independent and a set of generators, the following theorem is an immediate consequence of Theorem 28.9.

Theorem 28.10. If E is a finitely generated K-vector space, then any two bases of E are finite and have the same number of vectors.

Definition. A module E is **n-dimensional** if it has a basis of n elements. A **finite-dimensional** module is a module that is n-dimensional for some natural number n. The **dimension** of a finite-dimensional K-vector space E is the unique natural number n such that E is n-dimensional.

The dimension of a finite-dimensional K-vector space E is denoted by $\dim_K E$ or simply $\dim E$.

If K is a ring with identity, we saw in Example 28.6 that the K-module K^n is n-dimensional. We did not define the "dimension" of a finite-dimensional module simply because it may have bases containing different numbers of elements. Indeed, there is a module that is n-dimensional for every strictly positive integer n (Exercise 32.15).

Theorem 28.11. If L is a linearly independent subset of a K-vector space E and if the subspace M generated by L is not E, then for every vector $b \notin M$, the set $L \cup \{b\}$ is linearly independent.

Proof. Suppose that

$$\sum_{k=1}^{n} \lambda_k x_k + \lambda b = 0$$

where $(x_k)_{1 \leq k \leq n}$ is a sequence of distinct vectors of L. If $\lambda \neq 0$, then

$$b = -\lambda^{-1} \left(\sum_{k=1}^{n} \lambda_k x_k \right) \in M,$$

a contradiction. Hence, $\lambda = 0$, so

$$\sum_{k=1}^{n} \lambda_k x_k = 0,$$

and therefore, $\lambda_1 = \ldots = \lambda_n = \lambda = 0$ as L is linearly independent.

Theorem 28.12. Let B be a subset of n vectors of an n-dimensional vector space E. The following statements are equivalent:

1° B is a basis of E.
2° B is linearly independent.
3° B is a set of generators for E.

Proof. Condition 2° implies 1°, for if B did not generate E, by Theorem 28.11, there would be a linearly independent subset of $n + 1$ vectors of E, a contradiction of Theorem 28.9. Condition 3° implies 1°, for B contains a basis B' of E by Theorem 28.7, but B' has n elements and hence is B by Theorem 28.10.

Theorem 28.13. If F is a subspace of a finite-dimensional vector space E, then F is finite-dimensional and dim $F \leq$ dim E. If F is a proper subspace of E, then dim $F <$ dim E.

Proof. Let $n =$ dim E. Every linearly independent subset of the vector space F is *a fortiori* a linearly independent subset of the vector space E and has, therefore, no more than n elements by Theorem 28.9. Consequently, the set of all natural numbers k such that F has a linearly independent subset of k vectors has a largest member m, and $m \leq n$. Let B be a linearly independent subset of F having m vectors. If the subspace generated by B were not F, then F would contain a linearly independent subset of $m + 1$ vectors by Theorem 28.11, a contradiction. Hence, B is a set of generators for F and is thus a basis of F, so F is finite-dimensional and dim $F \leq$ dim E. If dim $F =$ dim E, then a basis of F is a basis of E by Theorem 28.12, and consequently, $F = E$.

In Example 28.2 we saw directly that a subspace of R^2 properly containing a one-dimensional subspace was all of R^2. This fact is also an immediate consequence of Theorem 28.13 since R^2 is two-dimensional.

Theorem 28.14. Let E be an n-dimensional vector space. Every set of generators for E has at least n elements, contains a basis of E, and is itself a basis of E if and only if it has exactly n elements. Every linearly independent subset of E has at most n elements, is contained in a basis of E, and is itself a basis of E if and only if it has exactly n elements.

Proof. For the first assertion, it suffices by Theorems 28.7, 28.10, and 28.12 to show that every infinite set G of generators for E contains a finite set of generators. Let $(a_k)_{1 \leq k \leq n}$ be an ordered basis of E. For each $k \in [1, n]$, there is a finite subset G_k of G such that a_k is a linear combination of G_k. Hence, $\bigcup_{k=1}^{n} G_k$ is a finite subset of G generating E, for the subspace it generates contains $\{a_1, \ldots, a_n\}$ and hence is E.

A linearly independent subset of E has at most n elements by Theorem 28.9 and is itself a basis if and only if it has exactly n elements by Theorems 28.10 and 28.12. It remains to show that a linearly independent subset L of E is contained in a basis. By hypothesis, there is a basis B of E having n

elements. Then $L \cup B$ is a finite set of generators, so by Theorem 28.7, there exists a basis C of E such that $L \subseteq C \subseteq L \cup B$.

EXERCISES

28.1. A nonempty subset M of a K-module E is a submodule if and only if for all $x, y \in M$ and all $\lambda \in K$, $\lambda x + y \in M$.

28.2. Determine if M is a subspace of the R-vector space R^3 where M is the set of all $(\lambda_1, \lambda_2, \lambda_3) \in R^3$ such that
(a) $\lambda_3 = 0$.
(b) $\lambda_1 = \lambda_2$.
(c) $\lambda_1 \lambda_2 = 0$.
(d) $\lambda_1 = 2\lambda_2 = 3\lambda_3$.
(e) $\lambda_1 + 1 = 2\lambda_3$.
(f) $2\lambda_1 + 5\lambda_2 - 7\lambda_3 = 0$.
(g) $\lambda_1^2 + \lambda_2^2 = 0$.
(h) $\lambda_1^2 + \lambda_2^2 \geq 0$.

28.3. Determine if M is a subspace of the R-vector space R^R where M is the set of all functions f from R into R satisfying
(a) $f(3) = 0$.
(b) $f(0) = 3$.
(c) $f(3) = 2f(4)$.
(d) $f(4) \geq 0$.
(e) $\lim\limits_{t \to 3} f(t)$ exists.
(f) $\lim\limits_{t \to +\infty} t^2 f(t) = 0$.
(g) $\sin f(2) = 0$.
(h) $f(-t) = -f(t)$ for all $t \in R$.
(i) $f(t) = 0$ for at least one real number t.
(j) $f(t + 3) = f(t^2)$ for all $t \in R$.
(k) $f \circ g = f \circ h$, where g and h are given members of R^R.
(l) f is differentiable and $(Df)(2) = \int_0^2 f(t)\,dt$.

28.4. If L, M, and N are submodules of a module E and if $L \supseteq M$, then $L \cap (M + N) = (L \cap M) + (L \cap N)$. Give an example of three subspaces L, M, and N of R^2 such that $L \cap (M + N) \supset (L \cap M) + (L \cap N)$.

28.5. If A is a subset of a commutative ring K and if E is a K-module, then $\{x \in E: A.x = \{0\}\}$ is a submodule of E.

28.6. Prove Theorem 28.2.

28.7. Determine if the following are bases of R^3, and if so, express each member of the standard basis of R^3 as a linear combination of it: $\{(1, -1, 3)$, $(-4, 2, 0)$, $(-3, -1, 15)\}$, $\{(1, 3, 0)$, $(2, -1, 7)$, $(-3, 5, 1)\}$, $\{(1, \pi, 8)$, $(\sqrt{1 + \sqrt{2}}, 0, 4)$, $(3, \log_3 2, -1)$, $(-17, 3, \sqrt[3]{17} + \sqrt[3]{3})\}$.

28.8. Find all the bases B of R^3 containing $\{(1, 0, 1)\}$ and contained in $\{(1, 0, 1)$, $(2, 1, 4)$, $(-5, 1, 7)$, $(5, 1, 7)\}$.

28.9. If $(E, +)$ is an abelian group, then a subset of E is a subgroup if and only if it is a submodule of the associated Z-module, and an element b of E has infinite order if and only if $\{b\}$ is a linearly independent subset of the associated Z-module.

28.10. Let K be a commutative ring with identity, and let $\mathfrak{a}$ be an ideal of K.
(a) A subset of K is a submodule of the K-module K if and only if it is an ideal of the ring K.
(b) An element b of $\mathfrak{a}$ is a zero-divisor in K if and only if $\{b\}$ is a linearly dependent subset of the K-module $\mathfrak{a}$.
(c) A linearly independent subset of the K-module $\mathfrak{a}$ contains at most one element; hence, if $\mathfrak{a}$ is the sum of principal ideals but is not itself a principal ideal, then the K-module $\mathfrak{a}$ is finitely generated but has no basis at all. (An example of such a ring is given in Exercise 34.9.)

28.11. Cite theorems from §12 and Appendix B formally justifying the assertions about various finite sets made in the proofs of Theorems 28.7, 28.9, 28.12, and 28.13. The proof of which theorem tacitly uses Theorem 12.7? Which tacitly uses Exercise A.14?

28.12. (a) If $(b_k)_{1 \leq k \leq n}$ is a linearly independent sequence of vectors of a vector space and if $c = \sum_{k=1}^{n} \lambda_k b_k$ where $\lambda_1 \neq 0$, then $\{c, b_2, \ldots, b_n\}$ is linearly independent.
(b) If (e_1, e_2) is the standard ordered basis of the Z_6-module $Z_6{}^2$ and if $b = 2e_1 + 3e_2$, then neither $\{b, e_1\}$ nor $\{b, e_2\}$ is linearly independent.

28.13. For what real numbers α is $\{(1 + \alpha, 1, 1), (1, 1 + \alpha, 1), (1, 1, 1 + \alpha)\}$ a basis of R^3? linearly independent but not a basis? linearly dependent?

28.14. Is $\{(1, \sqrt{2}), (\sqrt{2}, 2)\}$ a linearly independent subset of the R-vector space R^2? of the Q-vector space R^2 obtained by restricting scalar multiplication? [Use Exercise 1.10.]

28.15. (a) Show that $\{(1, \alpha, \alpha^2), (1, \beta, \beta^2), (1, \gamma, \gamma^2)\}$ is a linearly independent subset of R^3 for all real numbers α, β, and γ. When is it a basis?
(b) If for each natural number n the function f_n is defined by $f_n(x) = e^{nx}$ for all $x \in R$, then (f_1, f_2, f_3) is a linearly independent sequence of the R-vector space R^R. [Differentiate twice and use (a).]

28.16. If for each real number α the function g_α is defined by $g_\alpha(x) = \cos(x + \alpha)$ for all $x \in R$, what is the dimension of the subspace of R^R generated by $\{g_\alpha : \alpha \in R\}$?

28.17. If E is a module over an integral domain K, then $M = \{x \in E : \{x\}$ is linearly dependent$\}$ is a submodule, and $M = \{0\}$ if E has a basis.

28.18. (a) If E is a K-module, the submodule generated by an element $b \in E$ is Kb.
(b) If K is a commutative ring with identity and if E is a nonzero K-module generated by a single element, then E has a basis if and only if the annihilator of E (Exercise 27.8) is $\{0\}$.

28.19. If B is a basis of a nonzero K-module E and if $b \in B$, then $B + b$ is linearly independent if and only if the element 2 of K is not a zero-divisor, and $B + b$ is a basis if and only if 2 is an invertible element of K.

28.20. If (b_1, b_2, b_3) is an ordered basis of a K-module E, then $(b_1 + b_2, b_2 + b_3, b_3 + b_1)$ is a linearly independent sequence if and only if the element 2 of K is not a zero-divisor, and $(b_1 + b_2, b_2 + b_3, b_3 + b_1)$ is a basis if and only if 2 is invertible.

28.21. If E is a finite-dimensional vector space, then E has exactly one basis if and only if either E is zero-dimensional or E is one-dimensional over a field isomorphic to the field Z_2.

28.22. What is the ratio of the number of ordered bases of the two-dimensional vector space $Z_5{}^2$ over Z_5 to the number of ordered couples of vectors of $Z_5{}^2$?

*28.23. Let K be a finite field of q elements, let E be an n-dimensional K-vector space, and let $m \in [1, n]$.

(a) How many linearly independent sequences of m vectors of E are there? [Use Theorem 28.11.] How many ordered bases of E are there?
(b) How many bases are there of a given m-dimensional subspace of E? Conclude that there are

$$\frac{1}{n!} \prod_{k=0}^{n-1} (q^n - q^k)$$

bases of E.
(c) Show that there are

$$\prod_{k=0}^{m-1} \frac{q^{n-k} - 1}{q^{m-k} - 1}$$

m-dimensional subspaces of E.

*28.24. Let E be a nonzero finitely generated module over an integral domain K. If the annihilator of E (Exercise 27.8) is $\{0\}$ and if every subgroup of the additive group E is a submodule of the K-module E, then K is isomorphic either to the ring Z of integers or to the field Z_p for some prime p.

28.25. If E is a one-dimensional module over a commutative ring with identity, then every basis of E has one element.

28.26. If E is an n-dimensional module over a finite ring, then every basis of E has n elements. [How many elements does E have?]

28.27. Let L be a quotient field of an integral domain K, and let E be an L-vector space. Every linearly independent subset of the K-module E (obtained by restricting scalar multiplication) is also a linearly independent subset of the L-vector space E.

*28.28. If E is an n-dimensional module over an integral domain K, then every linearly independent subset of E is finite and contains at most n elements, every set of generators for E contains at least n elements, and every basis of E contains exactly n elements. [Let L be a quotient field of K; use Exercise 28.27, and show that every set of generators for the K-module K^n is also a set of generators for the L-vector space L^n.]

28.29. Give an example of a set S of three elements of the Z-module Z^2 such that S is a set of generators for Z^2 containing no basis of Z^2, and every subset of S having two elements is linearly independent but is contained in no basis. [Use Exercise 28.28.]

*28.30. If K is a commutative ring with identity and if B is a basis of a nonzero K-module E, then for each $\lambda \in K$, λB is linearly independent if and only if λ is not a zero-divisor, and λB is a basis of E if and only if λ is invertible.

28.31. If E is an n-dimensional module over an integral domain K and if $n \geq 1$, then every linearly independent subset of n elements of E is a basis if and only if K is a field. [Use Exercise 28.30.]

*28.32. Let K be a ring with identity, and let E be a nonzero K-module.
(a) If every set of generators for E contains a basis of E, then for every $\lambda \in K$, either λ or $1 - \lambda$ is invertible.
(b) If E is one-dimensional, then every set of generators for E contains a basis of E if and only if the set $\mathfrak{a}$ of all elements of K having no left inverse (Exercise 4.11) is a left ideal (Exercise 19.15). [Use Exercise 19.17.]

*28.33. Let $(G, +)$ be a finite abelian group, and let s be the sum of all the elements of G. If G has exactly one element a of order 2, then $s = a$; otherwise, $s = 0$. [See Exercise 19.13; use Exercise 27.10.]

29. Linear Transformations

Analogous to the definition of a homomorphism from one algebraic structure into another is the following definition of a homomorphism from one K-algebraic structure into another.

Definition. Let K be a ring. A **homomorphism** from a K-algebraic structure $(E, +, .)$ with one composition into another $(F, +, .)$ is a function u from E into F that is a homomorphism from the algebraic structure $(E, +)$ into $(F, +)$ and satisfies

$$u(\lambda.x) = \lambda.u(x)$$

for all $x \in E$ and all $\lambda \in K$. Similarly, a **homomorphism** from a K-algebraic structure $(E, +, \cdot, .)$ with two compositions into another $(F, +, \cdot, .)$ is a function u from E into F that is a homomorphism from the algebraic structure

$(E, +, \cdot)$ into $(F, +, \cdot)$ and satisfies $u(\lambda.x) = \lambda.u(x)$ for all $x \in E$ and all $\lambda \in K$. A **monomorphism** from a K-algebraic structure E into a K-algebraic structure F is an injective homomorphism from E into F, and an **epimorphism** from E onto F is a surjective homomorphism from E onto F. An **endomorphism** of a K-algebraic structure is a homomorphism from itself into itself.

Thus, an isomorphism from a K-algebraic structure E onto a K-algebraic structure F is simply a bijective homomorphism. The proof of the following theorem is easy.

Theorem 29.1. Let E, F, and G be K-algebraic structures with the same number of compositions. If u is a homomorphism from E into F and if v is a homomorphism from F into G, then $v \circ u$ is a homomorphism from E into G.

If u is an epimorphism from a K-module E onto a K-algebraic structure F, it is easy to verify that F is a K-module. Thus, *a homomorphic image of a K-module is a K-module*, and if K is a division ring, *a homomorphic image of a K-vector space is a K-vector space*.

A homomorphism from one module into another is usually called a **linear transformation**. A linear transformation from a module E into itself is often called a **linear operator** on E. The **kernel** of a linear transformation u from a K-module E into a K-module F is the subset $u^*(\{0\})$ of E. Since a linear transformation u from a K-module E into a K-module F is, in particular, a homomorphism from the additive group E into the additive group F, $u(0) = 0$ and $u(-x) = -u(x)$ for all $x \in E$ by Theorem 20.5. The following theorem is, consequently, easy to prove.

Theorem 29.2. Let u be a linear transformation from a K-module E into a K-module F. If M is a submodule of E, then $u_*(M)$ is a submodule of F. If N is a submodule of F, then $u^*(N)$ is a submodule of E. In particular, the range of u is a submodule of F, and the kernel of u is a submodule of E.

If E is a K-module, a function u from E into a K-module F is a linear transformation if and only if

$$(1) \qquad u(\lambda x + \mu y) = \lambda u(x) + \mu u(y)$$

for all $x, y \in E$ and all $\lambda, \mu \in K$. Indeed, any linear transformation clearly satisfies (1); conversely, if u satisfies (1), we obtain $u(x + y) = u(x) + u(y)$ upon setting $\lambda = \mu = 1$, and $u(\lambda x) = \lambda u(x)$ upon setting $\mu = 0$. The satisfaction of (1) is often expressed by the word "linearity."

Example 29.1. Let us find all linear operators on the R-vector space R^2. If u is a linear operator on R^2 and if α_{11}, α_{12}, α_{21}, and α_{22} are the real numbers satisfying

$$u(e_1) = \alpha_{11}e_1 + \alpha_{21}e_2$$
$$u(e_2) = \alpha_{12}e_1 + \alpha_{22}e_2,$$

then by linearity,

$$u(\lambda_1, \lambda_2) = u(\lambda_1 e_1 + \lambda_2 e_2) = \lambda_1 u(e_1) + \lambda_2 u(e_2)$$
$$= (\lambda_1\alpha_{11} + \lambda_2\alpha_{12})e_1 + (\lambda_1\alpha_{21} + \lambda_2\alpha_{22})e_2$$
$$= (\lambda_1\alpha_{11} + \lambda_2\alpha_{12}, \lambda_1\alpha_{21} + \lambda_2\alpha_{22}).$$

Conversely, if α_{11}, α_{12}, α_{21}, and α_{22} are any real numbers, it is easy to verify that the function u defined by

$$u(\lambda_1, \lambda_2) = (\lambda_1\alpha_{11} + \lambda_2\alpha_{12}, \lambda_1\alpha_{21} + \lambda_2\alpha_{22})$$

is a linear operator on R^2. Thus, each linear operator on R^2 is completely determined by an ordered quadruple $(\alpha_{11}, \alpha_{12}, \alpha_{21}, \alpha_{22})$ of real numbers.

Certain familiar geometric transformations of the plane are linear, some examples of which follow.

Example 29.2. Let r_α be the rotation of the plane about the origin through α degrees (Figure 16); i.e., let r_α be the function such that for all $x \in R^2$, $r_\alpha(x)$ is the point into which that rotation carries x. If $(\lambda_1, \lambda_2) = (\rho \cos \sigma, \rho \sin \sigma)$, then

$$r_\alpha(\lambda_1, \lambda_2) = (\rho \cos(\alpha + \sigma), \rho \sin(\alpha + \sigma))$$
$$= (\rho \cos \alpha \cos \sigma - \rho \sin \alpha \sin \sigma, \rho \sin \alpha \cos \sigma + \rho \cos \alpha \sin \sigma)$$
$$= (\lambda_1 \cos \alpha - \lambda_2 \sin \alpha, \lambda_1 \sin \alpha + \lambda_2 \cos \alpha).$$

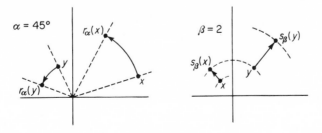

Figure 16

Consequently, by Example 29.1, r_α is a linear operator, the ordered quadruple determining r_α being $(\cos \alpha, -\sin \alpha, \sin \alpha, \cos \alpha)$.

Example 29.3. If E is a vector space over a field K, then for each $\beta \in K$, the function s_β defined by

$$s_\beta(x) = \beta x$$

for all $x \in E$ is a linear operator on E, since $\beta(x + y) = \beta x + \beta y$ and $\beta(\lambda x) = \lambda(\beta x)$. If $\beta \neq 0$, then s_β is an automorphism of E and

$$\overset{\leftarrow}{s_\beta} = s_{\beta^{-1}},$$

for

$$(s_{\beta^{-1}} \circ s_\beta)(x) = \beta^{-1}(\beta x) = x = \beta(\beta^{-1} x) = (s_\beta \circ s_{\beta^{-1}})(x).$$

The linear operators s_β where $\beta \neq 0$ are called *similitudes* of E. If E is the $\boldsymbol{R}$-vector space $\boldsymbol{R}^2$, then s_{-1} is the rotation r_{180} of the plane about the origin through $180°$ by Example 29.2. If $\beta \geq 1$, s_β is called a *stretching*, and if $0 < \beta \leq 1$, s_β is called a *contraction*. If $\beta < 0$, then as $\beta x = -|\beta| \, x$ we have

$$s_\beta = s_{-1} \circ s_{|\beta|},$$

and hence, s_β is a stretching or contraction followed by a rotation through $180°$. Also in this case, as $\beta x = |\beta|(-x)$, we have

$$s_\beta = s_{|\beta|} \circ s_{-1},$$

and hence, s_β is a rotation through $180°$ followed by a stretching or contraction.

Example 29.4. If M is a line in the plane passing through the origin, the *reflection* s_M of $\boldsymbol{R}^2$ in M is the rotation of the plane in space through $180°$ about M as axis (Figure 17). Geometrically, it is apparent that $s_M \circ s_M = 1_{R^2}$ and hence that $\overset{\leftarrow}{s_M} = s_M$. If M is the X-axis, then $s_M(\lambda_1, \lambda_2) = (\lambda_1, -\lambda_2)$, and if M is the Y-axis, then $s_M(\lambda_1, \lambda_2) = (-\lambda_1, \lambda_2)$. In general, s_M is a linear operator for every line M through the origin (Exercise 29.6).

Example 29.5. If M and N are distinct lines in the plane passing through the origin, the *projection* on M along N is the function $p_{M,N}$

such that for all $x \in R^2$, $p_{M,N}(x)$ is the intersection of M with the line through x parallel to N (Figure 17). Geometrically, it is apparent that M and N are respectively the range and kernel of $p_{M,N}$ and that $p_{M,N}(x) = x$ if and only if $x \in M$. If M is the X-axis and N the Y-axis, then $p_{M,N}(\lambda_1, \lambda_2) = (\lambda_1, 0)$, and if M is the Y-axis and N the X-axis, then $p_{M,N}(\lambda_1, \lambda_2) = (0, \lambda_2)$. In general, any such projection is a linear operator (Exercise 29.7).

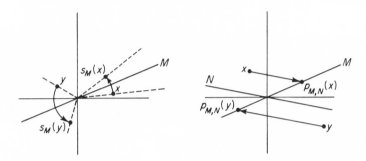

Figure 17

Example 29.6. The differentiation operator D on the vector space $P(R)$ of all polynomial functions on R satisfies $D(p + q) = Dp + Dq$ and $D(\lambda p) = \lambda Dp$ for all $p, q \in P(R)$ and all $\lambda \in R$, and therefore, D is a linear operator. If S is the function defined by

$$[S(p)](x) = \int_0^x p(t)\, dt,$$

for all $p \in P(R)$ and all $x \in R$, then S is a linear operator on $P(R)$ by theorems of calculus.

Theorem 29.3. Let u and v be linear transformations from a K-module E into a K-module F. The set

$$H = \{x \in E \colon u(x) = v(x)\}$$

is a submodule of E. If A is a set of generators for the K-module E and if $u(x) = v(x)$ for all $x \in A$, then $u = v$.

The proof is easy.

Theorem 29.4. If $(a_k)_{1 \le k \le n}$ is an ordered basis of a K-module E and if $(b_k)_{1 \le k \le n}$ is a sequence of elements of a K-module F, then there is one and only one linear transformation u from E into F satisfying

$$u(a_k) = b_k$$

for all $k \in [1, n]$.

Proof. By Theorem 28.4, the function u defined by

$$u\left(\sum_{k=1}^{n} \lambda_k a_k\right) = \sum_{k=1}^{n} \lambda_k b_k$$

is a well-defined function from E into F, and clearly, $u(a_k) = b_k$ for all $k \in [1, n]$. It is easy to verify that u is linear. By Theorem 29.3, u is the only linear transformation whose value at a_k is b_k for all $k \in [1, n]$.

Corollary. If $(a_k)_{1 \le k \le m}$ is a linearly independent sequence of vectors of a finite-dimensional K-vector space E and if $(b_k)_{1 \le k \le m}$ is a sequence of vectors of a K-vector space F, then there is a linear transformation u from E into F satisfying

$$u(a_k) = b_k$$

for all $k \in [1, m]$.

Proof. The assertion follows from Theorem 29.4, since $\{a_1, \ldots, a_m\}$ is contained in a basis by Theorem 28.14.

Theorem 29.5. Let B be a basis of a K-module E, and let u be a linear transformation from E into a K-module F.

1° u is a monomorphism if and only if $u_*(B)$ is linearly independent and $u(b) \ne u(c)$ whenever b and c are distinct members of B.
2° u is an epimorphism if and only if $u_*(B)$ generates F.
3° u is an isomorphism if and only if $u_*(B)$ is a basis of F and $u(b) \ne u(c)$ whenever b and c are distinct members of B.

The proof is easy.

Theorem 29.6. Let u be a linear transformation from an n-dimensional vector space E into an n-dimensional vector space F. The following statements are equivalent:

1° u is an isomorphism.
2° u is a monomorphism.
3° u is an epimorphism.
4° For every basis B of E, $u_*(B)$ is a basis of F.
5° For some basis B of E, $u_*(B)$ is a basis of F.

Proof. Clearly 1° implies 2° and 3°. Let B be a basis of E; then B has n

elements. If 2° holds, then $u_*(B)$ is a linearly independent set of n elements by 1° of Theorem 29.5, so $u_*(B)$ is a basis by Theorem 28.12; if 3° holds, then $u_*(B)$ is a set of generators for F having at most n elements by 2° of Theorem 29.5, so $u_*(B)$ is a basis by Theorem 28.14. Thus, each of 2° and 3° implies 4°. Clearly, 4° implies 5°. Finally, if 5° holds, then both B and $u_*(B)$ have n elements, so $u(b) \neq u(c)$ whenever b and c are distinct members of B, and therefore u is an isomorphism by 3° of Theorem 29.5. Thus, 5° implies 1°.

Definition. Let u be a linear transformation from one vector space into another. If the range of u is finite-dimensional, its dimension is called the **rank** of u and is denoted by $\rho(u)$. If the kernel of u is finite-dimensional, its dimension is called the **nullity** of u and is denoted by $\nu(u)$.

Theorem 29.7. Let E and F be K-modules, and let u be a nonzero linear transformation from E into F. If E is n-dimensional and if $(a_k)_{1 \leq k \leq n}$ is any ordered basis of E such that $\{a_k : r + 1 \leq k \leq n\}$ is a basis of the kernel of u, then $(u(a_k))_{1 \leq k \leq r}$ is an ordered basis of the range of u.

Proof. The sequence $(u(a_k))_{1 \leq k \leq r}$ is linearly independent, for if

$$\sum_{k=1}^{r} \lambda_k u(a_k) = 0,$$

then

$$u\left(\sum_{k=1}^{r} \lambda_k a_k\right) = 0,$$

so $\sum_{k=1}^{r} \lambda_k a_k$ belongs to the kernel of u and hence is also a linear combination of $\{a_k : r + 1 \leq k \leq n\}$, whence $\lambda_k = 0$ for all $k \in [1, r]$, since $(a_k)_{1 \leq k \leq n}$ is linearly independent. Every element of $u_*(E)$ is a linear combination of the sequence $(u(a_k))_{1 \leq k \leq r}$, for if $x \in E$ and if

$$x = \sum_{k=1}^{n} \mu_k a_k,$$

then

$$u(x) = \sum_{k=1}^{n} \mu_k u(a_k) = \sum_{k=1}^{r} \mu_k u(a_k),$$

since $u(a_k) = 0$ for all $k \in [r + 1, n]$. Therefore, $(u(a_k))_{1 \leq k \leq r}$ is an ordered basis of $u_*(E)$.

Corollary. If u is a linear transformation from an n-dimensional vector space E into a vector space F, then the range of u is finite-dimensional, and

$$\rho(u) + \nu(u) = n.$$

Proof. The assertion is clear if $u = 0$, so we shall assume that u is a nonzero linear transformation. By Theorems 28.13 and 28.14, there is an ordered basis $(a_k)_{1 \le k \le n}$ of E such that for some $r \in N_n$, $\{a_k : r + 1 \le k \le n\}$ is a basis of the kernel of u. Consequently, $\nu(u) = n - r$, and by Theorem 29.7, $\rho(u) = r$.

We shall apply this Corollary to derive a familiar result of elementary algebra: A system of m homogeneous linear equations in n unknowns has non-trivial solutions if $m < n$.

Theorem 29.8. Let K be a field, and let (α_{ij}) be a family of elements of K indexed by $[1, m] \times [1, n]$. The set H of all $(x_1, \ldots, x_n) \in K^n$ such that

$$\alpha_{11}x_1 + \alpha_{12}x_2 + \ldots + \alpha_{1n}x_n = 0$$

$$\alpha_{21}x_1 + \alpha_{22}x_2 + \ldots + \alpha_{2n}x_n = 0$$

$$\cdot \qquad \cdot \qquad \qquad \cdot \qquad \cdot$$
$$\cdot \qquad \cdot \qquad \qquad \cdot \qquad \cdot$$
$$\cdot \qquad \cdot \qquad \qquad \cdot \qquad \cdot$$

$$\alpha_{m1}x_1 + \alpha_{m2}x_2 + \ldots + \alpha_{mn}x_n = 0$$

is a subspace of the K-vector space K^n. If $m < n$, then H is a nonzero subspace, and in fact, dim $H \ge n - m$.

Proof. Let $u \colon K^n \to K^m$ be defined by

$$u(x_1, \ldots, x_n) = \left(\sum_{j=1}^{n} \alpha_{1j}x_j, \sum_{j=1}^{n} \alpha_{2j}x_j, \ldots, \sum_{j=1}^{n} \alpha_{mj}x_j \right).$$

It is easy to verify that u is a linear transformation from K^n into K^m. Clearly, $H = \ker u$, and $\rho(u) \le \dim K^m = m$. Hence, if $m < n$,

$$\dim H = \nu(u) = n - \rho(u) \ge n - m > 0.$$

EXERCISES

29.1. A linear operator u on R^2 satisfies $u(2, 1) = (-1, 7)$ and $u(3, 2) = (0, 5)$. What is $u(e_1)$? $u(e_2)$? $u(\lambda_1, \lambda_2)$? Show that u is an automorphism of R^2. What is $u^{\leftarrow}(e_1)$? $u^{\leftarrow}(e_2)$? $u^{\leftarrow}(\lambda_1, \lambda_2)$?

29.2. A linear operator u on R^3 satisfies $u(2, 0, 1) = (5, 3, 0)$, $u(3, 1, 0) = (4, 2, 1)$, and $u(0, 3, -1) = (1, 1, -1)$. What is $u(e_1)$? $u(e_2)$? $u(e_3)$? $u(\lambda_1, \lambda_2, \lambda_3)$?

29.3. Find a basis for the subspace of R^3 consisting of all $(\lambda_1, \lambda_2, \lambda_3)$ such that:

(a) $\lambda_1 + 2\lambda_2 + 3\lambda_3 = 0$.
(b) $2\lambda_1 - 3\lambda_2 + 7\lambda_3 = 0$.
(c) $3\lambda_1 - 4\lambda_3 = 0$.

Find a basis for the subspace of R^4 consisting of all $(\lambda_1, \lambda_2, \lambda_3, \lambda_4)$ such that:

(d) $\lambda_1 - 2\lambda_2 + \lambda_3 - \lambda_4 = 0$.
(e) $2\lambda_1 - 3\lambda_2 + 5\lambda_3 - 7\lambda_4 = 0$.
(f) $\lambda_1 + \lambda_2 - 2\lambda_4 = 0$.

29.4. Prove that if f is an epimorphism from a K-module E onto a K-algebraic structure F, then F is a K-module.

29.5. (a) Prove Theorem 29.1.
(b) Prove Theorem 29.2.

29.6. Let M be a line in the plane R^2 passing through the origin but other than the X-axis or Y-axis, and let m be its slope. Determine explicitly $s_M(\lambda_1, \lambda_2)$, and infer that s_M is a linear operator.

29.7. Let M and N be two different lines in the plane R^2 passing through the origin, let m and n be respectively the slopes of M and N, and let $p_{M,N}$ be the projection of the plane on M along N. Determine explicitly
(a) $p_{M,N}(\lambda_1, \lambda_2)$ in terms of m and n if neither is the Y-axis;
(b) $p_{M,N}(\lambda_1, \lambda_2)$ in terms of m if N is the Y-axis;
(c) $p_{M,N}(\lambda_1, \lambda_2)$ in terms of n if M is the Y-axis.
Conclude that in all cases, $p_{M,N}$ is a linear operator.

*29.8. If u is a linear operator on R^2 of rank 1, there exist a projection p on some line through the origin along another, a similitude s, and a rotation r such that either $u = s \circ p$ or $u = r \circ s \circ p$, according as the kernel and range of u are distinct or identical.

29.9. (a) Prove Theorem 29.3.
(b) Complete the proof of Theorem 29.4 (and, in particular, the discussion of Example 29.1) by verifying that u is indeed linear.
(c) Prove Theorem 29.5.

29.10. Let E and F be n-dimensional K-modules, and let u be a linear transformation from E into F.
(a) Show that conditions $1°$, $4°$, and $5°$ of Theorem 29.6 are equivalent.
(b) Give an example to show that $2°$ does not always imply $3°$. [Let E and F be the Z-module Z.]
(c) If E is m-dimensional as well as n-dimensional where $m < n$, show that there is a surjective linear operator on E that is not an automorphism of E.

29.11. Give an example of a linear operator on the two-dimensional R-vector space C that is not a linear operator on the one-dimensional C-vector space C.

29.12. Let $n > 0$, $m > 0$, and $q \geq 2$.

(a) How many homomorphisms are there from the additive group Z_q^n into Z_q^m?

(b) If K is a field of q elements and if E and F are respectively an n-dimensional and an m-dimensional vector space over K, how many linear transformations are there from E into F? If $n \leq m$, how many of them are injective? [Use Exercise 28.23.]

29.13. If E and F are finite-dimensional vector spaces, for every linear transformation u from E into F there exists a linear transformation v from F into E such that $u \circ v \circ u = u$. [Use Theorem 29.7.]

*29.14. Let u and v be linear operators on a finite-dimensional vector space E.

(a) There exists a linear operator w on E such that $u = v \circ w$ if and only if $u_*(E) \subseteq v_*(E)$.

(b) There exists a linear operator w on E such that $u = w \circ v$ if and only if $\ker u \supseteq \ker v$.

29.15. Let K be a ring with identity, let E and F be K-modules, and let u be a homomorphism from $(E, +)$ into $(F, +)$.

(a) Let $L = \{\lambda \in K : u(\lambda x) = \lambda u(x)$ for all $x \in E\}$. Then L is a subring of K, and for every $\lambda \in L$, if λ is invertible in K, then $\lambda^{-1} \in L$.

(b) If K is a prime field, then u is a linear transformation from the K-vector space E into the K-vector space F.

30. Direct Sums and Quotient Spaces

How can a vector space or module be described as put together in some natural way from certain of its subspaces or submodules? This is the analogue of a question raised concerning groups in §24. There the method of "putting together" was that of forming cartesian products; a group was called the direct product (or direct sum, depending on the notation used for the composition) of certain of its subgroups if it was isomorphic in a natural way to their cartesian product. We have already observed in Example 27.5 that the cartesian product of K-modules may be made into a K-module in a natural way, and therefore, we make the following definition.

Definition. Let $(M_k)_{1 \leq k \leq n}$ be a sequence of submodules of a K-module E. We shall say that E is the **direct sum** of $(M_k)_{1 \leq k \leq n}$ if the function

$$A : (x_k) \mapsto \sum_{k=1}^{n} x_k$$

is an isomorphism from the K-module $\prod_{k=1}^{n} M_k$ onto the K-module E.

Let $(M_k)_{1 \le k \le n}$ be a sequence of submodules of a K-module E, and let A be the function of the preceding definition. Then A is a linear transformation from $\prod_{k=1}^{n} M_k$ into E whose range is the submodule $M_1 + \ldots + M_n$, for

$$A((x_k) + (y_k)) = A((x_k + y_k)) = \sum_{k=1}^{n} (x_k + y_k)$$

$$= \sum_{k=1}^{n} x_k + \sum_{k=1}^{n} y_k = A((x_k)) + A((y_k))$$

(Theorem A.7) and

$$A(\lambda(x_k)) = A((\lambda x_k)) = \sum_{k=1}^{n} \lambda x_k$$

$$= \lambda \sum_{k=1}^{n} x_k = \lambda A((x_k))$$

by Theorem 27.2.

Theorem 30.1. A K-module E is the direct sum of a sequence $(M_k)_{1 \le k \le n}$ of submodules if and only if the following two conditions hold:

1° $E = M_1 + \ldots + M_n$

2° If $\sum_{k=1}^{n} x_k = 0$ where $x_k \in M_k$ for each $k \in [1, n]$, then $x_k = 0$ for all $k \in [1, n]$.

Proof. Clearly, A is surjective if and only if 1° holds. Therefore, by Theorem 20.9, A is an isomorphism if and only if 1° and 2° hold.

Theorem 30.2. If M and N are submodules of a K-module E, then the K-module $M + N$ is the direct sum of M and N if and only if $M \cap N = \{0\}$.

Proof. By Theorem 30.1, it suffices to show that $M \cap N = \{0\}$ if and only if $x + y = 0$ implies that $x = y = 0$, whenever $x \in M$, $y \in N$. Necessity: If $x + y = 0$ where $x \in M$, $y \in N$, then $x = -y \in N$, so $x \in M \cap N = \{0\}$, whence $x = 0$ and thus also $y = 0$. Sufficiency: Given $x \in M \cap N$, let $y = -x$; then $x \in M$, $y \in N$, and $x + y = 0$, so $x = 0$.

Corollary. A K-module E is the direct sum of submodules M and N if and only if $E = M + N$ and $M \cap N = \{0\}$.

Example 30.1. Let $(e_k)_{1 \le k \le n}$ be a basis of a K-module E, and let M_k be the one-dimensional submodule Ke_k for each $k \in [1, n]$. By the definition of a basis, $(M_k)_{1 \le k \le n}$ satisfies 1° and 2° of Theorem 30.1, and consequently, E is the direct sum of $(M_k)_{1 \le k \le n}$.

Example 30.2. If H is an $(n - 1)$-dimensional subspace and M a one-dimensional subspace of an n-dimensional vector space E and if M is not contained in H, then E is the direct sum of M and H. Indeed, as M is not contained in H, $H \subset M + H$ and $M \cap H \subset M$, so $n - 1 < \dim(M + H)$ and $\dim(M \cap H) < 1$; therefore,

$$\dim(M + H) = n \qquad \text{and} \qquad \dim(M \cap H) = 0,$$

so $M + H = E$ and $M \cap H = \{0\}$.

Definition. A **supplement** (or **complement**) of a submodule M of a K-module E is a submodule N of E such that E is the direct sum of M and N. A submodule M of E is a **direct summand** of E if there is a supplement of M in E.

By the Corollary of Theorem 30.2, if N is a supplement of M, then M is a supplement of N; in this case, we shall say that M and N are supplementary submodules. An important property of finite-dimensional vector spaces is that every subspace has a supplement:

Theorem 30.3. A subspace M of a finite-dimensional vector space E is a direct summand of E.

Proof. Since E and $\{0\}$ are supplementary subspaces, we may assume that M is a nonzero proper subspace. By Theorems 28.13 and 28.14, there is an ordered basis $(a_k)_{1 \leq k \leq n}$ of E such that $(a_k)_{1 \leq k \leq m}$ is an ordered basis of M for some $m \in [1, n - 1]$. Let N be the subspace generated by $\{a_k : m + 1 \leq k \leq n\}$. Clearly, $M + N = E$ and $M \cap N = \{0\}$, so M and N are supplementary subspaces.

Theorem 30.4. If E and F are finite-dimensional K-vector spaces, then

$$\dim(E \times F) = \dim E + \dim F.$$

Proof. Let $\{b_1, \ldots, b_m\}$ be a basis of E, $\{c_1, \ldots, c_n\}$ a basis of F. We shall show that $\{(b_1, 0), \ldots, (b_m, 0), (0, c_1), \ldots, (0, c_n)\}$ is a basis of $E \times F$, from which the assertion follows. If $(x, y) \in E \times F$, there exist $\lambda_1, \ldots, \lambda_m$, $\mu_1, \ldots, \mu_n \in K$ such that $x = \sum_{i=1}^{m} \lambda_i b_i$, $y = \sum_{j=1}^{n} \mu_j c_j$, whence

$$(x, y) = (x, 0) + (0, y) = \left(\sum_{i=1}^{m} \lambda_i b_i, 0 \right) + \left(0, \sum_{j=1}^{n} \mu_j c_j \right)$$

$$= \sum_{i=1}^{m} \lambda_i (b_i, 0) + \sum_{j=1}^{n} \mu_j (0, c_j).$$

If

$$\sum_{i=1}^{m} \beta_i(b_i, 0) + \sum_{j=1}^{n} \gamma_j(0, c_j) = (0, 0),$$

then

$$\left(\sum_{i=1}^{m} \beta_i b_i, \sum_{j=1}^{n} \gamma_j c_j \right) = (0, 0),$$

so $\sum_{t=1}^{m} \beta_i b_i = 0$ and $\sum_{j=1}^{n} \gamma_j c_j = 0$, whence $\beta_1 = \ldots = \beta_m = 0$ and $\gamma_1 = \ldots = \gamma_n = 0$. Thus, $\{(b_1, 0), \ldots, (b_m, 0), (0, c_1), \ldots, (0, c_n)\}$ is a basis of $E \times F$.

Corollary. If a finite-dimensional vector space E is the direct sum of subspaces M and N, then dim E = dim M + dim N.

Theorem 30.5. If M and N are subspaces of a finite-dimensional vector space E, then

$$\text{dim}(M + N) + \text{dim}(M \cap N) = \text{dim } M + \text{dim } N.$$

Proof. Clearly, $H = \{(z, z) \in M \times N : z \in M \cap N\}$ is a subspace of the vector space $M \times N$. Moreover, dim H = dim$(M \cap N)$ since $z \mapsto (z, z)$ is clearly an isomorphism from the vector space $M \cap N$ onto H. The function $u: (x, y) \mapsto x - y$ from $M \times N$ into $M + N$ is easily seen to be an epimorphism whose kernel is H. Thus, by the Corollary of Theorem 29.7 and Theorem 30.4,

$$\text{dim}(M + N) + \text{dim}(M \cap N) = \text{dim}(M + N) + \text{dim } H$$
$$= \rho(u) + \nu(u) = \text{dim}(M \times N)$$
$$= \text{dim } M + \text{dim } N.$$

Let M be a submodule of a K-module E. In §18 we introduced a composition on E/M by defining $(x + M) + (y + M)$ to be $x + y + M$. We may introduce a scalar multiplication on E/M by defining

(1) $$\lambda.(x + M) = \lambda x + M.$$

for all $\lambda \in K$, $x + M \in E/M$. To show that scalar multiplication is well-defined, we must show that if $x + M = x' + M$, then $\lambda x + M = \lambda x' + M$ for all $\lambda \in K$. But if $x + M = x' + M$, then $x - x' \in M$, so $\lambda x - \lambda x' = \lambda(x - x') \in M$, whence $\lambda x + M = \lambda x' + M$. Thus scalar multiplication is well-defined by (1). It is easy to verify that $(E/M, +, .)$ is a K-module. Indeed,

$$\lambda.((x + M) + (y + M)) = \lambda.(x + y + M) = \lambda(x + y) + M,$$
$$\lambda.(x + M) + \lambda.(y + M) = (\lambda x + M) + (\lambda y + M) = (\lambda x + \lambda y) + M,$$

so (VS 1) holds,

$$(\lambda + \mu).(x + M) = (\lambda + \mu)x + M,$$
$$\lambda.(x + M) + \mu.(x + M) = (\lambda x + M) + (\mu x + M) = (\lambda x + \mu x) + M,$$

so (VS 2) holds,

$$(\lambda\mu).(x + M) = (\lambda\mu)x + M,$$
$$\lambda.(\mu.(x + M)) = \lambda.(\mu x + M) = \lambda(\mu x) + M,$$

so (VS 3) holds, and

$$1.(x + M) = (1.x) + M = x + M,$$

so (VS 4) holds. The module $(E/M, +, .)$ is called the **quotient module** of E defined by M. Clearly $\varphi_M: x \mapsto x + M$ is an epimorphism from the module E onto the module E/M.

The following analogue of Theorem 20.9 is easy to prove:

Theorem 30.6. (Factor Theorem for Modules) Let f be an epimorphism from a K-module E onto a K-module F, and let N be the kernel of f. Then

$$g: x + N \mapsto f(x)$$

is a well-defined isomorphism from the K-module E/N onto F satisfying $g \circ \varphi_N = f$, where φ_N is the canonical epimorphism from E onto E/N. Moreover, f is an isomorphism if and only if $N = \{0\}$.

Theorem 30.7. If M and N are supplementary submodules of a K-module E, then

$$f: x \mapsto x + N, \qquad x \in M,$$

is an isomorphism from M onto E/N.

Proof. Clearly, f is a linear transformation. If $z \in E$, then there exist $x \in M$, $y \in N$ such that $z = x + y$, whence $z + N = x + y + N = x + N = f(x)$;

thus f is surjective. The kernel of f is clearly $M \cap N = \{0\}$. Therefore, f is an isomorphism.

Definition. If M is a subspace of a vector space E and if E/M is finite-dimensional, the **codimension** (or **deficiency**) of M in E is the dimension of E/M.

Theorem 30.8. If M is a subspace of a finite-dimensional vector space E, then

$$\dim M + \operatorname{codim} M = \dim E.$$

Proof. By Theorem 30.3, M admits a supplement N, which is isomorphic to E/M by Theorem 30.7. Thus, by the Corollary of Theorem 30.4,

$$\dim M + \operatorname{codim} M = \dim M + \dim N = \dim E.$$

EXERCISES

30.1. Verify the statements of the proof of Theorem 30.5.

30.2. Let M and N be supplementary submodules of a K-module E. The **projection of E on M along** N is the function p from E into E defined by $p(x + y) = x$ for all $x \in M$, $y \in N$. Show that the projection p of E on M along N is an endomorphism of the K-module E satisfying $p \circ p = p$; moreover, $M = \{z \in E : p(z) = z\}$, $N = \ker p$, and $1_E - p$ is the projection of E on N along M.

30.3. If E is a K-module and if p is an endomorphism of E such that $p \circ p = p$, then E is the direct sum of M and N where $M = \{z \in E : p(z) = z\}$, $N = \ker p$, and p is the projection of E on M along N.

30.4. State and prove the analogue for modules of Theorem 20.8.

30.5. Prove Theorem 30.6.

30.6. State and prove the analogue for modules of Example 20.4.

30.7. State and prove the analogue for modules of Exercise 20.11.

30.8. State and prove the analogue for modules of Exercise 20.12.

30.9. If M and N are subspaces of a finite-dimensional vector space E, then

$$\operatorname{codim}(M + N) + \operatorname{codim}(M \cap N) = \operatorname{codim} M + \operatorname{codim} N.$$

[Use Theorem 30.5.]

30.10. If M is a submodule of a module E, then the module E/M is finite-dimensional if and only if M has a finite-dimensional supplement in E.

*30.11. If M is an m-dimensional subspace of an n-dimensional vector space E over a finite field of q elements, then M has $q^{m(n-m)}$ supplements in E.

30.12. Let E be an n-dimensional vector space and F an r-dimensional vector space. If N is a subspace of E of codimension r, there is a bijection from the set of all linear transformations from E onto F with kernel N onto the set of all isomorphisms from E/N onto F.

*30.13. Let K be a finite field of q elements. Let E be an n-dimensional K-vector space, let F be an m-dimensional K-vector space, and let r be an integer such that $1 \le r \le \min\{m, n\}$.
(a) Show that E has

$$\prod_{j=1}^{r} \frac{q^{n-j+1} - 1}{q^j - 1}$$

subspaces of codimension r. [Use Exercise 28.23(c).]
(b) Show that there are

$$q^{r(r-1)/2} \prod_{j=1}^{r} \frac{(q^{n-j+1} - 1)(q^{m-j+1} - 1)}{q^j - 1}$$

linear transformations from E into F of rank r. [Use Exercises 30.12 and 28.23(c).]

30.14. A submodule M of a K-module E is a **maximal submodule** of E if M is a maximal element of the set of all proper submodules of E, ordered by $\subseteq$. A K-module F is **simple** (or **irreducible**) if F is a nonzero module whose only submodules are F and $\{0\}$. A submodule M of a K-module E is maximal if and only if E/M is a simple module. [Use Exercise 30.8.]

30.15. (a) The set of even integers is a submodule of the Z-module Z admitting no supplement.
(b) If K is an integral domain, the following statements are equivalent:
1° K is a field.
2° The K-module K is simple (Exercise 30.14).
3° Every submodule of the K-module K has a supplement.

31. Duality*

Linear equations are used to describe lines in plane analytic geometry and both planes and lines in solid analytic geometry. One of our purposes here is to relate concepts of plane and solid analytic geometry to vector space concepts. For example, if K is a field, we shall see in general that linear

---

* Subsequent chapters do not depend on this section.

equations in n unknowns with coefficients from K are "dual" to hyperplanes—cosets of $(n - 1)$-dimensional subspaces—of the n-dimensional vector space K^n.

Let E and F be K-modules. We shall denote by $\mathrm{Hom}_K(E, F)$ the set of all homomorphisms (i.e., linear transformations) from E into F. We shall denote by $\mathrm{End}_K(E)$ the set of all endomorphisms of E (i.e., linear operators on E). We recall from Example 27.4 that if $u, v \in F^E$, $u + v$ is the function from E into F defined by $(u + v)(x) = u(x) + v(x)$ for all $x \in E$, and if $\lambda \in K$, λu is the function from E into F defined by $(\lambda u)(x) = \lambda u(x)$ for all $x \in E$.

Theorem 31.1. Let D, E, F, G be K-modules.

$1°$ If $u, v \in \mathrm{Hom}_K(E, F)$, then $u + v \in \mathrm{Hom}_K(E, F)$.
$2°$ If $u, v \in \mathrm{Hom}_K(E, F)$ and if $w \in \mathrm{Hom}_K(D, E)$, then

$$(u + v) \circ w = u \circ w + v \circ w.$$

$3°$ If $u, v \in \mathrm{Hom}_K(E, F)$ and if $t \in \mathrm{Hom}_K(F, G)$, then

$$t \circ (u + v) = t \circ u + t \circ v.$$

Proof. $1°$ If $x, y \in E$,

$$(u + v)(x + y) = u(x + y) + v(x + y) = u(x) + u(y) + v(x) + v(y)$$
$$= u(x) + v(x) + u(y) + v(y)$$
$$= (u + v)(x) + (u + v)(y),$$

and if $\lambda \in K$,

$$(u + v)(\lambda x) = u(\lambda x) + v(\lambda x) = \lambda u(x) + \lambda v(x)$$
$$= \lambda[u(x) + v(x)] = \lambda(u + v)(x).$$

$2°$ If $x \in D$,

$$[(u + v) \circ w](x) = (u + v)(w(x)) = u(w(x)) + v(w(x))$$
$$= (u \circ w)(x) + (v \circ w)(x) = [u \circ w + v \circ w](x).$$

$3°$ If $x \in E$,

$$[t \circ (u + v)](x) = t((u + v)(x)) = t(u(x) + v(x))$$
$$= t(u(x)) + t(v(x)) = (t \circ u)(x) + (t \circ v)(x)$$
$$= [t \circ u + t \circ v](x).$$

Corollary. If E is a K-module, $(\mathrm{End}_K(E), +, \circ)$ is a ring.

The assertion follows from Theorems 31.1 and 29.1.

If E and F are modules over a *commutative* ring with identity K, then $\mathrm{Hom}_K(E, F)$ is a submodule of the K-module F^E of all functions from E into F:

Theorem 31.2. If K is a commutative ring with identity and if E and F are K-modules, then $\mathrm{Hom}_K(E, F)$ is a submodule of the K-module F^E.

Proof. By Theorem 31.1, we need only show that if $\alpha \in K$ and $u \in \mathrm{Hom}_K(E, F)$, then $\alpha u \in \mathrm{Hom}_K(E, F)$. If $x, y \in E$ and $\lambda \in K$, then

$$(\alpha u)(x + y) = \alpha u(x + y) = \alpha(u(x) + u(y))$$

$$= \alpha u(x) + \alpha u(y) = (\alpha u)(x) + (\alpha u)(y),$$

$$(\alpha u)(\lambda x) = \alpha u(\lambda x) = \alpha \lambda u(x)$$

$$= \lambda \alpha u(x) = \lambda(\alpha u)(x),$$

so $\alpha u \in \mathrm{Hom}_K(E, F)$.

A frequently used convention in mathematical notation is the **Kronecker delta convention:** If Γ is some (index) set and if K is a ring with identity occurring in a given context, then for every $(\alpha, \beta) \in \Gamma \times \Gamma$, $\delta_{\alpha\beta}$ is defined by

$$\delta_{\alpha\beta} = \begin{cases} 1 \text{ if } \alpha = \beta, \\ 0 \text{ if } \alpha \neq \beta, \end{cases}$$

where "1" and "0" denote respectively the multiplicative and additive neutral elements of K.

Theorem 31.3. If K is a commutative ring with identity and if E and F are respectively an n-dimensional and an m-dimensional K-module, then $\mathrm{Hom}_K(E, F)$ is an nm-dimensional module. Indeed, if $(a_k)_{1 \leq k \leq n}$ is an ordered basis of E, if $(b_k)_{1 \leq k \leq m}$ is an ordered basis of F, and if, for each $i \in [1, n]$ and each $j \in [1, m]$, u_{ij} is the unique linear transformation from E into F satisfying

$$u_{ij}(a_k) = \delta_{ik}b_j$$

for all $k \in [1, n]$, then $\{u_{ij} : i \in [1, n] \text{ and } j \in [1, m]\}$ is a basis of $\mathrm{Hom}_K(E, F)$.

Proof. Let $B = \{u_{ij} : i \in [1, n] \text{ and } j \in [1, m]\}$. If

$$\sum_{j=1}^{m} \sum_{i=1}^{n} \lambda_{ij} u_{ij} = 0,$$

then for each $k \in [1, n]$,

$$0 = \sum_{j=1}^{m} \sum_{i=1}^{n} \lambda_{ij} u_{ij}(a_k) = \sum_{j=1}^{m} \lambda_{kj} b_j,$$

so $\lambda_{kj} = 0$ for all $j \in [1, m]$. Hence, B is linearly independent. If $u \in \text{Hom}_K(E, F)$ and if $(\alpha_{ij})_{1 \leq j \leq m}$ is the sequence of scalars satisfying

$$u(a_i) = \sum_{j=1}^{m} \alpha_{ij} b_j$$

for each $i \in [1, n]$, then

$$u(a_k) = \left(\sum_{j=1}^{m} \sum_{i=1}^{n} \alpha_{ij} u_{ij} \right)(a_k)$$

for all $k \in [1, n]$ by a calculation similar to the preceding, so

$$u = \sum_{j=1}^{m} \sum_{i=1}^{n} \alpha_{ij} u_{ij}$$

by Theorem 29.3. Thus, B is a set of generators for $\text{Hom}_K(E, F)$.

Definition. Let E be a module over a commutative ring with identity K. A **linear form** on E is a linear transformation from E into the K-module K. The K-module $\text{Hom}_K(E, K)$ of all linear forms on E is usually denoted by E^* and is called the **algebraic dual** of E. Linear forms are also called **linear functionals.**

If E is a module over a commutative ring with identity K, it is customary to denote $t'(x)$ by

$$\langle x, t' \rangle$$

for all $x \in E$, $t' \in E^*$. Thus, the function

$$(x, t') \mapsto \langle x, t' \rangle$$

from $E \times E^*$ into K satisfies

$$\langle x + y, t' \rangle = \langle x, t' \rangle + \langle y, t' \rangle,$$
$$\langle x, s' + t' \rangle = \langle x, s' \rangle + \langle x, t' \rangle,$$
$$\langle \lambda x, t' \rangle = \lambda \langle x, t' \rangle = \langle x, \lambda t' \rangle$$

for all $x, y \in E, s', t' \in E^*, \lambda \in K$.

If $(a_k)_{1 \leq k \leq n}$ is an ordered basis of an n-dimensional module E over a commutative ring with identity K, by Theorem 31.3, there is an ordered basis $(a'_k)_{1 \leq k \leq n}$ of E^* satisfying

$$\langle a_i, a'_j \rangle = \delta_{ij}$$

for all $i, j \in [1, n]$, since $\{1\}$ is a basis of the K-module K; the ordered basis $(a'_k)_{1 \leq k \leq n}$ of E^* is called the **ordered basis of** E^* **dual to** $(a_k)_{1 \leq k \leq n}$, or simply the **ordered dual basis** of E^*.

We shall denote the algebraic dual of E^* by E^{**}. If E is n-dimensional, then so are both E^* and E^{**} by Theorem 31.3. For each $x \in E$, we define $x^\wedge$ to be the function from E^* into K satisfying

$$x^\wedge(t') = t'(x)$$

for all $t' \in E^*$. It is easy to verify that $x^\wedge \in E^{**}$ and that

$$J: x \mapsto x^\wedge$$

is a linear transformation from E into E^{**}. We shall call J the **evaluation linear transformation** from E into E^{**}.

Theorem 31.4. If E is a finite-dimensional module over a commutative ring with identity K, then the evaluation linear transformation J is an isomorphism from E onto E^{**}.

Proof. It is easy to verify that if $(a_k)_{1 \leq k \leq n}$ is an ordered basis of E, then $(J(a_k))_{1 \leq k \leq n}$ is the ordered basis of E^{**} dual to the ordered basis of E^* dual to $(a_k)_{1 \leq k \leq n}$. By Theorem 29.5, therefore, J is an isomorphism.

If E is finite-dimensional, therefore, we may call J the **evaluation isomorphism** from E onto E^{**}.

Theorem 31.5. If E is a finite-dimensional module over a commutative ring with identity K and if $(a'_k)_{1 \leq k \leq n}$ is an ordered basis of E^*, then there is an ordered basis $(a_k)_{1 \leq k \leq n}$ of E such that $(a'_k)_{1 \leq k \leq n}$ is the ordered basis of E^* dual to $(a_k)_{1 \leq k \leq n}$.

Proof. Let $(a''_k)_{1 \leq k \leq n}$ be the ordered basis of E^{**} dual to $(a'_k)_{1 \leq k \leq n}$, and let $a_k = J^{\leftarrow}(a''_k)$ for all $k \in [1, n]$. Then $(a_k)_{1 \leq k \leq n}$ is an ordered basis of E by Theorems 31.4 and 29.5, and

$$\langle a_i, a'_j \rangle = J(a_i)(a'_j) = a''_i(a'_j) = \delta_{ij}$$

for all $i, j \in [1, n]$, so $(a'_k)_{1 \leq k \leq n}$ is the ordered basis of E^* dual to $(a_k)_{1 \leq k \leq n}$.

Definition. Let E be a module over a commutative ring with identity K. For each submodule M of E, the **annihilator** of M is the subset $M^{\circ \rightarrow}$ of E^* defined by

$$M^{\circ \rightarrow} = \{t' \in E^*: \langle x, t' \rangle = 0 \text{ for all } x \in M\},$$

and for each submodule N of E^*, the **annihilator** of N is the subset $N^{\leftarrow \circ}$ of E defined by

$$N^{\leftarrow \circ} = \{x \in E: \langle x, t' \rangle = 0 \text{ for all } t' \in N\}.$$

It is easy to verify that $M^{\circ \rightarrow}$ is a submodule of E^* and that $N^{\leftarrow \circ}$ is a submodule of E.

Theorem 31.6. Let E be an n-dimensional vector space over a field K. If M is an m-dimensional subspace of E, then $M^{\circ \rightarrow}$ is an $(n - m)$-dimensional subspace of E^*, and $M^{\circ \rightarrow \leftarrow \circ} = M$. If N is a p-dimensional subspace of E^*, then $N^{\leftarrow \circ}$ is an $(n - p)$-dimensional subspace of E, and $N^{\leftarrow \circ \circ \rightarrow} = N$. The function

$$M \mapsto M^{\circ \rightarrow}$$

is a bijection from the set of all m-dimensional subspaces of E onto the set of all $(n - m)$-dimensional subspaces of E^*, and its inverse is the function

$$N \mapsto N^{\leftarrow \circ}.$$

Proof. Let $(a_k)_{1 \leq k \leq n}$ be an ordered basis of E such that $(a_k)_{1 \leq k \leq m}$ is an ordered basis of M, and let $(a'_k)_{1 \leq k \leq n}$ be the ordered dual basis of E^*. If $t' \in M^{\circ \rightarrow}$ and if $t' = \sum_{k=1}^{n} \lambda_k a'_k$, then for all $j \in [1, m]$,

$$\lambda_j = \sum_{k=1}^{n} \lambda_k \langle a_j, a'_k \rangle = \left\langle a_j, \sum_{k=1}^{n} \lambda_k a'_k \right\rangle = \langle a_j, t' \rangle = 0,$$

so t' is a linear combination of $\{a_k': m + 1 \leq k \leq n\}$. But clearly, $a_k' \in M^{\circ\rightarrow}$ for each $k \in [m + 1, n]$. Therefore, $M^{\circ\rightarrow}$ has dimension $n - m$. Similarly, let $(a_k')_{1 \leq k \leq n}$ be an ordered basis of E^* such that $(a_k')_{1 \leq k \leq p}$ is an ordered basis of N; by Theorem 31.5, $(a_k')_{1 \leq k \leq n}$ is the ordered basis of E^* dual to an ordered basis $(a_k)_{1 \leq k \leq n}$ of E, and an argument similar to the preceding shows that $N^{\leftarrow c}$ has dimension $n - p$.

Clearly, $M \subseteq M^{\circ\rightarrow\leftarrow\circ}$. As $\dim M^{\circ\rightarrow\leftarrow\circ} = n - (n - m) = m$, therefore, $M = M^{\circ\rightarrow\leftarrow\circ}$. Similarly, $N^{\leftarrow\circ\circ\rightarrow} = N$. The final assertion now follows from Theorem 5.3.

Theorem 31.7. Let K be a field, and let M be a subspace of the n-dimensional vector space K^n. The following statements are equivalent:

1° $\dim M = n - 1$.
2° M is the kernel of a nonzero linear form.
3° There exists a sequence $(\alpha_k)_{1 \leq k \leq n}$ of scalars not all of which are zero such that

$$M = \{(\lambda_1, \ldots, \lambda_n) \in K^n : a_1\lambda_1 + \ldots + a_n\lambda_n = 0\}.$$

Further, if 1°–3° hold and if $(\beta_k)_{1 \leq k \leq n}$ is a sequence of scalars such that

$$M = \{(\lambda_1, \ldots, \lambda_n) \in K^n : \beta_1\lambda_1 + \ldots + \beta_n\lambda_n = 0\},$$

then there is a nonzero scalar γ such that

$$\beta_k = \gamma\alpha_k$$

for all $k \in [1, n]$.

Proof. To show that 1° implies 2°, let $N = M^{\circ\rightarrow}$. By Theorem 31.6, N is one-dimensional, and $M = N^{\leftarrow\circ}$. Let u be a nonzero member of N. Then N is the set of all scalar multiples of u, so

$$M = N^{\leftarrow\circ} = \{x \in E : \langle x, \lambda u \rangle = 0 \text{ for all } \lambda \in K\}$$
$$= \{x \in E : \langle x, u \rangle = 0\} = \ker u.$$

By the corollary of Theorem 29.7, 2° implies 1°.

Also, 2° and 3° are equivalent, for if $(\alpha_k)_{1 \leq k \leq n}$ is any sequence of scalars and if $u = \sum_{k=1}^{n} \alpha_k e_k'$ where $(e_k')_{1 \leq k \leq n}$ is the ordered basis of $(K^n)^*$ dual to the standard ordered basis of K^n, a simple calculation shows that

$$\ker u = \{(\lambda_1, \ldots, \lambda_n) : \alpha_1\lambda_1 + \ldots + \alpha_n\lambda_n = 0\},$$

and that $u \neq 0$ if and only if $\alpha_k \neq 0$ for some $k \in [1, n]$. If also M is the kernel of $v = \sum_{k=1}^{n} \beta_k e'_k$, then $v \in M^{\circ \rightarrow}$, and so $v = \gamma u$ for some nonzero scalar γ, since $M^{\rightarrow}$ is one-dimensional and since $v \neq 0$, and therefore $\beta_k = \gamma \alpha_k$ for all $k \in [1, n]$.

A **hyperplane** of a vector space is a coset of a subspace of codimension 1. Theorem 31.7 allows us to characterize hyperplanes of the K-vector space K^n, where K is a field. Indeed, let $H = M + x_0$, where codim $M = 1$, and let $x_0 = (\mu_1, \ldots, \mu_n)$. By Theorem 31.7, there exist $\alpha_1, \ldots, \alpha_n \in K$, not all zero, such that M is the set of all $(\lambda_1, \ldots, \lambda_n) \in K^n$ satisfying

$$(1) \qquad\qquad \alpha_1 \lambda_1 + \ldots + \alpha_n \lambda_n = 0.$$

Let $\beta = \alpha_1 \mu_1 + \ldots + \alpha_n \mu_n$. We shall show that H is the set of all $(\zeta_1, \ldots, \zeta_n) \in K^n$ such that

$$(2) \qquad\qquad \alpha_1 \zeta_1 + \ldots + \alpha_n \zeta_n = \beta.$$

Indeed, if (2) holds and if $\lambda_k = \zeta_k - \mu_k$ for each $k \in [1, n]$, then

$$\alpha_1 \lambda_1 + \ldots + \alpha_n \lambda_n = \alpha_1 \zeta_1 + \ldots + \alpha_n \zeta_n - (\alpha_1 \mu_1 + \ldots + \alpha_n \mu_n)$$
$$= \beta - \beta = 0,$$

so $(\zeta_1, \ldots, \zeta_n) = (\lambda_1, \ldots, \lambda_n) + (\mu_1, \ldots, \mu_n) \in M + x_0$; conversely, if $(\lambda_1, \ldots, \lambda_n) \in M$ and if $(\zeta_1, \ldots, \zeta_n) = (\lambda_1, \ldots, \lambda_n) + (\mu_1, \ldots, \mu_n)$, then

$$\alpha_1 \zeta_1 + \ldots + \alpha_n \zeta_n = \alpha_1 \lambda_1 + \ldots + \alpha_n \lambda_n + \alpha_1 \mu_1 + \ldots + \alpha_n \mu_n$$
$$= 0 + \beta = \beta.$$

On the other hand, let $\alpha_1, \ldots, \alpha_n$ be scalars, not all zero, and let $\beta \in K$. Let M be the set of all $(\lambda_1, \ldots, \lambda_n) \in K^n$ satisfying (1). By Theorem 31.7, M is a subspace of K^n of codimension 1. If $x_0 = (\mu_1, \ldots, \mu_n)$ is a vector such that

$$\alpha_1 \mu_1 + \ldots + \alpha_n \mu_n = \beta$$

(if $\alpha_i \neq 0$, $(0, \ldots, 0, \beta/\alpha_i, 0, \ldots, 0)$ is such a vector), then by the above argument $M + x_0$ is the set of all $(\zeta_1, \ldots, \zeta_n) \in K^n$ satisfying (2). In sum, *a subset H of K^n is a hyperplane if and only if there exist a scalar β and a sequence $(\alpha_k)_{1 \leq k \leq n}$ of scalars, not all of which are zero, such that H is the set of all $(\lambda_1, \ldots, \lambda_n) \in K^n$ satisfying*

$$\alpha_1 \lambda_1 + \ldots + \alpha_n \lambda_n = \beta,$$

in which case H is a coset of the $(n-1)$-dimensional subspace of all $(\lambda_1, \ldots, \lambda_n) \in K^n$ satisfying

$$\alpha_1 \lambda_1 + \ldots + \alpha_n \lambda_n = 0.$$

In plane analytic geometry, a line L is either defined or derived to be the set of all $(x_1, x_2) \in R^2$ satisfying

(3) $$\alpha_1 x_1 + \alpha_2 x_2 = \beta$$

where α_1, α_2, and β are given real numbers and not both α_1 and α_2 are zero, and a line L' is shown to be parallel to L if and only if for some $\beta' \in R$ the line L' is the set of all $(x_1, x_2) \in R^2$ satisfying

$$\alpha_1 x_1 + \alpha_2 x_2 = \beta'.$$

In solid analytic geometry, a plane P is either defined or derived to be the set of all $(x_1, x_2, x_3) \in R^3$ satisfying

(4) $$\alpha_1 x_1 + \alpha_2 x_2 + \alpha_3 x_3 = \gamma$$

where α_1, α_2, α_3, and γ are given real numbers and not all of α_1, α_2, α_3 are zero, and a plane P' is shown to be parallel to P if and only if for some $\gamma' \in R$ the plane P' is the set of all $(x_1, x_2, x_3) \in R^3$ satisfying

$$\alpha_1 x_1 + \alpha_2 x_2 + \alpha_3 x_3 = \gamma'.$$

(A line or plane is considered parallel to itself, so β' may be β and γ' may be γ.) A line or plane is called **homogeneous** if it contains the origin, which is the zero vector. Clearly, the line of R^2 defined by (3) is homogeneous if and only if $\beta = 0$, and the plane of R^3 defined by (4) is homogeneous if and only if $\gamma = 0$. Consequently, we see from the preceding that *the lines of plane analytic geometry are precisely the cosets of one-dimensional subspaces of R^2*, i.e., the hyperplanes of R^2, and that two lines are parallel if and only if they are cosets of the same subspace. Similarly, *the planes of solid analytic geometry are precisely the cosets of two-dimensional subspaces of R^3*, i.e., the hyperplanes of R^3, and two planes are parallel if and only if they are cosets of the same subspace.

In solid analytic geometry, a line is either defined or derived to be the intersection of two nonparallel planes. Let M and N be two distinct two-dimensional subspaces of R^3, and let L be the intersection of the cosets $P = M + x_0$ and $Q = N + y_0$ of M and N respectively. Then since $M \neq N$, the subspace $M + N$ properly contains the two-dimensional subspace M, and hence, $M + N = R^3$; therefore, there exist $m \in M$ and $n \in N$ such that $x_0 - y_0 = m + n$. Let $z_0 = x_0 - m$. Then

$$P = M + x_0 = M + m + z_0 = M + z_0,$$

and also as $z_0 = y_0 + n$,

$$Q = N + y_0 = N + n + y_0 = N + z_0.$$

Consequently,

$$L = P \cap Q = (M + z_0) \cap (N + z_0),$$

which is easily seen to be $(M \cap N) + z_0$, and furthermore, $M \cap N$ is one-dimensional by Theorem 30.5. Thus, a line of solid analytic geometry is a coset of a one-dimensional subspace of R^3. Conversely, if L is the coset $D + z_0$ of a one-dimensional subspace D of R^3, it follows easily from Theorem 28.7 that there exist two-dimensional subspaces M and N of R^3 such that $D = M \cap N$, whence

$$L = (M \cap N) + z_0 = (M + z_0) \cap (N + z_0),$$

and consequently, L is the intersection of two nonparallel planes. In summary, *the lines of solid analytic geometry are precisely the cosets of one-dimensional subspaces of R^3.*

If M is a one-dimensional subspace of R^2, that is, a homogeneous line of the plane, by the preceding discussion R^2/M is the set of all lines in the plane parallel to M, and the canonical epimorphism φ_M from R^2 onto R^2/M is the function associating to each point of R^2 the line containing it parallel to M. If N is a homogeneous line distinct from M, then R^2 is the direct sum of M and N by Example 30.2. By Theorem 30.7, therefore, the restriction of φ_M to N is a bijection from N onto R^2/M; this is just the vector space translation of the geometric assertion that each line in the plane parallel to M intersects N in one and only one point. Similarly, vector space concepts admit a geometric description if the vector space is R^3 (Exercise 31.1).

EXERCISES

31.1. Let L be a one-dimensional subspace and P a two-dimensional subspace of R^3. Describe in geometrical language
   (a) the set R^3/P and the canonical epimorphism φ_P;

(b) the set R^3/L and the canonical epimorphism φ_L;

(c) the set P/L and the canonical epimorphism from P onto P/L, if $L \subseteq P$;

(d) the restriction of φ_P to L and the restriction of φ_L to P, if $L \nsubseteq P$.

31.2. Let E be an n-dimensional vector space where $n \geq 1$. A **linear variety** of E is a coset of a subspace of E. A linear variety is **proper** if it is a proper subset of E. Two linear varieties are **parallel** if they are cosets of subspaces one of which is contained in the other.

(a) A linear variety V is the coset of one and only one subspace M of E; for every $x_0 \in V$, $M = (-x_0) + V$ and $V = M + x_0$. The dimension of the subspace M is called the **dimension** of the linear variety V.

(b) Is the relation "____ is parallel to . . . ." an equivalence relation on the set of all linear varieties of E? Is the relation it induces on the set of all p-dimensional linear varieties of E an equivalence relation?

(c) If V and W are parallel linear varieties, then either $V \subseteq W$ or $W \subseteq V$ or $V \cap W = \emptyset$.

(d) Let V and W be cosets of subspaces M and N respectively. If $V \cap W \neq \emptyset$, then $V \cap W$ is a coset of $M \cap N$. If $M + N = E$, then $V \cap W \neq \emptyset$ and hence $V \cap W$ is a coset of $M \cap N$.

(e) If H is a hyperplane and if V is a proper linear variety, either $V \cap H \neq \emptyset$ or V is parallel to H, and in the latter case, there is one and only one hyperplane H' containing V parallel to H. In particular, every point of E belongs to one and only one hyperplane parallel to a given hyperplane H.

(f) The intersection of a hyperplane and a p-dimensional linear variety not parallel to it is a $(p - 1)$-dimensional linear variety.

(g) If $\mathscr{P}$ is a partition of E all of whose members are hyperplanes, then any two members of $\mathscr{P}$ are parallel, and consequently, there is a subspace H of codimension 1 such that $\mathscr{P} = E/H$.

(h) If M and N are subspaces of E, then $M + N = E$ (respectively, $M \cap N = \{0\}$, E is the direct sum of M and N) if and only if the intersection of any linear variety parallel to M with any linear variety parallel to N is a nonempty set (respectively, a set containing at most one point, a set containing exactly one point).

*31.3. Let E be an n-dimensional vector space where $n \geq 1$.

(a) A nonempty subset V of E is a linear variety if and only if for every sequence $(x_k)_{1 \leq k \leq r}$ of elements of V and every sequence $(\lambda_k)_{1 \leq k \leq r}$ of scalars such that

$$\sum_{k=1}^{r} \lambda_k = 1,$$

the vector

$$\sum_{k=1}^{r} \lambda_k x_k$$

belongs to V.

(b) If the intersection of a set of linear varieties of E is nonempty, that intersection is a linear variety. In particular, every nonempty subset A of

E is contained in a smallest linear variety, called the **linear variety generated** by A.

(c) A proper linear variety of dimension p is the intersection of $n - p$ hyperplanes.

(d) If V and W are linear varieties such that $V \cap W \neq \emptyset$ and if Z is the linear variety generated by $V \cup W$, then

$$\dim Z = \dim V + \dim W - \dim(V \cap W).$$

(e) If V and W are parallel linear varieties such that $V \cap W = \emptyset$ and if Z is the linear variety generated by $V \cup W$, then

$$\dim Z = \max\{\dim V, \dim W\} + 1.$$

31.4. State as many axioms and theorems of solid geometry as you can that are restatements in geometric language for the vector space R^3 of assertions about linear varieties made in Exercises 31.2 and 31.3.

31.5. Let E be a module over a commutative ring with identity K.
(a) The function $J: E \to E^{**}$ is a linear transformation.
(b) If M is a submodule of E, then $M^{\circ\to}$ is a submodule of E^*.
(c) If N is a submodule of E^*, then $N^{\leftarrow\circ}$ is a submodule of E.

31.6. If E and G are isomorphic K-modules, if F and H are isomorphic K-modules, and if K is commutative, then $\operatorname{Hom}_K(E, F)$ and $\operatorname{Hom}_K(G, H)$ are isomorphic K-modules.

31.7. If E and F are modules over a commutative ring K and if M is a submodule of E, then $\{u \in \operatorname{Hom}_K(E, F): u(M) = \{0\}\}$ is a submodule of $\operatorname{Hom}_K(E, F)$. What is its dimension if K is a field and if E, F, and M have dimensions n, m, and p respectively?

31.8. Let n be a strictly positive integer, and let $P_n(R)$ be the n-dimensional subspace of $P(R)$ consisting of all polynomial functions of degree $< n$. For each real number a, the function $a^\wedge: p \mapsto p(a)$ is a linear form on $P_n(R)$. If $(a_k)_{1 \leq k \leq n}$ is a sequence of distinct real numbers, then $(a_k{}^\wedge)_{1 \leq k \leq n}$ is an ordered basis of $P_n(R)^*$. To what ordered basis $(p_k)_{1 \leq k \leq n}$ of $P_n(R)$ is $(a_k{}^\wedge)_{1 \leq k \leq n}$ dual? [Express each p_k as a product of polynomial functions of degree 1.]

31.9. Let E and F be modules over a commutative ring with identity K. For each $u \in \operatorname{Hom}_K(E, F)$, the **transpose** of u is the function u^t from F^* into E^* defined by

$$u^t(y') = y' \circ u$$

for all $y' \in F^*$.
(a) Show that

$$\langle x, u^t(y') \rangle = \langle u(x), y' \rangle$$

for all $x \in E$, $y' \in F^*$, $u \in \operatorname{Hom}_K(E, F)$.

(b) Show that $u^t \in \mathrm{Hom}_K(F^*, E^*)$.

(c) Show that $\ker u^t = u_*(E)^{\circ\rightarrow}$.

(d) Show that $u_*^t(F^*) \subseteq (\ker u)^{\circ\rightarrow}$.

(e) If K is a field and if E and F are n-dimensional vector spaces, then $u_*^t(F^*) = (\ker u)^{\circ\rightarrow}$, $\rho(u) = \rho(u^t)$, $\nu(u) = \nu(u^t)$.

31.10. Let E, F, and G be modules over a commutative ring K.

(a) If $u \in \mathrm{Hom}_K(E, F)$ and if $v \in \mathrm{Hom}_K(F, G)$, then $(v \circ u)^t = u^t \circ v^t$.

(b) If u is an isomorphism from E onto F, then u^t is an isomorphism from F^* onto E^*, and $(u^t)^{\leftarrow} = (u^{\leftarrow})^t$.

(c) If $u \in \mathrm{Hom}_K(E, F)$ and if M is a submodule of E, then $u_*(M)^{\circ\rightarrow} = (u^t)^*(M^{\circ\rightarrow})$.

31.11. Let E and F be finite-dimensional vector spaces over a field.

(a) The function $T: u \mapsto u^t$ is an isomorphism from the vector space $\mathrm{Hom}_K(E, F)$ onto the vector space $\mathrm{Hom}_K(F^*, E^*)$.

(b) The function $T: u \mapsto u^t$ is an anti-isomorphism (Exercise 15.8) from the ring $\mathrm{End}_K(E)$ onto the ring $\mathrm{End}_K(E^*)$.

*31.12. Let E be a finite-dimensional vector space over a field, and let M and N be subspaces of E.

(a) Show that $M \subseteq N$ if and only if $M^{\circ\rightarrow} \supseteq N^{\circ\rightarrow}$.

(b) Show that $(M + N)^{\circ\rightarrow} = M^{\circ\rightarrow} \cap N^{\circ\rightarrow}$.

(c) Show that $(M \cap N)^{\circ\rightarrow} = M^{\circ\rightarrow} + N^{\circ\rightarrow}$. [Use (b) and Theorem 30.5.]

(d) Generalize (b) and (c) to any finite number of subspaces.

(e) If $u, u_1, \ldots, u_n \in E^*$, then $\displaystyle\bigcap_{k=1}^{n} \ker u_k \subseteq \ker u$ if and only if u is a linear combination of $(u_k)_{1 \leq k \leq n}$.

*31.13. Let E be a module over an integral domain K.

(a) If the annihilator of E in K (Exercise 27.8) is not the zero ideal, then $E^* = \{0\}$.

(b) If E is a quotient field of K considered as a K-module, then $E^* \neq \{0\}$ if and only if $E = K$.

31.14. If E is a finite-dimensional vector space over a field and if N is a subspace of E^*, then $N^{\circ\rightarrow} = J_*(N^{\leftarrow\circ})$.

31.15. Let E be a vector space over a field K.

(a) E is a module over the ring $\mathrm{End}_K(E)$, where scalar multiplication is defined by $u.x = u(x)$ for all $u \in \mathrm{End}_K(E)$, $x \in E$; moreover, the annihilator of the $\mathrm{End}_K(E)$-module E (Exercise 27.8) is $\{0\}$.

(b) If E is finite-dimensional, then the $\mathrm{End}_K(E)$-module E is generated by each nonzero element of E.

(c) If, in addition, $\dim_K E \geq 2$, then the $\mathrm{End}_K(E)$-module E has no basis. Conclude that the statement obtained from Exercise 28.18(b) by omitting the hypothesis that K be commutative is not in general true.

32. Matrices*

Here we shall show how to describe a linear transformation from one finite-dimensional module into another in a particularly concrete way by means of matrices—rectangular arrays of scalars—relative to ordered bases of the domain and codomain. If m and n are strictly positive integers and if K is a set, an m by n **matrix** over K is a function from $[1, m] \times [1, n]$ into K. We shall denote the set of all m by n matrices over K by $\mathscr{M}_K(m, n)$.

Thus, by definition $\mathscr{M}_K(m, n)$ is the set $K^{[1, m] \times [1, n]}$. Usually, matrices are denoted by means of indices: If $\alpha_{ij} \in K$ for all $(i, j) \in [1, m] \times [1, n]$, the m by n matrix whose value at each (i, j) is α_{ij} is denoted by $(\alpha_{ij})_{(i,j) \in [1,m] \times [1,n]}$ [or simply (α_{ij}) if it is clear what the set of indices is]. If $A = (\alpha_{ij})$ is an m by n matrix over K, for each $i \in [1, m]$ the ordered n-tuple $(\alpha_{i1}, \alpha_{i2}, \ldots, \alpha_{in})$ is called the *ith row* of A, and for each $j \in [1, n]$ the ordered m-tuple $(\alpha_{1j}, \alpha_{2j}, \ldots, \alpha_{mj})$ is called the *jth column* of A. This terminology arises from the fact that an m by n matrix (α_{ij}) is often denoted by the rectangular array

$$
\begin{bmatrix}
\alpha_{11} & \alpha_{12} & \cdot & \cdot & \cdot & \alpha_{1n} \\
\alpha_{21} & \alpha_{22} & & \cdot & \cdot & \alpha_{2n} \\
\cdot & \cdot & & & & \cdot \\
\cdot & \cdot & & & & \cdot \\
\alpha_{m1} & \alpha_{m2} & \cdot & \cdot & \cdot & \alpha_{mn}
\end{bmatrix}
$$

in which α_{ij} is the entry appearing in the ith row and the jth column. For a specific matrix it is often possible to dispense with indices. For example,

$$
\begin{bmatrix}
2 & 3 & -1 \\
0 & 7 & 5
\end{bmatrix}
$$

is the 2 by 3 matrix (α_{ij}) where $\alpha_{11} = 2$, $\alpha_{12} = 3$, $\alpha_{13} = -1$, $\alpha_{21} = 0$, $\alpha_{22} = 7$, and $\alpha_{23} = 5$.

Definition. If $A = (\alpha_{ij})$ and $B = (\beta_{ij})$ are m by n matrices over a ring K, then the **sum** $A + B$ of A and B is the m by n matrix $(\alpha_{ij} + \beta_{ij})$, and for each $\lambda \in K$, the **scalar product** λA of λ and A is the m by n matrix $(\lambda \alpha_{ij})$.

Thus, by definition,

$$(\alpha_{ij}) + (\beta_{ij}) = (\alpha_{ij} + \beta_{ij}),$$
$$\lambda(\alpha_{ij}) = (\lambda \alpha_{ij}).$$

* Subsequent chapters do not depend on this section.

If K is a ring with identity, then with these definitions of addition and scalar multiplication, $\mathscr{M}_K(m, n)$ is a K-module, namely, the K-module $K^{[1,\,m]\times[1,\,n]}$ of Example 27.4, where F of that example is the K-module K and E is the set $[1, m] \times [1, n]$. The neutral element for addition is called the **zero matrix,** for all of its entries are zero. The additive inverse of the matrix (α_{ij}) is therefore $(-\alpha_{ij})$.

Definition. If $A = (\alpha_{ij})$ is an m by n matrix over a ring K and if $B = (\beta_{ij})$ if an n by p matrix over K, the **product** AB of A and B is the m by p matrix (γ_{ij}) over K where

$$\gamma_{ij} = \sum_{k=1}^{n} \alpha_{ik}\beta_{kj}$$

for all $(i, j) \in [1, m] \times [1, p]$.

Thus, the product of two matrices over the same ring is defined if and only is the number of columns of the first is the same as the number of rows of the second; in that case, the entry in the ith row and jth column of the product is the result of multiplying pairwise the entries in the ith row of the first by the entries in the jth column of the second and then adding. The following are some examples of matric multiplication.

$$\begin{bmatrix} a_{11} & a_{12} \\ a_{21} & a_{22} \end{bmatrix}\begin{bmatrix} b_{11} & b_{12} \\ b_{21} & b_{22} \end{bmatrix} = \begin{bmatrix} a_{11}b_{11} + a_{12}b_{21} & a_{11}b_{12} + a_{12}b_{22} \\ a_{21}b_{11} + a_{22}b_{21} & a_{21}b_{12} + a_{22}b_{22} \end{bmatrix}$$

$$\begin{bmatrix} 2 & 1 & 0 \\ 3 & 0 & 7 \end{bmatrix}\begin{bmatrix} 2 & 3 & 5 & 8 \\ 4 & 8 & 6 & 1 \\ -1 & 7 & 0 & 7 \end{bmatrix} = \begin{bmatrix} 8 & 14 & 16 & 17 \\ -1 & 58 & 15 & 73 \end{bmatrix}$$

$$\begin{bmatrix} 5 & 3 & 1 \end{bmatrix}\begin{bmatrix} 2 \\ 0 \\ 6 \end{bmatrix} = \begin{bmatrix} 16 \end{bmatrix} \qquad \begin{bmatrix} 2 \\ 0 \\ 6 \end{bmatrix}\begin{bmatrix} 5 & 3 & 1 \end{bmatrix} = \begin{bmatrix} 10 & 6 & 2 \\ 0 & 0 & 0 \\ 30 & 18 & 6 \end{bmatrix}$$

$$\begin{bmatrix} 1 & 3 & -2 \\ -2 & -6 & 4 \\ 4 & 12 & -8 \end{bmatrix}\begin{bmatrix} 3 & -1 & 2 \\ 3 & 5 & -4 \\ 6 & 7 & -5 \end{bmatrix} = \begin{bmatrix} 0 & 0 & 0 \\ 0 & 0 & 0 \\ 0 & 0 & 0 \end{bmatrix}$$

Definition. Let $(a_j)_{1 \leq j \leq n}$ and $(b_i)_{1 \leq i \leq m}$ be ordered bases respectively of an

n-dimensional module E and m-dimensional module F over a commutative ring with identity K. For each $u \in \operatorname{Hom}_K(E, F)$, the **matrix of u relative to** $(a_j)_{1 \leq j \leq n}$ **and** $(b_i)_{1 \leq i \leq m}$ is the m by n matrix (α_{ij}) where

$$u(a_j) = \sum_{i=1}^{m} \alpha_{ij} b_i$$

for all $(i, j) \in [1, m] \times [1, n]$.

For example, the matrix of the differentiation linear transformation $D: P \mapsto Dp$ from $P_4(\mathbf{R})$ into $P_3(\mathbf{R})$ relative to the ordered bases $(p_j)_{0 \leq j \leq 3}$ and $(p_i)_{0 \leq i \leq 2}$ (Example 28.7) is

$$\begin{bmatrix} 0 & 1 & 0 & 0 \\ 0 & 0 & 2 & 0 \\ 0 & 0 & 0 & 3 \end{bmatrix}.$$

To avoid cumbersome notation, we shall sometimes denote an ordered basis $(a_k)_{1 \leq k \leq n}$ of a module simply by $(a)_n$ and the matrix of a linear transformation u relative to ordered bases $(a)_n$ and $(b)_m$ by $[u; (b)_m, (a)_n]$, or simply by $[u]$ if it is understood in the context which ordered bases are in use. [Note the order: In our notation for the matrix of u relative to $(a)_n$ and $(b)_m$, the notation for the given ordered basis $(a)_n$ of the domain of u occurs *last* and is preceded by the notation for the given ordered basis $(b)_m$ of the codomain.] The entries in the jth *column* of $[u; (b)_m, (a)_n]$ are thus the scalars occurring in the expression of $u(a_j)$ as a linear combination of the sequence $(b_1, \ldots, b_m)$. The reason for this choice is given in the following theorem.

Theorem 32.1. Let D, E, and F be finite-dimensional modules with ordered bases $(a)_p$, $(b)_n$, and $(c)_m$ respectively over a commutative ring K with identity. Then

$$M: u \mapsto [u; (c)_m, (b)_n]$$

is an isomorphism from the K-module $\operatorname{Hom}_K(E, F)$ onto the K-module $\mathscr{M}_K(m, n)$, and for each $u \in \operatorname{Hom}_K(D, E)$ and each $v \in \operatorname{Hom}_K(E, F)$,

$$[v \circ u; (c)_m, (a)_p] = [v; (c)_m, (b)_n][u; (b)_n, (a)_p].$$

Proof. The proof that M is an isomorphism is straightforward, so we shall consider only the last assertion. If $(\alpha_{ij}) = [u; (b)_n, (a)_p]$ and if $(\beta_{ij}) =$

$[v; (c)_m, (b)_n]$, then

$$(v \circ u)(a_j) = v(u(a_j)) = v\left(\sum_{k=1}^{n}\alpha_{kj}b_k\right)$$

$$= \sum_{k=1}^{n}\alpha_{jk}v(b_k) = \sum_{k=1}^{n}\alpha_{kj}\left(\sum_{i=1}^{m}\beta_{ik}c_i\right)$$

$$= \sum_{k=1}^{n}\left(\sum_{i=1}^{m}\alpha_{kj}\beta_{ik}c_i\right) = \sum_{i=1}^{m}\left(\sum_{k=1}^{n}\alpha_{kj}\beta_{ik}c_i\right)$$

$$= \sum_{i=1}^{m}\left(\sum_{k=1}^{n}\beta_{ik}\alpha_{kj}\right)c_i,$$

so $[v \circ u; (c)_m, (a)_p] = [v; (c)_m, (b)_n][u; (b)_n, (a)_p]$.

Multiplication of matrices is purposely defined just to obtain the above equality relating the product of the matrices of two linear transformations with the matrix of their composite.

If E is a K-module, we shall denote by $\mathcal{M}_K(n)$ the set $\mathcal{M}_K(n, n)$ of all n by n matrices over K. An n by n matrix is called a **square matrix of order n.** Since the product of two square matrices of order n over a ring is a square matrix of order n, matric multiplication is a composition on $\mathcal{M}_K(n)$.

When representing a linear operator on a finite-dimensional module E by a matrix, one usually selects the same ordered basis for E considered both as the domain and the codomain of the linear operator. If $u \in \mathrm{End}_K(E)$ and if $(a_k)_{1 \leq k \leq n}$ is an ordered basis of E, we shall abbreviate $[u; (a)_n, (a)_n]$ to $[u; (a)_n]$. The following theorem is an immediate consequence of Theorem 32.1.

Theorem 32.2. If $(a_k)_{1 \leq k \leq n}$ is an ordered basis of an n-dimensional module E over a commutative ring with identity K, then

$$M: u \mapsto [u; (a)_n]$$

is an isomorphism from the ring $(\mathrm{End}_K(E), +, \circ)$ onto $(\mathcal{M}_K(n), +, \cdot)$. In particular, $\mathcal{M}_K(n)$ is a ring with identity.

By Theorems 31.1 and 32.1, if K is a commutative ring with identity, if $U, V \in \mathcal{M}_K(m, n)$, if $W, Z \in \mathcal{M}_K(n, p)$, and if $Y \in \mathcal{M}_K(p, q)$, then

$$(U + V)W = UW + VW,$$

$$U(W + Z) = UW + UZ,$$

$$(UW)Y = U(WY).$$

Actually, a simple calculation shows that these three identities hold even if K lacks an identity element or if K is not commutative (Exercise 32.12). In particular, therefore, *for every ring K and every $n \in N^*$, under matric addition and matric multiplication $\mathcal{M}_K(n)$ is a ring.*

If K is a ring with identity 1, it is a simple matter to show that the n by n matrix

$$\begin{bmatrix} 1 & 0 & 0 & . & . & . & 0 \\ 0 & 1 & 0 & . & . & . & 0 \\ 0 & 0 & 1 & . & . & . & 0 \\ . & . & . & & & & . \\ . & . & . & & & & . \\ 0 & 0 & 0 & . & . & . & 1 \end{bmatrix}$$

is the multiplicative identity of the ring $\mathcal{M}_K(n)$. The **(principal) diagonal** of a square matrix (α_{ij}) of order n is the sequence $\alpha_{11}, \alpha_{22}, \ldots, \alpha_{nn}$ of elements occurring on the diagonal joining the upper left and lower right corners of the matrix, written as a square array. Thus, if K is a ring with identity 1, the multiplicative identity of the ring $\mathcal{M}_K(n)$, which is denoted by I_n and called the **identity matrix** is the matrix whose diagonal entries are all 1 and whose nondiagonal entries are all 0. In terms of the Kronecker delta notation,

$$I_n = (\delta_{ij})_{(i,j) \in [1,n] \times [1,n]}.$$

An *invertible matrix* of order n is, of course, an invertible element of the ring $\mathcal{M}_K(n)$; if K is a commutative ring with identity, and if E is an n-dimensional K-module, then the invertible elements of $\mathrm{End}_K(E)$ are precisely the automorphisms of E, so that invertible matrices correspond to automorphisms of E under any isomorphism of the rings $\mathcal{M}_K(n)$ and $\mathrm{End}_K(E)$.

EXERCISES

32.1. Determine the eight products XYZ where each of X, Y, and Z is either

$$\begin{bmatrix} 2 & 0 \\ 3 & -1 \end{bmatrix} \text{ or } \begin{bmatrix} 4 & 2 \\ 2 & -4 \end{bmatrix}.$$

32.2. Make the eight computations of Exercise 32.1 if the numerals denote elements of Z_5.

32.3. Determine $BA - I_2$, $CB + A$, and $C^2 + AB$ if

$$A = \begin{bmatrix} 2 & 4 \\ 1 & 0 \\ 3 & -2 \end{bmatrix}, \quad B = \begin{bmatrix} 2 & 1 & -2 \\ 0 & 2 & 1 \end{bmatrix}, \quad \text{and} \quad C = \begin{bmatrix} 2 & 1 & 3 \\ 0 & 2 & -2 \\ 2 & 2 & 3 \end{bmatrix}.$$

32.4. Make the computations of Exercise 32.3 if the numerals denote elements of Z_5.

32.5. Determine the real numbers λ for which the following matrices over R are invertible, and for each such λ determine the inverse.

$$\begin{bmatrix} \lambda & 1 & 0 \\ 0 & \lambda & 1 \\ 1 & \lambda & 0 \end{bmatrix} \qquad \begin{bmatrix} 1 & 0 & \lambda \\ 0 & 1 & 0 \\ \lambda & 0 & 1 \end{bmatrix} \qquad \begin{bmatrix} 1 & \lambda & 0 \\ \lambda & 1 & \lambda \\ 0 & \lambda & 1 \end{bmatrix}$$

32.6. Determine the matrices R_α, S_M, and $P_{M,N}$ of the linear operators r_α, s_M, and $p_{M,N}$ of §29 relative to the standard ordered basis of R^2. By matric computation, show that $r_\alpha \circ r_\beta = r_{\alpha+\beta}$, $P_{M,N} \circ P_{M,N} = P_{M,N}$, and $s_M \circ s_M = 1_{R^2}$.

32.7. What are the matrices of r_α, s_M, and $p_{M,N}$ relative to the ordered basis $((1, 1), (0, -1))$ of R^2? relative to $((1, -1), (2, 4))$?

32.8. Let u be the linear operator on $P_5(R)$ satisfying

$$[u(p)](x) = p(x + 1)$$

for all $p \in P_5(R)$ and all $x \in R$. What is the matrix of u relative to the ordered basis $(p_k)_{0 \le k \le 4}$?

32.9. Determine the matrices of the linear operators defined in Exercises 29.1 and 29.2 relative to the standard ordered bases.

*32.10. If u is a linear operator on an n-dimensional vector space E such that $u^n = 0$ but $u^{n-1} \ne 0$ [where u^n denotes the linear operator $u \circ u \circ \ldots \circ u$ (n factors)], then for some $a \in E$, $(u^k(a))_{0 \le k \le n-1}$ is an ordered basis of E. What is the matrix of u with respect to this ordered basis? What is such an ordered basis if u is the differential linear operator D on the n-dimensional vector space $P_n(R)$?

32.11. Let K be a ring with identity, and let $(a_j)_{1 \le j \le n}$ and $(b_i)_{1 \le i \le m}$ be ordered bases respectively of an n-dimensional K-module E and an m-dimensional K-module F. For each $u \in \operatorname{Hom}_K(E, F)$, the **matrix of u relative to** $(a_j)_{1 \le j \le n}$ **and** $(b_i)_{1 \le i \le m}$ is the matrix (α_{ij}) *over the reciprocal ring L of K* (Exercise 15.8) where

$$u(a_j) = \sum_{i=1}^{m} \alpha_{ij} b_i$$

for all $(i, j) \in [1, m] \times [1, n]$.

(a) Prove the statement obtained by deleting "commutative" from the assertion of Theorem 32.1.

(b) The function $M: u \mapsto [u; (a)_n]$ of Theorem 32.2 is an isomorphism from the ring $\text{End}_K(E)$ onto $\mathscr{M}_L(n)$.

32.12. Let K be a ring, let $U, V \in \mathscr{M}_K(m, n)$, let $W, Z \in \mathscr{M}_K(n, p)$, and let $Y \in \mathscr{M}_K(p, q)$.
(a) Verify that $(UW)Y = U(WY)$, $(U + V)W = UW + VW$, and $U(W + Z) = UW + UZ$.
(b) Infer that under matric addition and multiplication, $\mathscr{M}_K(n)$ is a ring.

32.13. Let E and F be vector spaces over a field of dimensions n and m respectively, let $(a_k)_{1 \leq k \leq n}$ and $(b_k)_{1 \leq k \leq m}$ be ordered bases of E and F respectively, and let $u \in \text{Hom}_K(E, F)$. If $(\alpha_{ij}) = [u; (b)_m, (a)_n]$ and if $(\beta_{ij}) = [u^t; (a')_n, (b')_m]$ (Exercise 31.9), where $(a'_k)_{1 \leq k \leq n}$ and $(b'_k)_{1 \leq k \leq m}$ are the ordered bases dual to $(a_k)_{1 \leq k \leq n}$ and $(b_k)_{1 \leq k \leq m}$ respectively, then $\beta_{ij} = \alpha_{ji}$ for all $i \in [1, n], j \in [1, m]$.

32.14. Let $(e_\alpha)_{\alpha \in A}$ be a family of distinct terms of a K-module E such that $\{e_\alpha : \alpha \in A\}$ is a basis of E.
(a) Extend Theorem 29.4 as follows: If $(b_\alpha)_{\alpha \in A}$ is a family of elements of a K-module F also indexed by A, there is one and only one linear transformation u from E into F satisfying $u(e_\alpha) = b_\alpha$ for all $\alpha \in A$.
(b) The K-module E^* is isomorphic to the K-module K^A.

*32.15. Let E be a vector space having a denumerable basis $\{e_n : n \in N^*\}$ (e.g., let $E = P(R)$ and let $e_n = p_{n-1}$ for all $n \in N^*$), and let K be the ring $\text{End}_K(E)$. Let u_1 and u_2 be the linear operators on E satisfying

$$u_1(e_{2n-1}) = 0, \qquad u_2(e_{2n-1}) = e_n,$$

$$u_1(e_{2n}) = e_n, \qquad u_2(e_{2n}) = 0$$

for all $n \in N^*$ (Exercise 32.14(a)).
(a) Show that $\{u_1, u_2\}$ is a basis of the K-module K.
(b) Let $v_k = u_2 u_1{}^k$ for each $k \in N$. Show that

$$v_k(e_{2^k(2q-1)}) = e_q$$

for all $q \in N^*$ and that

$$v_k(e_r) = 0$$

if r is not an odd multiple of 2^k.
(c) For each $m \in N^*$, $\{v_0, v_1, \ldots, v_{m-1}, u_1{}^m\}$ is a basis of $m + 1$ elements of the K-module K.
(d) For each $k \in N$, let s_k be the linear operator on E satisfying

$$s_k(e_j) = e_{2^k(2j-1)}$$

for all $j \in N^*$. Prove that $\tau: w \mapsto (ws_k)_{k \geq 0}$ is an isomorphism from the K-module K onto the K-module K^N.

CHAPTER VI

POLYNOMIALS

Polynomial functions on the field of real numbers, i.e., functions f satisfying

$$f(x) = a_n x^n + a_{n-1} x^{n-1} + \ldots + a_1 x + a_0$$

for all real numbers x, where $a_0, a_1, \ldots, a_{n-1}, a_n$ are given real numbers, and the ways of combining them by addition and multiplication are familiar from elementary algebra. The definition of "polynomial function" is easily generalized by replacing the field of real numbers with any commutative ring with identity. In modern algebra it is convenient to study objects called "polynomials" that are closely related to but not identical with polynomial functions. Such objects are fundamental to the study of fields, and this chapter is primarily devoted to their study.

33. Algebras

Vector spaces and modules, the most important examples of K-algebraic structures with one composition, were investigated in Chapter V. The most important examples of K-algebraic structures with two compositions are called simply algebras.

Definition. Let K be a commutative ring with identity. A **K-algebra**, or an **algebra over K**, is a K-algebraic structure $(A, +, \cdot, .)$ with two compositions such that

(A 1) $\qquad\qquad\qquad (A, +, .)$ is a K-module,

(A 2) $\qquad\qquad\qquad (A, +, \cdot)$ is a ring,

(A 3) $\qquad \alpha(xy) = (\alpha x)y = x(\alpha y)$ for all $x, y \in A$ and all $\alpha \in K$.

If, in addition, $(A, +, \cdot)$ is a division ring, then $(A, +, \cdot, .)$ is called a **division algebra** over K.

Example 33.1. Let E be a module over a commutative ring with identity K. For every $\alpha \in K$ and every $u \in \text{End}_K(E)$, αu belongs to $\text{End}_K(E)$ by Theorem 31.2. One easily verifies that

$$\alpha(uv) = (\alpha u)v = u(\alpha v)$$

for all $\alpha \in K$ and all $u, v \in \text{End}_K(E)$. Thus, $\text{End}_K(E)$ is a K-algebra.

Example 33.2. If K is a commutative ring with identity, then $\mathscr{M}_K(n)$ is a K-algebra. If E is an n-dimensional K-module, the K-algebra $\mathscr{M}_K(n)$ is isomorphic to the K-algebra $\text{End}_K(E)$ (Theorem 32.2).

Example 33.3. If $(A, +, \cdot)$ is a ring with identity and if K is a subring of the center of A containing the identity element of A, by the K-algebra A we shall mean $(A, +, \cdot, .)$ where . is the restriction to $K \times A$ of the given multiplication $\cdot$ on A. It is easy to verify that $(A, +, \cdot, .)$ is indeed a K-algebra since K is contained in the center of A. In particular, a commutative ring with identity may be regarded as a one-dimensional algebra over itself.

Example 33.4. If $(A, +, \cdot)$ is a ring, then $(A, +, \cdot, .)$ is a Z-algebra where . is the function defined in §11 and §14. Indeed, $(A, +, .)$ is a Z-module by Theorem 14.8, and (A 3) holds by (1) of §15. Thus, every ring can be made into a Z-algebra in a natural way.

Example 33.5. If A is a K-algebra and if E is a set, the compositions and scalar multiplication induced on the set A^E of all functions from E into A by the compositions and scalar multiplication of the K-algebra A (Examples 15.5 and 27.4) convert A^E into a K-algebra.

Example 33.6. If $(A, +, \cdot, .)$ is a K-algebraic structure such that $(A, +, .)$ is a K-module and $(A, +, \cdot)$ is a trivial ring, then

$$(\alpha x)y = \alpha(xy) = x(\alpha y) = 0$$

for all $x, y \in A$ and all $\alpha \in K$, so $(A, +, \cdot, .)$ is a K-algebra. Such algebras are called **trivial algebras.** A particularly trivial algebra is a **zero algebra,** one containing only one element.

Definition. If B is a stable subset of a K-algebraic structure $(A, +, \cdot, .)$, then $(B, +_B, \cdot_B, ._B)$ is a **subalgebra** of $(A, +, \cdot, .)$ if it is itself a K-algebra.

A stable subset B of $(A, +, \cdot, .)$ is also called a subalgebra if the K-algebraic structure $(B, +_B, \cdot_B, \cdot_B)$ is a subalgebra in the sense just defined. One consequence of Theorem 28.1 is the following theorem.

Theorem 33.1. If A is a K-algebra, a nonempty subset B of A is a subalgebra of A if and only if for all $x, y \in B$ and all $\alpha \in K$, the elements $x + y$, xy, and αx all belong to B, i.e., if and only if B is a stable subset of A.

Theorem 33.2. Let A be a K-algebra. The intersection of any set of subalgebras of A is a subalgebra; consequently, if S is a subset of A, the intersection of the set of all subalgebras of A containing S is the smallest subalgebra of A containing S.

The proof is easy (see Theorem 28.2).

Theorem 33.2 enables us to make the following definition:

Definition. If S is a subset of a K-algebra A, the **subalgebra generated** by S is the smallest subalgebra B of A containing S, S is called a **set of generators** of B, and B is said to be **generated** by S.

Theorem 33.3. Let f and g be homomorphisms from a K-algebra A into a K-algebra B. The set

$$H = \{x \in A : f(x) = g(x)\}$$

is a subalgebra of A. If S is a set of generators for the algebra A and if $f(x) = g(x)$ for all $x \in S$, then $f = g$.

The proof is easy (see Theorem 29.3).

If $(A, +, \cdot, .)$ is a K-algebra, by the *ring A* we mean $(A, +, \cdot)$, by the *module A* (or *vector space A* if K is a field) we mean $(A, +, .)$, by the *(additive) group A* we mean $(A, +)$, and by the *multiplicative semigroup A* we mean $(A, \cdot)$. An algebra A is *commutative* if multiplication is a commutative composition, and A is an *algebra with identity* if the ring A is a ring with identity.

Definition. An **ideal** of an algebra A is a subset of A that is both an ideal of the ring A and a submodule of the module A.

Thus, a nonempty subset $\mathfrak{a}$ of a K-algebra A is an ideal if and only if for all $a, b \in \mathfrak{a}$, for all $x \in A$, and for all $\lambda \in K$, the elements $a + b$, xa, ax, and λa all belong to $\mathfrak{a}$. If there is an identity element e for multiplication, then an ideal $\mathfrak{a}$ of the ring A is also an ideal of the algebra A, for if $a \in \mathfrak{a}$ and if $\lambda \in K$, then

$$\lambda a = \lambda(ea) = (\lambda e)a \in \mathfrak{a}.$$

Otherwise, however, there may exist ideals of the ring A that are not ideals of the algebra A. For example, the R-vector space R becomes a trivial R-algebra when multiplication is defined by $xy = 0$; Q is an ideal of the trivial ring R but is not an ideal of the trivial R-algebra R.

If $\mathfrak{a}_1, \ldots, \mathfrak{a}_n$ are ideals of A, then

$$\mathfrak{a}_1 + \ldots + \mathfrak{a}_n,$$

$$\mathfrak{a}_1 \cap \ldots \cap \mathfrak{a}_n$$

are ideals of A by Theorems 19.3 and 28.2.

If $\mathfrak{a}$ is an ideal of a K-algebra A, the compositions and scalar multiplication of A induce compositions and a scalar multiplication on the quotient set $A/\mathfrak{a}$, given by

$$(x + \mathfrak{a}) + (y + \mathfrak{a}) = (x + y) + \mathfrak{a},$$

$$(x + \mathfrak{a})(y + \mathfrak{a}) = xy + \mathfrak{a},$$

$$\alpha(x + \mathfrak{a}) = \alpha x + \mathfrak{a},$$

as we saw in our discussion of quotient rings and modules. The canonical surjection $\varphi_\mathfrak{a}: x \mapsto x + \mathfrak{a}$ is thus an epimorphism, so by the following theorem, $A/\mathfrak{a}$ is a K-algebra, called the **quotient algebra** defined by $\mathfrak{a}$.

Theorem 33.4. If f is an epimorphism from a K-algebra A onto a K-algebraic structure B with two compositions, then B is a K-algebra. Moreover, if A is commutative, so is B.

Proof. By Theorem 21.1 and the remark following Theorem 29.1, it suffices to make the easy verification that $\lambda(zw) = (\lambda z)w = z(\lambda w)$ for all $\lambda \in K$, $z, w \in B$.

Corollary. If f is a homomorphism from a K-algebra A into a K-algebraic structure $(B, +, \cdot, .)$, then the range $f_*(A)$ of f is a subalgebra of $(B, +, \cdot, .)$. Moreover, if A is commutative, so is $f_*(A)$.

The proof is similar to that of the Corollary of Theorem 21.1.

Theorem 33.5. Let f be an epimorphism from a K-algebra A onto a K-algebra B. The kernel $\mathfrak{a}$ of f is an ideal of A, and there is one and only one isomorphism

g from the K-algebra $A/\mathfrak{a}$ onto B such that

$$g \circ \varphi_{\mathfrak{a}} = f.$$

Furthermore, f is an isomorphism if and only if $\mathfrak{a} = \{0\}$.

The assertion is a consequence of Theorems 21.4 and 30.6.

Theorem 33.6. If A is a K-algebra with identity e, then

$$h: \alpha \mapsto \alpha e$$

is a homomorphism from the K-algebra K into A, and h is a monomorphism if and only if $\{e\}$ is linearly independent.

Proof. That h is a homomorphism follows from the identities

$$(\alpha + \beta)e = \alpha e + \beta e,$$
$$(\alpha\beta)e = \alpha(\beta e),$$
$$(\alpha e)(\beta e) = \alpha(e(\beta e)) = \alpha(\beta e^2)$$
$$= \alpha(\beta e) = (\alpha\beta)e.$$

Furthermore, by the definition of linear independence, $\{e\}$ is linearly independent if and only if the kernel of h is $\{0\}$.

In view of Theorem 33.6, if A is a K-algebra with identity e such that $\{e\}$ is linearly independent (in particular, if K is a field), K is frequently "identified" with the subalgebra Ke of A; that is, the subalgebra Ke of A is denoted simply by "K," and for each $\alpha \in K$, the element αe of A is denoted simply by "α."

Theorem 33.7. If K is a field and if A is a finite-dimensional K-algebra with identity, then every cancellable element of A is invertible.

Proof. Let e be the identity element of A, and let a be a cancellable element. Then $L_a: x \mapsto ax$ and $R_a: x \mapsto xa$ are injective linear operators on the K-vector space A, so L_a and R_a are permutations of A by Theorem 29.6. In particular, there exist $b, c \in A$ such that

$$ab = L_a(b) = e = R_a(c) = ca,$$

whence as

$$c = c(ab) = (ca)b = b,$$

a is invertible.

Corollary. If K is a field, if A is a finite-dimensional K-algebra with identity, and if every nonzero element of A is cancellable, then A is a division algebra.

EXERCISES

33.1. Complete the verifications needed in Examples 33.1, 33.3, and 33.5.

33.2. (a) Prove Theorem 33.2.
(b) Prove Theorem 33.3.
(c) Complete the proof of Theorem 33.4.

33.3. Let K be a commutative ring with identity, let $(A, +, \cdot, .)$ be a K-algebraic structure with two compositions, and for each $a \in A$, let L_a be the function from A into A defined by $L_a(x) = ax$ for all $x \in A$. Then $(A, +, \cdot, .)$ is a K-algebra if and only if $(A, +, .)$ is a K-module, L_a is a linear operator on the K-module A for all $a \in A$, and $L: a \mapsto L_a$ is a homomorphism from $(A, +, \cdot, .)$ into the K-algebra $\operatorname{End}_K(A)$.

33.4. Let E be a module over a commutative ring K. If $(e_i)_{1 \le i \le m}$ is an ordered basis of E and if (z_{ij}) is a family of elements of E indexed by $[1, m] \times [1, m]$, there is one and only one composition $\cdot$ on E that is distributive over addition, satisfies $e_i e_j = z_{ij}$ for all $i, j \in [1, m]$, and in addition satisfies (A 3); furthermore, $\cdot$ is associative if and only if $e_i(e_j e_k) = (e_i e_j)e_k$ for all $i, j, k \in [1, m]$, and $\cdot$ is commutative if and only if $e_i e_j = e_j e_i$ for all $i, j \in [1, m]$.

33.5. If $(A, +, .)$ is a two-dimensional vector space over a field K and if $\cdot$ is a composition on A that is distributive over $+$ and admits an identity element, then $\cdot$ is associative and commutative, and hence $(A, +, \cdot, .)$ is a commutative algebra.

33.6. If A is a K-algebra, the left regular representation (Exercise 21.6) of the ring A is also a homomorphism from the K-algebra A into the K-algebra $\operatorname{End}_K(A)$.

33.7. Let A be a K-algebra with identity element e.
(a) If x is an invertible element of A and if λ is an invertible element of K, then λx is invertible and $(\lambda x)^{-1} = \lambda^{-1} x^{-1}$.
(b) If λx is an invertible element of A, then x is invertible, and if, in addition, $\{e\}$ is linearly independent, then λ is cancellable. Need λ be invertible in K under these circumstances?

33.8. If n is an integer > 1 such that the (additive) order of every element of a ring $(A, +, \cdot)$ divides n, then there is one and only one scalar multiplication $.$ from $Z_n \times A$ into A such that $(A, +, \cdot, .)$ is a Z_n-algebra. In particular, if A is an integral domain whose characteristic is a prime p, then there is one and only one scalar multiplication from $Z_p \times A$ into A such that $(A, +, \cdot, .)$ is a Z_p-algebra.

33.9. If A is a K-algebra, a modular ideal (Exercise 19.5) of the ring A is also an ideal of the algebra A.

*33.10. If A is a nontrivial K-algebra possessing no nonzero proper ideals, then there are no nonzero proper ideals of the ring A. [Show that a nonzero ideal of the ring A contains the submodule generated by $A \cdot A$.]

33.11. Generalize Theorem 33.7 as follows: If K is a field and if A is a finite-dimensional K-algebra, then every cancellable element of A is invertible. [Use Exercise 7.13.]

33.12. If A is a one-dimensional algebra over a field K, then either A is a trivial algebra or A is isomorphic to the K-algebra K. [If $u^2 = \alpha u \neq 0$, consider $\alpha^{-1} u$.]

*33.13. An element a of a ring or algebra A is a **square** if there exists $x \in A$ satisfying $x^2 = a$. Let A be a two-dimensional algebra with identity element e over a field K whose characteristic is not 2.
(a) The algebra A is commutative (Exercise 33.5).
(b) There is a basis $\{e, v\}$ of A such that $v^2 = \alpha e$ for some $\alpha \in K$.
(c) If α is not a square of K, then A is a division algebra. [Use Theorem 33.7.]
(d) If α is a nonzero square of K, then A has a basis $\{e_1, e_2\}$ where $e_1{}^2 = e_1$, $e_2{}^2 = e_2$, $e_1 e_2 = e_2 e_1 = 0$. [Use Exercise 33.12.]
(e) If $\alpha = 0$, then there is a one-dimensional ideal $\mathfrak{a}$ of A such that $\mathfrak{a}$ is a trivial algebra and $A/\mathfrak{a}$ is isomorphic to the K-algebra K.

*33.14. Let $\{e, v\}$ be a basis of a two-dimensional vector space $(A, +, .)$ over a field K whose characteristic is not 2.
(a) For each scalar α that is not a square of K, there is one and only one composition $\cdot_\alpha$ on A such that $(A, +, \cdot_\alpha, .)$ is a division algebra over K whose identity element is e and $v \cdot_\alpha v = \alpha e$. [Use Exercises 33.4, 33.5, and 33.13.]
(b) If α and β are scalars that are not squares of K, then $(A, +, \cdot_\alpha, .)$ and $(A, +, \cdot_\beta, .)$ are isomorphic K-algebras if and only if α/β (or equivalently, $\alpha\beta$) is a square of K; in this case, there are exactly two isomorphisms.
(c) Any two-dimensional division algebra over $\boldsymbol{R}$ is isomorphic to the division algebra $\boldsymbol{C}$ of complex numbers, and the only automorphisms of the $\boldsymbol{R}$-algebra $\boldsymbol{C}$ are the identity automorphism and the conjugation automorphism $z \mapsto \bar{z}$.

*33.15. Let K be a field, and let A be a two-dimensional K-algebra containing a one-dimensional ideal $\mathfrak{a}$ such that the algebra $\mathfrak{a}$ is a trivial algebra and $A/\mathfrak{a}$ is isomorphic to the K-algebra K. Then there is a basis $\{e, u\}$ of A such that $e^2 = e$ and $u^2 = 0$. [Infer from Exercise 33.11 that either A has an identity element or every element of A is a zero-divisor; in the latter case, show that $a^3 - a^2 = 0$ where $a + \mathfrak{a}$ is the identity element of $A/\mathfrak{a}$, and conclude that $a^4 = a^2$.] Furthermore, either (a) $eu = ue = u$, or (b) $eu = u$, $ue = 0$, or (c) $eu = 0$, $ue = u$, or (d) $eu = ue = 0$.

*33.16. (a) For each of the four possible algebras described in Exercise 33.15, determine the matrices of L_e and L_u relative to the ordered basis (e, u), where L is the left regular representation (Exercise 33.6). For which of the four is L injective?

(b) Show that the three algebras determined by (a), (b), and (c) of Exercise 33.15 are isomorphic to the subalgebras of $\mathscr{M}_K(2)$ consisting respectively of all the matrices

$$\begin{bmatrix} \alpha & 0 \\ \beta & \alpha \end{bmatrix} \qquad \begin{bmatrix} \alpha & \beta \\ 0 & 0 \end{bmatrix} \qquad \begin{bmatrix} \alpha & 0 \\ \beta & 0 \end{bmatrix}$$

where $\alpha, \beta \in K$.

(c) Show that the algebra determined by (d) of Exercise 33.15 is not isomorphic to any subalgebra of $\mathscr{M}_K(2)$. [First find the form of all matrices $X, Y \in \mathscr{M}_K(2)$ satisfying $X^2 = X$, $Y^2 = 0$.]

(d) Show that the subalgebra determined by (d) of Exercise 33.15 is isomorphic to the subalgebra of $\mathscr{M}_K(3)$ consisting of all the matrices

$$\begin{bmatrix} \alpha & 0 & 0 \\ 0 & 0 & 0 \\ 0 & \beta & 0 \end{bmatrix}$$

where $\alpha, \beta \in K$.

*33.17. Let K be a field, and let A be a two-dimensional K-algebra containing a one-dimensional ideal $\mathfrak{a}$ such that both the algebras $\mathfrak{a}$ and $A/\mathfrak{a}$ are trivial algebras.

(a) Either A is a trivial algebra, or else there exists $e \in A$ such that $\{e, e^2\}$ is a basis of A and $e^3 = 0$. [If x and x^2 are linearly dependent for all $x \in A$, show that every square is zero, then that $Au = \{0\}$ for all $u \in \mathfrak{a}$, and finally that $A^2 = \{0\}$.]

(b) Show that the two algebras described in (a) are isomorphic respectively to the subalgebras of $\mathscr{M}_K(3)$ consisting of all the matrices

$$\begin{bmatrix} 0 & 0 & 0 \\ \alpha & 0 & \beta \\ 0 & 0 & 0 \end{bmatrix} \qquad \begin{bmatrix} 0 & 0 & \alpha \\ \alpha & 0 & \beta \\ 0 & 0 & 0 \end{bmatrix}$$

where $\alpha, \beta \in K$.

33.18. Let K be a field, and let A be a two-dimensional K-algebra none of whose one-dimensional ideals is a trivial algebra. Either A is a division algebra, or A has a basis $\{e, u\}$ satisfying $e^2 = e$, $u^2 = u$, $eu = ue = 0$ and is isomorphic to the subalgebra of $\mathcal{M}_K(2)$ consisting of all the diagonal matrices

$$\begin{bmatrix} \alpha & 0 \\ 0 & \beta \end{bmatrix}$$

where $\alpha,\ \beta \in K$.

*33.19. No two of the seven two-dimensional algebras that are not division algebras described in Exercises 33.15–33.18 are isomorphic. Thus, to within isomorphism there are exactly seven two-dimensional algebras over a field that are not division algebras.

*33.20. The five sets of all the matrices

$$\begin{bmatrix} \alpha & \beta \\ \alpha & \beta \end{bmatrix} \quad \begin{bmatrix} \alpha & \beta \\ \beta & \alpha \end{bmatrix} \quad \begin{bmatrix} 0 & \alpha \\ 0 & \beta \end{bmatrix} \quad \begin{bmatrix} \alpha + \beta & \beta \\ -\beta & \alpha - \beta \end{bmatrix} \quad \begin{bmatrix} \alpha & \beta \\ 0 & \alpha + \beta \end{bmatrix}$$

where α and β are any elements of a field K are two-dimensional subalgebras of $\mathcal{M}_K(2)$. For each, determine which of the algebras discussed in Exercises 33.15–33.18 is isomorphic to it.

33.21. Let K be a commutative ring with identity, A a K-algebra with identity. An ordered quadruple $(1, i, j, k)$ is a **quaternionic basis** of A if $(1, i, j, k)$ is an ordered basis of A, if 1 is the multiplicative identity of A, and if the following equalities hold:

$$i^2 = j^2 = k^2 = -1,$$
$$ij = k,\ jk = i,\ ki = j,$$
$$ji = -k,\ kj = -i,\ ik = -j.$$

The algebra A is an **algebra of quaternions** over K if A possesses a quaternionic basis.

(a) There is an algebra of quaternions over K. [Use Exercise 33.4.]

(b) If A and A' are algebras of quaternions over K, then A and A' are isomorphic.

(c) If $(1, i, j, k)$ is a quaternionic basis of A and if the characteristic of K is not 2, then $\{1, -1, i, -i, j, -j, k, -k\}$ is a group under multiplication that is isomorphic to the quaternionic group.

33.22. Let A be an algebra of quaternions over an integral domain K, and let $(1, i, j, k)$ be a quaternionic basis of A. For each $z \in A$, if $z = \alpha_1 + \alpha_2 i + \alpha_3 j + \alpha_4 k$, let $z^* = \alpha_1 - \alpha_2 i - \alpha_3 j - \alpha_4 k$; z^* is called the **conjugate** of z.

(a) If $z^2 = -1$ and if $(1, z)$ is a linearly independent sequence, then $z^* = -z$. [Assume first that the characteristic of K is not 2.]

(b) If $(1, i', j', k')$ is another quaternionic basis of A, then $(\alpha_1 + \alpha_2 i' + \alpha_3 j' + \alpha_4 k')^* = \alpha_1 - \alpha_2 i' - \alpha_3 j' - \alpha_4 k'$ for all $\alpha_1,\ \alpha_2,\ \alpha_3,\ \alpha_4 \in K$. [Use (a).] Conclude that the definition of z^* does not depend upon the choice of quaternionic basis.

(c) For all $z, w \in A$, $(zw)^* = w^*z^*$.

(d) For each $z \in A$, zz^* is a scalar multiple of the identity element 1; we denote that scalar by $N(z)$, so that $zz^* = N(z).1$; $N(z)$ is called the **norm** of z.

(e) For all $z, w \in A$, $N(zw) = N(z)N(w)$.

(f) For all $z \in A$, z is invertible in A if and only if $N(z)$ is invertible in K.

(g) If K is a totally ordered field, then A is a noncommutative division algebra.

34. The Algebra of Polynomials

Let K be a commutative ring with identity. Addition and scalar multiplication by the K-module K^N of all sequences of elements of K indexed by N (Example 27.4) are given by

$$(\alpha_n) + (\beta_n) = (\alpha_n + \beta_n),$$

$$\lambda(\alpha_n) = (\lambda\alpha_n)$$

for all (α_n), $(\beta_n) \in K^N$ and all $\lambda \in K$. We define a composition, called **multiplication**, on K^N by

$$(\alpha_n)(\beta_n) = (\gamma_n)$$

where

$$\gamma_n = \sum_{k=0}^{n} \alpha_k \beta_{n-k}$$

for all $n \in N$.

Let us verify that $(K^N, +, \cdot, .)$ is a commutative K-algebra. To show that multiplication is associative, let (α_n), (β_n), and (γ_n) be elements of K^N, let

$$(\lambda_n) = [(\alpha_n)(\beta_n)](\gamma_n),$$

$$(\mu_n) = (\alpha_n)[(\beta_n)(\gamma_n)],$$

and let $n \in N$. Then

$$\lambda_n = \sum_{m=0}^{n} \left(\sum_{i=0}^{m} \alpha_i \beta_{m-i} \right) \gamma_{n-m} = \sum_{m=0}^{n} \left(\sum_{i=0}^{m} \alpha_i \beta_{m-i} \gamma_{n-m} \right),$$

$$\mu_n = \sum_{m=0}^{n} \alpha_m \left(\sum_{j=0}^{n-m} \beta_j \gamma_{n-m-j} \right) = \sum_{m=0}^{n} \left(\sum_{j=0}^{n-m} \alpha_m \beta_j \gamma_{n-m-j} \right).$$

Let

$$R = \{(i, j, k) \in N^3 : i + j + k = n\},$$

and for each $m \in [0, n]$ let

$$S_m = \{(i, j, k) \in R: i + j = m\},$$
$$T_m = \{(i, j, k) \in R: j + k = n - m\}.$$

Then $\{S_0, S_1, \ldots, S_n\}$ and $\{T_0, T_1, \ldots, T_n\}$ are clearly partitions of R, so (by Theorem A.10),

$$\lambda_n = \sum_{m=0}^{n} \left(\sum_{(i,j,k)\in S_m} \alpha_i \beta_j \gamma_k \right) = \sum_{(i,j,k)\in R} \alpha_i \beta_j \gamma_k$$
$$= \sum_{m=0}^{n} \left(\sum_{(i,j,k)\in T_m} \alpha_i \beta_j \gamma_k \right) = \mu_n.$$

Multiplication is commutative, for

$$\sum_{i=0}^{n} \alpha_i \beta_{n-i} = \alpha_0 \beta_n + \alpha_1 \beta_{n-1} + \ldots + \alpha_{n-1} \beta_1 + \alpha_n \beta_0$$
$$= \beta_n \alpha_0 + \beta_{n-1} \alpha_1 + \ldots + \beta_1 \alpha_{n-1} + \beta_0 \alpha_n$$
$$= \beta_0 \alpha_n + \beta_1 \alpha_{n-1} + \ldots + \beta_{n-1} \alpha_1 + \beta_n \alpha_0$$
$$= \sum_{i=0}^{n} \beta_i \alpha_{n-i}.$$

It is easy to verify that multiplication is distributive over addition and that

$$\lambda[(\alpha_n)(\beta_n)] = [\lambda(\alpha_n)](\beta_n) = (\alpha_n)[\lambda(\beta_n)].$$

Consequently, $(K^N, +, \cdot, .)$ is a commutative K-algebra, called the **algebra of formal power series over** K.

The subset $K^{(N)}$ of K^N defined by

$$K^{(N)} = \{(\alpha_n) \in K^N: \alpha_n \neq 0 \text{ for only finitely many } n \in N\}$$

is a subalgebra of the algebra of formal power series. Indeed, if (α_n) and (β_n) belong to $K^{(N)}$ and if $\alpha_n = 0$ for all $n \geq r$ and $\beta_n = 0$ for all $n \geq s$, then

$$\alpha_n + \beta_n = 0 \text{ for all } n \geq \max\{r, s\},$$
$$\lambda \alpha_n = 0 \text{ for all } n \geq r,$$
$$\sum_{k=0}^{n} \alpha_k \beta_{n-k} = 0 \text{ for all } n \geq r + s.$$

An element of $K^{(N)}$ is called a **polynomial over** K, and the subalegbra $K^{(N)}$ of the algebra of formal power series over K is called the **algebra of polynomials over K.**

Using the Kronecker index notation, we see from the discussion of Example 28.8 that

$$\{(\delta_{r,n})_{n \geq 0} : r \in N\}$$

is a basis of the K-module $K^{(N)}$. Let us denote temporarily the polynomial $(\delta_{r,n})_{n \geq 0}$ by e_r. Then

(1) $$e_r e_s = e_{r+s}$$

for all $r, s \in N$, for if $e_r e_s = (\alpha_n)_{n \geq 0}$, then

$$\alpha_n = \sum_{j=0}^{n} \delta_{r,j} \delta_{s,n-j} = \delta_{r+s,n}$$

since $\delta_{r,j} \delta_{s,n-j} = 0$ unless $j = r$ and $n - j = s$, in which case $\delta_{r,j} \delta_{s,n-j} = 1$. We shall denote the polynomial $e_1 = (\delta_{1,n})_{n \geq 0}$ by a special symbol, usually "X" or some other capital letter. By (1), the function $r \mapsto e_r$ is a monomorphism from $(N, +)$ into $(K^{(N)}, \cdot)$. Consequently, $e_n = X^n$ for all $n \geq 1$, as we observed before Theorem 20.3. Also $e_0 = (\delta_{0,n})_{n \geq 0}$ is the multiplicative identity, for if $(\alpha_n) \in K^N$, then

$$\sum_{k=0}^{n} \delta_{0,k} \alpha_{n-k} = \alpha_n$$

for all $n \geq 0$. In accordance with the notation introduced in §11, if x is an element of a ring with identity, then x^0 denotes the identity element; therefore, $X^m = e_m$ for $m = 0$ as well as for all $m \in N^*$. The polynomials X^m where $m \geq 0$ are called **monomials.** Thus, the set of all monomials is a basis of the K-module of polynomials. If (α_n) is a polynomial such that $\alpha_n = 0$ for all $n > m$, then

$$(\alpha_n) = \alpha_m X^m + \alpha_{m-1} X^{m-1} + \ldots + \alpha_1 X + \alpha_0 X^0$$

as we saw in Example 28.8.

The algebra of polynomials over K is usually denoted by $K[X]$ (though if another symbol is used to denote $(\delta_{1,n})_{n \geq 0}$, that symbol replaces "$X$" in the expression "$K[X]$").

Theorem 34.1. The K-algebra $K[X]$ of polynomials over K is a commutative algebra with identity. The set of monomials is a basis of $K[X]$, and the function $\alpha \mapsto \alpha X^0$ is an isomorphism from the K-algebra K onto the subalgebra of $K[X]$ generated by the identity element.

The final assertion is a consequence of Theorem 33.6 and the linear independence of $\{X^0\}$. In view of that assertion, it is customary to "identify" K with the subalgebra of $K[X]$ generated by the identity element. Thus, for any $\alpha \in K$, the symbol "α" is also used to denote the polynomial αX^0. When we speak of K as a subalgebra of $K[X]$, it is the subalgebra of $K[X]$ generated by X^0 that we have in mind. A polynomial $f \in K[X]$ is called a **constant** polynomial if $f = \alpha X^0$ for some $\alpha \in K$.

Corollary. The K-algebra $K[X]$ is generated by $\{X^0, X\}$. Consequently, if φ and ψ are homomorphisms from $K[X]$ into a K-algebra B such that $\varphi(X^0) = \psi(X^0)$ and $\varphi(X) = \psi(X)$, then $\varphi = \psi$.

The assertion follows from Theorems 34.1 and 33.3.

Theorem 34.2. Let K and L be commutative rings with identity, and let φ be a homomorphism from K into L such that $\varphi(1) = 1$. Then

$$\bar{\varphi}: \sum_{k=0}^{m} \alpha_k X^k \mapsto \sum_{k=0}^{m} \varphi(\alpha_k) Y^k$$

is a homomorphism from the ring $K[X]$ into the ring $L[Y]$ and is, furthermore, the only homomorphism ψ from $K[X]$ into $L[Y]$ satisfying

$$\psi(X) = Y,$$

$$\psi(\alpha X^0) = \varphi(\alpha) Y^0$$

for all $\alpha \in K$. If φ is an isomorphism from K onto L, then $\bar{\varphi}$ is an isomorphism from $K[X]$ onto $L[Y]$.

Proof. It is easy to verify that $\bar{\varphi}$ is indeed a homomorphism from the ring $K[X]$ into the ring $L[Y]$ satisfying $\bar{\varphi}(X) = Y$ and $\bar{\varphi}(\alpha X^0) = \varphi(\alpha) Y^0$ for all $\alpha \in K$ and that $\bar{\varphi}$ is, moreover, an isomorphism from $K[X]$ onto $L[Y]$ if φ is an isomorphism from K onto L. If ψ is a homomorphism from the ring $K[X]$ into the ring $L[Y]$ such that $\psi(X) = Y$ and $\psi(\alpha X^0) = \varphi(\alpha) Y^0$ for all $\alpha \in K$, then for every polynomial $f = \sum_{k=0}^{m} \alpha_k X^k \in K[X]$,

$$\psi(f) = \psi\left(\sum_{k=0}^{m} (\alpha_k X^0) X^k \right) = \sum_{k=0}^{m} \psi(\alpha_k X^0) \psi(X)^k$$

$$= \sum_{k=0}^{m} \varphi(\alpha_k) Y^k = \bar{\varphi}(f),$$

so $\psi = \bar{\varphi}$.

The homomorphism $\bar{\varphi}$ of Theorem 34.2 is called the **homomorphism induced by the given homomorphism** φ from K into L. The symbol "X" in such expressions as $\alpha_m X^m + \alpha_{m-1} X^{m-1} + \cdots + \alpha_1 X + \alpha_0$ is often called an "indeterminate," but there is nothing indeterminate about what it denotes in any given context: "X" denotes the polynomial $(\delta_{1,n})_{n \geq 0}$ over the scalar ring in question. If K and L are commutative rings with identity, the polynomial X of $K[X]$ is identical with the polynomial Y of $L[Y]$ if and only if the zero elements of K and L are the same and the identity elements of K and L are also the same. For example, if K is a subring of L, then $K[X]$ is a subring of $L[Y]$, but the polynomial X of $K[X]$ is identical with the polynomial Y of $L[Y]$ if and only if the identity element of K is also the identity element of L (which is, indeed, the case if L is an integral domain by Theorem 15.2). In other words, *if L is a commutative ring with identity element 1 and if K is a subring of L containing 1, then $K[X]$ is a subring of $L[X]$, and the polynomial X of $K[X]$ is identical with the polynomial X of $L[X]$.*

Definition. Let f be the polynomial (α_n) over a commutative ring with identity K. The **coefficient** of X^n in f is the scalar α_n for each natural number n. If $f \neq 0$, the **degree** of f is the largest integer m such that the coefficient α_m of X^m in f is not zero, and the degree of f is denoted by $\deg f$. If $\deg f = m$, then α_m is called the **leading coefficient** of f. If the leading coefficient of f is 1, then f is called a **monic** polynomial. The coefficient of X^0 in f is called the **constant coefficient** of f.

The degree of a polynomial f is thus defined only if $f \neq 0$. However, it is customary to say that "f is a polynomial of degree $\leq m$ (respectively, of degree $< m$)" either if $f = 0$ or if f is a nonzero polynomial whose degree n satisfies $n \leq m$ (respectively, $n < m$). The constant polynomials, for example, are the polynomials of degree ≤ 0. Often if a polynomial f is denoted by $\sum_{k=0}^{m} \alpha_k X^k$, it is tacitly assumed that m is the degree of f; this assumption is not always valid, however, and whether m is to be regarded as the degree of a polynomial denoted by $\sum_{k=0}^{m} \alpha_k X^k$ depends on the context. A polynomial is often called a **linear** (respectively, **quadratic**, **cubic**, **quartic**, or **biquadratic**, **quintic**) polynomial if its degree is 1 (respectively, 2, 3, 4, 5).

Theorem 34.3. Let f and g be nonzero polynomials over a commutative ring with identity K. If $\deg f \neq \deg g$, then $f + g \neq 0$, and

$$\deg(f + g) = \max\{\deg f, \deg g\}.$$

If $\deg f = \deg g$ and if $f + g \neq 0$, then

$$\deg(f + g) \leq \deg f.$$

If either the leading coefficient of f or the leading coefficient of g is not a zero-divisor, then $fg \neq 0$, and

$$\deg fg = \deg f + \deg g.$$

In particular, a monic polynomial over K is not a zero-divisor of $K[X]$.

Proof. The assertions concerning $f + g$ are immediate. Let α_n and β_m be the leading coefficients of f and g respectively. Then $\alpha_n\beta_m$ is clearly the coefficient of X^{n+m} in fg and is, moreover, the leading coefficient of fg whenever it is not zero. Consequently, if either α_n or β_m is not a zero-divisor, then $\alpha_n\beta_m \neq 0$, so $fg \neq 0$ and

$$\deg fg = n + m = \deg f + \deg g.$$

Theorem 34.4. If K is an integral domain, then $K[X]$ is an integral domain, and the only invertible elements of $K[X]$ are the constant polynomials determined by invertible elements of K.

Proof. By Theorem 34.3, the product of nonzero polynomials over K is a nonzero polynomial. If $fg = X^0$, then

$$0 = \deg X^0 = \deg f + \deg g,$$

so $\deg f = \deg g = 0$, and hence, both f and g are constant polynomials. If $f = \alpha X^0$ and $g = \beta X^0$, then

$$X^0 = fg = (\alpha X^0)(\beta X^0) = \alpha\beta X^0,$$

so $\alpha\beta = 1$ and hence α and β are invertible elements of K.

Analogous to the Division Algorithm for the ring of integers is the following theorem for the ring of polynomials over a field.

Theorem 34.5. (Division Algorithm) If K is a field and if g is a nonzero polynomial over K, then for every polynomial $f \in K[X]$, there exist unique polynomials q and r over K satisfying

(2) $$f = qg + r,$$

(3) $$\text{either } r = 0 \text{ or } \deg r < \deg g.$$

Proof. Let $n = \deg g$. To prove the existence of polynomials q and r satisfying (2) and (3) we shall proceed by induction on the degree of f. Let S

be the set of all $m \in N$ such that for every polynomial $f \in K[X]$ of degree $< m$ there exist polynomials q and r over K satisfying (2) and (3). If $m \leq n$, then $m \in S$, for if f is a polynomial of degree $< n$, then the polynomials $q = 0$ and $r = f$ satisfy (2) and (3). Suppose that $m \in S$ and that $m + 1 > n$, whence $m \geq n$. To show that $m + 1 \in S$ it suffices to show that for every polynomial f of degree m there exist polynomials q and r satisfying (2) and (3). Let

$$f_1 = f - \frac{\alpha}{\beta} X^{m-n} g$$

where α and β are the leading coefficients of f and g respectively. Then either $f_1 = 0$ or $\deg f_1 < m$, so as $m \in S$, there exist polynomials q_1 and r_1 over K such that $f_1 = q_1 g + r_1$ and either $r_1 = 0$ or $\deg r_1 < n$. Then

$$q = \frac{\alpha}{\beta} X^{m-n} + q_1$$

and $r = r_1$ satisfy (2) and (3). Consequently, $m + 1 \in S$, so by induction $S = N$.

To prove uniqueness, suppose that

$$f = qg + r = q_1 g + r_1$$

where r and r_1 are of degree $< n$. Then

$$(q - q_1)g = r_1 - r.$$

If $q - q_1 \neq 0$, then $r_1 - r \neq 0$ by Theorem 34.4 and

$$\deg(q - q_1)g = \deg(q - q_1) + \deg g \geq n > \deg(r_1 - r)$$

by Theorem 34.3, a contradiction. Hence, $q - q_1 = 0$, whence also $r_1 - r = (q - q_1)g = 0$.

The polynomials q and r of Theorem 34.5 are called respectively the **quotient** and **remainder** obtained by dividing g into f.

The Division Algorithm for Z enabled us to describe all ideals of Z (Theorem 19.8); similarly, the Division Algorithm for $K[X]$ permits us to describe the ideals of $K[X]$.

Theorem 34.6. Let K be a field. Every ideal of the K-algebra $K[X]$ is a principal ideal. In fact, if $\mathfrak{a}$ is a nonzero ideal of $K[Y]$, then there is one and only one monic polynomial $g \in K[X]$ such that $\mathfrak{a} = (g)$.

Proof. Let $\mathfrak{a}$ be a nonzero ideal of $K[X]$, and let n be the smallest member of $\{m \in N$: there is a polynomial of degree m belonging to $\mathfrak{a}\}$. If g_1 is a polynomial of degree n belonging to $\mathfrak{a}$ and if α is the leading coefficient of g_1, then $g = \alpha^{-1}g_1$ is a monic polynomial of degree n belonging to $\mathfrak{a}$. Let $f \in \mathfrak{a}$. By Theorem 34.5, there exist $q, r \in K[X]$ such that $f = qg + r$ and either $r = 0$ or $\deg r < n$. Then $r = f - qg \in \mathfrak{a}$, so $r = 0$ by the definition of n. Hence, $f = qg \in (g)$. Thus, $\mathfrak{a} = (g)$. The polynomial g is the only monic polynomial of degree n belonging to $\mathfrak{a}$, for if h were another, then $g - h$ would be a nonzero polynomial belonging to $\mathfrak{a}$ of degree $< n$, a contradiction of the definition of n.

The monic polynomial g satisfying $\mathfrak{a} = (g)$ is called the **monic generator** of $\mathfrak{a}$.

Theorem 34.7. If K is a field, then $K[X]$ is a principal ideal domain.

Proof. By Theorem 34.6, we need only verify (N) (page 198). Let $\mathfrak{A}$ be a nonempty set of ideals of $K[X]$. If $\mathfrak{A}$ contains only the zero ideal, then that ideal is a maximal element of $\mathfrak{A}$. In the contrary case, let n be the smallest of those natural numbers m such that $\mathfrak{A}$ contains a principal ideal generated by a polynomial of degree m. Let g be a polynomial of degree n such that $(g) \in \mathfrak{A}$. If $(h) \supseteq (g)$ and $(h) \in \mathfrak{A}$, then $g = uh$ for some $u \in K[X]$, and therefore, $\deg g = \deg u + \deg h$; but then

$$n \leq \deg h \leq \deg g = n,$$

so $\deg h = n$ and thus $\deg u = 0$; therefore, u is a nonzero constant polynomial and hence is a unit of $K[X]$, so $(h) = (g)$. Thus, (g) is a maximal element of $\mathfrak{A}$ by Theorem 34.6.

If K is a field, the set P of all monic irreducible polynomials is a representative set of irreducible elements of the principal ideal domain $K[X]$, for the units of $K[X]$ are the nonzero constant polynomials by Theorem 34.4. Consequently, we shall call a polynomial p over a field K an **irreducible polynomial** over K if it is an irreducible element of $K[X]$ and a **prime polynomial** over K if it is a monic irreducible element of $K[X]$.

The following theorem will prove useful in Chapter VIII.

Theorem 34.8. Let L be a commutative ring with identity, let K be a subring of L containing the identity element of L, and let $g \in L[X]$. If there exists a polynomial $h \in K[X]$ whose leading coefficient is an invertible element of K (in particular, if h is monic) such that $gh \in K[X]$, then $g \in K[X]$.

Proof. Let $g = \sum_{k=0}^{n} \alpha_k X^k$ and $h = \sum_{k=0}^{m} \beta_k X^k$ where n and m are the degrees respectively of g and h, and let $S = \{k \in [0, n]: \alpha_{n-j} \in K$ for all $j \in [0, k]\}$. Then $0 \in S$, for the coefficient of X^{m+n} in gh is $\alpha_n \beta_m$, which belongs to K, and as β_m has an inverse β_m^{-1} in K by hypothesis,

$$\alpha_n = (\alpha_n \beta_m)\beta_m^{-1} \in K.$$

Suppose that $k \in S$ where $k < n$. The coefficient of $X^{m+n-k-1}$ in gh is either

$$\alpha_{n-k-1}\beta_m + \alpha_{n-k}\beta_{m-1} + \ldots + \alpha_n \beta_{m-k-1}$$

or

$$\alpha_{n-k-1}\beta_m + \alpha_{n-k}\beta_{m-1} + \ldots + \alpha_{n+m-k-1}\beta_0$$

according as $k + 1 \leq m$ or $k + 1 > m$. By our inductive hypothesis, all terms after the first belong to K. Since the sum is a coefficient of gh and hence belongs to K, we have $\alpha_{n-k-1}\beta_m \in K$, whence

$$\alpha_{n-k-1} = (\alpha_{n-k-1}\beta_m)\beta_m^{-1} \in K.$$

Consequently, $k + 1 \in S$, so by induction $S = [0, n]$. Therefore $g \in K[X]$.

EXERCISES

34.1. Complete the verification of the statement that $(K^N, +, \cdot, .)$ is a K-algebra.

34.2. Find the quotient and remainder obtained by dividing $2X^2 + 4X - 1$ into $X^4 + 2X^3 - 3X^2 + 4X + 1$ where the scalar field is Q (respectively, Z_5, Z_7).

34.3. For what prime numbers p is $X^6 + X^5 + 4X^4 + X^3 - 7X^2 + 4$ divisible by $X^2 - 2$ in $Z_p[X]$? $X^5 + 2X^2 - 1$ divisible by $X^2 + 3$ in $Z_p[X]$?

34.4. If $(f_k)_{0 \leq k \leq n}$ is a sequence of nonzero polynomials over a field K such that $\deg f_k = k$ for each $k \in [0, n]$, then $(f_k)_{0 \leq k \leq n}$ is an ordered basis of the subspace of $K[X]$ of all polynomials of degree $\leq n$.

*34.5. Let K be a commutative ring with identity, let g be a nonzero polynomial over K of degree n, and let α be the leading coefficient of g.
(a) If f is a polynomial over K of degree m and if k is the larger of $m - n + 1$ and 0, then there exist polynomials q and r over K such that

$$\alpha^k f = qg + r,$$

either $r = 0$ or $\deg r < n$,

and furthermore, q and r are unique if α is not a zero-divisor. [Modify the proof of Theorem 34.5.]

(b) If α is invertible, then for every polynomial f over K there exist unique polynomials q and r over K satisfying (2) and (3).

*34.6. Let K be a commutative ring with identity, and let $f = \sum_{k=0}^{m} \alpha_k X^k$ be a polynomial over K of degree m.

(a) If $g = \sum_{k=0}^{n} \beta_k X^k$ is a polynomial over K of degree $n > 0$ and if $fg = 0$, then there is a nonzero polynomial h over K of degree $< n$ such that $fh = 0$. [Assume $\beta_0 \neq 0$. If $\alpha_k g = 0$ for all $k \in [0, m-1]$, consider $h = \beta_0 X^0$; if $\alpha_k g = 0$ for all $k \in [0, p-1]$ where $p < m$ and if $\alpha_p g \neq 0$, show that $\alpha_p \beta_0 = 0$ and consider $h = \sum_{k=0}^{n-1} \alpha_p \beta_{k+1} X^k$.]

(b) If f is a zero-divisor of $K[X]$, then $\lambda f = 0$ for some nonzero scalar λ.

34.7. Let K be an integral domain, and let $\leq$ be an ordering on K compatible with its ring structure. Show that P is the set of all positive elements for an ordering $\leq$ on $K[X]$ that is compatible with its ring structure and induces on K its given ordering if P is the set of all $f = \sum \alpha_k X^k$ such that

(a) $\alpha_k \geq 0$ for all $k \in N$;

(b) $\alpha_k \geq 0$ if k is even and $\alpha_k \leq 0$ if k is odd;

(c) either $f = 0$ or the leading coefficient of f is positive;

(d) either $f = 0$ or $\alpha_m > 0$ where m is the smallest of the natural numbers k such that $\alpha_k \neq 0$.

34.8. If K is a totally ordered integral domain, which of the orderings on $K[X]$ defined in Exercise 34.7 are total? If the set of positive elements of K is well-ordered, for which of those orderings on $K[X]$ is P well-ordered?

34.9. (a) The ideal $(2) + (X)$ is a maximal ideal but not a principal ideal of $Z[X]$.

(b) The constant polynomial 1 is a greatest common divisor of 2 and X in $Z[X]$, but there are no polynomials $g, h \in Z[X]$ such that $1 = 2g + Xh$.

(c) The $Z[X]$-module $(2) + (X)$ is finitely generated but has no basis (Exercise 28.10).

(d) The polynomial X is an irreducible element of $Z[X]$ and (X) is a prime ideal (Exercise 19.6), but (X) is not a maximal ideal of $Z[X]$.

34.10. If K is an integral domain, then $K[X]$ is a principal ideal domain if and only if K is a field. [Consider an ideal like that of Exercise 34.9(a) if K is not a field.]

34.11. Let a and b be nonzero polynomials over a field.

(a) Verify the assertions obtained from Exercise 22.9(a) and (b) by replacing "$0 \leq r < b$" with "either $r = 0$ or $\deg r < \deg b$" and "positive integers" with "polynomials."

(b) Calculate the monic greatest common divisor and least common multiple of the following pairs of polynomials over Q:

$$2X^4 + 9X^3 - 11X^2 + 5X - 1, \quad 2X^3 - 3X^2 + 5X - 2;$$
$$X^6 - 2X^5 + X^4 + 2X^3 + X^2 - 8X + 5, \quad X^5 - 2X^4 + X^3 - 6X + 3.$$

35. Substitution

Substituting numbers for the indeterminate of a polynomial is a familiar operation of elementary algebra, and we shall next investigate it in a general setting.

Definition. Let A be an algebra with identity element e over a commutative ring with identity K. For every polynomial

$$f = \alpha_0 + \alpha_1 X + \ldots + \alpha_n X^n$$

over K and every $c \in A$, we define $f(c)$ by

$$f(c) = \alpha_0 e + \alpha_1 c + \ldots + \alpha_n c^n.$$

The element $f(c)$ is said to be obtained by **substituting** c in f for the indeterminate X.

Theorem 35.1. Let A be an algebra with identity element e over a commutative ring with identity K. For every $c \in A$,

$$S_c : f \mapsto f(c)$$

is a homomorphism from the K-algebra $K[X]$ into A whose range is the subalgebra of A generated by e and c.

Proof. Let $f = \sum_{i=0}^{n} \alpha_i X^i$ and $g = \sum_{j=0}^{m} \beta_j X^j$, and let $\alpha_i = 0$ for all $i > n$, $\beta_j = 0$ for all $j > m$. Then

$$f(c)g(c) = \left(\sum_{i=0}^{n} \alpha_i c^i \right) \left(\sum_{j=0}^{m} \beta_j c^j \right) = \sum_{i=0}^{n} \sum_{j=0}^{m} \alpha_i \beta_j c^{i+j}$$
$$= \sum_{k=0}^{n+m} \left(\sum_{j=0}^{k} \alpha_j \beta_{k-j} \right) c^k = (fg)(c)$$

(Theorem A.10). It is easy to verify also that

$$(f + g)(c) = f(c) + g(c),$$
$$(\alpha f)(c) = \alpha f(c)$$

for every scalar α. Hence, S_c is a homomorphism from $K[X]$ into A. By the corollary of Theorem 33.4, the range of S_c is, therefore, a subalgebra of A containing e and c, for $e = X^0(c)$ and $c = X(c)$. An easy inductive proof shows, however, that a subalgebra of A containing e and c also contains $f(c)$ for all $f \in K[X]$. Thus, the range of S_c is the subalgebra of A generated by e and c.

The homomorphism S_c of Theorem 35.1 is called the **substitution homomorphism** determined by c, and its range is denoted by $K[c]$. Thus, $K[c]$ is the smallest subalgebra of A containing e and c. Since $K[X]$ is a commutative K-algebra, $K[c]$ *is a commutative subalgebra* of A.

For examples, let $f = 2X^2 - 3X + 4 \in R[X]$. Since R is an R-algebra with identity, upon substituting 2 for X in f, we obtain

$$f(2) = 8 - 6 + 4 = 6.$$

Since C is an R-algebra with identity, upon substituting $1 + i$ for X in f, we obtain

$$f(1 + i) = 2(1 + i)^2 - 3(1 + i) + 4 = 1 + i.$$

Since $\mathscr{M}_R(2)$ is an R-algebra with identity, upon substituting

$$\begin{bmatrix} 1 & -2 \\ 2 & 3 \end{bmatrix}$$

for X in f, we obtain

$$f\left(\begin{bmatrix} 1 & -2 \\ 2 & 3 \end{bmatrix}\right) = \begin{bmatrix} -6 & -16 \\ 16 & 10 \end{bmatrix} - \begin{bmatrix} 3 & -6 \\ 6 & 9 \end{bmatrix} + \begin{bmatrix} 4 & 0 \\ 0 & 4 \end{bmatrix}$$

$$= \begin{bmatrix} -5 & -10 \\ 10 & 5 \end{bmatrix}.$$

Since R^R is an R-algebra whose identity element is the constant function taking each real number into 1, upon substituting cosine for X in f and evaluating the resulting function at $\pi/3$, we obtain

$$[f(\cos)]\left(\frac{\pi}{3}\right) = 2\cos^2\frac{\pi}{3} - 3\cos\frac{\pi}{3} + 4 = 3.$$

Let $\mathscr{C}^\infty$ be the vector space over R of all real-valued functions on R having derivatives of all orders. As $\mathrm{End}_R(\mathscr{C}^\infty)$, the algebra of all linear operators on $\mathscr{C}^\infty$, is an R-algebra whose identity element is the identity linear operator on $\mathscr{C}^\infty$, upon substituting the differentiation linear operator D for X in f and evaluating the resulting linear transformation at the function cosine, we obtain

$$[f(D)](\cos) = 2D^2\cos - 3D\cos + 4\cos = 2\cos + 3\sin,$$

whence

$$[f(D)(\cos)]\left(\frac{\pi}{6}\right) = \sqrt{3} + \frac{3}{2}.$$

Because $R[X]$ itself is an R-algebra with identity, upon substituting X^2 for X in f, we obtain

$$f(X^2) = 2(X^2)^2 - 3X^2 + 4 = 2X^4 - 3X^2 + 4,$$

and upon substituting f for X in f, we obtain

$$f(f) = 2(2X^2 - 3X + 4)^2 - 3(2X^2 - 3X + 4) + 4$$
$$= 8X^4 - 24X^3 + 44X^2 - 39X + 24.$$

In general, if K is any commutative ring with identity and if f is any polynomial over K, upon substituting X for itself in f, we obtain simply f itself, so

$$f(X) = f.$$

For this reason, a polynomial f may equally well be denoted by $f(X)$.

Let A be an algebra with identity over a field K, and let $c \in A$. Since S_c is a homomorphism from $K[X]$ into A, the kernel $\mathfrak{a}$ of S_c is an ideal of $K[X]$ and hence is either the zero ideal or is the ideal generated by a unique monic polynomial g by Theorem 34.6.

Definition. Let K be a field, let c be an element of a K-algebra A with identity, and let $\mathfrak{a}$ be the kernel of the homomorphism S_c from $K[X]$ into A. We shall say that c is an **algebraic** element of A if $\mathfrak{a}$ is not the zero ideal, and that c is a **transcendental** element of A if $\mathfrak{a}$ is the zero ideal, If c is algebraic, the unique monic polynomial $g \in K[X]$ such that $\mathfrak{a} = (g)$ is called the **minimal polynomial** of c, and its degree is called the **degree** of c.

Thus, c is algebraic if and only if there is a nonzero polynomial f over K such that $f(c) = 0$; if c is algebraic and if g is its minimal polynomial, then a polynomial f over K satisfies $f(c) = 0$ if and only if $g \mid f$.

Since $S_c(X^0)$ is the identity element of A, $X^0 \notin \mathfrak{a}$, and therefore, *the minimal polynomial of an algebraic element is a nonconstant polynomial.*

Theorem 35.2. If A is a finite-dimensional algebra with identity over a field K, then every element of A is algebraic.

Proof. If c were a transcendental element of A, then S_c would be a monomorphism from the K-algebra $K[X]$ into A satisfying $S_c(X^n) = c^n$ for all $n \in N$; as the set of all monomials is a basis of $K[X]$, the set of all powers of c would be an infinite linearly independent subset of A by Theorem 29.5, in contradiction to Theorem 28.9.

Theorem 35.3. Let K be a field, and let c be an element of a K-algebra with identity A. Then c is an algebraic element of A if and only if $K[c]$ is a finite-dimensional subalgebra of A. If c is algebraic and if n is the degree of the minimal polynomial g of c, then $(1, c, c^2, \ldots, c^{n-1})$ is an ordered basis of the K-vector space $K[c]$, and in particular

$$n = \dim_K K[c].$$

Proof. By Theorem 35.2, if $K[c]$ is finite-dimensional, then c is algebraic. We shall show that if c is algebraic and if n is the degree of the minimal polynomial g of c, then $(1, c, c^2, \ldots, c^{n-1})$ is an ordered basis of $K[c]$, whence in particular, $K[c]$ is finite-dimensional. To show that $\{1, c, c^2, \ldots, c^{n-1}\}$ generates the K-vector space $K[c]$, let $z \in K[c]$. Then there exists a polynomial $f \in K[X]$ such that $z = f(c)$. By Theorem 34.5, there exist polynomials q and r over K satisfying

$$f = gq + r,$$

either $r = 0$ or $\deg r < n$.

As $g(c) = 0$, we have

$$z = f(c) = g(c)q(c) + r(c) = r(c)$$

by Theorem 35.1. Hence, $z = 0$ if $r = 0$, and $z = \sum_{k=0}^{m} \alpha_k c^k$ if r is the polynomial

$\sum_{k=0}^{m} \alpha_k X^k$ of degree $m < n$. Also, $(1, c, c^2, \ldots, c^{n-1})$ is a linearly independent sequence, for if

$$\sum_{k=0}^{n-1} \beta_k c^k = 0$$

and if $h = \sum_{k=0}^{n-1} \beta_k X^k$, then $h(c) = 0$, so $h \in (g)$ and hence $h = gu$ for some polynomial u. But if $h \neq 0$, then

$$n > \deg h = \deg g + \deg u \geq \deg g = n,$$

a contradiction, so $h = 0$, and therefore, $\beta_0 = \ldots = \beta_{n-1} = 0$.

Theorem 35.4. Let K be a field, and let c be an element of a K-algebra with identity A. Then $K[c]$ is a field if and only if c is algebraic and its minimal polynomial is irreducible.

Proof. If c is transcendental, then $K[c]$ is isomorphic to $K[X]$ and hence is not a field by Theorem 34.4. If c is algebraic and if g is the minimal polynomial of c, then $K[c]$ and $K[X]/(g)$ are isomorphic K-algebras by Theorems 35.1 and 33.5. But $K[X]/(g)$ is a field if and only if (g) is a maximal ideal of the ring $K[X]$ by Theorem 19.6, or equivalently, if and only if g is irreducible by Theorem 22.7.

Theorem 35.5. If A is a division algebra over a field K, then the minimal polynomial of every algebraic element of A is irreducible.

Proof. If c is an algebraic element of A, then $K[c]$ is a finite-dimensional subdomain of A by Theorem 35.3, so $K[c]$ is a field by the corollary of Theorem 33.7 applied to the K-algebra $K[c]$, and therefore, the minimal polynomial of c is irreducible by Theorem 35.4.

Theorem 35.6. If K is a commutative ring with identity and if A is a K-algebra with identity element e, then for all $f, g \in K[X]$ and for every $c \in A$,

$$[f(g)](c) = f(g(c)).$$

Proof. Both $S_c \circ S_g$ and $S_{g(c)}$ are homomorphisms from $K[X]$ into A by Theorem 35.1, and

$$(S_c \circ S_g)(X) = S_c(g) = g(c) = S_{g(c)}(X),$$
$$(S_c \circ S_g)(X^0) = S_c(X^0) = e = S_{g(c)}(X^0).$$

Therefore $S_c \circ S_g = S_{g(c)}$ by the corollary of Theorem 34.1. Consequently for every $f \in K[X]$,

$$[f(g)](c) = S_c(f(g)) = S_c(S_g(f))$$
$$= S_{g(c)}(f) = f(g(c)).$$

Corollary. If $f \in K[X]$, if $\alpha \in K$, and if $g = f(X + \alpha)$, then $f = g(X - \alpha)$.

Proof. If $g = f(X + \alpha)$, then

$$g(X - \alpha) = [f(X + \alpha)](X - \alpha)$$
$$= f((X - \alpha) + \alpha)$$
$$= f(X) = f$$

by Theorem 35.6.

Definition. Let A be a K-algebra with identity. An element $c \in A$ is a **root** of a polynomial $f \in K[X]$ if $f(c) = 0$.

In discussing the roots of a polynomial, it is important to keep in mind what algebra is being considered. For example, the polynomial $X^2 + 1$ has no roots in the R-algebra R, but has two roots in the R-algebra C.

Theorem 35.7. If f is a nonzero polynomial over a commutative ring with identity K and if $\alpha \in K$, then α is a root of f if and only if $X - \alpha$ divides f in $K[X]$.

Proof. Necessity: Let $g = f(X + \alpha)$. Then $f = g(X - \alpha)$ by the corollary of Theorem 35.6, and

$$g(0) = [f(X + \alpha)](0) = f(\alpha) = 0$$

by Theorem 35.6. Since $g(0)$ is the coefficient of X^0 in g, therefore, there exist $\beta_1, \ldots, \beta_n \in K$ such that $g = \sum_{k=1}^{n} \beta_k X^k$. Hence

$$f = g(X - \alpha) = \sum_{k=1}^{n} \beta_k (X - \alpha)^k$$
$$= (X - \alpha) \sum_{k=0}^{n-1} \beta_{k+1} (X - \alpha)^k,$$

so $X - \alpha$ divides f in $K[X]$. Sufficiency: If $f = (X - \alpha)h$ where $h \in K[X]$, then $f(\alpha) = (\alpha - \alpha)h(\alpha) = 0$ by Theorem 35.1.

Let f be a nonzero polynomial over a commutative ring with identity K, let α be a root of f in K, and let P be the set of all positive integers p such that $(X - \alpha)^p$ divides f in $K[X]$. By Theorem 35.7, $1 \in P$. If $p \in P$ and if $f = (X - \alpha)^p g$ where $g \in K[X]$, then $\deg f = p + \deg g$ by Theorem 34.3, so $p \leq \deg f$. Thus, P is a nonempty subset of the integer interval $[1, \deg f]$ and hence has a largest member.

Definition. Let K be a commutative ring with identity, and let $\alpha \in K$ be a root of a nonzero polynomial f over K. We shall call the largest of those numbers p such that $(X - \alpha)^p$ divides f in $K[X]$ the **multiplicity** of α in f. If the multiplicity of α is 1, α is called a **simple root** of f; if the multiplicity of α is greater than 1, α is called a **multiple** or **repeated root** of f.

In the definition we specified that $(X - \alpha)^p$ divide f in $K[X]$. Let L be a commutative ring with identity, and let K be a subring of L containing the identity element of L. *If an element α of K is a root of a polynomial f over K, then the multiplicity of α in f, regarded as a polynomial over L, is identical with the multiplicity of α in f, regarded as a polynomial over K.* Indeed, if $(X - \alpha)^p$ divides f in $L[X]$, then $(X - \alpha)^p$ divides f in $K[X]$ by Theorem 34.8, as $(X - \alpha)^p$ is a monic polynomial.

Theorem 35.8. Let f be a nonzero polynomial over an integral domain K. If $(\alpha_k)_{1 \leq k \leq n}$ is a sequence of distinct roots in K of f and if the multiplicity of α_k in f is m_k for each $k \in [1, n]$, then there exists a polynomial $g \in K[X]$ such that

$$(1) \qquad f = (X - \alpha_1)^{m_1} \ldots (X - \alpha_n)^{m_n} g,$$

$$g(\alpha_k) \neq 0$$

for all $k \in [1, n]$.

Proof. Let S be the set of all positive integers n such that for every sequence $(\alpha_k)_{1 \leq k \leq n}$ of distinct roots of f, (1) holds for some polynomial $g \in K[X]$ satisfying $g(\alpha_k) \neq 0$ for all $k \in [1, n]$. If α is a root of f of multiplicity m, then there is a polynomial $g \in K[X]$ such that $f = (X - \alpha)^m g$ by the definition of multiplicity. If $g(\alpha) = 0$, then by Theorem 35.7, there would exist a polynomial $h \in K[X]$ such that $g = (X - \alpha)h$, whence $f = (X - \alpha)^{m+1}h$, a contradiction of the definition of m. Hence, $g(\alpha) \neq 0$, so $1 \in S$. Suppose that $n \in S$, let $(\alpha_k)_{1 \leq k \leq n+1}$ be a sequence of $n + 1$ distinct roots of f in K, and let m_k be the multiplicity of α_k in f for each $k \in [1, n + 1]$. Since $n \in S$ and as m_{n+1} is the multiplicity of α_{n+1} in f, there exist polynomials $g, u \in K[X]$ such that

$$(2) \qquad (X - \alpha_1)^{m_1} \ldots (X - \alpha_n)^{m_n} g = f = (X - \alpha_{n+1})^{m_{n+1}} u,$$

$$g(\alpha_k) \neq 0$$

for all $k \in [1, n]$. As

$$0 = f(\alpha_{n+1}) = (\alpha_{n+1} - \alpha_1)^{m_1} \ldots (\alpha_{n+1} - \alpha_n)^{m_n} g(\alpha_{n+1})$$

by Theorem 35.1 and as K is an integral domain, α_{n+1} is a root of g. Let p be the multiplicity of α_{n+1} in g. As before, there exists $h \in K[X]$ such that

$$g = (X - \alpha_{n+1})^p h,$$

$$h(\alpha_{n+1}) \neq 0.$$

Also, for all $k \in [1, n]$ we have $h(\alpha_k) \neq 0$, since $g(\alpha_k) \neq 0$. As $(X - \alpha_{n+1})^p$ divides g and hence divides f, we have $p \leq m_{n+1}$ by the definition of m_{n+1}. Because $K[X]$ is an integral domain by Theorem 34.4, upon cancelling $(X - \alpha_{n+1})^p$ in (2), we obtain

$$(X - \alpha_1)^{m_1} \ldots (X - \alpha_n)^{m_n} h = (X - \alpha_{n+1})^{m_{n+1}-p} u.$$

Consequently, $m_{n+1} - p = 0$, since

$$(\alpha_{n+1} - \alpha_1)^{m_1} \ldots (\alpha_{n+1} - \alpha_n)^{m_n} h(\alpha_{n+1}) \neq 0.$$

Therefore, from (2), we obtain

$$(X - \alpha_1)^{m_1} \ldots (X - \alpha_n)^{m_n} (X - \alpha_{n+1})^{m_{n+1}} h = f,$$

$$h(\alpha_k) \neq 0$$

for all $k \in [1, n + 1]$. Thus, $n + 1 \in S$. By induction, therefore, the proof is complete.

Corollary 35.8.1. If f is a polynomial of degree n over an integral domain K and if $(\alpha_k)_{1 \leq k \leq r}$ is a sequence of distinct roots of f in K, then

$$(3) \qquad \sum_{k=1}^{r} m_k \leq n$$

where m_k is the multiplicity of α_k in f for each $k \in [1, r]$, and in particular, $r \leq n$.

An immediate consequence of Corollary 35.8.1 is the following apparently more general statement: If K is a subdomain of an integral domain A and if f is a polynomial of degree n over K, then (3) holds for any sequence $(\alpha_k)_{1 \leq k \leq r}$ of distinct roots of f in A where m_k is the multiplicity of α_k in f for each

$k \in [1, r]$. Indeed, we need only apply Corollary 35.8.1 to f, regarded as a polynomial over A.

Corollary 35.8.2. If f and g are polynomials of degrees $\leq n$ over an integral domain K and if there is a sequence $(\alpha_k)_{1 \leq k \leq n+1}$ of distinct elements of K such that $f(\alpha_k) = g(\alpha_k)$ for all $k \in [1, n + 1]$, then $f = g$.

Proof. The polynomial $f - g$ is of degree $\leq n$ but has at least $n + 1$ roots in K, so $f - g$ is the zero polynomial by Corollary 35.8.1.

Definition. Let S be a subset of an integral domain K, and let f be a polynomial over K. The **number of roots of f in** S is, of course, the number of elements in $\{\alpha \in S : f(\alpha) = 0\}$. The **number of roots of f in** S, **multiplicities counted,** is the sum of the multiplicities in f of all the roots of f in S.

By Corollary 35.8.1, *if f is a nonzero polynomial over an integral domain K, then the number of roots of f in K, multiplicities counted, does not exceed the degree of f.*

Definition. Let A be a K-algebra with identity, and let $f \in K[X]$. The **polynomial function on A defined by** f is the function $f^{\sim}$ from A into A defined by

$$f^{\sim} : x \mapsto f(x).$$

From the definition of $f^{\sim}$ and Theorem 35.1 it follows easily that the function $S : f \mapsto f^{\sim}$ is a homomorphism from the K-algebra $K[X]$ into the K-algebra A^A of all functions from A into A. The following theorem is an immediate consequence of Corollary 35.8.2.

Theorem 35.9. If K is an infinite integral domain, then the function $S : f \mapsto f^{\sim}$ is a monomorphism from the K-algebra $K[X]$ into the K-algebra K^K of all functions from K into K.

The range of the homomorphism S is, of course, the K-algebra $P(K)$ of all polynomial functions on K, which we have previously discussed informally. In our discussion of Example 28.7, we used calculus to show that $\{p_n : n \in N\}$ is a linearly independent subset of $P(\boldsymbol{R})$. Since p_n is the polynomial function defined by X^n, however, by Theorem 35.9, the fact that $\boldsymbol{R}$ is infinite is all that is really needed to establish this assertion.

If K is a finite field, the homomorphism S from $K[X]$ into K^K is not a monomorphism since $K[X]$ is infinite and K^K is finite. For example, if $f_n = X^n + X \in \boldsymbol{Z}_2[X]$, then $f_n^{\sim}$ is the zero function on $\boldsymbol{Z}_2$ for all $n \in N^*$.

Often it is important to know whether a root of a polynomial is a multiple root, and an important criterion for a root to be a multiple root may be given in terms of the derivative of a polynomial.

Definition. Let $f = \sum\limits_{k=0}^{n} \alpha_k X^k$ be a polynomial over a commutative ring with identity K. The **derivative** of f is the polynomial Df defined by

$$Df = \sum_{k=1}^{n} k.\alpha_k X^{k-1}.$$

The function $D : f \mapsto Df$ from $K[X]$ into $K[X]$ is called the **differential operator** on $K[X]$.

If $f \in R[X]$, the polynomial function on R defined by Df is identical with the derivative of the polynomial function on R defined by f considered in calculus. Of course, we could not use calculus in attempting to frame a suitable definition of the derivative of a polynomial over an arbitrary commutative ring with identity.

Theorem 35.10. If K is a commutative ring with identity, the differential operator D on $K[X]$ is a linear operator on the K-module $K[X]$ and, in addition, satisfies

(4) $$D(fg) = f \cdot Dg + Df \cdot g$$

for all $f, g \in K[X]$.

Proof. It is easy to verify that D is indeed a linear operator on $K[X]$. Let $f = \sum\limits_{k=0}^{n} \alpha_k X^k \in K[X]$, and let H be the set of all the polynomials $g \in K[X]$ for which (4) holds. Certainly, the constant polynomial X^0 belongs to H, for $DX^0 = 0$. For every $m \in N^*$, X^m belongs to H, for

$$D(fX^m) = D\left(\sum_{k=0}^{n} \alpha_k X^{k+m}\right) = \sum_{k=0}^{n} \alpha_k D(X^{k+m})$$

$$= \sum_{k=0}^{n} (k+m).\alpha_k X^{k+m-1}$$

$$= \sum_{k=0}^{n} m.\alpha_k X^{k+m-1} + \sum_{k=1}^{n} k.\alpha_k X^{k-1+m}$$

$$= f \cdot DX^m + Df \cdot X^m.$$

For every polynomial $h \in K[X]$, the function $L_h : f \mapsto fh$ is clearly a linear operator on the K-module $K[X]$. By definition, H is the set of all polynomials g such that

$$[D \circ L_f - L_f \circ D - L_{Df}](g) = 0$$

and is thus the kernel of a linear operator on $K[X]$. Thus, H is a subspace of $K[X]$ containing X^m for all $m \in N$. Consequently, $H = K[X]$, so (4) holds for all $g \in K[X]$.

Corollary. If f is a polynomial over a commutative ring with identity, then for every $n \in N^*$,

$$Df^n = n.f^{n-1}Df.$$

The assertion follows by induction from Theorem 35.10.

Theorem 35.11. If K is an integral domain, an element α of K is a multiple root of a nonzero polynomial f over K if and only if α is a root of both f and Df.

Proof. Let m be the multiplicity of α in f. By Theorem 35.8, there exists a polynomial $g \in K[X]$ such that $f = (X - \alpha)^m g$ and $g(\alpha) \neq 0$. By Theorem 35.10 and its corollary,

$$Df = m.(X - \alpha)^{m-1}g + (X - \alpha)^m Dg.$$

Hence, if $m \geq 2$, then $(Df)(\alpha) = 0$, but if $m = 1$, then $(Df)(\alpha) = g(\alpha) \neq 0$.

EXERCISES

35.1. Verify in detail that a subalgebra of a K-algebra A with identity element e that contains e and c also contains $K[c]$.

35.2. Determine $h(-2)$, $h(1 - i)$, $h\left(\begin{bmatrix} 2 & 1 \\ -1 & 2 \end{bmatrix}\right)$, $[h(\sin)](\pi/6)$, $[h(D)(\sin)](\pi/6)$, $[h(D)(h(\sin))](\pi/6)$, $h(X^2 + 1)$, $h(h)$, $h(h(h))$, and $[h(h)](h)$ where h is the polynomial over R defined by (a) $h = 2X + 1$; (b) $h = X^2 - 2$; (c) $h = X^3 - X + 1$.

35.3. Prove that the function $f \mapsto f^{\sim}$ is a homomorphism from the K-algebra $K[X]$ into the K-algebra A^A where A is a K-algebra with identity.

35.4. If A is a K-algebra with identity element e and if u is a homomorphism from the K-algebra $K[X]$ into A such that $u(1) = e$, then u is the substitution homomorphism S_c for some $c \in A$.

35.5. (a) If K is a field, then u is an automorphism of the K-algebra $K[X]$ if and only if u is the substitution endomorphism S_g for some linear polynomial $g \in K[X]$.

(b) If K is a field, then the group of all automorphisms of the K-algebra $K[X]$ is isomorphic to the group $(K^* \times K, \circ)$ whose composition is defined by

$$(\alpha, \beta) \circ (\gamma, \delta) = (\alpha\gamma, \gamma\beta + \delta).$$

(c) If K is a commutative ring with identity containing a nonzero element α such that $\alpha^2 = 0$ and if $g = \alpha X^2 + X$, then S_g is an automorphism of the K-algebra $K[X]$. [Compute $S_g(X - \alpha X^2)$.]

*35.6. If K is a field and if u is an automorphism of the ring $K[X]$, then the restriction φ of u to K is an automorphism of the field K, and there is a linear polynomial g over K such that $u = S_g \circ \bar\varphi$ (with the notation of Theorem 34.2). [Use Theorem 34.4 and Exercise 35.5(a).]

35.7. If A and B are K-algebras with identity elements e and e' respectively and if u is a homomorphism from A into B such that $u(e) = e'$, then $u(f(a)) = f(u(a))$ for every $a \in A$ and every polynomial $f \in K[X]$.

35.8. (Remainder Theorem) If f is a polynomial over a field K, then for each $\alpha \in K$ the remainder obtained by dividing $X - \alpha$ into f is the constant polynomial $f(\alpha)$.

35.9. How many roots does $X^2 - 1$ have in Z_8? in Z_{12}?

35.10. If K is a finite field of q elements, the restriction of the function $f \mapsto f^\sim$ to the subspace H of $K[X]$ of all polynomials of degree $< q$ is an isomorphism from H onto the vector space K^K of all functions from K into K.

35.11. If K is a finite field of q elements, give an example of a polynomial f over K of degree q such that the polynomial function $f^\sim \in K^K$ is the zero function. [Express f as a product of linear polynomials.]

35.12. Verify that the differential operator D on $K[X]$ is linear.

*35.13. Let K be an integral domain.
(a) What are the kernel and range of the differential operator D if the characteristic of K is zero?
(b) What are the kernel and range of D if the characteristic of K is a prime p?

35.14. (Taylor's Formula) If A is an algebra with identity over a field K whose characteristic is zero, then for every polynomial f over K of degree n and for all elements a and h of A,

$$f(a + h) = \sum_{k=0}^{n} \frac{1}{k!} (D^k f)(a) h^k.$$

[Prove the equality first for $f = X^n$.]

35.15. (Chain Rule) If K is a commutative ring with identity, for all polynomials f, g over K,

$$D[f(g)] = (Df)(g) \cdot Dg.$$

[Prove the equality first for $f = X^n$.]

35.16. If f is a nonzero polynomial over an integral domain K, then an element $\alpha \in K$ is a root of f of multiplicity n if and only if $(D^k f)(\alpha) = 0$ for all $k \in N_n$ and $(D^n f)(\alpha) \neq 0$.

35.17. Let K be a commutative ring with identity. The following statements are equivalent:

1° K is an integral domain.
2° For every positive integer n, every polynomial over K of degree n has at most n roots in K.
3° Every linear polynomial over K has at most one root in K.

*35.18. Let λ be a euclidean stathm (Exercise 22.13) on an integral domain A. The following statements are equivalent:

1° For all $a, b \in A^*$, if $a + b \neq 0$, then $\lambda(a + b) \leq \max\{\lambda(a), \lambda(b)\}$.
2° For all $a, b \in A^*$, there exist unique elements q and r of A such that $a = bq + r$ and either $r = 0$ or $\lambda(r) < \lambda(b)$.
3° Either A is a field, or there exist an isomorphism φ from A onto the ring of polynomials over a field K and a strictly increasing function f from N into N such that $\lambda(a) = f(\deg \varphi(a))$ for all $a \in A^*$.

[To show that 2° implies 1°, if $\lambda(a + b) > \lambda(a)$ and $\lambda(a + b) > \lambda(b)$, divide $a + b$ into $a^2 - b^2 + b$ in two ways to obtain remainders b and $-a$. To show that 1° and 2° imply 3°, use Exercise 22.13(b) and, if the set N of nonzero noninvertible elements of A is not empty, consider $c \in N$ where $\lambda(c)$ is the smallest member of $\lambda_*(N)$.]

36. Irreducibility Criteria

If A is a commutative ring with identity, we shall say that a nonzero polynomial f is **irreducible** over A if f is an irreducible element of the ring $A[X]$, and that f is **reducible** over A if $f \in A[X]$ and if f is neither a unit nor an irreducible element of $A[X]$. An important but often difficult problem of algebra is to determine whether a given polynomial over A is irreducible.

If f is a nonconstant polynomial over a field K and if $n = \deg f$, then f is *irreducible over K if and only if the degree of every divisor of f in $K[X]$ is either n or 0*, for the units of $K[X]$ are the polynomials of degree zero by Theorem 34.4, and therefore, the associates of f are those polynomials of degree n that divide f by Theorem 34.3. Every linear polynomial over a field K clearly has a root in K. Thus, *every linear polynomial over a field K is irreducible and has a* *root in K.* By Theorem 35.7, however, *a polynomial of degree* ≥ 2 *that is irreducible over K has no roots in K.* A polynomial may be reducible over K and yet have no roots in K; the polynomial $X^4 + 2X^2 + 1$ over R is such a polynomial. However, *if f is a quadratic or cubic polynomial over a field K, then f is irreducible over K if and only if f has no roots in K,* for if $f = gh$ where g and h are nonconstant polynomials over K, then either g or h is linear and hence has a root in K, which necessarily is also a root of f. Thus, $X^2 + 1$ is irreducible over R as it has no roots in R, but $X^2 + 1$ is, of course, reducible over C.

The polynomial $f = 2X + 6$ is irreducible over Q but reducible over Z, for the polynomial $2X^0$ divides f in $Z[X]$ but is not a unit of $Z[X]$. A criterion for irreducibility of nonconstant polynomials over an integral domain that includes the above criterion for irreducibility of nonconstant polynomials over a field is the following: *If f is a nonconstant polynomial over an integral domain A and if n = deg f, then f is irreducible over A if and only if the nonzero coefficients of f are relatively prime in A and the degree of every divisor of f in A[X] is either n or 0.* Indeed, the units of $A[X]$ are those constant polynomials determined by the units of A by Theorem 34.4. The condition is, therefore, necessary, for if α is a common divisor of the nonzero coefficients of a nonconstant polynomial f irreducible over A, then αX^0 divides f and clearly is not an associate of f, so αX^0 is a unit of $A[X]$ and hence α is a unit of A. Conversely, if the condition holds and if $f = gh$, then either g or h is a constant polynomial αX^0 by hypothesis, and since αX^0 divides f in $A[X]$, clearly α is a common divisor of the nonzero coefficients of f, so, by hypothesis, α is a unit of A, and therefore, αX^0 is a unit of $A[X]$. Consequently, *if A is a subdomain of a field K, if f is a nonconstant polynomial over A that is irreducible over K, and if the nonzero coefficients of f are relatively prime in A, then f is irreducible over A.* One of our most important results is the converse for the case where A is a unique factorization domain and K a quotient field of A (Theorem 36.3).

Definition. A nonzero polynomial f over an integral domain A is **primitive** over A if its nonzero coefficients are relatively prime in A (or, in case f has only one nonzero coefficient α, if α is a unit).

Lemma 36.1. *If f is a nonzero polynomial over a unique factorization domain A, then α is a greatest common divisor of the nonzero coefficients of f if and only if there is a primitive polynomial f_1 over A such that $f = \alpha f_1$.*

Proof. Necessity: Since α is a greatest common divisor of the nonzero coefficients of f, clearly there exists $f_1 \in A[X]$ such that $\alpha f_1 = f$. If β is a common divisor of the nonzero coefficients of f_1, then $\alpha\beta$ is a common divisor of the nonzero coefficients of f, so $\alpha\beta \mid \alpha$, and therefore, β is a unit. Hence, f_1 is primitive.

Sufficiency: By Theorem 22.10 the nonzero coefficients of f admit a greatest common divisor λ, and as observed above, there is a polynomial $f_2 \in A[X]$ such that $f = \lambda f_2$. Since $\alpha f_1 = f$, α is a common divisor of the nonzero coefficients of f, and therefore, there exists $\beta \in A$ such that $\alpha\beta = \lambda$. Thus, $\alpha\beta f_2 = f = \alpha f_1$, so $\beta f_2 = f_1$, and therefore, β is a common divisor of the nonzero coefficients of f_1. But since f_1 is primitive, β is, therefore, a unit, so α and λ are associates, and consequently, α is also a greatest common divisor of the nonzero coefficients of f.

Lemma 36.2. Let A be a unique factorization domain, let

$$g = \sum_{k=0}^{m} \beta_k X^k, \qquad h = \sum_{k=0}^{q} \gamma_k X^k$$

be nonzero polynomials over A, let

$$gh = \sum_{k=0}^{m+q} \alpha_k X^k,$$

and let π be an irreducible element of A such that πX^0 divides neither g nor h in $A[X]$. If r is the smallest integer in $[0, m]$ such that $\pi \nmid \beta_r$ and if s is the smallest integer in $[0, q]$ such that $\pi \nmid \gamma_s$, then $\pi \nmid \alpha_{r+s}$.

Proof. By hypothesis, $\pi \mid \beta_i$ for all $i < r$ and $\pi \mid \gamma_{r+s-i}$ for all $i > r$, so $\pi \mid \beta_i \gamma_{r+s-i}$ for all $i \neq r$. If π were a divisor of α_{r+s}, then π would divide

$$\alpha_{r+s} - \sum_{i \neq r} \beta_i \gamma_{r+s-i} = \beta_r \gamma_s,$$

whence π would divide either β_r or γ_s by (UFD 3), a contradiction.

Theorem 36.1. (Gauss's Lemma) If f and g are primitive polynomials over a unique factorization domain A, then fg is primitive over A.

Proof. If π is an irreducible element of A, then πX^0 divides neither f nor g in $A[X]$ by hypothesis, so by Lemma 36.2, there is a coefficient of fg not divisible in A by π. Consequently, by (UFD 1), the units of A are the only divisors of all the nonzero coefficients of fg, so fg is a primitive polynomial.

Theorem 36.2. Let A be a unique factorization domain, let f be a nonzero polynomial over A, and let K be a quotient field of A. If there exist polynomials g, $h \in K[X]$ such that $f = gh$, then there exist polynomials g_0, $h_0 \in A[X]$ that are multiples by nonzero scalars of K of g and h respectively such that $f = g_0 h_0$.

Proof. As the nonzero coefficients of g and h are quotients of elements of A, there exist nonzero elements α and β of A such that αg and βh are polynomials over A. By Theorem 22.10, the nonzero coefficients of αg and βh

admit greatest common divisors α_1 and β_1 respectively in A, and by Lemma 36.1, there exist primitive polynomials g_1 and h_1 over A such that $\alpha g = \alpha_1 g_1$ and $\beta h = \beta_1 h_1$. By Theorem 36.1, $g_1 h_1$ is primitive, so by Lemma 36.1, $\alpha_1 \beta_1$ is a greatest common divisor of the nonzero coefficients of $\alpha_1 \beta_1 g_1 h_1 = \alpha \beta g h = \alpha \beta f$. Consequently, $\alpha \beta$ divides $\alpha_1 \beta_1$ in A, so there exists $\gamma \in A$ such that $\alpha \beta \gamma = \alpha_1 \beta_1$, whence $f = \gamma g_1 h_1$. The polynomials $g_0 = \gamma g_1$ and $h_0 = h_1$, therefore, have the desired properties.

Theorem 36.3. Let A be a unique factorization domain, and let K be a quotient field of A. If f is a nonconstant polynomial over A, then f is irreducible over A if and only if f is irreducible over K and the nonzero coefficients of f are relatively prime in A.

Proof. It follows at once from Theorem 36.2 that the condition is necessary, and we saw in our earlier discussion that it is also sufficient.

Thus, a nonconstant polynomial irreducible over a unique factorization domain is also irreducible over its quotient field. The following theorem is useful in determining the irreducibility of many polynomials.

Theorem 36.4. (Eisenstein's Criterion) Let K be a quotient field of a unique factorization domain A, and let $f = \sum_{k=0}^{n} \alpha_k X^k$ be a nonconstant polynomial over A of degree n. If there exists an irreducible element π of A such that $\pi \nmid \alpha_n$, $\pi \mid \alpha_k$ for all $k \in [0, n-1]$, and $\pi^2 \nmid \alpha_0$, then f is irreducible over K.

Proof. By Theorem 36.2, it suffices to prove that if $f = g_0 h_0$ where g_0 and h_0 are polynomials over A, then either g_0 or h_0 is a constant polynomial. Let

$$g_0 = \sum_{k=0}^{m} \beta_k X^k, \qquad h_0 = \sum_{k=0}^{q} \gamma_k X^k.$$

As $\beta_0 \gamma_0 = \alpha_0$ and as $\pi^2 \nmid \alpha_0$, either $\pi \nmid \beta_0$ or $\pi \nmid \gamma_0$, but as $\pi \mid \alpha_0$, either $\pi \mid \beta_0$ or $\pi \mid \gamma_0$ by (UFD 3). We shall assume that $\pi \nmid \beta_0$ and that $\pi \mid \gamma_0$. As $\beta_m \gamma_q = \alpha_n$ and as $\pi \nmid \alpha_n$, we conclude that $\pi \nmid \gamma_q$. By Lemma 36.2, $\pi \nmid \alpha_s$ where s is the smallest integer in $[0, q]$ such that $\pi \nmid \gamma_s$. By hypothesis, therefore,

$$s = n = m + q \geq q \geq s,$$

so $m = 0$ and g_0 is a constant polynomial.

Example 36.1. The polynomials $2X^5 + 18X^3 + 30X^2 - 24$ and $4X^7 - 20X^5 + 100X - 10$ are irreducible over $\mathbf{Q}$, as we see by applying Eisenstein's Criterion where $\pi = 3$ and $\pi = 5$ respectively. Neither polynomial, however, is irreducible over Z. Dividing each of these polynomials by the constant polynomial 2, we obtain polynomials that are irreducible over Z by Theorem 36.3.

Example 36.2. If p is a prime number and if n is an integer ≥ 2, then $\sqrt[n]{p}$ is irrational. Indeed, $X^n - p$ is irreducible over Q by Eisenstein's Criterion and, consequently, has no roots in Q. The root $\sqrt[n]{p}$ of $X^n - p$ is, therefore, irrational.

Example 36.3. If u is an automorphism of an integral domain A, then an element a of A is clearly irreducible if and only if $u(a)$ is irreducible, for $a = bc$ if and only if $u(a) = u(b)u(c)$, and b is invertible if and only if $u(b)$ is. In particular, if f is a nonzero polynomial over an integral domain K and if $\alpha \in K$, then $f(X)$ is irreducible if and only if $f(X + \alpha)$ is, for the substitution endomorphism defined by the polynomial $X + \alpha$ is an automorphism of the ring $K[X]$ (corollary of Theorem 35.6). Thus, if $f = X^3 + 3X + 2$, then $f(X + 1) = X^3 + 3X^2 + 6X + 6$, so $f(X + 1)$ is irreducible over Q by Eisenstein's Criterion, and therefore, f is also irreducible over Q.

Eisenstein's Criterion is not applicable to many polynomials over Z that are actually irreducible, such as $X^5 + X^3 + 1$ (Exercise 36.14). One may always apply the method of finding factors learned in beginning algebra to polynomials over Z, however, to determine whether they are irreducible.

Example 36.4. We shall show that $f = X^5 + X^3 + 1$ is irreducible over Z. Suppose that $f = gh$ where g and h are polynomials over Z of degree < 5. We may assume that both g and h are monic, for the product of their leading coefficients is 1, and hence their leading coefficients are either both 1 or both -1; in the latter case, we may replace g and h by the monic polynomials $-g$ and $-h$. Let c and e be the constant terms of g and h respectively. Then $ce = 1$, so either $c = e = 1$ or $c = e = -1$. Since neither 1 nor -1 is a root of f, neither g nor h is linear. We may suppose, therefore, that g is cubic and h quadratic. Let $g = X^3 + aX^2 + bX + c$, $h = X^2 + dX + e$. Then

(1) $$d + a = 0,$$

(2) $$e + ad + b = 1,$$

(3) $$ea + bd + c = 0,$$

(4) $$be + cd = 0,$$

(5) $$ce = 1.$$

If $c = e = 1$, we obtain $a = b = -d$ from (1) and (4), whence $-d^2 - d = 0$ and $-d - d^2 + 1 = 0$ from (2) and (3), which is impossible. If $c = e = -1$, we obtain $a = b = -d$ from (1) and (4), whence $-d^2 - d = 2$ and $d - d^2 - 1 = 0$ by (2) and (3); consequently, $-2 - d = d^2 = d - 1$, so $2d = -1$, which is impossible as d is an integer. Therefore, $X^5 + X^3 + 1$ is irreducible over Z and hence over Q.

EXERCISES

36.1. Let α be the leading coefficient of a linear polynomial f over an integral domain A.

(a) The polynomial f has a root in A if and only if $f = \alpha g$ for some monic linear polynomial g.

(b) The polynomial f has a root in A and is irreducible over A if and only if α is invertible in A.

36.2. (a) If a and b are integers, then $X^3 + aX^2 + bX + 1$ is reducible over Z if and only if either $a = b$ or $a + b = -2$.

(b) Determine necessary and sufficient conditions on integers a and b for $X^3 + aX^2 + bX - 1$ to be reducible over Z.

36.3. (a) For what integers b is $3X^2 + bX + 5$ reducible over Z?

(b) If a and c are nonzero integers, give an upper bound in terms of the number of positive divisors of a and of c on the number of integers b for which $aX^2 + bX + c$ is reducible over Z.

36.4. Determine all quadratic and cubic irreducible polynomials over Z_2.

36.5. Determine all quartic and quintic irreducible polynomials over Z_2. [Use Exercise 36.4.]

36.6. (a) Let $\mathfrak{a}$ be a proper ideal of a commutative ring with identity K, and let φ be the canonical epimorphism from K onto $K/\mathfrak{a}$. If $f = \sum\limits_{k=0}^{n} \alpha_k X^k$ is a monic polynomial over K such that $\sum\limits_{k=0}^{n} \varphi(\alpha_k) X^k$ is irreducible over $K/\mathfrak{a}$, then f is irreducible over K.

(b) Show that the following polynomials are irreducible over Z:

$$X^5 + 6X^4 + 5X^2 - 2X + 9$$

$$X^5 + 7X^4 - 3X^3 - 6X^2 + 5X + 21$$

$$X^5 + 2X^4 + 3X^3 + 4X^2 + 5.$$

[Use Exercise 36.5.]

36.7. If $\alpha_n X^n + \alpha_{n-1} X^{n-1} + \ldots + \alpha_1 X + \alpha_0$ is an irreducible polynomial of degree n over a field K, then so is $\alpha_0 X^n + \alpha_1 X^{n-1} + \ldots + \alpha_{n-1} X + \alpha_n$.

36.8. Let a and n be integers ≥ 2.
  (a) If $\sqrt[n]{a} \in Q$, then $\sqrt[n]{a} \in Z$.
  (b) The real number $\sqrt[n]{a}$ is rational if and only if $a = b^n$ for some positive integer b. [Express a as a product of powers of primes.]

36.9. Show that the following polynomials are irreducible over Q:
  (a) $X^8 - 21X^5 + 98X^3 - 56$
  (b) $7X^6 - 52X^4 + 65X^2 - 104$
  (c) $210X^4 + 207X^2 - 184$.

36.10. Use the technique of Example 36.3 to show that the following polynomials are irreducible over Q:
  (a) $X^3 + 4X^2 + 3X - 1$ (b) $X^3 + 6X^2 + 1$
  (c) $X^3 - X^2 + 7X + 2$ (d) $X^3 + 5X^2 - 3X - 1$.

36.11. Determine all integers b such that $X^5 - bX - 1$ is reducible over Z.

36.12. Determine which of the following polynomials are irreducible over Q:
  (a) $X^3 - 2X + 3$ (b) $X^3 - 3X^2 + 6X - 6$
  (c) $X^4 - 3X^2 + 1$ (d) $X^4 + 4X^2 + 10$
  (e) $X^4 - X - 1$ (f) $X^5 - X + 1$
  (g) $X^5 + X + 1$ (h) $X^6 - 6X^4 + 12$
  (i) $X^5 - 5X^4 + 7X^3 - X^2 + 2X - 1$.

36.13. If K is a field and if $a \in K$, then $X^4 - a$ is reducible over K if and only if either $a = b^2$ for some $b \in K$ or $a = -4c^4$ for some $c \in K$.

36.14. If $f = \sum_{k=0}^{m} \alpha_k X^k$ is a polynomial over Z of odd degree $m > 1$ such that $\alpha_m = \alpha_{m-2} = 1$ and $\alpha_{m-1} = 0$, then for no integer n does $f(X + n)$ satisfy the hypotheses of Eisenstein's Criterion. [Consider the coefficients of X^{m-1} and X^{m-2} in $f(X + n)$.]

36.15. Let K be a quotient field of a unique factorization domain A, and let $f = \sum_{k=0}^{n} \alpha_k X^k$ be a nonconstant polynomial over A of degree $n \geq m$. If there exists an irreducible element π of A such that $\pi \nmid \alpha_n$, $\pi \nmid \alpha_m$, $\pi \mid \alpha_k$ for all $k \in [0, m-1]$, and $\pi^2 \nmid \alpha_0$, then there is an irreducible factor of f of degree $\geq m$. [Modify the proof of Eisenstein's Criterion.]

*36.16. Let K be a finite field of q elements.
  (a) There are exactly $\frac{1}{2}(q^2 - q)$ prime quadratic polynomials over K.
  (b) There are exactly $\frac{1}{3}(q^3 - q)$ prime cubic polynomials over K.

36.17. Let f and g be nonzero polynomials over a unique factorization domain A.
  (a) If f is primitive and if $g \mid f$, then either g is a unit of $A[X]$, or g is an associate of f, or $0 < \deg g < \deg f$.

(b) The integral domain $A[X]$ satisfies (UFD 1). [Proceed by induction on the degree of a polynomial.]

(c) If g is a primitive polynomial over A and if $g \mid \alpha f$ where $\alpha \in A^*$ and $f \in A[X]$, then $g \mid f$.

*36.18. Let A be a unique factorization domain, and let p, g, $h \in A[X]$ be polynomials such that p is irreducible, $\deg p > 0$, $p \mid gh$, but $p \nmid g$. Let λ be a nonzero element of the ideal $(p) + (g)$ of smallest possible degree.

(a) There exists $\alpha \in A^*$ such that $\lambda \mid \alpha g$ and $\lambda \mid \alpha p$. [Apply Exercise 34.5 to λ and g and also to λ and p.]

(b) The polynomial λ is a constant polynomial. [Let $\lambda = \beta \lambda_1$ where λ_1 is primitive, and apply Exercise 36.17(c).]

(c) The polynomial p divides h in $A[X]$. [Let $\lambda = up + vg$, and apply Exercise 36.17(c).]

36.19. Let A be a unique factorization domain.

(a) If α is an irreducible element of A and if g and h are polynomials over A such that αX^0 divides gh but not g, then αX^0 divides h.

(b) The integral domain $A[X]$ is a unique factorization domain. [Use Exercises 36.17–36.18.]

(c) The ring $Z[X]$ is a unique factorization domain that is not a principal ideal domain. [Use Exercise 34.10.]

37. Adjoining Roots

Let $\mathscr{L}$ be a set of subfields of a field K, and let $L = \cap \mathscr{L}$. The multiplicative identity of each member of $\mathscr{L}$ is that of K by Theorem 15.2, so L contains the multiplicative identity 1 of K. The sum and product of any two elements of L again belong to each member of $\mathscr{L}$ and hence to L, and similarly, the multiplicative inverse of each nonzero element of L belongs to each member of $\mathscr{L}$ and hence to L. Thus, we have proved the following theorem:

Theorem 37.1. Let K be a field. If $\mathscr{L}$ is a set of subfields of K, then $\cap \mathscr{L}$ is the largest subfield of K contained in each member of $\mathscr{L}$. If S is a subset of K, the intersection of all the subfields of K containing S is the smallest subfield of K containing S.

Theorem 37.1 enables us to make the following definition:

Definition. Let K be a field, and let S be a subset of K. The **subfield generated** by S is the smallest subfield L of K containing S, and S is called a **set of generators** or a **generating set** for L.

If f is a polynomial over a field K, f may have no roots in K but yet may possess roots in some larger field. For example, $X^2 - 2$ has no roots in Q but has roots in R, and $X^2 + 1$ has no roots in R but has roots in C. Our principal goal is to prove that for every nonconstant polynomial f over K, there is (to within isomorphism) a unique field L such that f is a product of linear polynomials in $L[X]$ and L is generated by K and the roots of f in L. First, however, we shall prove that, for every prime polynomial f over K, there is (to within isomorphism) a unique field that contains K and a root of f and is generated by K and that root.

Definition. A field L is an **extension field** of a field K if K is a subfield of L.

If C is a subset of an extension field L of K, the smallest subfield of L containing $K \cup C$ is called the **extension field of K generated by** C and is denoted by $K(C)$. If $C = \{c_1, \ldots, c_n\}$, the extension field of K generated by C is also called the extension field of K generated by $c_1, \ldots, c_n$ and is denoted by $K(c_1, \ldots, c_n)$. If c is an element of an extension field L of K, the subalgebra of the K-algebra L generated by 1 and c is clearly the smallest subdomain of L containing $K \cup \{c\}$; thus, by Theorem 35.1, $K[c]$ is the smallest *subdomain* of L containing $K \cup \{c\}$, whereas $K(c)$ is the smallest *subfield* of L containing $K \cup \{c\}$.

Theorem 37.2. If B and C are subsets of an extension field L of a field K, then $K(B \cup C) = [K(B)](C)$.

Proof. The field $K(B \cup C)$ contains the field $K(B)$ and the set C, so $[K(B)](C) \subseteq K(B \cup C)$. But $[K(B)](C)$ is a field containing K and $B \cup C$, so $K(B \cup C) \subseteq [K(B)](C)$.

Theorem 37.3. Let σ and τ be monomorphisms from a field L into a field F. The set

$$H = \{x \in L : \sigma(x) = \tau(x)\}$$

is a subfield of L. If S is a set of generators for the field L and if $\sigma(x) = \tau(x)$ for all $x \in S$, then $\sigma = \tau$.

The proof is easy.

Corollary. Let L be an extension field of the field K, and let C be a subset of L such that $L = K(C)$. If σ and τ are monomorphisms from L into a field F such that $\sigma(x) = \tau(x)$ for all $x \in K \cup C$, then $\sigma = \tau$.

Definition. A field L is a **finite-dimensional extension field** of K, or simply a **finite extension** of K, if L is an extension field of K and if the K-vector space L is finite-dimensional.

If L is a finite extension of a field K, the dimension of the K-vector space L is often denoted by $[L : K]$ and is called the **degree** of L over K.

Theorem 37.4. Let L be a finite extension of a field K, and let E be a finite-dimensional L-vector space. If $(\alpha_i)_{1 \le i \le n}$ is an ordered basis of the K-vector space L and if $(b_j)_{1 \le j \le m}$ is an ordered basis of the L-vector space E, then

$$B = \{\alpha_i b_j : i \in [1, n] \text{ and } j \in [1, m]\}$$

is a basis of the K-vector space E obtained by restricting scalar multiplication.

Proof. The set B is linearly independent, for if

$$\sum \lambda_{ij} \alpha_i b_j = 0$$

where $\lambda_{ij} \in K$ for all $(i, j) \in [1, n] \times [1, m]$, then

$$\sum_{j=1}^{m} \left(\sum_{i=1}^{n} \lambda_{ij} \alpha_i \right) b_j = 0,$$

whence

$$\sum_{i=1}^{n} \lambda_{ij} \alpha_i = 0$$

for each $j \in [1, m]$, as $(b_j)_{1 \le j \le m}$ is an ordered basis of the L-vector space E, and therefore, $\lambda_{ij} = 0$ for all $i \in [1, n], j \in [1, m]$, as $(\alpha_i)_{1 \le i \le n}$ is an ordered basis of the K-vector space L. Also, B is a set of generators for the K-vector space E, for if $x \in E$, there exist $\beta_1, \ldots, \beta_m$ in L such that

$$x = \sum_{j=1}^{m} \beta_j b_j,$$

and, for each β_j, there exist $\lambda_{1j}, \lambda_{2j}, \ldots, \lambda_{nj}$ in K such that

$$\beta_j = \sum_{i=1}^{n} \lambda_{ij} \alpha_i,$$

whence

$$x = \sum_{j=1}^{m} \sum_{i=1}^{n} \lambda_{ij} \alpha_i b_j.$$

Corollary. If K and L are subfields of a field E and if $K \subseteq L$, then E is a finite extension of K if and only if E is a finite extension of L and L is a finite extension of K, in which case

$$[E : K] = [E : L][L : K].$$

We recall that a *prime* polynomial over a field K is a monic irreducible polynomial over K.

Definition. If f is a prime polynomial over a field K and if L is an extension field of K containing a root c of f such that $L = K(c)$, then L is called a **stem field** of f over K, or a field **obtained by adjoining a root** of f to K.

For example, the field C of complex numbers is a stem field of $X^2 + 1$ over R, for $C = R(i)$ and i is a root of $X^2 + 1$.

If L is an extension field of a field K, we shall say that an element c of L is **algebraic** or **transcendental** over K according as c is an algebraic or a transcendental element of the K-algebra L. Thus, if $c \in L$, then c is algebraic over K if and only if there is a nonzero polynomial $f \in K[X]$ such that $f(c) = 0$. If c is algebraic over K, the **degree** of c over K is the degree of c as an element of the K-algebra L, and the **minimal polynomial** of c over K is the minimal polynomial of c as an element of the K-algebra L. By Theorem 35.5, the minimal polynomial of an element c algebraic over K is a prime polynomial.

Theorem 37.5. If f is a prime polynomial of degree n over a field K and if L is a field obtained by adjoining a root c of f to K, then

    1° the element c is algebraic over K and f is the minimal polynomial of c over K;

    2° $L = K[c]$;

    3° the substitution homomorphism S_c is an epimorphism from $K[X]$ onto L with kernel (f);

    4° $(1, c, c^2, \ldots, c^{n-1})$ is an ordered basis of the K-vector space L, and, in particular, the degree of c over K is n.

Proof. Since $f(c) = 0$, c is algebraic over K. The minimal polynomial g of c over K then divides f as $(f) \subseteq (g)$, but since f is a prime polynomial and since g is a nonconstant monic polynomial, we conclude that $g = f$. The remaining assertions follow from Theorems 35.3 and 35.4.

By 3° of Theorem 37.5, a stem field over K of a prime polynomial $f \in K[X]$ is isomorphic to $K[X]/(f)$; this suggests that to construct an extension field of K containing a root of a prime polynomial f over K, we should construct a field isomorphic to $K[X]/(f)$.

Theorem 37.6. (Kronecker) If f is a prime polynomial over a field K, then there is a stem field of f over K.

Proof. By Theorems 22.7 and 19.6, the ring $K[X]/(f)$ is a field. Let φ be the function from K into $K[X]/(f)$ defined by

Clearly, φ is a homomorphism, and because the zero polynomial is the only constant polynomial belonging to (f), φ is not the zero homomorphism and hence is a monomorphism by Theorem 21.5. By the corollary of Theorem 17.3, therefore, there is an algebraic structure $(L, +, \cdot)$ containing the field K algebraically and an isomorphism Φ from L onto $K[X]/(f)$ extending φ. Since L is isomorphic to the field $K[X]/(f)$, L is itself a field. Let

$$c = \Phi^{\leftarrow}(X + (f)).$$

For every polynomial $g = \sum_{k=0}^{n} \beta_k X^k$ over K,

$$\Phi(g(c)) = \sum_{k=0}^{n} \varphi(\beta_k)\Phi(c)^k = \sum_{k=0}^{n} [\beta_k X^0 + (f)][X + (f)]^k$$

$$= \sum_{k=0}^{n} \beta_k X^k + (f) = g + (f).$$

In particular,

$$\Phi(f(c)) = f + (f) = (f),$$

the zero element of $K[X]/(f)$, so

$$f(c) = 0$$

as Φ is injective. Also, for every $b \in L$, there exists $g \in K[X]$ such that $\Phi(b) = g + (f)$, whence $b = g(c)$ as $g + (f) = \Phi(g(c))$. Therefore, c is a root of f and $L = K[c]$, so L is a stem field of f over K.

We shall next show that a stem field of a prime polynomial is unique to within isomorphism. Actually, we shall prove a somewhat more general result:

Theorem 37.7. Let φ be an isomorphism from a field K onto a field $\bar{K}$, and let $\bar{\varphi}: g \mapsto \bar{g}$ be the induced isomorphism from the ring $K[X]$ onto $\bar{K}[Y]$. If f is a prime polynomial over K and if L and $\bar{L}$ are fields obtained by adjoining roots c and $\bar{c}$ respectively of f and $\bar{f}$ to K and $\bar{K}$, then

$$\sigma: g(c) \mapsto \bar{g}(\bar{c})$$

is a well-defined isomorphism from L onto $\bar{L}$ and is, furthermore, the only isomorphism that extends φ and satisfies $\sigma(c) = \bar{c}$.

Proof. Since $\bar{\varphi}$ is an isomorphism from $K[X]$ onto $\bar{K}[Y]$, the image $\bar{f}$ of f under $\bar{\varphi}$ is clearly a prime polynomial over $\bar{K}$. To show that σ is well-defined and injective, we need to show that $g(c) = h(c)$ if and only if $\bar{g}(\bar{c}) = \bar{h}(\bar{c})$ for all g, $h \in K[X]$. But by 3° of Theorem 37.5, $g(c) = h(c)$ if and only if f divides $g - h$ in $K[X]$, and $\bar{g}(\bar{c}) = \bar{h}(\bar{c})$ if and only if $\bar{f}$ divides $\bar{g} - \bar{h}$ in $\bar{K}[Y]$. As $\bar{\varphi}$ is as isomorphism, f divides $g - h$ in $K[X]$ if and only if $\bar{f}$ divides $\bar{g} - \bar{h}$ in $\bar{K}[Y]$. Therefore, σ is well-defined and injective. As $L = K[c]$ and as $\bar{L} = \bar{K}[\bar{c}]$ by Theorem 37.5, the domain of σ is L and its range is $\bar{L}$. Since

$$\sigma(g(c) + h(c)) = \sigma((g + h)(c)) = \bar{\varphi}(g + h)(\bar{c})$$
$$= (\bar{g} + \bar{h})(\bar{c}) = \bar{g}(\bar{c}) + \bar{h}(\bar{c})$$

and similarly,

$$\sigma(g(c)h(c)) = \bar{g}(\bar{c})\bar{h}(\bar{c}),$$

σ is an isomorphism from L onto $\bar{L}$. Furthermore,

$$\sigma(c) = \sigma(X(c)) = \bar{X}(\bar{c}) = Y(\bar{c}) = \bar{c},$$

and for all $\alpha \in K$,

$$\sigma(\alpha) = \sigma((\alpha X^0)(c)) = \bar{\varphi}(\alpha X^0)(\bar{c})$$
$$= (\varphi(\alpha) Y^0)(\bar{c}) = \varphi(\alpha).$$

If τ is an isomorphism from L onto $\bar{L}$ extending φ and satisfying $\tau(c) = \bar{c}$, then $\tau = \sigma$ by the corollary of Theorem 37.3.

Definition. Let L and L' be extension fields of a field K. An isomorphism (monomorphism) σ from L onto (into) L' is a *K*-**isomorphism** (*K*-**monomorphism**) if $\sigma(x) = x$ for all $x \in K$. A *K*-**automorphism** of L is a *K*-isomorphism from L onto L.

Thus, a *K*-isomorphism (*K*-monomorphism) from L onto (into) L' is simply an isomorphism (monomorphism) from the *K*-algebra L onto (into) the *K*-algebra L', and a *K*-automorphism of L is an automorphism of the *K*-algebra L. By Theorem 27.1, *the set of all K-automorphisms of an extension field L of K is a subgroup of the group of all permutations of L.* From Theorem 37.7, we obtain the following corollary:

Corollary. If f is a prime polynomial over a field K and if L and $\bar{L}$ are fields obtained by adjoining roots c and $\bar{c}$ respectively of f to K, then there is one and only one *K*-isomorphism σ from L onto $\bar{L}$ satisfying $\sigma(c) = \bar{c}$.

Example 37.1. Let f be the polynomial $X^3 + X^2 + 1$ over Z_2. Then f is a cubic polynomial having no roots in Z_2, so f is a prime polynomial over Z_2. Let L be a field obtained by adjoining a root c of f to Z_2. Then by Theorem 37.5, $(1, c, c^2)$ is an ordered basis of the Z_2-vector space L, which has eight elements as it is three-dimensional over a field of two elements. Since the prime subfield of L is Z_2, the characteristic of L is 2. Since $c^3 + c^2 + 1 = 0$, we may express c^3 and c^4 as linear combinations of 1, c, and c^2, and thus, we obtain the following multiplication table for elements of that basis, where each entry is expressed as a linear combination of 1, c, and c^2:

	1	c	c^2
1	1	c	c^2
c	c	c^2	$1 + c^2$
c^2	c^2	$1 + c^2$	$1 + c + c^2$

In general, if A is a finite-dimensional K-algebra and if $(a_k)_{1 \le k \le n}$ is an ordered basis of A, multiplication on A is completely determined by the products of pairs of basis elements; indeed, if it is known that $a_i a_j = \sum_{k=1}^{n} \beta_{ijk} a_k$ for all $i, j \in [1, n]$, then the product xy of any elements $x = \sum_{i=1}^{n} \lambda_i a_i$ and $y = \sum_{j=1}^{n} \mu_j a_j$ may be calculated by

$$xy = \sum_{i=1}^{n} \sum_{j=1}^{n} \lambda_i \mu_j a_i a_j = \sum_{i=1}^{n} \sum_{j=1}^{n} \lambda_i \mu_j \left(\sum_{k=1}^{n} \beta_{ijk} a_k \right)$$
$$= \sum_{k=1}^{n} \left(\sum_{i=1}^{n} \sum_{j=1}^{n} \lambda_i \mu_j \beta_{ijk} \right) a_k.$$

For this reason, a multiplication table expressing each product $a_i a_j$ as a linear combination of $(a_k)_{1 \le k \le n}$ gives a complete description of the composition. In the case where $f = X^n + \alpha_{n-1} X^{n-1} + \ldots + \alpha_1 X + \alpha_0$ is a prime polynomial over a field K and L is a field obtained by adjoining a root c of f to K, we may select $1, c, c^2, \ldots, c^{n-1}$ as ordered basis of the K-algebra L, and to complete the table we need only express the powers c^m of c where $m \in [n, 2n - 2]$ as linear combinations of $1, c, c^2, \ldots, c^{n-1}$, as in Example 37.1. But this is easily done by repeated use of the equality

A stem field L of a prime polynomial over a field K may contain more than one root of the polynomial. In Example 37.1, c^2 and c^4 are also easily seen to be roots of f. As f is a cubic polynomial, therefore, by Theorem 35.8 we have in $L[X]$ the factorization

$$f = (X - c)(X - c^2)(X - c^4)$$

of f into linear polynomials over L. By the corollary of Theorem 37.7, there exist Z_2-automorphisms σ_1 and σ_2 of L such that $\sigma_1(c) = c^2$ and $\sigma_2(c) = c^4$. But if σ is any Z_2-automorphism of L, $\sigma(c)$ must be a root of f, since clearly,

$$f(\sigma(c)) = \sigma(f(c)) = \sigma(0) = 0.$$

Thus, as Z_2-automorphisms of L having the same value at c are identical by Theorem 37.3, the identity automorphism 1_L, σ_1, and σ_2 are the only Z_2-automorphisms of L. For the same reason, since we have by direct calculation

$$\sigma_1{}^2(c) = \sigma_1(c^2) = c^4 = \sigma_2(c),$$

$$\sigma_1{}^3(c) = \sigma_1(c^4) = c^8 = c,$$

we conclude that $\sigma_1{}^2 = \sigma_2$ and that $\sigma_1{}^3 = 1_L$. Thus, the group of Z_2-automorphisms of L is $\{1_L, \sigma_1, \sigma_1{}^2\}$, a cyclic group of order 3.

In our preceding discussion, c was an arbitrarily chosen root of f, and that discussion pertains equally well, therefore, if some other root of f in L is chosen instead. Thus, if $c_1 = c^4$, then as before $c_1{}^2$ and $c_1{}^4$ are the other roots of f, and therefore, it is no surprise to learn by direct calculation that $c_1{}^2 = c$ and $c_1{}^4 = c^2$.

Definition. Let f be a nonconstant polynomial over a field K, and let L be an extension field of K. The polynomial f **splits over** L if f is a product of linear polynomials in $L[X]$. The extension field L of K is a **splitting field of** f **over** K if f splits over L and if L is the extension field of K generated by the roots of f in L.

Thus, if f is a nonconstant polynomial over K and if L is an extension field of K, then f splits over L if and only if in $L[X]$ we have a factorization

(1) $$f = \beta(X - c_1)(X - c_2) \ldots (X - c_n)$$

of f into linear polynomials over L. If so, the field $K(c_1, \ldots, c_n)$ is a splitting field of f over K (the sequence $(c_k)_{1 \leq k \leq n}$ need not be a sequence of *distinct* elements of L, of course). Indeed, $c_1, \ldots, c_n$ are all the roots of f in L and *a fortiori* all the roots of f in $K(c_1, \ldots, c_n)$, for if $c \in L$ is a root of f, then

$$0 = f(c) = \beta(c - c_1)(c - c_2) \ldots (c - c_n),$$

whence $c = c_k$ for some $k \in [1, n]$ (the assertion follows also from Corollary 35.8.1). Consequently, every extension field of K over which f splits contains a splitting field of f over K; in fact, this splitting field is the smallest subfield containing K over which f splits. Conversely, if L is a splitting field of f over K and if $c_1, \ldots, c_n$ are elements of L such that (1) holds, then as we have just seen, $c_1, \ldots, c_n$ are all the roots of f in L, so $L = K(c_1, \ldots, c_n)$.

If E is a splitting field of f over K and if L is a subfield of E containing K, then E is also a splitting field of f over L, for if $c_1, \ldots, c_n$ are the roots of f in E, then

$$E = K(c_1, \ldots, c_n) \subseteq L(c_1, \ldots, c_n) \subseteq E,$$

so $E = L(c_1, \ldots, c_n)$.

If L is an extension of a field K and if $c_1, \ldots, c_m$ are all the roots in L of a nonconstant polynomial f over K, then a splitting field E of f over $K(c_1, \ldots, c_m)$ is also a splitting field of f over K. Indeed, if $c_1, \ldots, c_m, c_{m+1}, \ldots, c_n$ are all the roots of f in E, then

$$E = [K(c_1, \ldots, c_m)](c_{m+1}, \ldots, c_n) = K(c_1, \ldots, c_m, c_{m+1}, \ldots, c_n)$$

by Theorem 37.2.

Any stem field of a prime quadratic polynomial f over a field K is also a splitting field of f over K, for if c is a root of f in a stem field L of f over K, then $f = (X - c)h$ where $h \in L[X]$, and since f is quadratic, h is linear. In particular, C is a splitting field of $X^2 + 1$ over R, and indeed, we have the factorization

$$X^2 + 1 = (X - i)(X + i)$$

in $C[X]$. In Example 37.1 we saw that a stem field of $X^3 + X^2 + 1$ over Z_2 is actually a splitting field of that polynomial.

Theorem 37.8. For every nonconstant polynomial f over a field there is a splitting field of f over that field.

Proof. Let S be the set of all $n \in N^*$ such that if f is any polynomial of degree n over a field, there is a splitting field of f over that field. If f is a linear polynomial over a field K, then K itself is a splitting field of f over K, so $1 \in S$. Suppose that $n \in S$, and let f be a polynomial over a field K of degree $n + 1$. Let g be a prime factor of f in $K[X]$. By Theorem 37.6, there is a field

K_1 obtained by adjoining a root c_0 of g to K. Consequently, c_0 is a root of f, so there exists a polynomial $h \in K_1[X]$ such that $f = (X - c_0)h$ by Theorem 35.7. Then deg $h = n$, so there is a splitting field L of h over K_1. The leading coefficient β of h is clearly the leading coefficient of f, so $\beta \in K$. Since L is a splitting field of h over K_1, there exists a sequence $c_1, \ldots, c_n$ of elements of L such that

$$h = \beta(X - c_1) \ldots (X - c_n).$$

Then

$$f = \beta(X - c_0)(X - c_1) \ldots (X - c_n)$$

and

$$L = K_1(c_1, \ldots, c_n) = [K(c_0)](c_1, \ldots, c_n) = K(c_0, c_1, \ldots, c_n)$$

by Theorem 37.2, so L is a splitting field of f over K. Therefore, $n + 1 \in S$, so by induction the proof is complete.

Finally we shall show that a splitting field of a nonconstant polynomial is unique to within isomorphism. Actually, we shall prove a somewhat more general result:

Theorem 37.9. Let φ be an isomorphism from a field K onto a field $\bar{K}$, and for each $g \in K[X]$, let $\bar{g} \in \bar{K}[Y]$ be the image of g under the induced isomorphism $\bar{\varphi}$ from the ring $K[X]$ onto the ring $\bar{K}[Y]$. Let L be a splitting field of a nonconstant polynomial f over K, and let $\bar{L}$ be a splitting field of the corresponding polynomial $\bar{f}$ over $\bar{K}$.

1° There is an isomorphism from L onto $\bar{L}$ extending φ.

2° If σ is any isomorphism from L onto $\bar{L}$ extending φ and if $c \in L$, then c is a root of f if and only if $\sigma(c)$ is a root of $\bar{f}$.

3° If each root of $\bar{f}$ in $\bar{L}$ is a simple root, the number of isomorphisms from L onto $\bar{L}$ extending φ is $[L : K]$.

Proof. Let $f = \sum_{k=0}^{n} \alpha_k X^k$. If F is any subfield of L containing K, for any monomorphism σ from F into $\bar{L}$ extending φ we shall denote by $\bar{\sigma}$ the induced isomorphism from the ring $F[X]$ onto the ring $\sigma_*(F)[Y]$. We first make two observations about such an extension $\bar{\sigma}$: (a) $\bar{\sigma}(f) = \bar{f}$, for

$$\bar{\sigma}(f) = \sum_{k=0}^{n} \sigma(\alpha_k) Y^k = \sum_{k=0}^{n} \varphi(\alpha_k) Y^k = \bar{f}.$$

(b) For every $c \in F$ and every polynomial g over F, c is a root of g if and only if $\sigma(c)$ is a root of $\bar{\sigma}(g)$. Indeed, if $g = \sum_{k=0}^{m} \beta_k X^k$, then

$$[\bar{\sigma}(g)](\sigma(c)) = \sum_{k=0}^{m} \sigma(\beta_k)\sigma(c)^k = \sigma\left(\sum_{k=0}^{m} \beta_k c^k\right)$$
$$= \sigma(g(c)),$$

so as σ is injective, $\sigma(c)$ is a root of $\bar{\sigma}(g)$ if and only if c is a root of g.

Let

$$f = \beta(X - c_1) \ldots (X - c_n)$$

be the factorization of f over L, where β is the leading coefficient α_n of f. Let $F_0 = K$, and for each $i \in [1, n]$ let $F_i = K(c_1, \ldots, c_i)$. We shall show that if τ is a monomorphism from F_i into $\bar{L}$ extending φ, where $i < n$, and if g is the minimal polynomial of c_{i+1} over F_i, then $\bar{\tau}(g)$ divides $\bar{f}$ in $\bar{L}[Y]$, $\bar{\tau}(g)$ splits over $\bar{L}$, and there are exactly r monomorphisms from F_{i+1} into $\bar{L}$ extending τ, where r is the number of roots of $\bar{\tau}(g)$ in $\bar{L}$. Indeed, g divides f in $F[X]$ since $f(c_{i+1}) = 0$, so $\bar{\tau}(g)$ divides $\bar{\tau}(f) = \bar{f}$ in $\tau_*(F)[Y]$ and a fortiori in $\bar{L}[Y]$; hence as $\bar{f}$ is the product of linear polynomials in $\bar{L}[Y]$, so also is $\bar{\tau}(g)$. Let $d_1, \ldots, d_r$ be the roots of $\bar{\tau}(g)$ in $\bar{L}$. As g is the minimal polynomial of c_{i+1} over F, by Theorem 37.7 for each $j \in [l, r]$ there exists a monomorphism τ_j from $F_i(c_{i+1}) = F_{i+1}$ into $\bar{L}$ extending τ such that $\tau_j(c_{i+1}) = d_j$. If σ is any monomorphism from F_{i+1} into $\bar{L}$ extending τ, then $\sigma(c_{i+1})$ is a root of $\bar{\sigma}(g) = \bar{\tau}(g)$ (by (b)) and hence is d_j for some $j \in [l, r]$, whence $\sigma = \tau_j$ by the Corollary of Theorem 37.3. Therefore there are exactly r monomorphisms from F_{i+1} into $\bar{L}$ extending τ.

Let S be the set of all $i \in [0, n]$ such that there is a monomorphism from F_i into $\bar{L}$ extending φ. Clearly $0 \in S$. If $i \in S$ and $i < n$, then $i + 1 \in S$ by the preceding, so by induction $S = [0, n]$.

In particular, as $L = F_n$, there is a monomorphism σ from L into $\bar{L}$ extending φ. By (a),

$$\bar{f} = \bar{\sigma}(f) = \sigma(\beta)\bar{\sigma}(X - c_1) \ldots \bar{\sigma}(X - c_n)$$
$$= \varphi(\beta)(Y - \sigma(c_1)) \ldots (Y - \sigma(c_n)).$$

Therefore $\bar{L} = \bar{K}(\sigma(c_1), \ldots, \sigma(c_n))$ as $\bar{L}$ is a splitting field of $\bar{f}$ over $\bar{K}$, so the range $\sigma_*(L)$ of σ is a subfield of $\bar{L}$ containing a set of generators for $\bar{L}$, and hence σ is surjective. Thus σ is an isomorphism from L onto $\bar{L}$ extending φ. Therefore 1° holds, and 2° follows from (a) and (b) applied to σ.

Suppose finally that every root of $\bar{f}$ in $\bar{L}$ is simple, and let S' be the set of all $i \in [0, n]$ such that the number of monomorphisms from F_i into $\bar{L}$ extending φ is $[F_i : K]$. Clearly $0 \in S'$. Assume that $i \in S$, where $i < n$.

Then there are $[F_i: K]$ monomorphisms from F_i into $\bar{L}$ extending φ; let τ be any one of them, and let g be the minimal polynomial of c_{i+1} over F_i. By the preceding, $\bar{\tau}(g)$ divides $\bar{f}$ in $\bar{L}[Y]$, so each root of $\bar{\tau}(g)$ in $\bar{L}$ is simple; as $\bar{\tau}(g)$ splits over $\bar{L}$, therefore, the number of roots of $\bar{\tau}(g)$ in $\bar{L} = \deg \bar{\tau}(g) = \deg g = [F_i(c_{i+1}): F_i]$ by Theorem 37.5. Thus by what we proved earlier, there are $[F_i(c_{i+1}): F_i]$ monomorphisms from $F_i(c_{i+1})$ into $\bar{L}$ extending τ. Hence as $F_i(c_{i+1}) = F_{i+1}$, there are $[F_{i+1}: F_i][F_i: K] = [F_{i+1}: K]$ monomorphisms from F_{i+1} into $\bar{L}$ extending φ. Therefore $i + 1 \in S'$. By induction, $n \in S'$. As we saw above, however, a monomorphism from L into $\bar{L}$ extending φ is actually an isomorphism. Thus there are $[F_n: K] = [L: K]$ isomorphisms from L onto $\bar{L}$ extending φ.

Corollary. If f is a nonconstant polynomial over a field K, then any two splitting fields of f over K are K-isomorphic.

Example 37.2. Let $f = (X^2 - 2)(X^2 - 3)$, a product of two prime polynomials over Q, and let L be the splitting field of f contained in C. Then $\sqrt{2}$ and $\sqrt{3}$ belong to L, so $L = Q(\sqrt{2}, \sqrt{3})$ as we have the factorization

$$f = (X - \sqrt{2})(X + \sqrt{2})(X - \sqrt{3})(X + \sqrt{3})$$

in $Q(\sqrt{2}, \sqrt{3})$. In $Q(\sqrt{2})$ we have the factorization

$$f = (X - \sqrt{2})(X + \sqrt{2})(X^2 - 3).$$

To determine whether $L = Q(\sqrt{2})$, or equivalently whether $X^2 - 3$ is reducible over $Q(\sqrt{2})$, we need to determine whether $X^2 - 3$ has a root in $Q(\sqrt{2})$. If $\alpha + \beta\sqrt{2}$ were a root of $X^2 - 3$ where $\alpha, \beta \in Q$, then by substitution we would have

$$(\alpha^2 + 2\beta^2 - 3) + 2\alpha\beta\sqrt{2} = 0,$$

whence

$$\alpha^2 + 2\beta^2 - 3 = 0,$$
$$2\alpha\beta = 0,$$

since $(1, \sqrt{2})$ is an ordered basis of the Q-vector space $Q(\sqrt{2})$; consequently, either $\beta = 0$ or $\alpha = 0$, and therefore, either $\alpha^2 - 3 = 0$ or $2\beta^2 - 3 = 0$, neither of which is possible as $X^2 - 3$ and $2X^2 - 3$ are irreducible over Q by Eisenstein's Criterion. Therefore, $X^2 - 3$ is irreducible over $Q(\sqrt{2})$, so L is obtained by adjoining a root to $Q(\sqrt{2})$ of the prime polynomial $X^2 - 3$. Hence, by Theorem 37.4,

$$[L: Q] = [L: Q(\sqrt{2})][Q(\sqrt{2}): Q] = 2 \cdot 2 = 4,$$

and a basis of the Q-vector space L is $\{1, \sqrt{2}, \sqrt{3}, \sqrt{6}\}$.

Thus, $Q(\sqrt{2}, \sqrt{3})$ is the splitting field of a reducible polynomial over Q. But $Q(\sqrt{2}, \sqrt{3})$ is also a stem field of a prime polynomial over Q. Indeed, $\sqrt{2} + \sqrt{3}$ belongs to $Q(\sqrt{2}, \sqrt{3})$, and it is easy to verify that $X^4 - 10X^2 + 1$ is a prime polynomial over Q of which $\sqrt{2} + \sqrt{3}$ is a root. Consequently, $Q(\sqrt{2} + \sqrt{3})$ is a four-dimensional vector space over Q by Theorem 37.5. Since $Q(\sqrt{2} + \sqrt{3})$ is contained in $Q(\sqrt{2}, \sqrt{3})$, which is also four-dimensional over Q, we conclude that $Q(\sqrt{2} + \sqrt{3}) = Q(\sqrt{2}, \sqrt{3})$.

Example 37.3. Let K be a field such that the polynomial $f = X^3 - 2$ is irreducible over K, and let L be a splitting field of f over K. Then L contains a root c of f, and dividing $X - c$ into f, we obtain the factorization

$$f = (X - c)(X^2 + cX + c^2)$$

in $K(c)$. Clearly, d is a root of the polynomial $g = X^2 + cX + c^2$ if and only if $c^{-1}d$ is a root of the polynomial $h = X^2 + X + 1$. Consequently, there are roots of h in L, and if e is any root of h, then ec is a root of g, so $L = K(c, ec) = K(c, e)$ (for an extension field of $K(c)$ contains e if and only if it contains ec). Let us suppose that $L = K(c)$, or equivalently, that h is reducible over $K(c)$. Then h has a root e in $K(c)$. If h were irreducible over K, then h would be the minimal polynomial of e over K; and hence, 2 would be a divisor of 3 since

$$3 = \deg f = [K(c): K] = [K(c): K(e)][K(e): K]$$
$$= 2[K(c): K(e)]$$

by Theorem 37.5 and the corollary of Theorem 37.4, which is impossible. Thus, if h is reducible over $K(c)$, then h is also reducible over K. If h is irreducible over K, therefore, h is irreducible over $K(c)$, so by Theorem 37.5,

$$[L: K] = [L: K(c)][K(c): K] = 2 \cdot 3 = 6,$$

and $\{1, c, c^2, e, ec, ec^2\}$ is a basis of the K-vector space L where c is any root of f and e any root of h.

If $K = Q$, then $h(1) \neq 0$ and $h(-1) \neq 0$, so h is irreducible over Q. It is customary to define ω by

$$\omega = \frac{1}{2}(-1 + i\sqrt{3}) = \cos\frac{2\pi}{3} + i\sin\frac{2\pi}{3},$$

which is a root of h in C; the other root of h in C is then

$$\omega^2 = \frac{1}{2}(-1 - i\sqrt{3}) = \cos\frac{4\pi}{3} + i\sin\frac{4\pi}{3}.$$

Thus, selecting $\sqrt[3]{2}$ for c, we obtain the basis $\{1, \sqrt[3]{2}, \sqrt[3]{4}, \omega, \omega\sqrt[3]{2}, \omega\sqrt[3]{4}\}$ of the Q-vector space L, and choosing e to be either ω or ω^2, we see from the above that the roots of f are $\sqrt[3]{2}$, $\omega\sqrt[3]{2}$, and $\omega^2\sqrt[3]{2}$.

EXERCISES

37.1. Prove Theorem 37.3.

37.2. For each quadratic or cubic prime polynomial f over Z_2 (Exercise 36.4), determine the number of elements in a stem field of f and construct a multiplication table like that of Example 37.1. If c is a root of f in a stem field, express $(1 + c)^{-1}$ as a linear combination of powers of c.

37.3. Let f be the prime polynomial $X^5 + X^2 + 1$ over Z_2. Determine the number of elements in a stem field of f and construct a multiplication table like that of Example 37.1. If c is a root of f in a stem field, express $(1 + c + c^4)^{-1}$ as a linear combination of powers of c.

37.4. Let c be a root in a stem field of the prime polynomial $f = X^4 + 2X^3 + 4X^2 - 10$ over Q. Construct a multiplication table like that of Example 37.1, and express c^{-1} as a linear combination of powers of c.

37.5. If f is a prime polynomial of degree n over a finite field of q elements, how many elements does a stem field of f over K have?

37.6. If c is an element of an extension field L of a field K such that $\dim_K K(c) = n$ and if c is a root of a polynomial $g \in K[X]$ of degree n, then g is irreducible over K.

*37.7. Let A be a two-dimensional algebra with identity element e over a field K, let $\{e, u\}$ be a basis of the K-vector space A, and let $\alpha, \beta \in K$ be such that $u^2 = \alpha u + \beta e$.

(a) If $f = X^2 - \alpha X - \beta$ is irreducible over K, then A is a stem field of f over K.

(b) If f has two simple roots in K, then the vector space A has a basis $\{e_1, e_2\}$ satisfying

$$e_1{}^2 = e_1, \, e_2{}^2 = e_2, \, e_1 e_2 = e_2 e_1 = 0.$$

(c) If f has one repeated root in K, then there exists $v \in A$ such that $\{e, v\}$ is a basis of the vector space A satisfying $v^2 = 0$.

37.8. Let K be a field whose characteristic is not 2, and let α and β be elements of K such that $X^2 - \alpha$ and $X^2 - \beta$ are irreducible over K. If L is a splitting field of $f = (X^2 - \alpha)(X^2 - \beta)$ over K, then $[L:K]$ is 2 or 4 according as $X^2 - \alpha\beta$ is reducible or irreducible over K.

37.9. Let λ be an element of a field K such that the polynomial $f = X^3 - \lambda$ is irreducible over K. If L is a splitting field of f over K and if c is a root of f in L and e a root of $h = X^2 + X + 1$ in L, then $L = K(c, e)$ and $[L:K]$ is 3 or 6 according as h is reducible or irreducible over K.

*37.10. Let p be a prime, and let $f = X^p - \lambda$ be an irreducible polynomial over a field K. If L is a splitting field of f over K, then $L = K(c, e)$ where c is a root of f and e a root of $h = X^{p-1} + X^{p-2} + \ldots + X + 1$, and $[L:K] = ps$ where s is the degree of any prime divisor of h in $K[X]$ (thus s divides $p - 1$). [Observe that $(X - 1)h = X^p - 1$ and that the roots of $X^p - 1$ in L form a multiplicative group; to show that f is irreducible over $K(e)$, use the corollary of Theorem 37.4.]

37.11. Is a stem field of the polynomial $X^3 - X + 1$ over Q also a splitting field of that polynomial?

*37.12. Let K be a field whose characteristic is not 2, let $f = X^4 + \alpha X^2 + \beta$ be an irreducible polynomial over K, and let L be a splitting field of f over K.
(a) $[L:K]$ is either 4 or 8. [Consider $g = X^2 + \alpha X + \beta$.]
(b) If K is a totally ordered field and if $\beta < 0$, then $[L:K] = 8$.

37.13. By use of the preceding exercises, describe the splitting fields contained in C of the following polynomials over Q:
(a) $(X^2 - 2)(X^2 + 1)$ (b) $(X^2 - 3)(X^2 - 12)$
(c) $(X^2 - 3)(X^2 + 12)$ (d) $X^5 - 17$
(e) $X^{11} + 31$ (f) $X^4 - 5X^2 - 1$
(g) $X^4 + 2X^2 - 8$ (h) $X^3 - X^2 + 2X - 1$.

38. Finite Fields

A finite field is of prime characteristic, for the prime subfield of a field of characteristic zero is isomorphic to the infinite field Q by Theorem 21.8.

Theorem 38.1. If K is a finite field whose characteristic is p, then K contains p^n elements for some strictly positive integer n, and its prime subfield is isomorphic to the field $\mathbf{Z}_p$.

Proof. The prime subfield P of K is isomorphic to the field $\mathbf{Z}_p$, which has p elements, by Theorem 21.8. As the P-vector space K is finite, it has a finite basis by Theorem 28.8. If the dimension of the P-vector space K is n, then K is isomorphic to the P-vector space P^n by Theorem 28.5, and hence, K has p^n elements.

Lemma 38.1. If p is a prime number and if $0 < k < p$, then p divides $\binom{p}{k}$.

Proof. If $1 \leq j \leq \max\{k, p - k\}$, then $j < p$, so p does not divide j. But p divides $p! = k! \, (p - k)! \binom{p}{k}$. By (UFD 3'), therefore, p divides $\binom{p}{k}$.

Theorem 38.2. If K is an integral domain whose characteristic is a prime p, then

$$\sigma: x \mapsto x^p$$

is a monomorphism from K into K.

Proof. Let $a, \ b \in K$. If $0 < k < p$, then p divides $\binom{p}{k}$, and hence, $\binom{p}{k}a^{p-k}b^k = 0$. Consequently,

$$(a + b)^p = \sum_{k=0}^{p}\binom{p}{k}a^{p-k}b^k = \binom{p}{0}a^p + \binom{p}{p}b^p$$
$$= a^p + b^p.$$

As $(ab)^p = a^p b^p$, therefore, σ is a homomorphism. But as K is an integral domain, $a^p = 0$ implies that $a = 0$, so σ is a monomorphism.

If σ_n is the composite of σ with itself n times, an inductive argument shows that

$$\sigma_n(x) = x^{p^n}$$

for all $x \in K$. From Theorem 38.2, therefore, we obtain the following corollaries.

Corollary 38.2.1. If K is an integral domain whose characteristic is a prime p, then for every natural number n,

$$\sigma_n: x \mapsto x^{p^n}$$

is a monomorphism from K into K.

Corollary 38.2.2. If K is a finite field whose characteristic is a prime p, then for every natural number n the function $\sigma_n \colon x \mapsto x^{p^n}$ is an automorphism of K.

Theorem 38.3. If K is a finite field of q elements, then every element of K is a root of $X^q - X$, and

(1) $$X^q - X = \prod_{a \in K} (X - a).$$

In particular, K is a splitting field of $X^q - X$ over the prime subfield P of K.

Proof. The multiplicative group K^* has $q - 1$ elements, so $a^{q-1} = 1$ for all $a \in K^*$ by Theorem 23.8. Hence, $a^q = a$ for all $a \in K$, so the polynomial $\prod_{a \in K} (X - a)$ divides $X^q - X$ in $K[X]$ by Theorem 35.8. As both polynomials are monic polynomials of degree q, therefore, (1) holds.

In particular, $a^{p-1} = 1$ for every nonzero element a of the field Z_p and $a^p = a$ for every element a of Z_p. Expressing these equalities as congruences, we obtain:

Corollary 38.3.1. (Fermat) If p is a prime, then for every integer b not divisible by p,

(2) $$b^{p-1} \equiv 1 \pmod{p},$$

and for every integer b,

$$b^p \equiv b \pmod{p}.$$

Corollary 38.3.2. If K is a finite field of q elements and if E is an extension of K of degree n, then E is a splitting field of $X^{q^n} - X$ over K.

Proof. As the K-vector space E is n-dimensional, E has q^n elements. By Theorem 38.3, E is, therefore, a splitting field of $X^{q^n} - X$ over K.

Corollary 38.3.3. If K is a finite field and if E_1 and E_2 are finite extensions of K having the same degree, then E_1 and E_2 are K-isomorphic.

The assertion follows from Corollary 38.3.2 and the corollary of Theorem 37.9.

Theorem 38.4. Finite fields having the same number of elements are isomorphic.

Proof. Let K_1 and K_2 be fields having q elements. By Theorem 38.1, there exist a prime p and a strictly positive integer n such that $q = p^n$, and there is an isomorphism φ from the prime subfield P_1 of K_1 onto the prime subfield P_2 of K_2. The induced isomorphism $\bar{\varphi}$ from $P_1[X]$ onto $P_2[Y]$ takes the polynomial $X^q - X$ of $P_1[X]$ into the polynomial $Y^q - Y$ of $P_2[Y]$. By Theorem 38.3, K_1 and K_2 are splitting fields of $X^q - X$ over P_1 and $Y^q - Y$ over P_2 respectively. Therefore, K_1 and K_2 are isomorphic by Theorem 37.9.

By Theorem 38.1, q is the number of elements in a finite field only if $q = p^n$ for some prime p and some strictly positive integer n. We next shall show that, for every prime p and for every $n \in N^*$, there is a finite field having p^n elements. Theorem 38.3 suggests that to prove the existence of such a field, we should consider a splitting field of $X^{p^n} - X$ over Z_p.

Theorem 38.5. (E. H. Moore) If p is a prime and if $n \in N^*$, there is a field possessing p^n elements.

Proof. Let $q = p^n$, let $f = X^q - X$, and let K be a splitting field of f over Z_p. Then there is a sequence $(a_k)_{1 \leq k \leq q}$ of elements of K such that

$$X^q - X = \prod_{k=1}^{q} (X - a_k)$$

and every root of f in K is one of $a_1, \ldots, a_q$. Since the function $\sigma : x \mapsto x^q$ is a monomorphism from K into itself by Corollary 38.2.1, the set L of roots of f in K is a subfield of K by Theorem 37.3. Consequently, L contains the prime subfield Z_p of K and hence is K itself by the definition of splitting field. Therefore, every element of K is a term of the sequence $(a_k)_{1 \leq k \leq q}$, so K has at most q elements. But

$$Df = q.X^{q-1} - 1 = -1$$

as the characteristic of K is p, so Df has no roots in K. Therefore, by Theorem 35.11, $(a_k)_{1 \leq k \leq q}$ is a sequence of distinct terms of K. Consequently, K has exactly q elements.

In view of Theorems 38.1, 38.4, and 38.5, it is customary to speak of *the* field of q elements where q is a power of a prime. Finite fields are often called *Galois fields* after the French mathematician Galois, and a finite field of q elements is correspondingly often denoted by $GF(q)$.

The only finite groups isomorphic to subgroups of the multiplicative group K^* of a field K are cyclic; in deriving this result, we shall use an important property of the Euler φ-function.

Theorem 38.6. For each $n \in N^*$,

$$\sum_{d \mid n} \varphi(d) = n,$$

the sum ranging over all positive divisors d of n.

Proof. Let G be a cyclic group of order n. By Theorem 23.9, for each positive divisor d of n, there exists a unique subgroup of G of order d; consequently, G contains exactly $\varphi(d)$ elements of order d. Since every element of G has order d for some divisor d of n by Theorem 23.8, the desired equality follows (Theorem B.1).

Theorem 38.7. If G is a finite subgroup of the multiplicative group K^* of a field K, then G is cyclic.

Proof. Let n be the order of G. If d is a positive divisor of n and if a is an element of G of order d, then every element of G of order d is a generator of the cyclic subgroup $[a]$ of G generated by a; indeed, $1, a, a^2, \ldots, a^{d-1}$ are d distinct roots of $X^d - 1$; if b has order d, then b is a root of $X^d - 1$, so $b \in [a]$ by Corollary 35.8.1, and hence, b is a generator of $[a]$. Thus, the number m_d of elements of G of order d is either zero or $\varphi(d)$. But every element of G has order d for some divisor d of n by Theorem 23.8, so (by Theorem B.1),

$$\sum_{d \mid n} m_d = n.$$

Since $m_d \leq \varphi(d)$ for every divisor d of n, we therefore have $m_d = \varphi(d)$ for every divisor d of n by Theorem 38.6. In particular, $m_n = \varphi(n) \geq 1$, so G is cyclic.

Corollary. If K is a finite field, the multiplicative group K^* is cyclic.

EXERCISES

In these exercises, p always denotes a prime number.

*38.1. If K is a field whose multiplicative group K^* is cyclic, then K is finite. [Consider -1 if the characteristic of K is zero; otherwise, show that K is a stem field of a prime polynomial over the prime subfield of K.]

*38.2. If K is a field and if f is the function from K into K defined by

$$f(x) = \begin{cases} x^{-1} \text{ if } x \neq 0, \\ 0 \text{ if } x = 0, \end{cases}$$

then f is an automorphism of K if and only if K has either 2, 3, or 4 elements.

38.3. If K is a finite field, for every $m \in N^*$, there is a prime polynomial over K of degree m. [Apply the corollary of Theorem 38.7 to an extension field of K, and use Theorem 35.4.]

38.4. An integer m is a **Mersenne number** if $m = 2^n - 1$ for some $n \in N^*$. A **Mersenne prime** is a Mersenne number that is a prime. If $2^n - 1$ is a Mersenne prime, then n is a prime.

*38.5. If K is a finite field, then $H \cup \{0\}$ is a subfield of K for every subgroup H of the multiplicative group K^* if and only if the order of K^* is either 1 or a Mersenne prime. [First show that the characteristic of K is 2; then use Theorem 23.9.]

38.6. Let K be a finite field having p^n elements.
(a) If L is a subfield of K, then there is a divisor r of n such that L has p^r elements. [Consider K as an L-vector space.]
(b) If $r \mid n$, then $X^{p^r} - X$ is a divisor of $X^{p^n} - X$ in $K[X]$. [If $rm = n$, show that $X^{p^n} - X = \sum_{k=0}^{m-1} (X^{p^r} - X)^{p^{rk}}$.]
(c) If $r \mid n$, then $\{x \in K : x^{p^r} = x\}$ is a subfield of K having p^r elements.
(d) For each positive divisor r of n there is one and only one subfield of K having p^r elements, and these are the only subfields of K.

*38.7. Let f be a prime polynomial of degree m over a field K having q elements, and let L be a field obtained by adjoining a root c of f to K.
(a) The sequence $(c^{q^k})_{0 \leq k \leq m-1}$ is a sequence of m distinct roots of f. [Use Exercise 35.7, Corollary 38.2.1, and Theorem 38.7.]
(b) In $L[X]$ we have the factorization

$$f = \prod_{k=0}^{m-1} (X - c^{q^k}),$$

and consequently, L is a splitting field of f over K.
(c) The polynomial f divides $X^{q^m} - X$ in $K[X]$. [Use Theorem 34.8.]
(d) The polynomial f divides $X^{q^s} - X$ in $K[X]$ if and only if $m \mid s$. [Apply Exercise 38.6(d) to a splitting field of $X^{q^s} - X$ over K.]

38.8. Let K be a finite field having q elements, and let $s \in N^*$.
(a) The polynomial $X^{q^s} - X$ is the product of all the prime polynomials over K whose degree divides s. [Use Exercise 38.7(d).]

(b) If $\Psi_q(m)$ denotes the number of prime polynomials over K of degree m, then

$$q^s = \sum_{m \mid s} m\Psi_q(m).$$

*38.9. Let K be a finite field having q elements, let E be an extension field of K of degree n, and let σ be the automorphism $x \mapsto x^q$ of E.

(a) The group G of K-automorphisms of E is cyclic of order n and is generated by σ. [If $u \in G$, consider $f(u(c))$ where c is a generator of E^* and where f is the minimal polynomial of c over K.]

(b) If $m \in N^*$ and if d is the positive greatest common divisor of m and n, then $\{x \in E : \sigma^m(x) = x\}$ is a subfield of E having q^d elements. [Use Exercise 38.6.]

*38.10. Let φ be the Euler φ-function. If p is a prime and if $n \in N^*$, then $n \mid \varphi(p^n - 1)$. [Use Exercise 38.9(a).]

*38.11. Let K be a finite field having q elements whose characteristic is not 2. A nonzero element a of K is a square of K (Exercise 33.13) if and only if $a^{(q-1)/2} = 1$, and a is not a square of K if and only if $a^{(q-1)/2} = -1$. [Use Theorem 38.7 to show that the kernel of the endomorphism $x \mapsto x^{(q-1)/2}$ of the group K^* is a proper subgroup of K^*, and consider also the range of the endomorphism $x \mapsto x^2$ of K^*.]

*38.12. Let K be a finite field having q elements whose characteristic is not 2, and let a_1 and a_2 be nonzero elements of K. The number n of ordered couples (x_1, x_2) of elements of K satisfying

$$a_1x_1^2 + a_2x_2^2 = b$$

is given as follows:

(a) if $b = 0$ and if $-a_1a_2$ is a square of K, then $n = 2q - 1$;
(b) if $b \neq 0$ and if $-a_1a_2$ is a square of K, then $n = 1$;
(c) if $b = 0$ and if $-a_1a_2$ is not a square of K, then $n = 1$;
(d) if $b \neq 0$ and if $-a_1a_2$ is not a square of K, then $n = q + 1$.

CHAPTER VII

THE REAL AND COMPLEX
NUMBER FIELDS*

To the Greek Pythagoreans is attributed the discovery, before the end of the fourth century B.C., that certain geometric magnitudes, such as the hypotenuse of an isosceles right triangle whose legs have unit length, cannot be expressed as rational numbers. The concept of a "real number" remained rather murky from that time until the nineteenth century, when Dedekind and Cantor showed that the real number system could formally be described as a certain kind of ordered field.

One property possessed by the totally ordered field R of real numbers that singles out R among all totally ordered fields is that every nonempty subset of R that is bounded above has a least upper bound. In honor of Dedekind, we shall call a totally ordered field possessing this property a "Dedekind ordered field." In §39 we shall relate this property with other properties that a totally ordered field may possess, and in §40 we shall present Cantor's method of constructing a Dedekind ordered field (Dedekind's method is outlined in the exercises). In §41 we shall obtain certain isomorphism theorems that establish the essential uniqueness of the ordered field of real numbers and enable us to construct in a natural way the logarithmic and exponential functions. Finally, in §42 we shall construct the field C of complex numbers and establish a further important property of R that is crucial to our proof in Chapter VIII of the "fundamental theorem of algebra": Every nonconstant polynomial over C has a root in C.

* Readers wishing to proceed directly to Galois theory may omit this chapter. Note need be made only of the definition of a real-closed, totally ordered field, given in §42, the theorem that the totally ordered field R of real numbers is real-closed (a special case of the Intermediate Value Theorem for continuous functions), and Rolle's Theorem for polynomials over R. These theorems are established in §42, but readers may have already encountered them in texts on advanced calculus or analysis.

369

39. Dedekind and Archimedean Ordered Fields

Definition. Let $(E, \leq)$ be an ordered structure, and let A be a nonempty subset of E. An element c of E is an **upper bound (lower bound)** of A if $x \leq c$ ($x \geq c$) for all $x \in A$. The set A is **bounded above (bounded below)** in E if there is an upper bound (lower bound) of A in E, and A is **bounded** in E if A is both bounded above and bounded below in E. An element c of E is a **supremum** (or **least upper bound**) of A in E if c is an upper bound of A and if $c \leq d$ for all upper bounds d of A in E; c is an **infimum** (or **greatest lower bound**) of A in E if c is a lower bound of A and if $c \geq d$ for all lower bounds d of A in E.

Theorem 39.1. If $(E, \leq)$ is an ordered structure and if A is a nonempty subset of E, then A has at most one supremum and at most one infimum in E.

Proof. If c and c' are suprema of A in E, then $c \leq c'$ as c' is an upper bound of A and as c is a supremum of A, and $c' \leq c$ as c is an upper bound of A and as c' is a supremum of A, so $c = c'$.

Consequently, if a nonempty subset A of an ordered structure admits a supremum (infimum) c, we may use the definite article and call c *the* supremum (infimum) of A. The supremum (infimum) of A is often denoted by sup A (inf A).

Fundamental to any discussion of the ordered field of real numbers is the concept of a Dedekind ordered field or group:

Definition. A **Dedekind ordered group** is a totally ordered group $(G, +, \leq)$ containing more than one element such that every nonempty subset of G that is bounded above admits a supremum. A **Dedekind ordered field** is a totally ordered field $(K, +, \cdot, \leq)$ such that $(K, +, \leq)$ is a Dedekind ordered group.

An element x of an ordered group $(G, +, \leq)$ is **positive** if $x \geq 0$, and x is **strictly positive** if $x > 0$. We recall also that the set of strictly positive elements of an ordered field K is denoted by K_+^*.

Theorem 39.2. Let $(G, +, \leq)$ be a totally ordered group containing more than one element. The following statements are equivalent:

1° $(G, +, \leq)$ is a Dedekind ordered group.
2° Every nonempty subset of G that is bounded below admits an infimum.
3° Every nonempty subset of G that contains a strictly positive element and is bounded above admits a supremum.

Proof. Since $x < y$ if and only if $-x > -y$ by Theorem 10.2, it is easy to verify that a is a supremum (infimum) of a nonempty subset A of G if and only if $-a$ is an infimum (supremum) of $\{-x: x \in A\}$. The equivalence of 1° and 2° readily follows. It remains for us to show that 3° implies 1°. Let A be a nonempty subset of G that has an upper bound b but contains no strictly positive element. If A has just one element, surely A admits a supremum. In the contrary case, there exist elements u and v of A such that $u < v$. Then $-u + b$ is an upper bound of the set $-u + A$, which contains the strictly positive element $-u + v$. By 3°, $-u + A$ admits a supremum c. It is easy to verify that $u + c$ is then the supremum of A.

Definition. An **archimedean ordered group** is a totally ordered group $(G, +, \leq)$ containing more than one element such that for all $a, b \in G$, if $a > 0$ and if $b > 0$, then there exists $n \in N^*$ such that $n.a > b$. An **archimedean ordered field** is a totally ordered field $(K, +, \cdot, \leq)$ such that $(K, +, \leq)$ is an archimedean ordered group.

Theorem 39.3. A Dedekind ordered group is archimedean ordered.

Proof. Let a and b be strictly positive elements of a Dedekind ordered group $(G, +, \leq)$, and let $A = \{n.a: n \in N^*\}$. If $n.a \leq b$ for all $n \in N^*$, then b would be an upper bound of A, so A would admit a supremum c. As $0 < a$, we have $-a < 0$ and hence $c - a < c$, so by the definition of c there would exist $n \in N^*$ such that $c - a < n.a$, whence

$$c = (c - a) + a < (n + 1).a \leq c,$$

a contradiction. Hence, $n.a > b$ for some $n \in N^*$.

The totally ordered field Q of rational numbers is archimedean ordered. Indeed, if $r, s \in Q_+^*$, then there exist $m, n, p, q \in N^*$ such that $r = m/n$ and $s = p/q$, whence $(pn + 1).r > s$.

Theorem 39.4. If K is a totally ordered field, then the characteristic of K is zero, and there is one and only one isomorphism from the ordered field Q of rationals onto the ordered prime subfield Q of K.

Proof. By 5° and 10° of Theorem 17.6, $n.1 > 0$ for all $n \in N^*$. Thus, the characteristic of K cannot be p for any $p > 0$. Hence, the characteristic of K is zero, so by Theorem 21.6, $f: n \mapsto n.1$ is the unique isomorphism from the ring Z onto the smallest subring B of K that contains 1. If $m < p$, then $p - m \in N^*$, so $p.1 - m.1 = (p - m).1 > 0$, whence $f(m) < f(p)$. Therefore, by Theorem 10.4, f is an isomorphism from the ordered ring Z onto the ordered ring B. The prime field Q of K is clearly the quotient field of B in K. The assertion, therefore, follows from Theorem 17.9.

Let φ be the isomorphism from the ordered field $\boldsymbol{Q}$ onto the ordered prime subfield Q of a totally ordered field K. For convenience we shall use a symbol denoting a given rational number r also to denote the corresponding element $\varphi(r)$ of the prime subfield Q. Thus, if m and n are integers and if $n \neq 0$, for example, we shall denote by m/n not only the rational number m/n but also the element $\dfrac{m.1}{n.1} = \dfrac{\varphi(m)}{\varphi(n)}$ of Q.

Definition. A subset S of a totally ordered field K is **dense** in K if for all $a, b \in K$, if $a < b$, then there exists $s \in S$ such that $a < s < b$.

Theorem 39.5. A totally ordered field K is archimedean ordered if and only if its prime subfield Q is dense in K.

Proof. Necessity: Let a and b be elements of K satisfying $a < b$.

Case 1: $0 \leq a < b$. By Theorem 17.6, $(b - a)^{-1} > 0$, so there exists $n \in N^*$ such that $n.1 > (b - a)^{-1}$, whence $1/n < b - a$. As K is archimedean ordered, $\{j \in N: j > n.a\}$ is not empty and hence contains a smallest element k. Then $k/n \in Q$, $k > 0$ as $n.a \geq 0$, and $a < k/n$. Consequently, $k - 1$ is a natural number, and hence, $k - 1 \leq n.a$ by the definition of k. Therefore,

$$a < \frac{k}{n} = \frac{k-1}{n} + \frac{1}{n} \leq a + \frac{1}{n} < a + (b - a) = b.$$

Case 2: $a < 0 < b$. Then 0 is the desired element s of Q satisfying $a < s < b$.

Case 3: $a < b \leq 0$. Then $0 \leq -b < -a$, so, by Case 1, there exists $s \in Q$ such that $-b < s < -a$; consequently, $-s \in Q$ and $a < -s < b$.

Sufficiency: Let $a, b \in K_+^*$. Then $a/b \in K_+^*$, so as Q is dense in K, there exist $m, n \in N^*$ such that $0 < m/n < a/b$, whence

$$b = (bn) \frac{1}{n} \leq (bn) \frac{m}{n} < (bn) \frac{a}{b} = n.a.$$

Definition. Let $(K, +, \cdot, \leq)$ be a totally ordered field. For each $a \in K$, the **absolute value** of a is the element $|a|$ of K_+ defined by

$$|a| = \begin{cases} a \text{ if } a \geq 0, \\ -a \text{ if } a < 0. \end{cases}$$

Theorem 39.6. Let a and b be elements of a totally ordered field K.

$1°$ $|a| = 0$ if and only if $a = 0$.

$2°$ $a \leq |a|$.

$3°$ $|-a| = |a|$.

$4°$ $|ab| = |a|\,|b|$.

$5°$ $|a + b| \leq |a| + |b|$.

$6°$ $|a - b| \geq |a| - |b|$.

$7°$ $|a| \leq b$ if and only if $-b \leq a \leq b$, and $|a| < b$ if and only if $-b < a < b$.

$8°$ If $a \neq 0$, then $|a^{-1}| = |a|^{-1}$.

Proof. Statements $1°$–$3°$ are immediate. Statement $4°$ follows from $8°$ and $9°$ of Theorem 17.6. Statement $5°$ follows from $2°$ and $3°$, for if $a + b \geq 0$, then

$$|a + b| = a + b \leq |a| + |b|$$

by $2°$, and if $a + b < 0$, then

$$|a + b| = -(a + b) = (-a) + (-b) \leq |-a| + |-b| = |a| + |b|$$

by $2°$ and $3°$. Statement $6°$ follows from $5°$, for

$$|a| = |(a - b) + b| \leq |a - b| + |b|$$

by $5°$, whence

$$|a| - |b| \leq |a - b|.$$

From $4°$ and $5°$ we obtain by induction the following corollary:

Corollary. If $(a_k)_{1 \leq k \leq n}$ is a sequence of elements of a totally ordered field K, then

$$\left| \sum_{k=1}^{n} a_k \right| \leq \sum_{k=1}^{n} |a_k|$$

and

$$\left| \prod_{k=1}^{n} a_k \right| = \prod_{k=1}^{n} |a_k|.$$

If a and b are elements of a totally ordered field K, the element $|a - b|$ of K may be thought of as the "distance" between a and b, for if K is the ordered field of real or rational numbers, then $|a - b|$ is the geometric distance between a and b. The following definition of "Cauchy sequence" singles out those sequences in K all of whose terms "sufficiently far out" are as "close" to each other (in terms of the "distance" defined by the absolute value) as desired.

Definition. Let K be a totally ordered field. A sequence $(a_n)_{n \geq k}$ of elements of K is a **Cauchy sequence** in K if for every $e \in K_+^*$ there exists $m \geq k$ such that $|a_n - a_p| < e$ for all $n \geq m$ and all $p \geq m$. A sequence $(a_n)_{n \geq k}$ of elements of K is a **bounded sequence** in K if there exists $c \in K_+^*$ such that $|a_n| \leq c$ for all $n \geq k$.

Theorem 39.7. If $(a_n)_{n \geq k}$ is a Cauchy sequence in a totally ordered field K, then $(a_n)_{n \geq k}$ is a bounded sequence in K.

Proof. By hypothesis, there exists $m \geq k$ such that $|a_n - a_p| < 1$ for all $n \geq m$ and all $p \geq m$. Let

$$c = \max\{|a_k|, \ldots, |a_m|\} + 1.$$

Then $|a_n| \leq c$ if $k \leq n \leq m$, and

$$|a_n| \leq |a_n - a_m| + |a_m| \leq c$$

if $n \geq m$.

Definition. Let $(a_n)_{n \geq k}$ be a sequence of elements of a totally ordered field K, and let $v \in K$. The sequence $(a_n)_{n \geq k}$ **converges** to v in K if, for every $e \in K_+^*$, there exists $m \geq k$ such that $|a_n - v| < e$ for all $n \geq m$. The sequence $(a_n)_{n \geq k}$ is a **convergent sequence** in K if there exists $v \in K$ such that $(a_n)_{n \geq k}$ converges to v in K.

Theorem 39.8. Let K be a totally ordered field. A sequence of elements of K converges in K to at most one element of K. A convergent sequence in K is a Cauchy sequence in K.

Proof. Suppose that a sequence $(a_n)_{n \geq k}$ of elements of K converged both to u and to v in K and that $u < v$. Then there would exist $m_1 \geq k$ such that

$$|a_n - u| < \tfrac{1}{2}(v - u)$$

for all $n \geq m_1$, and there would exist $m_2 \geq k$ such that

$$|a_n - v| < \tfrac{1}{2}(v - u)$$

for all $n \geq m_2$. Let $n = \max\{m_1, m_2\}$. Then

$$v - u = (v - a_n) + (a_n - u) \leq |a_n - v| + |a_n - u|$$
$$< \tfrac{1}{2}(v - u) + \tfrac{1}{2}(v - u) = v - u,$$

a contradiction.

Let $(a_n)_{n \geq k}$ be a sequence of elements of K that converges in K to an element u, and let $e \in K_+^*$. By definition, there exists $m \geq k$ such that

$$|a_n - u| < \frac{e}{2}$$

for all $n \geq m$. Consequently, if $n \geq m$ and if $p \geq m$, then

$$|a_n - a_p| = |(a_n - u) + (u - a_p)|$$

$$\leq |a_n - u| + |a_p - u| < \frac{e}{2} + \frac{e}{2} = e.$$

Therefore, $(a_n)_{n \geq k}$ is a Cauchy sequence.

If $(a_n)_{n \geq k}$ is a sequence converging in K to an element u, then u is called the **limit** of $(a_n)_{n \geq k}$ and is often denoted by $\lim\limits_{n \to \infty} a_n$ (the use of the definite article is justified by the first assertion of Theorem 39.8).

Definition. A totally ordered field K is **complete** if every Cauchy sequence in K is a convergent sequence in K.

A useful property of archimedean ordered fields is the following: If $e > 0$, then there exists $n \in N$ such that $2^{-n} < e$. Indeed, there exists $n \in N^*$ such that $0 < 1/n < e$ by Theorem 39.5, and as easy inductive argument (or Exercise 12.10 and Theorem B.7) shows that $m < 2^m$ for all $m \in N$, whence $0 < 2^{-n} < 1/n < e$. Consequently, in determining whether a sequence in an archimedean ordered field K is a Cauchy sequence or a convergent sequence, it suffices to consider only those strictly positive elements of K of the form 2^{-n} where $n \in N$.

The totally ordered field Q is not complete. We shall show that the sequence $(a_n)_{n \geq 0}$ of rationals defined by

$$a_n = \sum_{k=0}^{n} 2^{-k(k+1)}$$

for all $n \in N$ is a Cauchy sequence that does not converge to any rational number. If $m > n$, then

$$0 < a_m - a_n = \sum_{k=n+1}^{m} 2^{-k(k+1)} = \sum_{j=0}^{m-n-1} 2^{-(j+(n+1))(j+(n+2))}$$

$$\leq \sum_{j=0}^{m-n-1} 2^{-[(n+1)(n+2)+j]} = 2^{-(n+1)(n+2)} \sum_{j=0}^{m-n-1} 2^{-j}$$

$$< 2^{-(n+1)(n+2)} \cdot 2 < 2^{-n(n+3)}.$$

Consequently, $(a_n)_{n \geq 0}$ is a Cauchy sequence. For each $n \in N$, let

$$b_n = \sum_{k=0}^{n} 2^{n(n+1)-k(k+1)}.$$

Then b_n is an integer and

$$a_n = b_n 2^{-n(n+1)}$$

for all $n \in N$. Suppose that $(a_n)_{n \geq 0}$ converged to the rational r. Then for every $n \in N$ and every $e \in Q_+^*$, there would exist $q > n$ such that $|r - a_q| < e$, whence

$$|r - a_n| \leq |r - a_q| + |a_q - a_n| < e + 2^{-n(n+3)};$$

consequently, as $|r - a_n| < e + 2^{-n(n+3)}$ for every $e \in Q^*$ and every $n \in N$, we would have

$$|r - a_n| \leq 2^{-n(n+3)}$$

for all $n \in N$. Let $r = p/q$ where $p \in Z$ and $q \in N^*$. Since

$$\left| \frac{p}{q} - 2^{-n(n+1)}b_n \right| = |r - a_n| \leq 2^{-n(n+3)},$$

we have

$$|2^{n(n+1)}p - qb_n| \leq q \cdot 2^{n(n+1)-n(n+3)} = 4^{-n}q.$$

But as $2^{n(n+1)}p - qb_n$ is an integer and as $4^{-n}q < 1$ for all $n \geq q$, we would, therefore, conclude that $2^{n(n+1)}p - qb_n = 0$ and hence that $r = a_n$ for all $n \geq q$, which is impossible since $a_{q+1} > a_q$.

Theorem 39.9. A totally ordered field K is Dedekind ordered if and only if K is complete and archimedean ordered.

Proof. Necessity: By Theorem 39.3, K is archimedean ordered. Let $(a_n)_{n \geq k}$ be a Cauchy sequence in K. By Theorem 39.7, there exists $z \in K_+^*$ such that $|a_n| \leq z$ and hence $-z \leq a_n \leq z$ for all $n \geq k$. For each $n \geq k$, let $A_n = \{a_m : m \geq n\}$. Then z is an upper bound of A_n, so A_n admits a supremum b_n. Let $B = \{b_n : n \geq k\}$. If $m \geq n$, then $A_m \subseteq A_n$, and hence $b_m \leq b_n$. Since $-z \leq a_n \leq b_n$ for all $n \geq k$, $-z$ is a lower bound of B, and consequently, B admits an infimum c. We shall prove that $(a_n)_{n \geq k}$ converges to c.

Let $e \in K_+^*$. Since $(a_n)_{n \geq k}$ is a Cauchy sequence, there exists $p \geq k$ such that

$$|a_m - a_n| < \frac{e}{3}$$

whenever $m \geq p$ and $n \geq p$. As $c < c + \frac{1}{3}e$ and as $(b_n)_{n \geq k}$ is a decreasing sequence, there exists $q \geq p$ such that

$$0 \leq b_q - c < \frac{e}{3}$$

by the definition of c. By the definition of b_q there exists $r \geq q$ such that

$$0 \leq b_q - a_r < \frac{e}{3}.$$

Hence, if $n \geq r$, then

$$|a_n - c| \leq |a_n - a_r| + |a_r - b_q| + |b_q - c| < \frac{e}{3} + \frac{e}{3} + \frac{e}{3} = e$$

by 5° of Theorem 39.6. Thus, $(a_n)_{n \geq k}$ converges to c.

Sufficiency: By Theorem 39.2, it suffices to show that if B is a nonempty subset of K that is bounded above and contains a strictly positive element, then B admits a supremum. For each $n \in N$, let

$$K_n = \{j \in N: 2^{-n}j \text{ is an upper bound of } B\}.$$

As K is archimedean ordered, K_n is not empty; let k_n be the smallest member of K_n. Then $k_n > 0$ as B contains a strictly positive element. If $n \geq m$, then

(1) $$0 \leq 2^{-m}k_m - 2^{-n}k_n \leq 2^{-m},$$

for $2^{-n}(2^{n-m}k_m) = 2^{-m}k_m$, an upper bound of B, so $2^{n-m}k_m \geq k_n$ and hence $2^{-m}k_m \geq 2^{-n}k_n$ by the definition of k_n; also if $2^{-m}k_m - 2^{-n}k_n$ were greater than 2^{-m}, we would conclude that $2^{-m}(k_m - 1) > 2^{-n}k_n$, an upper bound of B, whence $k_m - 1 \in K_m$ as $k_m - 1 \in N$, a contradiction of the definition of k_m. By (1), the sequence $(2^{-n}k_n)_{n \geq 0}$ is a Cauchy sequence, since K is archimedean ordered, and therefore, $(2^{-n}k_n)_{n \geq 0}$ converges to an element b of K as K is complete. Let

$$S = \{2^{-n}k_n: n \in N\}.$$

We shall first show that b is a lower bound of S. For if b were greater than $2^{-m}k_m$ for some $m \in N$ and if $e = b - 2^{-m}k_m$, then e would be a strictly positive element of K, and adding e to the terms of the inequality (1), we would obtain

$$e \leq b - 2^{-n}k_n$$

for all $n \geq m$, which is impossible as $(2^{-n}k_n)_{n \geq 0}$ converges to b. Moreover, b

is the infimum of S, for if c were a lower bound of S such that $b < c$ and if $e' = c - b$, then e' would be a strictly positive element of K, and adding e' to the terms of the inequality (1), we would obtain

$$e' \leq c - 2^{-n}k_n + 2^{-m}k_m - b \leq 2^{-m}k_m - b$$

for all $m \geq 0$, which again is impossible as $(2^{-m}k_m)_{m \geq 0}$ converges to b. Thus, $b = \inf S$.

Finally, we shall show that b is the desired supremum of B. Every element x of B is a lower bound of S, and hence $x \leq b$ as $b = \inf S$. Thus, b is an upper bound of B. It remains for us to show that if $d < b$, then d is not an upper bound of B. Since K is archimedean ordered, there exists $m \in N$ such that $b - d > 2^{-m}$ and hence

$$d + 2^{-m} < b.$$

By the definition of k_m, there exists $x \in B$ such that $2^{-m}(k_m - 1) < x$, whence

$$2^{-m}k_m - x < 2^{-m}.$$

Therefore,

$$\begin{aligned}
d - x &= (d - 2^{-m}k_m) + (2^{-m}k_m - x) \\
&< d - 2^{-m}k_m + 2^{-m} \\
&< b - 2^{-m}k_m \leq 0,
\end{aligned}$$

so d is not an upper bound of B. Thus, $b = \sup B$, and the proof is complete.

The totally ordered field Q of rationals is thus an archimedean ordered field that is not Dedekind ordered.

EXERCISES

39.1. Make the verifications needed to complete the proof of Theorem 39.2.

39.2. Complete the proof of Theorem 39.6 and its corollary.

39.3. Let Q be the prime subfield of a totally ordered field K. A strictly positive element x of K is **infinitely large** if $x > q$ for all $q \in Q_+^*$, and x is **infinitely small** or an **infinitesimal** if $x < q$ for all $q \in Q_+^*$.
(a) If $x \in K_+^*$, then x is infinitely large if and only if x^{-1} is an infinitesimal, and x is an infinitesimal if and only if x^{-1} is infinitely large.
(b) The totally ordered field K is archimedean ordered if and only if there are no infinitesimals in K.

(c) If $A = \{x \in K: |x|$ is not infinitely large$\}$ and if $\mathfrak{a} = \{y \in K:$ either $y = 0$ or $|y|$ is an infinitesimal$\}$, then A is a subring of K and $\mathfrak{a}$ is a maximal ideal of A.

39.4. Let $\leq$ be the total ordering on $Q[X]$ defined in Exercise 34.7(c). The ordering on a quotient field K of $Q[X]$ induced by $\leq$ converts K into a nonarchimedean totally ordered field. What polynomials are infinitely large elements of K?

39.5. Let $(a_n)_{n \geq k}$ be a sequence of elements of a totally ordered field K, and let $A = \{a_n: n \geq k\}$. If $(a_n)_{n \geq k}$ is an increasing (decreasing) sequence, then $(a_n)_{n \geq k}$ converges to an element of K if and only if A admits a supremum (infimum) in K, in which case $\lim_{n \to \infty} a_n = \sup A$ ($\lim_{n \to \infty} a_n = \inf A$).

39.6. Let K be a totally ordered field.
(a) Every bounded increasing sequence in K is a Cauchy sequence in K if and only if every bounded decreasing sequence in K is a Cauchy sequence in K.
(b) Every bounded increasing sequence in K is a convergent sequence in K if and only if every bounded decreasing sequence in K is a convergent sequence in K.

*39.7. A totally ordered field K is archimedean ordered if and only if every bounded increasing sequence in K is a Cauchy sequence in K.

*39.8. A totally ordered field K is Dedekind ordered if and only if every bounded increasing sequence in K is a convergent sequence in K. [Use Exercise 39.7; examine the proof of Theorem 39.9.]

39.9. (a) Let $S_2 = \{x \in Q_+^*: x^2 < 2\}$. Show that S_2 is bounded above but does not admit a supremum in the totally ordered field Q. [If s is an upper bound of S_2, consider $s - r$ where $r = (s^2 - 2)/2s$.]
(b) Let $m \in N^*$, and let $S_m = \{x \in Q_+^*: x^2 < m\}$. Show that if m is not the square of an integer, then S_m is bounded above but does not admit a supremum in the totally ordered field Q.

The following terminology is used in the remaining exercises. Let $(E, \leq)$ be a totally ordered structure, and let $a, b \in E$. We make the following definitions:

$$[a, b] = \{x \in E: a \leq x \leq b\}, \qquad [a, \to[= \{x \in E: a \leq x\},$$
$$]a, b[= \{x \in E: a < x < b\}, \qquad]a, \to[= \{x \in E: a < x\},$$
$$[a, b[= \{x \in E: a \leq x < b\}, \qquad]\leftarrow, a] = \{x \in E: x \leq a\},$$
$$]a, b] = \{x \in E: a < x \leq b\}, \qquad]\leftarrow, a[= \{x \in E: x < a\}.$$

A nonempty subset J of E is an **interval** if J is either E or one of the eight sets defined above for some $a, b \in E$; J is a **bounded interval** if J is both a bounded set and an interval; J is a **closed interval** if J is either E or an interval of the form $[a, b]$, $[a, \to[$, or $]\leftarrow, a]$; J is an **open interval** if J is either E or an interval of the form $]a, b[$, $]a, \to[$, or $]\leftarrow, a[$. A subset H of E is **convex** if $[x, y] \subseteq H$ for all

$x, y \in H$. A subset C of E is **compact** if for every set $\mathscr{G}$ of open intervals such that $\cup \mathscr{G} \supseteq C$, there is a finite subset $\mathscr{H}$ of $\mathscr{G}$ such that $\cup \mathscr{H} \supseteq C$. A subset G of E is **open** if, for every $x \in G$, there is an open interval that contains x and is contained in G. Finally, a subset D of E is **connected** if, whenever G and H are open subsets of E such that $G \cap H = \emptyset$ and $D \subseteq G \cup H$, either $D \subseteq G$ or $D \subseteq H$.

*39.10. A totally ordered field K is Dedekind ordered if and only if K is archimedean ordered and for every sequence $(J_n)_{n \geq 0}$ of closed bounded intervals of K such that $J_{n+1} \subseteq J_n$ for all $n \in N$, the intersection $\cap \{J_n : n \in N\}$ is not empty. [Necessity: Use Exercise 39.8. Sufficiency: If B is bounded above, consider intervals containing elements of B and upper bounds of B.]

*39.11. A totally ordered field K is Dedekind ordered if and only if every nonempty bounded convex subset of K is a bounded interval. [Sufficiency: If B is bounded above and if $a \in B$, consider the union of $\{[a, x] : x \in B$ and $x \geq a\}$.]

*39.12. A totally ordered field K is Dedekind ordered if and only if every closed bounded interval of K is compact. [Necessity: Consider the supremum of $\{x \in [a, b] : [a, x]$ is contained in the union of a finite subset of $\mathscr{G}\}$. Sufficiency: If c is an upper bound of a set B admitting no supremum and if $b \in B$, show that $[b, c]$ is not compact by considering $\{]b - 1, x[: x \in B\} \cup \{]y, c + 1[: y$ is an upper bound of $B\}$.]

*39.13. A totally ordered field K is Dedekind ordered if and only if every closed bounded interval of K is connected. [Necessity: If $[a, b] \subseteq G \cup H$ and if $a \in G$, consider $\sup \{x \in [a, b] : [a, x] \subseteq G\}$. Sufficiency: If $a \in B$ and if b is an upper bound of B, let $G = \cup \{]a - 1, x[: x \in B\}, H = \cup \{[y, b + 1[: y$ is an upper bound of $B\}$.]

40. The Construction of a Dedekind Ordered Field

Two general methods of constructing a Dedekind ordered field are available, one due to Dedekind, the other to Cantor. Each method is applicable to many other problems besides that of constructing a Dedekind ordered field. We shall present Cantor's method and sketch Dedekind's procedure in the exercises (Exercises 40.6–40.10).

With the compositions induced on Q^N by addition and multiplication on Q, Q^N is a ring (Example 15.5). We shall denote by $\mathscr{C}(Q)$ the set of all sequences of rational numbers indexed by N that are Cauchy sequences in Q. Sequences in Q indexed by N will frequently be denoted by single letters; if $a \in Q^N$, for each $n \in N$ we shall denote by a_n the value of a at n, so that a is simply the sequence $(a_n)_{n \geq 0}$.

Lemma 40.1. The set $\mathscr{C}(Q)$ is a subring of Q^N containing the multiplicative identity of Q^N.

Proof. Let $a, b \in \mathcal{C}(\boldsymbol{Q})$. We shall show that $a - b$ and ab belong to $\mathcal{C}(\boldsymbol{Q})$, By Theorem 39.7, there exists a rational $c \geq 1$ such that $|a_n| \leq c$ and $|b_n| \leq c$ for all $n \geq 0$. Let $e \in \boldsymbol{Q}_+^*$. As a and b are Cauchy sequences, there exists $m \in \boldsymbol{N}$ such that $|a_n - a_p| < e/2c$ and $|b_n - b_p| < e/2c$ for all $n \geq m$ and all $p \geq m$. Hence, if $n \geq m$ and if $p \geq m$, then

$$\begin{aligned} |(a - b)_n - (a - b)_p| &= |(a_n - b_n) - (a_p - b_p)| \\ &\leq |a_n - a_p| + |b_n - b_p| \\ &< \frac{e}{2c} + \frac{e}{2c} \leq e \end{aligned}$$

and

$$\begin{aligned} |(ab)_n - (ab)_p| &= |a_n b_n - a_p b_p| = |(a_n b_n - a_p b_n) + (a_p b_n - a_p b_p)| \\ &\leq |a_n - a_p| |b_n| + |a_p| |b_n - b_p| \\ &< \frac{e}{2c} \cdot c + c \cdot \frac{e}{2c} = e. \end{aligned}$$

Clearly, $\mathcal{C}(\boldsymbol{Q})$ contains the multiplicative identity of $\boldsymbol{Q}^N$, which is the sequence whose terms are all 1.

Definition. A **null sequence** in a totally ordered field K is a sequence converging to zero.

We shall denote by $\mathcal{N}(\boldsymbol{Q})$ the set of all null sequences of elements of $\boldsymbol{Q}$ indexed by $\boldsymbol{N}$.

Lemma 40.2. If a belongs to $\mathcal{C}(\boldsymbol{Q})$ but not to $\mathcal{N}(\boldsymbol{Q})$, then there exists $m \in \boldsymbol{N}$ such that either $a_n > 2^{-m}$ for all $n \geq m$ or $a_n < -2^{-m}$ for all $n \geq m$.

Proof. As a is not a null sequence, there exists $e \in \boldsymbol{Q}_+^*$ such that for every $n \in \boldsymbol{N}$ there is a natural number $p \geq n$ for which $|a_p| \geq e$. Since a is a Cauchy sequence, there is a natural number m such that $2^{-m} < e/2$ and $|a_n - a_p| < e/2$ for all $n \geq m, p \geq m$. Then $|a_n| \geq e/2$ for any $n \geq m$, for since there exists $p \geq n$ such that $|a_p| \geq e$, we have

$$|a_n| = |a_p - (a_p - a_n)| \geq |a_p| - |a_p - a_n|$$

$$\geq e - \frac{e}{2} = \frac{e}{2}$$

by 6° of Theorem 39.6. If there existed $n \geq m$ and $p \geq m$ such that $a_n \geq e/2$

and $a_p \le -e/2$, then

$$\frac{e}{2} > |a_n - a_p| = a_n - a_p \ge e,$$

a contradiction. Hence, as $e/2 > 2^{-m}$, either $a_n > 2^{-m}$ for all $n \ge m$ or $a_n < -2^{-m}$ for all $n \ge m$.

Let $\le$ be the relation on $\mathscr{C}(Q)$ satisfying $a \le b$ if and only if $a_n \le b_n$ for all $n \in N$. It is easy to verify that $\le$ is an ordering on $\mathscr{C}(Q)$ compatible with its ring structure.

Lemma 40.3. The set $\mathscr{N}(Q)$ is a maximal ideal of the ring $\mathscr{C}(Q)$. If $h \in \mathscr{N}(Q)$, if $k \in \mathscr{C}(Q)$, and if $0 \le k \le h$, then $k \in \mathscr{N}(Q)$.

Proof. The second assertion is easy to prove. The multiplicative identity of $\mathscr{C}(Q)$ clearly does not belong to $\mathscr{N}(Q)$, but the zero element of $\mathscr{C}(Q)$ does, so $\mathscr{N}(Q)$ is a nonempty proper subset of $\mathscr{C}(Q)$.

Let $a, b \in \mathscr{N}(Q)$ and let $c \in \mathscr{C}(Q)$. We shall show that $a - b$ and ac belong to $\mathscr{N}(Q)$. By Theorem 39.7, there exists a rational $s \ge 1$ such that $|c_n| \le s$ for all $n \ge 0$. Let $e \in Q_+^*$, and let $m \in N$ be such that $|a_n| < e/2s$ and $|b_n| < e/2s$ for all $n \ge m$. Then

$$|a_n - b_n| \le |a_n| + |b_n| < \frac{e}{2s} + \frac{e}{2s} \le e$$

for all $n \ge m$, and

$$|a_n c_n| = |a_n|\,|c_n| < \frac{e}{2s} \cdot s < e$$

for all $n \ge m$. Hence, $a - b$ and ac converge to zero. Thus, $\mathscr{N}(Q)$ is a proper ideal of $\mathscr{C}(Q)$.

To show that $\mathscr{N}(Q)$ is a maximal ideal of $\mathscr{C}(Q)$, let $\mathfrak{a}$ be an ideal of $\mathscr{C}(Q)$ properly containing $\mathscr{N}(Q)$. We shall show that $\mathfrak{a} = \mathscr{C}(Q)$. As $\mathfrak{a} \supset \mathscr{N}(Q)$, there is a sequence $a \in \mathfrak{a}$ that is not a null sequence. By Lemma 40.2, there exists $m \in N$ such that $2^{-m} < |a_n|$ for all $n \ge m$. Let h be the sequence defined by

$$h_n = \begin{cases} 0 \text{ if } a_n \ne 0, \\ 1 \text{ if } a_n = 0. \end{cases}$$

Then $h_n = 0$ for all $n \ge m$, so $h \in \mathscr{N}(Q)$. Let $b = a + h$. Then $b \in \mathfrak{a}$ since $\mathscr{N}(Q) \subset \mathfrak{a}$. Also, $b_n \ne 0$ for all $n \ge 0$, so b is invertible in the ring Q^N, and its inverse is the sequence $(b_n^{-1})_{n \ge 0}$. Moreover, $|b_n| > 2^{-m}$, and hence,

$|b_n|^{-1} < 2^m$ for all $n \geq m$. Consequently, the inverse b^{-1} of b belongs to $\mathscr{C}(\boldsymbol{Q})$, for if $e \in \boldsymbol{Q}_+^*$, then there exists $q \geq m$ such that $|b_n - b_p| < 4^{-m}e$ for all $n \geq q, p \geq q$, whence

$$\begin{aligned}
|b_n^{-1} - b_p^{-1}| &= |b_n^{-1}(b_p - b_n)b_p^{-1}| \\
&= |b_n|^{-1}\,|b_p - b_n|\,|b_p|^{-1} \\
&< 2^m(4^{-m}e)2^m = e
\end{aligned}$$

for all $n \geq q, p \geq q$. Therefore, $\mathfrak{a}$ contains the invertible element b of the ring $\mathscr{C}(\boldsymbol{Q})$, so $\mathfrak{a} = \mathscr{C}(\boldsymbol{Q})$ by Theorem 19.4. Thus, $\mathscr{N}(\boldsymbol{Q})$ is a maximal ideal of $\mathscr{C}(\boldsymbol{Q})$.

Let $\mathscr{R}$ be the quotient ring $\mathscr{C}(\boldsymbol{Q})/\mathscr{N}(\boldsymbol{Q})$. By Lemma 40.3 and Theorem 19.6, $\mathscr{R}$ is a field. Let $\leq$ be the relation on $\mathscr{R}$ satisfying $\alpha \leq \beta$ if and only if there exist Cauchy sequences $a \in \alpha$ and $b \in \beta$ such that $a \leq b$.

Lemma 40.4. Let a and b be elements of $\mathscr{C}(\boldsymbol{Q})$ belonging respectively to the elements α and β of $\mathscr{R}$. Then $\alpha \leq \beta$ if and only if there exists $h \in \mathscr{N}(\boldsymbol{Q})$ such that $a \leq b + h$. If there exists $m \in \boldsymbol{N}$ such that $a_n \leq b_n$ for all $n \geq m$, then $\alpha \leq \beta$.

Proof. Necessity: As $\alpha \leq \beta$, there exist $a' \in \alpha$ and $b' \in \beta$ such that $a' \leq b'$. Then $a - a'$ and $b' - b$ are null sequences, so if $h = b' - b + a - a'$, then $h \in \mathscr{N}(\boldsymbol{Q})$ and

$$a = a' + (a - a') \leq b' + (a - a') = b + h.$$

Sufficiency: If $a \leq b + h$ where $h \in \mathscr{N}(\boldsymbol{Q})$, then $\alpha \leq \beta$ since $a \in \alpha$ and $b + h \in \beta$.

Suppose, finally, that $a_n \leq b_n$ for all $n \geq m$. If h is defined by

$$h_n = \begin{cases} a_n - b_n \text{ for all } n < m, \\ 0 \text{ for all } n \geq m, \end{cases}$$

then $h \in \mathscr{N}(\boldsymbol{Q})$ and $a \leq b + h$, so $\alpha \leq \beta$ by what we have just proved.

Lemma 40.5. The relation $\leq$ on $\mathscr{R}$ is an ordering.

Proof. Clearly, $\leq$ is reflexive. Suppose that $\alpha \leq \beta$ and $\beta \leq \alpha$. Since $\alpha \leq \beta$, there exist $a \in \alpha$ and $b \in \beta$ such that $a \leq b$. Since $\beta \leq \alpha$, by Lemma 40.4, there exists $h \in \mathscr{N}(\boldsymbol{Q})$ such that $b \leq a + h$. Hence, $0 \leq b - a \leq h$, so $b - a \in \mathscr{N}(\boldsymbol{Q})$ by Lemma 40.3, and therefore, $\alpha = \beta$. Next, suppose that

384   THE REAL AND COMPLEX NUMBER FIELDS   Chap. VII

$\alpha \leq \beta$ and that $\beta \leq \gamma$, and let a, b, and c be members of α, β, and γ respectively such that $a \leq b$. By Lemma 40.4, there exists $h \in \mathcal{N}(Q)$ such that $b \leq c + h$, so $a \leq c + h$, and therefore, $\alpha \leq \gamma$ again by Lemma 40.4. Thus, $\leq$ is an ordering on $\mathcal{R}$.

Lemma 40.6. If $\alpha \in \mathcal{R}$ and if $a \in \alpha$, then $\alpha > 0$ if and only if there exists $m \in N$ such that $a_n > 2^{-m}$ for all $n \geq m$.

Proof. Necessity: The sequence a is not a null sequence since $\alpha \neq 0$, and therefore, by Lemma 40.2, there exists $m \in N$ such that either $a_n > 2^{-m}$ for all $n \geq m$ or $a_n < -2^{-m}$ for all $n \geq m$. The latter condition would imply that $\alpha \leq 0$ by Lemma 40.4, a contradiction of our hypothesis.

Sufficiency: Clearly, a is not a null sequence, so $\alpha \neq 0$. But $\alpha \geq 0$ by Lemma 40.4. Therefore, $\alpha > 0$.

Lemma 40.7. $(\mathcal{R}, +, \cdot, \leq)$ is a totally ordered field.

Proof. If α, β, $\gamma \in \mathcal{R}$ and if $\alpha \leq \beta$, then there exist $a \in \alpha$ and $b \in \beta$ such that $a \leq b$, whence $a + c \leq b + c$ for any $c \in \gamma$, and therefore, $\alpha + \gamma \leq \beta + \gamma$. If $\alpha > 0$ and $\beta > 0$ and if $a \in \alpha$ and $b \in \beta$, then by Lemma 40.6 there exist natural numbers m and q such that $a_n \geq 2^{-m}$ for all $n \geq m$ and $b_n \geq 2^{-q}$ for all $n \geq q$, whence $a_n b_n \geq 2^{-(m+q)}$ for all $n \geq m + q$, and therefore $\alpha\beta > 0$ again by Lemma 40.6.

By Lemmas 40.2 and 40.6, if $\alpha \neq 0$, then either $\alpha > 0$ or $-\alpha > 0$. Consequently, as $\leq$ is compatible with the ring structure of $\mathcal{R}$, $\leq$ is a total ordering on $\mathcal{R}$.

For each rational r, the sequence $(r_n)_{n\geq 0}$ defined by $r_n = r$ for all $n \geq 0$ clearly belongs to $\mathcal{C}(Q)$; we shall denote the corresponding element $(r_n)_{n\geq 0} + \mathcal{N}(Q)$ of $\mathcal{R}$ by $[r]$. It is easy to verify that the function $r \mapsto [r]$ is the unique isomorphism from the totally ordered field Q onto the totally ordered prime subfield of $\mathcal{R}$ (Theorem 39.4).

Lemma 40.8. If α, $\beta \in \mathcal{R}$ and if $0 \leq \alpha < \beta$, then there exist natural numbers k and q such that $\alpha < [2^{-q}k] < \beta$.

Proof. Since $\alpha \geq 0$, there exist $h \in \mathcal{N}(Q)$ and $a_1 \in \alpha$ such that $h \leq a_1$. Let $a = a_1 - h$. Then $a \in \alpha$ and $a \geq 0$. Let $b \in \beta$. By Lemma 40.6, there exists $m \in N$ such that $b_n - a_n > 2^{-m}$ for all $n \geq m$. As a and b are Cauchy sequences, there exists $q \geq m + 2$ such that $|a_n - a_p| < 2^{-(m+2)}$ and $|b_n - b_p| < 2^{-(m+2)}$ for all $n \geq q$, $p \geq q$. By Theorem 39.7, $\{j \in N: a_n < 2^{-q}(j-1)$ for all $n \geq q\}$ is not empty; let k be its smallest member. Then $k - 1 \geq 1$ as $2^{-q}(k-1) > 0$, and $\alpha \leq [2^{-q}(k-1)]$ by Lemma 40.4. Consequently, $k - 1 \in N$, and therefore, $2^{-q}(k-2) \leq a_r$ for some $r \geq q$

by the definition of k. If $n \geq q$, then

$$b_n = b_r - (b_r - b_n) \geq b_r - |b_r - b_n| \geq b_r - 2^{-(m+2)}$$
$$> a_r + 2^{-m} - 2^{-(m+2)} \geq 2^{-q}(k-2) + 3 \cdot 2^{-(m+2)}$$
$$\geq 2^{-q}(k-2) + 3 \cdot 2^{-q} = 2^{-q}(k+1).$$

By Lemma 40.4, therefore, $[2^{-q}(k+1)] \leq \beta$. Consequently,

$$\alpha \leq [2^{-q}(k-1)] < [2^{-q}k] < [2^{-q}(k+1)] \leq \beta.$$

Theorem 40.1. $(\mathscr{R}, +, \cdot, \leq)$ is a Dedekind ordered field.

Proof. By Theorem 39.2, it suffices to show that a subset B of $\mathscr{R}$ that is bounded above and contains a strictly positive element admits a supremum in $\mathscr{R}$. By Lemma 40.8, there exist natural numbers p and j such that $[2^{-p}j]$ is an upper bound of B. For each $n \in N$ let

$$K_n = \{k \in N: [2^{-n}k] \text{ is an upper bound of } B\}.$$

Since $[2^{-n}j] > [2^{-p}j]$ if $n < p$ and since $[2^{-n}(2^{n-p}j)] = [2^{-p}j]$ if $n \geq p$, we conclude that $K_n \neq \emptyset$ for all $n \in N$; let k_n be the smallest member of K_n. Since B contains a strictly positive element, $k_n > 0$ and consequently $k_n - 1 \in N$. If $[2^{-n}k_n] \in B$ for some $n \geq 0$, then $[2^{-n}k_n]$ is clearly the supremum of B. Hence, we may assume that $[2^{-n}k_n] > \beta$ for all $\beta \in B$ and all $n \geq 0$. As in the proof of Theorem 39.9, if $n \geq m$, then

$$(1) \qquad\qquad 0 \leq 2^{-m}k_m - 2^{-n}k_n \leq 2^{-m},$$

so the sequence $(2^{-n}k_n)_{n \geq 0}$ is a Cauchy sequence. Let $a = (2^{-n}k_n)_{n \geq 0}$, and let α be the corresponding element $a + \mathscr{N}(\boldsymbol{Q})$ of $\mathscr{R}$. We shall first show that α is an upper bound of B. Let $\beta \in B$, and let $b \in \beta$. If $\alpha < \beta$, by Lemma 40.6, there would exist a natural number m such that $b_n - a_n > 2^{-m}$ for all $n \geq m$; as $\beta < [2^{-m}k_m]$, by Lemma 40.6, there exists $q \in N$ such that $2^{-m}k_m - b_n > 2^{-q}$ and, in particular, $b_n < 2^{-m}k_m$ for all $n \geq q$; hence, if $n = \max\{m, q\}$, we would have

$$2^{-m} < b_n - a_n = b_n - 2^{-n}k_n < 2^{-m}k_m - 2^{-n}k_n \leq 2^{-m}$$

by (1), a contradiction. Consequently, α is an upper bound of B.

To complete the proof, we shall show that if γ is an upper bound of B, then $\alpha \leq \gamma$. If $\gamma < \alpha$, then by Lemma 40.8, there would exist natural numbers k and q such that $\gamma < [2^{-q}k] < \alpha$, and so by Lemma 40.6, there would exist $m \geq q$ such that $2^{-n}k_n - 2^{-q}k > 2^{-m}$ and hence $2^{-n}k_n - 2^{-m} > 2^{-q}k$ for all $n \geq m$. We would then have

$$2^{-m}(k_m - 1) = 2^{-m}k_m - 2^{-m} > 2^{-q}k,$$

so $[2^{-m}(k_m - 1)] > [2^{-q}k] > \gamma$, an upper bound of B, whence $k_m - 1 \in K_m$ as $k_m - 1 \in N$, a contradiction of the definition of k_m. Thus, $\alpha = \sup B$, and the proof is complete.

Our construction of a Dedekind ordered field depends upon the existence of the rational field, which in turn depends upon our fundamental postulate that there exists a naturally ordered semigroup. Actually, the existence of a Dedekind ordered field conversely implies the existence of a naturally ordered semigroup (Exercise 40.5). In sum, there exists a Dedekind ordered field if and only if there exists a naturally ordered semigroup.

Since there is an isomorphism $f: r \mapsto [r]$ from the ordered field Q onto the ordered prime subfield of $\mathcal{R}$, by the corollary of Theorem 17.3, there exist a field R containing the field Q algebraically and an isomorphism g from R onto $\mathcal{R}$ extending f. We define an ordering $\leq$ on R by $x \leq y$ if and only if $g(x) \leq g(y)$. Equipped with this ordering, R is clearly a totally ordered field, and g is an isomorphism from the ordered field R onto the ordered field $\mathcal{R}$. Consequently, R is a Dedekind ordered field. For all $x, y \in Q$, $x \leq y$ if and only if $f(x) \leq f(y)$ as f is an isomorphism from the ordered field Q onto the ordered prime subfield of $\mathcal{R}$. Therefore, the total ordering of R induces on Q its given ordering. In §41 we shall see that any two Dedekind ordered fields are isomorphic and, moreover, that there is just one isomorphism from one onto the other. In view of this, there is nothing arbitrary in our choice if we simply call R the **ordered field of real numbers** and its elements **real numbers**. Then Q is the prime subfield of R, and by Theorem 39.5, Q is dense in R.

EXERCISES

40.1. (a) Prove that the relation $\leq$ on $\mathscr{C}(Q)$ satisfying $a \leq b$ if and only if $a_n \leq b_n$ for all $n \geq 0$ is an ordering compatible with the ring structure of $\mathscr{C}(Q)$.
(b) Prove the second assertion of Lemma 40.3.

40.2. Verify that $r \mapsto [r]$ is the (unique) isomorphism from the ordered field Q onto the ordered prime subfield of $\mathcal{R}$.

40.3. If Q is the prime subfield of a totally ordered field K, then K is archimedean ordered if and only if every convergent sequence in the totally ordered field

Q is also a convergent sequence in the totally ordered field K. [Consider $(n^{-1})_{n \geq 1}$.]

40.4. Give an example of a null sequence a in Q such that $a_n > 0$ for all $n \geq 0$.

*40.5. Let K be a Dedekind ordered field. A subset A of K is called a **Peano set** if $0 \in A$ and if for all $x \in K$, $x \in A$ implies that $x + 1 \in A$. Let P be the intersection of all the Peano sets in K.

(a) The set P is a Peano set, P contains both 0 and 1, and 0 is the smallest element of P.

(b) If $x \in P$ and if $x \geq 1$, then $x - 1 \in P$. [Show that $\{x \in P: \text{if } x \geq 1, \text{then } x - 1 \in P\}$ is a Peano set.]

(c) If $0 < x < 1$, then $x \notin P$. [Consider $\{x \in P: \text{either } x = 0 \text{ or } x \geq 1\}$.]

(d) For all $x, y \in K$, if $x \in P$ and if $x < y < x + 1$, then $y \notin P$.

(e) For all $x, y \in P$, if $x \leq y$, then $y - x \in P$.

(f) The set P is a subsemigroup of $(K, +)$.

(g) $(P, +, \leq)$ is a naturally ordered semigroup. [If A were a nonempty subset of P containing no smallest element, consider $\{x \in P: \text{for all } y \in P, \text{if } y \leq x, \text{then } y \notin A\}$.] Conclude that there exists a Dedekind ordered field if and only if there exists a naturally ordered semigroup.

40.6. A **Dedekind cut** is a nonempty proper subset A of Q_+^* having the following two properties:

$1°$ For all $x, y \in Q_+^*$, if $x < y$ and if $y \in A$, then $x \in A$.

$2°$ For all $x \in Q_+^*$, if $x \in A$, then there exists $y \in A$ such that $y > x$.

Let $\mathscr{C}$ be the set of all Dedekind cuts. For each $r \in Q_+^*$ we define L_r by

$$L_r = \{x \in Q_+^*: x < r\}.$$

(a) If $A \in \mathscr{C}$, then every strictly positive rational not belonging to A is greater than every member of A.

(b) If $r \in Q_+^*$, then $L_r \in \mathscr{C}$.

(c) If $A, B \in \mathscr{C}$, then $A + B$ and AB belong to $\mathscr{C}$. Thus, equipped with either of the compositions induced by $+$ and $\cdot$ on the set $\mathscr{C}$ of all Dedekind cuts, $\mathscr{C}$ is a commutative semigroup.

(d) The Dedekind cut L_1 is the multiplicative identity element of $\mathscr{C}$.

(e) Multiplication on $\mathscr{C}$ is distributive over addition.

*40.7. Every element of A of $\mathscr{C}$ is invertible for multiplication. $\Big[$ Let $B = \{y \in Q_+^*: y < u^{-1} \text{ for some } u \notin A\}$. To show that if $z < 1$, then $z \in AB$, let $a \in A$ and consider integral multiples of $\dfrac{a}{n}$ where $z < \dfrac{n}{n+1}$. $\Big]$

40.8. (a) The ordering $\subseteq$ on $\mathscr{C}$ is a total ordering compatible with both addition and multiplication.

(b) If $\mathscr{A}$ is a nonempty subset of $\mathscr{C}$ and if D is an element of $\mathscr{C}$ satisfying $X \subseteq D$ for all $X \in \mathscr{A}$, then $\cup \mathscr{A} \in \mathscr{C}$, and $\cup \mathscr{A}$ is the supremum of $\mathscr{A}$ in $\mathscr{C}$.

*40.9. (a) If $A, B \in \mathscr{C}$, then $A \subset A + B$. [Consider fractions whose denominator n satisfies $1/n \in B$.]

(b) If A and C are elements of $\mathscr{C}$ and if $A \subset C$, then there exists $B \in \mathscr{C}$ such that $A + B = C$. [Let $B = \{c' - c : c' \in C, \ c' > c, \ \text{and} \ c \notin A\}$. If $c \notin C$ and $c \notin A$, to show that $c \in A + B$, consider fractions whose denominator n satisfies $c + (1/n) \in C$.]

(c) Every element of $\mathscr{C}$ is cancellable for addition.

40.10. Let $(R, +)$ be an inverse-completion (§17, Exercises) of the semigroup $(\mathscr{C}, +)$. There exists a composition $\cdot$ and a total ordering $\leq$ on R such that $(R, +, \cdot, \leq)$ is a Dedekind ordered field. [Use Exercise 17.16 and 17.17.]

41. Isomorphisms of Archimedean Ordered Groups*

Here we shall show that every archimedean ordered group is isomorphic to an ordered subgroup of the ordered group of real numbers. From this we shall prove that any Dedekind ordered field is isomorphic to R and derive the existence of the logarithmic and exponential functions.

Definition. A **monomorphism** from an ordered semigroup $(E, +, \leq)$ into an ordered semigroup $(E', +', \leq')$ is a monomorphism f from the semigroup $(E, +)$ into the semigroup $(E', +')$ satisfying the condition

$$x \leq y \text{ if and only if } f(x) \leq' f(y)$$

for all $x, y \in E$.

If f is a function from an ordered semigroup E into an ordered semigroup E', then clearly, f is a monomorphism if and only if the function obtained by restricting the codomain of f to its range is an isomorphism from E onto the ordered semigroup $f_*(E)$. Consequently, by Theorem 10.4, *if E is a totally ordered semigroup, a function f from E into an ordered semigroup E' is a monomorphism if and only if f is strictly increasing and is a homomorphism from the semigroup E into the semigroup E'.*

We begin with three preliminary theorems.

Theorem 41.1. If a is a strictly positive element of an ordered group $(G, +, \leq)$, then

$$h_a : n \mapsto n.a$$

is an isomorphism from the ordered group of integers onto the ordered cyclic subgroup of G generated by a. In particular, $n.a > 0$ if and only if $n > 0$.

---

* Readers may omit this section and proceed directly to §42.

Theorem 41.2. If $(G, +, \leq)$ is a totally ordered commutative group, then for every strictly positive integer n,

$$g_n \colon x \longmapsto n.x$$

is a monomorphism from the ordered group G into itself.

Theorem 41.3. If A and B are nonempty subsets of an ordered group $(G, +, \leq)$ each of which admits a supremum, then $\sup A + \sup B$ is the supremum of $A + B$.

Proof. If $a \in A$ and $b \in B$, then

$$a + b \leq a + \sup B \leq \sup A + \sup B.$$

Thus, $\sup A + \sup B$ is an upper bound of $A + B$. Let c be an upper bound of $A + B$. For each $b \in B$, $a + b \leq c$, and hence, $a \leq c - b$ for all $a \in A$, so $\sup A \leq c - b$, and therefore, $b \leq -\sup A + c$. Thus, $\sup B \leq -\sup A + c$, so $\sup A + \sup B \leq c$. Consequently, $\sup A + \sup B = \sup(A + B)$.

In determining all archimedean ordered groups, we begin by considering those having a smallest strictly positive element.

Theorem 41.4. The ordered group of integers is a Dedekind ordered group having a smallest strictly positive element.

Proof. Let A be a subset of Z that contains a strictly positive integer and is bounded above. The set B of upper bounds of A is, therefore, a nonempty subset of N and hence possesses a smallest member b, which clearly is the supremum of A. By Theorem 39.2, therefore, Z is a Dedekind ordered group.

Next we shall show, that to within isomorphism, the ordered group of integers is the only archimedean ordered group (and, in particular, the only Dedekind ordered group) having a smallest strictly positive element.

Theorem 41.5. If $(G, +, \leq)$ is an archimedean ordered group possessing a smallest strictly positive element a, then

$$h \colon n \longmapsto n.a$$

is an isomorphism from the ordered group of integers onto G and is, furthermore, the only isomorphism. In particular, an archimedean ordered group with a smallest strictly positive element is Dedekind ordered.

Proof. By Theorem 41.1, it suffices to show that G is generated by a. Let $b > 0$. As G is archimedean ordered, $\{k \in N: b < k.a\}$ is not empty; let n be its smallest element. Then $n > 0$ as $b > 0$, so $n - 1 \in N$, and hence,

$$(n - 1).a \leq b < n.a.$$

Therefore, by Theorem 10.3,

$$0 \leq (1 - n).a + b < a,$$

so $(1 - n).a + b = 0$ as a is the smallest strictly positive element, whence $b = (n - 1).a$. If $b < 0$, then $-b > 0$, so $-b = m.a$ for some $m > 0$ by what we have just proved, whence $b = -(m.a)$. Therefore, G is generated by a. Any isomorphism from the ordered group Z onto the ordered group G clearly takes the smallest strictly positive element 1 of Z into the smallest strictly positive element a of G and hence is h by the corollary of Theorem 20.5.

Theorem 41.6. If a is a strictly positive element of a commutative archimedean ordered group $(G, +, \leq)$, there is one and only one monomorphism f_a from $(G, +, \leq)$ into $(R, +, \leq)$ such that $f_a(a) = 1$.

Proof. For each $x \in G$, let L_x be the set of all rational numbers r for which there exist $m \in Z$ and $n \in N^*$ such that

$$r = \frac{m}{n},$$

$$m.a < n.x.$$

We need three facts about L_x. First, if $r \in L_x$ and if $p/q \leq r$ where $p \in Z$ and $q \in N^*$, then $p.a < q.x$. Indeed, if $r = m/n$ where $m.a < n.x$, then $pn \leq qm$, so

$$n.(p.a) = (np).a \leq (qm).a = q.(m.a)$$

$$< q.(n.x) = n.(q.x)$$

by Theorems 41.1 and 41.2, whence $p.a < q.x$ by Theorem 41.2. Second, $L_x \neq \emptyset$, for there exists $m \in N^*$ such that $-x < m.a$, since G is archimedean ordered (if $-x \leq 0$, we may take $m = 1$), whence $(-m).a = -(m.a) < x$ and therefore $-m \in L_x$. Third, L_x is bounded above, for since G is archimedean ordered, there exists $n \in N^*$ such that $n.a > x$, and by our first observation no rational number greater than n belongs to L_x, so n is an upper bound of L_x.

We shall next show that

$$L_{x+y} = L_x + L_y$$

for all $x, y \in G$. Let $m/n \in L_{x+y}$, where $m \in Z$ and $n \in N^*$. Then

$$m.a < n.(x + y)$$

by the first observation above, so

$$0 < (-m).a + n.(x + y),$$

and therefore, since G is archimedean ordered, there exists $q \in N^*$ such that

$$a < q.[(-m).a + n.(x + y)] = (-qm).a + (nq).(x + y),$$

whence

$$(mq + 1).a < (nq).(x + y).$$

Since $L_{(nq).x}$ is bounded above, it possesses a largest integer j. Then

$$j.a < (nq).x \le (j + 1).a,$$

and, in particular, $j/nq \in L_x$. Let $k = mq - j$. Then

$$
\begin{aligned}
(nq).x + (nq).y = (nq).(x + y) &> (mq + 1).a \\
&= (j + k + 1).a = (j + 1).a + k.a \\
&\ge (nq).x + k.a,
\end{aligned}
$$

so $(nq).y > k.a$. Therefore, $k/nq \in L_y$, so

$$\frac{m}{n} = \frac{j}{nq} + \frac{k}{nq} \in L_x + L_y.$$

Thus, $L_{x+y} \subseteq L_x + L_y$. Conversely, if $m/n \in L_x$ and $p/q \in L_y$ where $m, p \in Z$ and $n, q \in N^*$, then $m.a < n.x$ and $p.a < q.y$, so

$$
\begin{aligned}
(mq + np).a = q.(m.a) + n.(p.a) \\
< q.(n.x) + n.(q.y) = (nq).(x + y),
\end{aligned}
$$

and therefore,

$$\frac{m}{n} + \frac{p}{q} = \frac{mq + np}{nq} \in L_{x+y}.$$

Thus, $L_x + L_y \subseteq L_{x+y}$.

Let f_a be the function from G into R defined by

$$f_a(x) = \sup L_x.$$

By Theorem 41.3,

$$f_a(x + y) = \sup L_{x+y} = \sup(L_x + L_y)$$
$$= \sup L_x + \sup L_y = f_a(x) + f_a(y)$$

for all $x, y \in G$. If $x < y$, then $y - x > 0$, so there exists $n \in N^*$ such that $a < n.(y - x)$, whence $1/n \in L_{y-x}$, and consequently,

$$f_a(y) = f_a(y - x) + f_a(x) \geq \frac{1}{n} + f_a(x) > f_a(x).$$

Therefore, f_a is a monomorphism from the ordered group G into the ordered additive group R. Furthermore, $f_a(a) = 1$, for $m/n \in L_a$ if and only if $m < n$ by Theorem 41.1, and therefore, L_a is the set of all rationals less than 1, whence $f_a(a) = \sup L_a = 1$.

Let g be a monomorphism from the ordered group G into R such that $g(a) = 1$. We shall show that $g = f_a$. Suppose that $g(x) < f_a(x)$ for some $x \in G$. Then, because Q is dense in R, there would exist a rational p/q such that $g(x) < p/q < f_a(x)$, whence by the definition of f_a and our first observation above, $p/q \in L_x$ and thus $p.a < q.x$. Therefore,

$$p = p.g(a) = g(p.a) < g(q.x)$$

$$= q.g(x) < q.\frac{p}{q} = p,$$

a contradiction. Suppose that $f_a(x) < g(x)$ for some $x \in G$. Again, there would exist a rational p/q such that $f_a(x) < p/q < g(x)$, so $p/q \notin L_x$ and thus $p.a \geq q.x$, whence

$$p = p.g(a) = g(p.a) \geq g(q.x)$$

$$= q.g(x) > q.\frac{p}{q} = p,$$

a contradiction. Therefore, $g = f_a$.

Thus, to within isomorphism, the only commutative archimedean ordered groups are the subgroups of the additive group of real numbers. The hypothesis of Theorem 41.6 that G be commutative is unnecessary, for every archimedean ordered group is commutative (Exercise 41.3).

We shall employ the following notation in our subsequent discussion: If s is a nonzero element of a totally ordered field K, we shall denote by M_s the function from K into K defined by

$$M_s(x) = sx.$$

If $s > 0$, then M_s is an automorphism of the ordered group $(K, +, \leq)$; If $s < 0$, then M_s is a strictly decreasing automorphism of the group $(K, +)$. Clearly,

$$M_s^{-1} = M_s^{\leftarrow}$$

for all $s \in K^*$.

Theorem 41.7. If K is an archimedean ordered field, there is one and only one monomorphism from the ordered field K into the ordered field $\mathbf{R}$.

Proof. By Theorem 41.6, there is one and only one monomorphism f from the ordered group $(K, +, \leq)$ into $(\mathbf{R}, +, \leq)$ such that $f(1) = 1$. Since any monomorphism from the field K into the field $\mathbf{R}$ must take the identity element 1 of K into the identity element 1 of $\mathbf{R}$, there is, therefore, at most one monomorphism from the ordered field K into $\mathbf{R}$, namely, f. It remains for us to show that $f(sx) = f(s)f(x)$ for all $s, x \in K$. First, let $s \in K^*_+$. Then $f(s) > 0$, and hence $f(s)^{-1} > 0$, so $M_{f(s)^{-1}} \circ f \circ M_s$ is a monomorphism from $(K, +, \leq)$ into $(\mathbf{R}, +, \leq)$ and

$$(M_{f(s)^{-1}} \circ f \circ M_s)(1) = M_{f(s)^{-1}}(f(s))$$
$$= f(s)^{-1}f(s) = 1.$$

As f is the only monomorphism such that $f(1) = 1$, therefore,

$$M_{f(s)^{-1}} \circ f \circ M_s = f,$$

so, since $M_{f(s)^{-1}} = M_{f(s)}^{\leftarrow}$, we conclude that $f \circ M_s = M_{f(s)} \circ f$. Thus, for all $x \in K$,

$$f(sx) = (f \circ M_s)(x) = (M_{f(s)} \circ f)(x) = f(s)f(x).$$

Finally, if $s < 0$, then $-s > 0$, so

$$f(sx) = -f((-s)x) = -f(-s)f(x) = f(s)f(x).$$

Therefore, f is a monomorphism from the ordered field K into the ordered field $\mathbf{R}$.

Thus, to within isomorphism, the only archimedean ordered fields are the subfields of the ordered field of real numbers.

Theorem 41.8. If $(G, +, \leq)$ is a commutative Dedekind ordered group possessing no smallest strictly positive element, then for each strictly positive element a, the unique monomorphism f_a from $(G, +, \leq)$ into $(\mathbf{R}, +, \leq)$ satisfying $f_a(a) = 1$ is surjective and hence is an isomorphism.

Proof. Let H be the range of f_a. Then $(H, +, \leq_H)$ is a Dedekind ordered group possessing no smallest strictly positive element as it is isomorphic to $(G, +, \leq)$. We shall show that $H = \mathbf{R}$. Let s be the infimum in $\mathbf{R}$ of all the strictly positive elements of H. First, we shall prove that $s = 0$. Suppose that $s > 0$. Then $s \notin H$, for otherwise, s would be the smallest strictly positive element of H. Hence, by the definition of s, there would exist $x \in H$ such that $s < x < 2s$ and also there would exist $y \in H$ such that $s < y < x$, whence $x - y \in H$ and $0 < x - y < s$, a contradiction. Therefore, $s = 0$.

Next, we shall prove that H is a dense subset of $\mathbf{R}$. Let u and v be real numbers such that $u < v$. Assume first that $u \geq 0$. By what we have just proved, there exists $x \in H$ such that $0 < x < v - u$. Let n be the smallest of those natural numbers k such that $u < k.x$. Then $n > 0$, so $n - 1 \in N$, and therefore,

$$(n - 1).x \leq u < n.x,$$

whence

$$u < n.x = x + (n - 1).x < v - u + u = v$$

and $n.x \in H$. If $u < 0 < v$, then 0 is a member of H between u and v. Finally, assume that $u < v \leq 0$. Then $0 \leq -v < -u$, so by what we have just proved there exists $x \in H$ such that $-v < x < -u$, whence $-x \in H$ and $u < -x < v$. Thus, H is dense in $\mathbf{R}$.

Finally, we shall prove that $H = \mathbf{R}$. Let $c \in \mathbf{R}$, and let $C = \{x \in H : x \leq c\}$. Then C is a nonempty subset of H possessing an upper bound in H, for by what we have just proved there exist $x, y \in H$ such that $c - 1 < x < c < y < c + 1$, so $x \in C$ and y is an upper bound of C in H. Let b be the supremum of C in H for the ordering $\leq_H$. If $b < c$ (respectively, $c < b$), there would be no elements x of H such that $b < x < c$ (respectively, $c < x < b$), a contradiction of the density of H in $\mathbf{R}$. Hence $c = b$ and thus $c \in H$.

Thus, to within isomorphism, the only (commutative) Dedekind ordered groups are the additive groups Z and $\mathbf{R}$.

Theorem 41.9. If K is a Dedekind ordered field, the unique monomorphism from the ordered field K into the ordered field R is surjective and hence is an isomorphism.

Proof. If a is a strictly positive element of K, then $a/2$ is also strictly positive and $a/2 < a$, so K has no smallest strictly positive element. The assertion, therefore, follows from Theorems 41.7 and 41.8.

Thus, to within isomorphism, R is the only Dedekind ordered field. By Theorem 41.9, the only monomorphism from the ordered field R into itself is the identity automorphism. Actually, the only monomorphism from the field R into itself is the identity automorphism (Exercise 41.4).

Clearly, $(R_+^*, \cdot, \leq)$ is a Dedekind ordered group, for the supremum in R_+^* of a nonempty subset of R_+^* that is bounded above is simply its supremum in R. The strictly positive elements of the ordered group $(R_+^*, \cdot, \leq)$ are, of course, the real numbers > 1. There is no smallest real number greater than 1, for if $c > 1$, then $1 < \frac{1}{2}(c + 1) < c$. Consequently, by Theorem 41.8, for each real number $a > 1$ there is one and only one isomorphism f_a from $(R_+^*, \cdot, \leq)$ onto $(R, +, \leq)$ satisfying $f_a(a) = 1$.

In our remaining discussion, we shall need the following easily proved facts about strictly increasing and strictly decreasing functions: (a) the inverse of a strictly increasing (strictly decreasing) function is strictly increasing (strictly decreasing); (b) the composite of two functions, one of which is strictly increasing and the other of which is strictly decreasing, is strictly decreasing; (c) the composite of two functions, both of which are either strictly increasing or strictly decreasing, is strictly increasing.

Definition. Let $a > 0$. If $a > 1$, the **logarithmic function to base** a is the unique isomorphism f_a from $(R_+^*, \cdot, \leq)$ onto $(R, +, \leq)$ such that $f_a(a) = 1$. If $0 < a < 1$, the **logarithmic function to base** a is $f_{a^{-1}} \circ J$, where J is the function from R_+^* onto R_+^* defined by

$$J(x) = x^{-1}.$$

The logarithmic function to base a is denoted by $\log_a$. Clearly, J is an automorphism of the multiplicative group R_+^*. Consequently, for each strictly positive real number a distinct from 1, $\log_a$ is an isomorphism from the multiplicative group R_+^*, onto the additive group R. Thus, for all x, $y \in R_+^*$ and all integers n,

$$\log_a xy = \log_a x + \log_a y,$$

$$\log_a x^n = n \log_a x,$$

and in particular,

$$\log_a x^{-1} = -\log_a x.$$

Furthermore,

$$\log_a a = 1,$$

for if $a > 1$, then $\log_a a = 1$ by definition, and if $0 < a < 1$, then

$$\log_a a = \log_{a^{-1}} a^{-1} = 1,$$

again by definition. If $a > 1$, then $\log_a$ is a strictly increasing function. Since J is a strictly decreasing function, $\log_a$ is strictly decreasing if $0 < a < 1$.

Definition. Let a be a strictly positive real number distinct from 1. The **exponential function to base a** is the function $\log_a^{\leftarrow}$ from R onto R_+^*.

We shall denote the exponential function to base a by $\exp_a$. As $\exp_a$ is the inverse of an isomorphism, $\exp_a$ is itself an isomorphism from the additive group R onto the multiplicative group R_+^*. Thus, for all $x, y \in R$ and all integers n,

$$\exp_a(x + y) = (\exp_a x)(\exp_a y),$$

$$\exp_a(nx) = (\exp_a x)^n,$$

and in particular,

$$\exp_a(-x) = (\exp_a x)^{-1}.$$

Furthermore, as $\log_a a = 1$,

$$\exp_a 1 = a,$$

so for all integers n,

$$\exp_a n = (\exp_a 1)^n = a^n.$$

For this reason, it is customary to denote $\exp_a x$ by a^x for all real numbers x. We also define 1^x to be 1 for all $x \in R$. With this notation,

$$a^{x+y} = a^x a^y,$$

$$a^{nx} = (a^x)^n,$$

and in particular,

$$a^{-x} = (a^x)^{-1}$$

for all $x, y \in R$, all integers n, and all $a \in R_+^*$.

If $a > 1$, then $\exp_a$ is strictly increasing since $\log_a$ is, and if $0 < a < 1$, then $\exp_a$ is strictly decreasing since $\log_a$ is.

Theorem 41.10. If $a \in R_+^*$ and if $a \neq 1$, then

(1) $$\log_a x^y = y \log_a x$$

for all $x > 0$ and all $y \in R$. If $a, b \in R_+^*$, then

(2)
$$a^{xy} = (a^x)^y$$

for all $x, y \in R$, and

(3)
$$(ab)^x = a^x b^x$$

for all $x \in R$.

Proof. First, let $a > 1$, let x be a strictly positive real number distinct from 1, and let $b = \log_a x$. If $x > 1$ (respectively, $0 < x < 1$), then $b > 0$ (respectively, $b < 0$) and hence also $b^{-1} > 0$ (respectively, $b^{-1} < 0$), so both $\exp_x$ and $M_{b^{-1}}$ are strictly increasing (respectively, strictly decreasing) functions. As $a > 1$, $\log_a$ is strictly increasing. Hence, the function g defined by

$$g = M_{b^{-1}} \circ \log_a \circ \exp_x$$

is a strictly increasing function from R onto R. But g is also an automorphism of the additive group R as it is the composite of isomorphisms. Thus, by Theorem 10.4, g is an automorphism of the ordered group $(R, +, \leq)$, and also

$$g(1) = b^{-1} \log_a x = 1.$$

But by Theorem 41.6, there is only one automorphism of the ordered group $(R, +, \leq)$ taking 1 into 1, and that automorphism is clearly the identity automorphism. Thus, g is the identity automorphism, so as $M_{b^{-1}} = M_b^{\leftarrow}$, we conclude that

$$\log_a \circ \exp_x = M_b,$$

whence

$$\log_a x^y = y \log_a x$$

for all $y \in R$. Also $\log_a 1^y = 0 = y \log_a 1$. Therefore, (1) holds if $a > 1$, $x > 0$, and $y \in R$. If $0 < a < 1$, $x > 0$, and $y \in R$, then by what we have just proved and the definition of $\log_a$,

$$\log_a x^y = \log_{a^{-1}}(x^y)^{-1} = \log_{a^{-1}}(x^{-y}) = (-y) \log_{a^{-1}} x$$
$$= (-y)\log_a x^{-1} = (-y)(-\log_a x) = y \log_a x.$$

To prove (2), let a be a strictly positive number distinct from 1. For all $x, y \in R$,

$$\log_a a^{xy} = (\log_a \circ \exp_a)(xy) = xy = y(\log_a \circ \exp_a)(x)$$
$$= y \log_a a^x = \log_a(a^x)^y$$

by (1). Therefore, since $\log_a$ is injective, $a^{xy} = (a^x)^y$. Also $1^{xy} = 1 = 1^y = (1^x)^y$ for all $x, y \in R$. Thus, (2) holds.

If a and b are strictly positive numbers, then

$$\log_2(ab)^x = x\log_2(ab) = x\log_2 a + x\log_2 b$$
$$= \log_2 a^x + \log_2 b^x = \log_2(a^x b^x)$$

for all $x \in R$ by (1), whence $(ab)^x = a^x b^x$ as $\log_2$ is injective.

Theorem 41.11. For every $a \in R_+$ and every $n \in N^*$, there is one and only one $x \in R_+$ satisfying $x^n = a$.

Proof. The assertion is clear if $a = 0$, so we shall assume that $a > 0$. If $x = a^{1/n}$, then $x^n = (a^{1/n})^n = a$ by (2) of Theorem 41.10. By Theorem 41.2, applied to the multiplicative group R_+^*, there is at most one strictly positive real number x satisfying $x^n = a$.

The unique positive real number x satisfying $x^n = a$ is often denoted by $\sqrt[n]{a}$ and is called the **positive nth root** of a.

Theorem 41.12. If $a \in R$, then $a \geq 0$ if and only if there is a real number x such that $x^2 = a$.

The assertion follows from Theorem 41.11 and $10°$ of Theorem 17.6.

EXERCISES

41.1. Prove Theorem 41.1.

41.2. Prove Theorem 41.2.

*41.3. Let $(G, +, \leq)$ be an archimedean ordered group that possesses no smallest strictly positive element.
(a) For every $c > 0$, there exists $d > 0$ such that $d + d < c$.
(b) If $a > 0$ and $b > 0$, then $a + b = b + a$. [If not, assume that $c = (a + b) + [-(b + a)] > 0$. If $d + d < c$, infer from the inequalities $m.d \leq a < (m + 1).d$ and $n.d \leq b < (n + 1).d$ that $c < d + d$.]
(c) The composition $+$ is commutative.

41.4. The only monomorphism from the field R into itself is the identity automorphism. [Use Theorem 41.12 to show that a monomorphism from R into itself is a monomorphism from the ordered field R into itself.]

41.5. If u is a function from R into R, then u is an automorphism of the ordered group $(R, +, \leq)$ if and only if there exists $b > 0$ such that $u(x) = bx$ for all $x \in R$.

41.6. If v is a function from R_+^* into R_+^*, then v is automorphism of the ordered group $(R_+^*, \cdot, \leq)$ if and only if there exists $b > 0$ such that $v(x) = x^b$ for all $x \in R_+^*$. [Use Exercise 41.5.]

41.7. If n is a positive odd integer, then $f_n: x \mapsto x^n$ is a strictly increasing permutation of R. [Use Theorems 41.2 and 41.11.]

41.8. (a) $\log_{10} 5$ is irrational.
(b) If a and b are integers > 1 and if there is a prime p dividing a but not b, then $\log_a b$ and $\log_b a$ are irrational.

41.9. Let a, b, and c be strictly positive real numbers distinct from 1. Show that $(\log_a b)(\log_b a) = 1$ and that $(\log_a b)(\log_b c) = \log_a c$.

41.10. Find all strictly positive real numbers x satisfying the following equalities:
(a) $\log_{8x} 2 + 3\log_{2x} 2 - 6\log_{4x} 2 = 0$.   (b) $(\log_2 x)(\log_4 x)(\log_8 x) = 36$.

(c) $x^{\sqrt{x}} = \sqrt{x^x}$.   (d) $x^{3x} = (3x)^x$.

41.11. Let K and L be subfields of R. If K and L are isomorphic ordered fields, then $K = L$.

42. The Field of Complex Numbers

If K is a field over which $X^2 + 1$ is irreducible, we shall denote by i a root of $X^2 + 1$ in a stem field $K(i)$ of $X^2 + 1$ over K. Then $(1, i)$ is an ordered basis of the K-vector space $K(i)$, and addition and multiplication in $K(i)$ are given by

$$(\alpha + \beta i) + (\gamma + \delta i) = (\alpha + \gamma) + (\beta + \delta)i,$$

$$(\alpha + \beta i)(\gamma + \delta i) = (\alpha\gamma - \beta\delta) + (\alpha\delta + \beta\gamma)i.$$

By the corollary of Theorem 37.7, there is a unique K-automorphism σ of $K(i)$ such that $\sigma(i) = -i$, since i and $-i$ are the roots of $X^2 + 1$ in $K(i)$. For each $z \in K(i)$, we shall denote $\sigma(z)$ by $\bar{z}$, which is called the **conjugate** of z. Thus, if $z = \alpha + \beta i$ where a, $\beta \in K$, then $\bar{z} = a - \beta i$. Clearly, $\bar{\bar{z}} = z$ for all $z \in K(i)$.

If K is a totally ordered field, then $X^2 + 1$ is irreducible over K, for $x^2 + 1 \geq 1 > 0$ for all $x \in K$ by $10°$ of Theorem 17.6. In particular, $X^2 + 1$ is irreducible over R. By the corollary of Theorem 37.7, any two stem fields of $X^2 + 1$ over R are R-isomorphic; hence, there is nothing arbitrary in our choice if we simply select some stem field C of $X^2 + 1$ over R and call its members "complex numbers." We also select one of the two roots of $X^2 + 1$

in C and denote it as above by i. The above equalities are then the familiar rules learned in elementary algebra for adding and multiplying complex numbers.

If $z = \alpha + \beta i$ where $\alpha, \beta \in R$, we recall that the **real part** of z is the real number α and is denoted by $\mathscr{R}z$, and that the **imaginary part** of z is the real number β and is denoted by $\mathscr{I}z$; thus, $z = \mathscr{R}z + i\mathscr{I}z$. As we saw in Example 20.3, $\mathscr{R}$ and $\mathscr{I}$ are epimorphisms from $(C, +)$ onto $(R, +)$.

A fundamental property of C is that every nonconstant polynomial over C has a root in C, a fact sometimes called the "fundamental theorem of algebra."

1

Definition. A field K is **algebraically closed** if every nonconstant polynomia over K has a root in K.

No finite field is algebraically closed. Indeed, if K is a finite field, the polynomial $1 + \prod_{a \in K} (X - a)$ is a nonconstant polynomial over K having no roots in K.

Theorem 42.1. The following conditions are equivalent for any field K:

$1°$ K is algebraically closed.
$2°$ The linear polynomials are the only irreducible polynomials over K.
$3°$ Every nonconstant polynomial over K is a product of linear polynomials.

Proof. Condition $1°$ implies $2°$ by Theorem 35.7. Since every nonconstant polynomial is a product of irreducible polynomials, $2°$ implies $3°$. Since every linear polynomial over K has a root in K, $3°$ implies $1°$.

In §47, we shall use Galois theory and Sylow's First Theorem to prove that C is algebraically closed. Basic to that proof is the fact that R is real-closed in the following sense:

Definition. Let K be a totally ordered field. If a, b, and c are elements of K, we shall say that c is **between** a and b if either $a < c < b$ or $b < c < a$. A polynomial f over K **changes signs** between elements a and b of K if 0 is between $f(a)$ and $f(b)$. Finally, we shall say that K is **real-closed** if for every polynomial f over K and for all elements a and b of K, if f changes signs between a and b, then f has a root in K between a and b.

For example, Q is not a real-closed field, for $X^2 - 2$ changes signs between 1 and 2 but has no root in Q between them.

Clearly, f changes signs between a and b is and only if $f(a)f(b) < 0$.

Theorem 42.2. (Bolzano's Theorem for Polynomials) The totally ordered field R of real numbers is real-closed.

Proof. Let a and b be real numbers such that $a < b$, and let f be a polynomial over R that changes signs between a and b. It suffices to consider the case where $f(a) < 0 < f(b)$, for if $f(b) < 0 < f(a)$, then $(-f)(a) < 0 < (-f)(b)$, and any root of $-f$ is also a root of f. Let B be the set of all $s \in R$ such that $a \leq s \leq b$ and $f(x) \leq 0$ for all real numbers x such that $a \leq x \leq s$. Then B is not empty since $a \in B$, and B is bounded above by b. Let $c = \sup B$, let g be the polynomial defined by

$$g = f(X + c) = \sum_{k=0}^{n} \alpha_k X^k,$$

and let α be the constant coefficient α_0 of g. Then $f = g(X - c)$ by the corollary of Theorem 35.6, and $f(c) = g(0) = \alpha$.

We shall arrive at a contradiction from the assumption that $\alpha \neq 0$. Let d be the minimum of all the numbers $|\alpha|/(2n|\alpha_k|)$ where $k \in [0, n]$ and $\alpha_k \neq 0$. Then $d \leq 1/2n < 1$, so if $|y| \leq d$, then

$$|\alpha_k| \, |y|^k \leq |\alpha_k| \, |y| \leq \frac{|\alpha|}{2n}$$

for each $k \in [1, n]$, whence

$$|g(y) - \alpha| = \left| \sum_{k=1}^{n} \alpha_k y^k \right| \leq \sum_{k=1}^{n} |\alpha_k| \, |y|^k \leq \frac{|\alpha|}{2},$$

and therefore,

$$-\frac{|\alpha|}{2} \leq g(y) - \alpha \leq \frac{|\alpha|}{2}.$$

Case 1: $\alpha > 0$. Then $c > a$, so if $x = c - \min\{c - a, d\}$, then $c > x \geq a$ and $-d \leq x - c < 0$, whence

$$f(x) - \alpha = g(x - c) - \alpha \geq \frac{-\alpha}{2},$$

and therefore,

$$f(x) \geq \frac{\alpha}{2} > 0,$$

a contradiction of the definition of c, since there exists $s \in B$ such that $c \geq s > x$.

Case 2: $\alpha < 0$. Let $c' = c + \min\{b - c, d\}$. Since $\alpha < 0$, we have $c < b$, and therefore, $c < c' \leq b$. If $c \leq x \leq c'$, then $0 \leq x - c \leq c' - c \leq d$, so

$$f(x) - \alpha = g(x - c) - \alpha \leq \frac{-\alpha}{2},$$

and therefore,

$$f(x) \leq \frac{\alpha}{2} < 0.$$

Consequently, $f(x) < 0$ for all x satisfying $a \leq x \leq c'$, a contradiction of the definition of c. Therefore, $\alpha = 0$, so $f(c) = 0$. Since $f(a) < 0$ and $f(b) > 0$, c is neither a nor b, and hence, $a < c < b$.

Students of calculus will recognize that in the proof we essentially established the continuity of the polynomial g at zero. Another familiar theorem of calculus, which when specialized to polynomials may be proved by algebraic methods, is the following:

Theorem 42.3. (Rolle's Theorem for Polynomials) Let f be a polynomial over a real-closed totally ordered field K. If a and b are roots of f in K and if f has no roots between a and b in K, then Df has a root in K between a and b.

Proof. Let m and n be the multiplicities respectively of the roots a and b of f. By Theorem 35.8, there exists a polynomial g over K such that

$$f = (X - a)^m (X - b)^n g,$$

$$g(a) \neq 0, \qquad g(b) \neq 0.$$

Now g does not change signs between a and b, for otherwise, g and hence f would have a root between a and b as K is real-closed. Let

$$h = m(X - b)g + n(X - a)g + (X - a)(X - b)Dg.$$

Then

$$h(a) = m(a - b)g(a),$$

$$h(b) = n(b - a)g(b),$$

so, as $g(a)$ and $g(b)$ are either both > 0 or both < 0, h changes signs between a and b and hence has a root c between them. But by Theorem 35.10 and its corollary,

$$(X - a)(X - b)g Df = fh.$$

Hence,

$$(c - a)(c - b)g(c)(Df)(c) = f(c)h(c) = 0.$$

Since $(c - a)(c - b)g(c) \neq 0$, therefore, $(Df)(c) = 0$.

Corollary. If f is a nonconstant polynomial over a real-closed totally ordered field K and if Df has n roots in K, then f has at most $n + 1$ roots in K.

EXERCISES

42.1. If a complex number z is a root of a polynomial f over R, then $\bar{z}$ is also a root of f. If f is a polynomial over R of degree n and if f has m real roots, multiplicities counted, then $n \equiv m \pmod 2$.

42.2. A complex number z is a **pure imaginary** number if $z = ai$ for some real number α. A monic cubic polynomial $X^3 + bX^2 + cX + d$ over R has one real and two pure imaginary roots if and only if $c > 0$ and $d = bc$.

*42.3. If f and g are polynomials over an algebraically closed field K such that the set of roots of f is identical with the set of roots of g and the set of roots of $f + 1$ is identical with the set of roots of $g + 1$, then $f = g$. [If f has r roots, if $f + 1$ has s roots, and if $\deg f = n$, show that Df has at least $2n - s - r$ roots, multiplicities counted, and infer that $f - g = 0$.]

*42.4. Let p be a prime, let $n > 1$, and let $f_p = p + \prod_{j=1}^{n}(X - 3jp)$.
 (a) The polynomial f_p has n real roots.
 (b) The polynomial f_p is irreducible over Q. [Use Eisenstein's Criterion.]
 (c) There exist infinitely many prime polynomials over Q of degree n having n real roots.

In the remaining exercises, K *always denotes a real-closed field.* If $x \in K$, then x is **negative** if $x \le 0$, and x is **strictly negative** if $x < 0$. We shall employ the notation for intervals introduced in the exercises of §39, and in addition we shall use the following terminology: Let $f \in K[X]$ and let $c \in K$. We shall say that f is **increasing** (**decreasing**) at c if there exist a, $b \in X$ such that $a < c < b$ and the restriction of the polynomial function $f^{\sim}$ to $[a, b]$ is a strictly increasing (strictly decreasing) function. We shall say that f has a **pure local minimum** (**pure local maximum**) at c if there exist a, $b \in K$ such that $a < c < b$, the restriction of $f^{\sim}$ to $[a, c]$ is strictly decreasing (strictly increasing), and the restriction of $f^{\sim}$ to $[c, b]$ is strictly increasing (strictly decreasing). Elements c and d of K are **consecutive roots** of f if c and d are roots of f, if $c < d$, and if f has no roots between c and d. If $x, y \in K$, we shall say that x and y **have the same sign** if either $x > 0$ and $y > 0$ or $x < 0$ and $y < 0$ (or equivalently, if $xy > 0$) and that x and y **have opposite signs** if either $x > 0$ and $y < 0$ or $x < 0$ and $y > 0$ (or equivalently, if $xy < 0$).

If $(a_k)_{0 \le k \le n}$ is a sequence of elements of K and if $\{k_0, k_1, \ldots, k_r\}$ is the set of all the indices $k \in [0, n]$ such that $a_k \ne 0$, where $k_0 < k_1 < \ldots < k_r$, then the **number of variations in sign** of $(a_k)_{0 \le k \le n}$ is the number of integers $s \in [1, r]$ such that $a_{k_{s-1}}$ and a_{k_s} have opposite signs. For example, the sequence 3, 0, 5, 0, -7, 1, 0, -3 has three variations in sign.

42.5. If a polynomial f over K changes sign (does not change signs) between elements a and b of K, then f has an odd (even) number of roots between a and b, multiplicities counted.

42.6. If a and b are consecutive roots of a polynomial f over K, then Df has an odd number of roots between a and b, multiplicities counted. [Modify the proof of Rolle's Theorem.]

42.7. (Mean Value Theorem for Polynomials) If f is a polynomial over K and if a and b are elements of K such that $a < b$, then there exists c between a and b such that $(b - a)Df(c) = f(b) - f(a)$. [Apply Rolle's Theorem to a suitable polynomial.]

42.8. If f is a polynomial over K such that $Df(c) > 0$ $(Df(c) < 0)$, then f is increasing (decreasing) at c. [Use Exercise 42.7.]

42.9. Let a and b be elements of K such that $a < b$, and let f be a nonconstant polynomial over K. The restriction of $f^{\sim}$ to $[a, b]$ is a strictly increasing (strictly decreasing) function if and only if $Df(x) \geq 0$ $(Df(x) \leq 0)$ for all $x \in [a, b]$.

42.10. If f is a nonconstant polynomial over K and if $c \in K$, then either f is increasing at c, or f is decreasing at c, or f has a pure local minimum at c, or f has a pure local maximum at c.

*42.11. Let f be a nonconstant polynomial over K, and let c be a root of f of multiplicity m.
(a) If g is the polynomial such that $f = (X - c)^m g$, then $(D^m f)(c) = m!\,g(c)$, and consequently, $(D^m f)(c)$ and $g(c)$ have the same sign.
(b) If m is odd, then f is increasing at c if $(D^m f)(c) > 0$, and f is decreasing at c if $(D^m f)(c) < 0$. If m is even, then f has a pure local minimum at c if $(D^m f)(c) > 0$, and f has a pure local maximum at c if $(D^m f)(c) < 0$.
(c) The polynomial $(Df)f$ is increasing at c.

42.12. Let f be a nonconstant polynomial over K, let $c \in K$, and let m be the smallest strictly positive integer such that $(D^m f)(c) \neq 0$. Show that the statements of Exercise 42.11(b) hold. [Apply Exercise 35.16 to $f - f(c)$.]

*42.13. Let f be a polynomial over K of degree n. For each $x \in K$, we define $V_f(x)$ to be the number of variations in sign of the sequence $((D^k f)(x))_{0 \leq k \leq n}$.
(a) If c is a root of f of multiplicity m, there exist $e > 0$ and $k_f(c) \in N$ such that

$$V_f(x) = V_f(c) + m + 2k_f(c)$$

for all $x \in \,]c - e, e[$ and

$$V_f(y) = V_f(c)$$

for all $y \in \,]c, c + e[$. [Proceed by induction on the degree of f, applying the inductive hypothesis to Df. Use Exercise 42.11(c).]
(b) If $f(c) \neq 0$, there exist $e > 0$ and $k_f(c) \in N$ such that

$$V_f(x) = V_f(c) + 2k_f(c)$$

for all $x \in \,]c - e, c[$ and

$$V_f(y) = V_f(c)$$

for all $y \in \,]c, c + e[$. [Show first that it suffices to consider the case where

$f(c) > 0$. Apply (a) to $g = f - f(c)$, and use Exercise 42.11(b) in considering the four cases determined by the sign of $(D^m f)(c)$ and the parity of m.]
(c) (Budan's Theorem) If $a < b$, the number of roots of f in $]a, b]$, multiplicities counted, is

$$V_f(a) - V_f(b) - 2k$$

for some $k \in N$. [For each root c of f or one of its nonzero derivatives, determine $V_f(x) - V_f(y)$ where x and y are close to c and $x < c < y$.]

*42.14. (Descartes' Rule of Signs) For each polynomial $f = \sum\limits_{k=0}^{n} \alpha_k X^k$ of degree $n > 0$ over K, we define $U_+(f)$ to be the number of variations in sign of the sequence $(\alpha_k)_{0 \leq k \leq n}$, and we define $U_-(f)$ to be the number of variations in the sign of the sequence $((-1)^k \alpha_k)_{0 \leq k \leq n}$. Then $U_-(f) = U_+(f(-X))$, the number of strictly positive roots of f, multiplicities counted, is

$$U_+(f) - 2j$$

for some $j \in N$, and the number of strictly negative roots of f, multiplicities counted, is

$$U_-(f) - 2k$$

for some $k \in N$. [Use Budan's Theorem and Exercise 35.14 where $a = 0$.]

*42.15. Let $f = \sum\limits_{k=0}^{n} \alpha_k X^k$ be a polynomial of degree n over K.
(a) $U_+(f) + U_-(f) \leq n$.
(b) If f has n roots in K, multiplicities counted, then f has $U_+(f)$ strictly positive roots, multiplicities counted, and $U_-(f)$ strictly negative roots, multiplicities counted. [Assume first that $f(0) \neq 0$.]
(c) If $\alpha_{k-1} = \alpha_k = 0$ for some $k \in [1, n-1]$, then f has fewer than n roots in K, multiplicities counted.
(d) If $\alpha_{k-1} = \alpha_k = \alpha_{k+1}$ for some $k \in [1, n-2]$, then f has fewer than n roots in K, multiplicities counted. [Apply (c) to $(X - 1)f$.]

42.16. Use Budan's Theorem and Descartes' Rule to determine the number of strictly positive and strictly negative roots, multiplicities counted, of the following polynomials, and locate the roots between consecutive integers.

(a) $X^3 + 4X + 6$.   (b) $X^3 + 3X^2 - 2X - 7$.

(c) $X^3 - 4X^2 + 6X - 1$.   (d) $X^4 + 3X^2 + 6X - 4$.

(e) $X^4 + 2X^3 + X^2 - 1$.   (f) $X^5 + X^4 - 4X^3 - 3X^2 + 3X + 1$.

42.17. Let f be a nonconstant polynomial over K. A **Sturm sequence** for f is a sequence $(f_k)_{0 \leq k \leq n+1}$ of nonzero polynomials having the following properties:

$1°$ $f_0 = f$ and $f_1 = Df$.

$2°$ For each $k \in [1, n]$ there exist $c_{k-1} > 0$ and a polynomial q_k such that $c_{k-1}f_{k-1} = f_k q_k - f_{k+1}$ and $\deg f_{k+1} < \deg f_k$.
$3°$ f_{n+1} divides f_n.

(a) There exists a Sturm sequence for f.
(b) If $(f_k)_{0 \leq k \leq n+1}$ and $(g_k)_{0 \leq k \leq m+1}$ are Sturm sequences for f, then $m = n$, and there exist strictly positive elements $b_0, \ldots, b_{n+1}$ of K such that $g_k = b_k f_k$ for all $k \in [0, n + 1]$.

*42.18. Let $(f_k)_{0 \leq k \leq n+1}$ be a Sturm sequence for a nonconstant polynomial f over K. For each $x \in K$ we define $W_f(x)$ to be the number of variations in sign of $(f_k(x))_{0 \leq k \leq n+1}$. Let $a < x < c < y < b$ where none of $f_0, f_1, \ldots, f_{n+1}$ has a root in $[a, c[\cup]c, b]$.
(a) The polynomial f_{n+1} is a greatest common divisor of f and Df, and consequently, the roots of f_{n+1} are precisely the multiple roots of f. Moreover, f_{n+1} is a greatest common divisor of $f_0, f_1, \ldots, f_n$.
(b) If c is a multiple root of f of multiplicity $m > 1$, then c is a root of each of $f_1, \ldots, f_{n+1}$ of multiplicity $m - 1$, and consequently, $W_f(y) = W_f(x) - 1$. [Use Exercise 42.9 to show that $f(x)$ and $Df(x)$ have opposite signs.]
(c) If c is a root of f_k where $k > 0$ but not a multiple root of f, then c is not a root of either f_{k-1} or f_{k+1}, and the number of variations in sign of the sequence $f_{k-1}(x), f_k(x), f_{k+1}(x)$ is identical with that of the sequence $f_{k-1}(y), f_k(y), f_{k+1}(y)$.
(d) If c is a simple root of f, then $W_f(y) = W_f(x) - 1$, but if c is not a root of f, then $W_f(y) = W_f(x)$.
(e) (Sturm's Theorem) If $a_0 < b_0$ and if neither a_0 nor b_0 is a root of f, then the number of roots of f in $[a_0, b_0]$ (multiplicities *not* counted) is $W_f(a_0) - W_f(b_0)$.

42.19. Use Sturm's Theorem to determine the number of real roots of the following polynomials, and locate the roots between consecutive integers.

(a) $X^3 - X + 4$.
(b) $X^3 + 3X^2 - 4X - 2$.
(c) $X^3 - 7X - 7$.
(d) $X^4 + 4X^3 - 4X + 1$.
(e) $7X^4 + 28X^3 + 34X^2 + 12X + 1$.
(f) $7X^4 + 28X^3 + 34X^2 + 12X + 5$.

CHAPTER VIII

GALOIS THEORY

In this chapter we shall continue our study of fields and present, in particular, an introduction to Galois theory. As we shall see, the theory of fields provides solutions to certain classical problems of geometry and algebra, such as the problem of trisecting an angle by ruler and compass and the problem of solving a polynomial equation "by radicals."

43. Algebraic Extensions

Throughout, K is a field and E is an extension field of K. We recall the following definitions given in §37. An element c of E is **algebraic** or **transcendental over** K according as c is an algebraic or transcendental element of the K-algebra E. If c is algebraic over K, the **minimal polynomial** of c over K is the minimal polynomial of the element c of the K-algebra E. The degree of the minimal polynomial of c over K is called the **degree** of c over K and is denoted by $\deg_K c$. The field E is an **algebraic extension** of the field K if every element of E is algebraic over K, and E is a **transcendental extension** of K if E is not an algebraic extension of K.

The simplest example of an element algebraic over K is any element of K itself. If $c \in K$, then the minimal polynomial of c over K is $X - c$. The complex number i is algebraic over $\boldsymbol{R}$, and its minimal polynomial is $X^2 + 1$. The real number $\sqrt[3]{2}$ is algebraic over $\boldsymbol{Q}$, and its minimal polynomial is $X^3 - 2$. Also $\sqrt{2 + \sqrt{2}}$ is algebraic over $\boldsymbol{Q}$, and its minimal polynomial is $X^4 - 4X^2 + 2$. On the other hand, there exist real numbers that are transcendental over $\boldsymbol{Q}$ (Exercise 43.12). Theorems of number theory assert, for example, that the real numbers π and e are transcendental over $\boldsymbol{Q}$.

Theorem 43.1. Let c be an element of E algebraic over K, and let g be a

407

monic polynomial over K satisfying $g(c) = 0$. The following conditions are equivalent:

1° g is the minimal polynomial of c over K.

2° For every nonzero polynomial f over K, if $f(c) = 0$, then $g \mid f$.

3° For every nonzero polynomial f over K, if $f(c) = 0$, then $\deg g \leq \deg f$.

4° g is irreducible over K.

Proof. Condition 1° implies 2° and 3° by the definition of minimal polynomial, and 1° implies 4° by Theorem 35.5. Conversely, since g is a multiple of the minimal polynomial f of c over K, condition 4° implies 1°, and each of 2° and 3° implies that $\deg g \leq \deg f$, whence $g = f$ as both are monic polynomials.

Theorem 43.2. *If c is an element of E algebraic over K whose degree over K is n, then*

1° $K(c) = K[c]$,

2° $[K(c):K] = n$,

3° $(1, c, c^2, \ldots, c^{n-1})$ *is an ordered basis of the K-vector space $K(c)$.*

The assertion is a consequence of Theorems 35.3 and 35.4.

Theorem 43.3. *If c is an element of E algebraic over K and if F is a subfield of E containing K, then c is algebraic over F, and the minimal polynomial of c over F divides the minimal polynomial of c over K in $F[X]$.*

Proof. If g is the minimal polynomial of c over K, then $g \in F[X]$ and $g(c) = 0$, so the assertion follows from Theorem 43.1.

We have determined the structure of the subfield $K(c)$ of an extension field E of K generated by K and an element c algebraic over K; what is the structure of $K(c)$ if c is transcendental over K? We first recall that $K[X]$ is an integral domain and hence admits a quotient field, which we shall denote by $K(X)$, whose elements are quotients of polynomials over K. The field $K(X)$ is often called the **field of rational functions** or the **field of rational fractions in one indeterminate** over K. Let c be an element of E transcendental over K. Because the kernel of S_c is the zero ideal, S_c is an isomorphism from $K[X]$ onto the subring $K[c]$, and the subfield $K(c)$ of E is a quotient field of $K[c]$, since $K(c)$ is the smallest subfield of E containing $K \cup \{c\}$ and *a fortiori* is the smallest subfield of E containing $K[c]$. By Theorem 17.5, therefore, there is one and only one isomorphism from $K(X)$ onto $K(c)$ extending S_c. Hence, *if c is transcendental over K, then $K(X)$ and $K(c)$ are K-isomorphic, and moreover, there is a unique isomorphism from $K(X)$ onto $K(c)$ whose* *restriction to K[X] is S_c.* In the sequel, we shall be almost exclusively concerned with algebraic extensions of fields.

Theorem 43.4. If E is an extension of a field K, then E is an algebraic extension of K if and only if every subring of E containing K is a field.

Proof. Necessity: Let A be a subring of E containing K. If c is a nonzero element of A, then $A \supseteq K[c]$, which is a field by Theorem 43.2, and hence, $c^{-1} \in A$. Therefore, A is a field. Sufficiency: Let c be a nonzero element of E. Then $K[c]$ is a subring of E containing K and hence is a field by hypothesis. Therefore, there is a polynomial $f \in K[X]$ such that $f(c) = c^{-1}$, whence $g(c) = 0$ where $g = Xf - 1$. Consequently, c is algebraic over K.

Theorem 43.5. If E is a finite extension of K, then E is an algebraic extension of K, and the degree over K of every element of E divides $[E:K]$.

Proof. By Theorem 35.2, E is an algebraic extension of K. For each $c \in E$,

$$[E:K] = [E:K(c)][K(c):K]$$

by the corollary of Theorem 37.4, so as $\deg_K c = [K(c):K]$, the degree of c over K divides $[E:K]$.

Corollary. If c is an element of E algebraic over K, then $K(c)$ is an algebraic extension of K.

The assertion is an immediate consequence of Theorems 43.2 and 43.5.

Definition. An extension E of a field K is a **simple extension** of K if there exists $c \in E$ such that $E = K(c)$.

The following theorem gives a necessary and sufficient condition for a finite extension to be simple.

Theorem 43.6. (Steinitz) If E is a finite extension of a field K, then E is a simple extension of K if and only if there are only a finite number of subfields of E containing K.

Proof. Necessity: By hypothesis, there exists $c \in E$ such that $E = K(c)$. Let f be the minimal polynomial of c over K. Let L be a subfield of E containing K, and let

$$g = X^m + \sum_{k=0}^{m-1} a_k X^k$$

be the minimal polynomial of c over L. By Theorem 43.3, g divides f in

$L[X]$ and hence in $E[X]$. Let

$$L_0 = K(a_0, \ldots, a_{m-1}).$$

Then $L_0 \subseteq L$, $g \in L_0[X]$, and g is certainly irreducible over L_0 as g is irreducible over L. Consequently, by Theorem 43.1, g is the minimal polynomial of c over L_0. Hence,

$$m = \deg g = [E: L_0] = [E: L][L: L_0] = m[L: L_0]$$

by Theorem 43.2 and the corollary of Theorem 37.4, so $[L: L_0] = 1$ and therefore, $L = L_0$. Thus, the only subfields of E containing K are the fields $K(a_0, \ldots, a_{m-1})$ where $X^m + \sum\limits_{k=0}^{m-1} a_k X^k$ is a divisor of f in $E[X]$. Since $E[X]$ is a principal ideal domain and hence a unique factorization domain, f has only a finite number of monic divisors in $E[X]$. Consequently, there are only a finite number of subfields of E containing K.

Sufficiency: If K is a finite field, then E is also a finite field since E is finite dimensional over K. The multiplicative group E^* is, therefore, cyclic by the corollary of Theorem 38.7, and consequently, $E = K(c)$ where c is a generator of E^*. We shall assume, therefore, that K is an infinite field. Let m be the largest of the integers $[K(a): K]$ where $a \in E$, and let $c \in E$ be such that $[K(c): K] = m$. We shall obtain a contradiction from the assumption that $K(c) \neq E$. Suppose that b were an element of E not belonging to $K(c)$. By hypothesis, the set of all the fields $K(bz + c)$ where $z \in K$ is finite. Since K is infinite, therefore, there exist $x, y \in K$ such that $x \neq y$ and $K(bx + c) = K(by + c)$. Let

$$d = bx + c.$$

Then $bx + c$ and $by + c$ belong to $K(d)$, so $b \in K(d)$ as

$$b = (x - y)^{-1}[(bx + c) - (by + c)],$$

and hence, also $c \in K(d)$ as

$$c = (bx + c) - bx.$$

Therefore, $K(d)$ is a field properly containing $K(c)$, whence

$$[K(d): K] = [K(d): K(c)][K(c): K] > m,$$

a contradiction of the definition of m. Therefore, $K(c) = E$, and the proof is complete.

Corollary. If E is a finite simple extension of K, then every subfield L of E containing K is also a finite simple extension of K.

Theorem 43.7. If $c_1, \ldots, c_n$ are elements of an extension E of K that are algebraic over K, then $K(c_1, \ldots, c_n)$ is a finite and hence algebraic extension of K.

Proof. Let S be the set of all integers $m \in [1, n]$ such that $K(c_1, \ldots, c_m)$ is a finite extension of K. Then $1 \in S$ by Theorem 43.2. If $m \in S$ and $m < n$, then $m + 1 \in S$, for c_{m+1} is algebraic over K and *a fortiori* over $K(c_1, \ldots, c_m)$, so, since $K(c_1, \ldots, c_{m+1}) = [K(c_1, \ldots, c_m)](c_{m+1})$, $K(c_1, \ldots, c_{m+1})$ is a finite extension of $K(c_1, \ldots, c_m)$ by Theorem 43.2 and hence is also a finite extension of K by the corollary of Theorem 37.4. By induction, therefore, $n \in S$.

Corollary. A splitting field of a polynomial over K is a finite extension of K.

Theorem 43.8. (Transitivity of Algebraic Extensions) If E is an extension of K and if F is a subfield of E containing K, then E is an algebraic extension of K if and only if E is an algebraic extension of F and F is an algebraic extension of K.

Proof. The condition is clearly necessary by Theorem 43.3. Sufficiency: Let $c \in E$, let

$$g = X^m + \sum_{k=0}^{m-1} a_k X^k$$

be the minimal polynomial of c over F, and let $L = K(a_0, \ldots, a_{m-1})$. Then $g \in L[X]$, so c is algebraic over L, and hence, $L(c)$ is a finite extension of L by Theorem 43.2. But L is a finite extension of K by Theorem 43.7. Hence, $L(c)$ is a finite extension of K by the corollary of Theorem 37.4, so c is algebraic over K by Theorem 43.5.

Theorem 43.9. Let E be an extension of K. The set A of all elements of E algebraic over K is an algebraic extension field of K, and every element of E algebraic over A belongs to A.

Proof. Let x and y be nonzero elements of A. By Theorem 43.7, $K(x, y)$ is an algebraic extension of K containing $x - y$, xy, and x^{-1}, so those elements are all algebraic over K and hence belong to A. Therefore, A is a field, and consequently, A is an algebraic extension of K. If c is an element of E algebraic over A, then $A(c)$ is an algebraic extension of A by the corollary of Theorem 43.5 and hence is also an algebraic extension of K by Theorem 43.8, so c is algebraic over K and therefore belongs to A.

We recall from §42 that a field K is **algebraically closed** if every nonconstant polynomial over K has a root in K.

Theorem 43.10. A field K is algebraically closed if and only if the only algebraic extension of K is K itself.

Proof. Necessity: Let E be an algebraic extension of K, and let c be any element of E. The minimal polynomial g of c over K is irreducible by Theorem 43.1 and hence is $X - a$ for some $a \in K$ by Theorem 42.1. Therefore, $c - a = g(c) = 0$, so $c = a \in K$. Consequently, $E = K$. Sufficiency: If K were not algebraically closed, there would exist a nonconstant polynomial f over K having no roots in K, and a splitting field of f over K would then be an algebraic extension of K properly containing K by the corollary of Theorem 43.7.

EXERCISES

43.1. Determine the minimal polynomials over Q of the following numbers:

(a) $\sqrt{2} + 5$ (b) $\sqrt[3]{2} - 5$

(c) $\sqrt{-1 + \sqrt{2}}$ (d) $\sqrt{2 + \sqrt{2}}$

(e) $\sqrt{2} + \sqrt[4]{2}$ (f) $\sqrt{2} + \sqrt[3]{3}$.

Find the other roots in C of the minimal polynomials. [Make a judicious guess, and determine the minimal polynomial of your guess.]

43.2. Determine the minimal polynomials over $Q(\sqrt{2})$ of the numbers of Exercise 43.1.

*__43.3.__ For every integer n, the cosine and sine of n degrees are algebraic over Q. [Use Theorem 43.8.] What is the minimal polynomial over Q of $2 \cos 15°$? of $2 \cos 12°$?

43.4. If B is a set of elements of an extension field E of K each of which is algebraic over K, then $K(B)$ is an algebraic extension of K.

43.5. Show that the following extensions of Q are simple: $Q(\sqrt{2}, \sqrt{5})$, $Q(\sqrt{3}, i)$, $Q(\sqrt{2}, \sqrt[3]{3})$.

43.6. If E is a splitting field of a polynomial of degree n over K, then $[E: K]$ divides $n!$.

*__43.7.__ If f is an irreducible polynomial of degree n over K and if E is a finite extension of K such that $[E: K]$ and n are relatively prime, then f is irreducible over E. [Compute $[E(c): K]$ in two ways where c is a root of f in a splitting field of f over E.]

*43.8. (a) If K is a subfield of R and if a and b are positive integers such that neither $\sqrt{a}$, $\sqrt{b}$, nor $\sqrt{ab}$ belongs to K, then $\sqrt{b}$ does not belong to $K(\sqrt{a})$.

(b) If $a_1, \ldots, a_n$ are integers > 1 that are not squares of other integers and if a_i and a_j are relatively prime whenever $i \neq j$, then

$$\sqrt{a_n} \notin Q(\sqrt{a_1}, \ldots, \sqrt{a_{n-1}}).$$

[Proceed by induction and use (a).]

(c) If $a_1, \ldots, a_n$ are integers > 1 that are not squares of other integers and if a_i and a_j are relatively prime whenever $i \neq j$, then the set of the 2^n numbers $(a_1^{s_i} \ldots a_n^{s_n})^{1/2}$ where s_j is either 0 or 1 for all $j \in [1, n]$ is linearly independent over Q.

(d) A strictly positive integer is **square-free** if it is not divisible by the square of any integer > 1. The set of all the numbers $\sqrt{a}$ where a is a square-free integer is linearly independent over Q.

(e) The field A of complex numbers algebraic over Q is not a finite extension of Q.

*43.9. If $f \in R[X]$ and if $a \in R$, there exists a real number M such that for all real numbers x, if $0 < |x - a| < 1$, then $|f(x) - f(a)| < M|x - a|$. [Consider $g = f(X + a)$.]

*43.10. If c is a real root of a polynomial f over Z of degree n, then there exists a real number M such that $f(r/s) = 0$ for all integers r and all strictly positive integers s satisfying $|c - (r/s)| < (Ms^n)^{-1}$. [Use Exercise 43.9.]

*43.11. A real number c is a **Liouville number** if there exist sequences $(r_k)_{k \geq 1}$ and $(s_k)_{k \geq 1}$ of integers and a positive number A satisfying the following conditions:

$1°$ $s_k \geq 2$ for all $k \geq 1$.

$2°$ $\left\{ \dfrac{r_k}{s_k} : k \geq 1 \right\}$ is an infinite set.

$3°$ $\left| c - \dfrac{r_k}{s_k} \right| < \dfrac{A}{s_k^{\,k}}$ for all but finitely many $k \in N^*$.

Prove that a Liouville number is transcendental over Q.

43.12. The numbers $\sum\limits_{k=1}^{\infty} 2^{-k^k}$ and $\sum\limits_{k=1}^{\infty} 2^{-k!}$ are Liouville numbers and hence are transcendental over Q.

*43.13. If $(a_k)_{k \geq 1}$ is the sequence of integers satisfying $a_1 = 1$ and $a_{k+1} = 2^{a_k}$ for all $k \geq 1$, then $\sum\limits_{k=1}^{\infty} a_k^{-1}$ is a Liouville number.

43.14. The set of real numbers transcendental over Q is dense in R.

44. Constructions by Ruler and Compass*

The theory of fields provides solutions to many geometric problems of antiquity. Among such problems are the following:

1° To construct, by ruler and compass, a square having the same area as a circle.

2° To construct, by ruler and compass, a cube having twice the volume of a given cube.

3° To trisect a given angle by ruler and compass.

4° To construct, by ruler and compass, a regular polygon having n sides.

The only figures constructible by ruler and compass are composed of lines, line segments, rays, circles, and arcs of circles. In the geometry of Euclid's day, the only use of a ruler was to draw the line or line segment joining two given points, and the only use of a compass was to draw the circle passing through one given point whose center was another given point. Consequently, a figure constructible by ruler and compass is completely determined by certain points.

To discuss the problem of determining what figures are constructible by ruler and compass in algebraic terms, we shall regard the plane as the coordinate plane R^2 of analytic geometry. If E is a subset of R^2, we shall say that a line (circle) is **constructible from** E if it is the line through two distinct points of E (the circle passing through one point of E whose center is another point of E). A point is **constructible from** E if it is a point common either to two distinct lines constructible from E, or to a line and a circle each constructible from E, or to two distinct circles constructible from E.

For each subset E of R^2, we define $s(E)$ to be the set of all points constructible from E. If E has at least two points, then $s(E) \supseteq E$, for if $p \in E$ and if q is another point of E, then the line through p and q intersects the circle of center q through p at p, so p is constructible from E. By Theorem 11.6, there is one and only one sequence $(E_n)_{n \geq 0}$ of subsets of R^2 such that $E_0 = \{(0,0),$ $(1,0)\}$ and $E_{n+1} = s(E_n)$ for all $n \geq 0$. From what we have just seen, $E_{n+1} \supseteq E_n$ for all $n \in N$, and consequently, $E_m \supseteq E_n$ whenever $m \geq n$. We shall say that a point of R^2 is **constructible** if it belongs to E_n for some $n \in N$; the set H of all constructible points is, therefore, $\bigcup_{n \in N} E_n$. A line or circle constructible from H is called simply **constructible**. Thus, H contains two initially given points, together with the set E_1 of all points constructible from them, together with the set E_2 of all points constructible from E_1, etc. The

---

* Readers may omit this section, together with Theorems 48.4, 48.5, and 48.6 and the discussion immediately preceding Theorem 48.4, which is devoted to the solution of certain classical geometric problems.

problem of deciding whether a geometric figure is constructible is, therefore, that of deciding whether the points determining the figure are constructible.

If (a, b) belongs to both G_1 and G_2 where each of G_1, G_2 is either a constructible line or a constructible circle and where $G_1 \neq G_2$, then (a, b) is a constructible point, for as $E_m \supseteq E_n$ whenever $m \geq n$, there exists $r \in N$ such that G_1 and G_2 are both constructible from E_r, whence $(a, b) \in E_{r+1}$, a subset of H. Thus, *every point constructible from H already belongs to H.*

To describe H we need the following facts about constructible points and lines:

(1) *The coordinate axes are constructible lines.* Indeed, the X-axis is the line through $(0, 0)$ and $(1, 0)$. The point $(-1, 0)$ is constructible, for the X-axis and the circle of center $(0, 0)$ through $(1, 0)$ intersect at $(-1, 0)$ and $(1, 0)$. The circle of center $(-1, 0)$ through $(1, 0)$ intersects the circle of center $(1, 0)$ through $(-1, 0)$ at $(0, \sqrt{3})$ and $(0, -\sqrt{3})$, and the line through those two points is the Y-axis.

(2) *If $a \neq 0$ and if any one of $(a, 0)$, $(-a, 0)$, $(0, a)$, $(0, -a)$ is constructible, then all four of those points are constructible,* for the circle of center $(0, 0)$ through any one of them intersects the X-axis at $(a, 0)$ and $(-a, 0)$ and the Y-axis at $(0, a)$ and $(0, -a)$.

(3) *If $(a, 0)$ is constructible, then so is (a, a),* for the circle of center $(a, 0)$ through $(0, 0)$ intersects the circle of center $(0, a)$ through $(0, 0)$ at (a, a) and $(0, 0)$.

If a is a real number, we shall say that a is **constructible** if the point $(a, 0)$ is a constructible point.

Theorem 44.1. Real numbers a and b are constructible if and only if (a, b) is a constructible point.

Proof. By (2), we may assume that $a \neq 0$ and $b \neq 0$. Necessity: The line through (a, a) and $(a, 0)$ intersects the line through (b, b) and $(0, b)$ at (a, b), so (a, b) is constructible by (2) and (3). Sufficiency: The circle of center (a, b) through $(0, 0)$ intersects the X-axis at $(2a, 0)$ and the Y-axis at $(0, 2b)$. The circle of center $(2a, 0)$ through $(0, 0)$ intersects the circle of center $(0, 0)$ through $(2a, 0)$ at $(a, \sqrt{3}a)$ and $(a, -\sqrt{3}a)$, and the line through those two points intersects the X-axis at $(a, 0)$. Similarly, $(\sqrt{3}b, b)$ and $(-\sqrt{3}b, b)$ are constructible, and the line through them intersects the Y-axis at $(0, b)$, so b is constructible by (2).

Theorem 44.2. The set K of constructible real numbers is a subfield of $\boldsymbol{R}$. If c is a positive constructible real number, then $\sqrt{c}$ is also constructible.

Proof. Let a and b be constructible real numbers. By (2), $-a \in K$. The line through $(0, a)$ and $(-a, 0)$ intersects the line through $(b, 0)$ and (b, b) at

$(b, a + b)$, so $a + b \in K$ by (2), (3), and Theorem 44.1. Consequently, K is a group under addition. Therefore, $b + 1 - a$ and $b + 1$ are also constructible numbers. The line through $(b + 1 - a, b + 1)$ and (b, b) intersects the X-axis at $(ab, 0)$, so ab is constructible by (3) and Theorem 44.1. Also if $a \neq 0$, the line through $(1, 0)$ and $(a, -1)$ intersects the line through $(0, 0)$ and $(1,1)$ at (a^{-1}, a^{-1}), so a^{-1} is constructible by Theorem 44.1. Thus, K is a subfield of R.

Let c be a positive constructible number. Since K is a subfield of R, the number $\frac{1}{2}(c + 1)$ is constructible. The circle of center $(\frac{1}{2}(c + 1), 0)$ through $(0, 0)$ intersects the line through (c, c) and $(c, 0)$ at $(c, \sqrt{c})$ and $(c, -\sqrt{c})$, so $\sqrt{c}$ is constructible by Theorem 44.1.

We shall say that a complex number $a + bi$ is a **constructible** complex number if the point (a, b) is constructible.

Theorem 44.3. The set L of constructible complex numbers is a subfield of C. Every root in C of a quadratic polynomial whose coefficients are constructible complex numbers is constructible.

Proof. Let K be the field of constructible real numbers. Then $L = K + Ki$ by Theorem 44.1, so as $\deg_K i = 2$, L is the subfield $K(i)$ of C by 3° of Theorem 43.2.

If $a \in K$ and if $z^2 = a$, then z is a constructible complex number; indeed, if $a \geq 0$, then z is either $\sqrt{a}$ or $-\sqrt{a}$ and hence is constructible by Theorem 44.2; if $a < 0$, then z is either $i\sqrt{-a}$ or $-i\sqrt{-a}$ and hence is constructible since i and $\sqrt{-a}$ are both constructible.

If w is a constructible complex number and if $z^2 = w$, then z is constructible. To see this, let $w = a + bi$ and $z = x + yi$. Then

$$x^2 - y^2 = a,$$

$$2xy = b.$$

If $b = 0$, then z is constructible by what we have just proved, so we shall assume that $b \neq 0$. Then $x \neq 0$ and $y = b/2x$, so

$$x^2 - \frac{b^2}{4x^2} = a$$

and hence,

$$x^4 - ax^2 - \frac{b^2}{4} = 0.$$

Consequently,

$$\left(x^2 - \frac{a}{2}\right)^2 = \frac{1}{4}(a^2 + b^2).$$

Therefore, $x^2 - \frac{1}{2}a$ is a constructible real number by Theorem 44.2 and hence, x^2 is also since a is constructible. Again by Theorem 44.2, x and hence also y are constructible. Therefore, z is constructible by Theorem 44.1.

To prove the final assertion, it suffices to show that if

$$z^2 + uz + v = 0$$

where u and v are constructible complex numbers, then z is constructible. But

$$\left(z + \frac{u}{2}\right)^2 = \frac{u^2}{4} - v,$$

so as $\frac{1}{4}u^2 - v$ is constructible, $z + \frac{1}{2}u$ and hence also z are constructible by what we have just proved.

Theorem 44.4. A complex number u is constructible if and only if there exists a sequence $(K_j)_{0 \leq j \leq m}$ of subfields of C such that

(4) $$Q = K_0 \subseteq K_1 \subseteq \ldots \subseteq K_m,$$

(5) $$[K_j : K_{j-1}] \leq 2 \text{ for each } j \in [1, m],$$

(6) $$u \in K_m.$$

Proof. A sequence $(K_j)_{0 \leq j \leq m}$ satisfying (4), (5), and (6) will be called an *admissible sequence* for u. By (5), for each $j \in [1, m]$, there exists $c_j \in K_j$ such that $K_j = K_{j-1}(c_j)$.

Necessity: First, we shall prove that if $(K_j)_{0 \leq j \leq m}$ and $(L_k)_{0 \leq k \leq n}$ are admissible sequences for u and v respectively, then there exists a sequence $(H_i)_{0 \leq i \leq m+n}$ admissible for both u and v. Indeed, for each $k \in [1, n]$ there exists $c_k \in L_k$ such that $L_k = L_{k-1}(c_k)$; let $H_j = K_j$ for each $j \in [0, m]$, and let $H_{m+k} = K_m(c_1, \ldots, c_k)$ for each $k \in [1, n]$. Then $H_{m+k} = H_{m+k-1}(c_k)$ for each $k \in [1, n]$, and since

$$L_{k-1} = Q(c_1, \ldots, c_{k-1}) \subseteq K_m(c_1, \ldots, c_{k-1}) = H_{m+k-1},$$

we have

$$[H_{m+k} : H_{m+k-1}] = \deg_{H_{m+k-1}} c_k \leq \deg_{L_{k-1}} c_k = [L_k : L_{k-1}] \leq 2.$$

By induction, therefore, if D is a finite set of complex numbers for each of which there is an admissible sequence of subfields, then there is a sequence of subfields that is admissible for all the elements of D.

For each $n \in N$, let D_n be the set of all real numbers x such that either (x, y) or (y, x) belongs to E_n for some real number y, and let $S = \{n \in N:$

there is an admissible sequence of subfields for each member of D_n}. Clearly, $0 \in S$, since $D_0 = \{0, 1\}$. Suppose that $n \in S$. To show that $n + 1 \in S$, we shall show that for every $(x, y) \in E_{n+1}$, there exist admissible sequences for both x and y. By the definition of E_{n+1}, (x, y) is a point common to both G_1 and G_2 where each of G_1 and G_2 is either a line or circle determined by two points of E_n and where $G_1 \neq G_2$; let F be the set consisting of the coordinates of two points of E_n determining G_1 and the coordinates of two points of E_n determining G_2. Then F is a finite subset of D_n, so by the preceding and since $n \in S$, there exists a sequence $(K_j)_{0 \leq j \leq m}$ of subfields of C admissible for all numbers in F. Algebraic calculations show that both x and y are of degree ≤ 2 over the subfield of C generated by F and hence also over K_m. Consequently, if K_{m+1} is $K_m(x)$ (respectively, $K_m(y)$), then $(K_j)_{0 \leq j \leq m+1}$ is an admissible sequence of subfields for x (for y). Therefore, $n + 1 \in S$, and hence, $S = N$ by induction.

Now let $u = x + yi$ be any constructible complex number. Then $(x, y) \in E_n$ for some $n \in N$, so by the preceding there exists a sequence $(K_j)_{0 \leq j \leq q}$ admissible for both x and y. Let $K_{q+1} = K_q(i)$; then $(K_j)_{0 \leq j \leq q+1}$ is clearly an admissible sequence for u.

Sufficiency: By induction, it suffices to prove that if K_{j-1} is a subfield of the field L of constructible complex numbers, then K_j is also a subfield of L. Let $c_j \in K_j$ be such that $K_j = K_{j-1}(c_j)$. Then by (5), the degree of c_j over K_{j-1} is 1 or 2, so c_j is constructible by Theorem 44.3, and hence, $K_{j-1}(c_j)$ is a subfield of L.

Theorem 44.5. Every constructible complex number is algebraic over Q, and its degree over Q is a power of 2.

Proof. Let $(K_j)_{0 \leq j \leq m}$ be a sequence of subfields of C admissible for a constructible complex number u. Then $u \in K_m$ and

$$[K_m : Q] = [K_m : K_{m-1}][K_{m-1} : K_{m-2}] \ldots [K_1 : K_0],$$

so, since $[K_j : K_{j-1}]$ is either 1 or 2 for each $j \in [1, m]$, $[K_m : Q]$ is a power of 2. By Theorem 43.5, the degree of u over Q divides $[K_m : Q]$ and hence is also a power of 2.

We shall say that a line segment is **constructible** if its endpoints are constructible points. But if (a_1, a_2) and (b_1, b_2) are distinct constructible points, by Theorem 44.2, the length $[(a_1 - b_1)^2 + (a_2 - b_2)^2]^{1/2}$ of the segment joining (a_1, a_2) to (b_1, b_2) is a constructible number. To show that a given line segment is not constructible, therefore, it suffices to show that its length is not a constructible number. The length of one side of a square whose area is that of a circle of unit radius is $\sqrt{\pi}$, and the length of an edge of a cube whose volume is twice that of a unit cube is $\sqrt[3]{2}$. But $\sqrt{\pi}$ is not even algebraic over Q, for if it were, its square π would be also, and by methods of analysis, one may prove that π is transcendental over Q. Also, $\sqrt[3]{2}$ is algebraic of degree 3 over Q as its minimal polynomial over Q is $X^3 - 2$, but 3 is not a power of 2. Consequently, it is not possible to construct by ruler and compass a square whose area is that of a circle of unit radius nor a cube whose volume is 2 units.

We shall say that an angle is **constructible** if its vertex is a constructible point and if each of its sides contains a constructible point other than the vertex. If (h, k) is the vertex of a constructible angle of α radians, the circle of center (h, k) and radius 1 is constructible, as it passes through the constructible point $(h + 1, k)$, and intersects each of the two sides of the angle. The chord joining those two points is, therefore, constructible and has the same length as the corresponding segment determined by the angle of α radians whose vertex is the origin and whose initial side is the positive X-axis. The length of that segment is $(2 - 2 \cos \alpha)^{1/2}$; but by Theorem 44.2, if $x \leq 1$, then x is constructible if and only if $(2 - 2x)^{1/2}$ is constructible. Therefore, if there exists a constructible angle of α radians, then $\cos \alpha$ is a constructible number. Conversely, if $\cos \alpha$ is a constructible number, then so is $\sin \alpha$ by Theorem 44.2, since $\sin \alpha$ is either $(1 - \cos^2 \alpha)^{1/2}$ or its negative, and hence the angle whose vertex is the origin, whose initial side is the positive X-axis, and whose terminal side is the ray from the origin through $(\cos \alpha, \sin \alpha)$ is a constructible angle of α radians. In sum, *there is a constructible angle of α radians if and only if $\cos \alpha$ is a constructible number, or, equivalently, if and only if $(\cos \alpha, \sin \alpha)$ is a constructible point.*

An angle of $\pi/3$ radians is, therefore, constructible since $\cos \pi/3 = \frac{1}{2}$. We shall show, however, that no angle of $\pi/9$ radians is constructible, and hence, an angle of $\pi/3$ radians cannot be trisected by ruler and compass. A trigonometric formula for $\cos 3\alpha$ in terms of $\cos \alpha$ yields

$$4 \cos^3 \frac{\pi}{9} - 3 \cos \frac{\pi}{9} - \cos \frac{\pi}{3} = 0,$$

so if $x = 2 \cos(\pi/9)$, then x is constructible if and only if $\cos(\pi/9)$ is, and

$$x^3 - 3x - 1 = 0.$$

Because neither 1 nor -1 is a root of $X^3 - 3X - 1$, that polynomial is irreducible over Q. Hence, by Theorem 43.1, x is algebraic over Q of degree 3, so by Theorem 44.5, x is not constructible.

The problem of constructing regular polygons by ruler and compass will be considered in §48.

EXERCISES

44.1. (a) What points belong to E_1?
(b) Determine an upper bound on the number of points in $s(E)$ if E has n points.

44.2. For each of the following, complete the proofs in detail by making the needed analytic verifications, and draw geometric diagrams appropriate to the proofs:
(a) Statements (1), (2), and (3);
(b) Theorem 44.1;
(c) Theorem 44.2.

44.3. Complete the proof of Theorem 44.4 by making the needed algebraic calculations.

44.4. The set A of all real numbers α such that $\cos \alpha$ is constructible is an additive subgroup of R, and if $\alpha \in A$, then $\frac{1}{2}\alpha \in A$.

*44.5. Show that $\cos(\pi/5)$ is a constructible number.

*44.6. If n is an integer, then there exists a constructible angle of n degrees if and only if n is an integral multiple of 3. [Use Exercises 44.4 and 44.5.]

44.7. The product of the lengths of the segments PA, PB, and PC is 2, where AB is a constructible line segment of length 4, C is the midpoint of AB, and P is a point on AB. Is P constructible?

44.8. There are two isosceles triangles inscribed in the circle of center $(0, 0)$ and radius 1, each one unit in area, such that $(1, 0)$ is the vertex common to the sides of equal length. Show that one of these triangles is constructible and that the other is not.

45. Galois Theory

Galois theory is concerned with relationships between subfields of a field and subgroups of its group of automorphisms. By means of Galois theory, certain problems about subfields of fields may be transformed into more amenable problems about subgroups of groups.

Definition. Let Λ be a set of monomorphisms from a field E into a field E'. The **fixed field** of Λ is

$$\{x \in E: \sigma(x) = \tau(x) \text{ for all } \sigma, \tau \in \Lambda\}.$$

By Theorems 37.3 and 37.1, the fixed field of Λ is, indeed, a subfield of E.

If E is a subfield of E' and if Λ contains the identity automorphism of E, then the fixed field of Λ is simply

$$\{x \in E: \sigma(x) = x \text{ for all } \sigma \in \Lambda\}.$$

Definition. Let E be a field, and let Ω be the group of all automorphisms of the field E. For each subfield K of E, the **automorphism group of E over K** is the group $K^{\blacktriangle}$ of all K-automorphisms of E. For each subgroup Λ of Ω, we denote the fixed field of Λ by $\Lambda^{\blacktriangledown}$. The **closure** of K in E is the subfield $K^{\blacktriangle\blacktriangledown}$ of E, and the **closure** of Λ is the subgroup $\Lambda^{\blacktriangledown\blacktriangle}$ of Ω. A subfield K of E is **closed** in E if $K^{\blacktriangle\blacktriangledown} = K$, and a subgroup Λ of Ω is **closed** if $\Lambda^{\blacktriangledown\blacktriangle} = \Lambda$.

By definition, $K^{\blacktriangle\blacktriangledown}$ is the set of all $y \in E$ such that $\sigma(y) = y$ for all automorphisms σ of E satisfying $\sigma(x) = x$ for all $x \in K$; thus, $K \subseteq K^{\blacktriangle\blacktriangledown}$. Similarly, $\Lambda^{\blacktriangledown\blacktriangle}$ is the set of all $\sigma \in \Omega$ such that $\sigma(x) = x$ for all $x \in E$ satisfying $\tau(x) = x$ for all $\tau \in \Lambda$; thus, $\Lambda \subseteq \Lambda^{\blacktriangledown\blacktriangle}$.

Theorem 45.1. Let K and L be subfields of a field E, and let Λ and Δ be subgroups of the group Ω of all automorphisms of E.

1° $K \subseteq K^{\blacktriangle\blacktriangledown}$, and $\Lambda \subseteq \Lambda^{\blacktriangledown\blacktriangle}$.
2° If $K \subseteq L$, then $K^{\blacktriangle} \supseteq L^{\blacktriangle}$; if $\Lambda \subseteq \Delta$, then $\Lambda^{\blacktriangledown} \supseteq \Delta^{\blacktriangledown}$.
3° $K^{\blacktriangle\blacktriangledown\blacktriangle} = K^{\blacktriangle}$, and $\Lambda^{\blacktriangledown\blacktriangle\blacktriangledown} = \Lambda^{\blacktriangledown}$.
4° K is the fixed field of a subgroup of Ω if and only if K is closed in E; Λ is the automorphism group of E over a subfield of E if and only if Λ is a closed subgroup of Ω.
5° The function $F \mapsto F^{\blacktriangle}$ is a bijection from the set $\mathscr{F}$ of all closed subfields of E onto the set $\mathscr{G}$ of all closed subgroups of Ω, and its inverse is the function $\Lambda \mapsto \Lambda^{\blacktriangledown}$.

Proof. We have already proved 1°, and 2° is also easy to prove. For 3°, we observe that by 1°, where $\Lambda = K^{\blacktriangle}$, we have $K^{\blacktriangle} \subseteq K^{\blacktriangle\blacktriangledown\blacktriangle}$, and by 2°, since $K \subseteq K^{\blacktriangle\blacktriangledown}$, we have $K^{\blacktriangle} \supseteq K^{\blacktriangle\blacktriangledown\blacktriangle}$. Therefore, $K^{\blacktriangle} = K^{\blacktriangle\blacktriangledown\blacktriangle}$. A similar argument shows that $\Lambda^{\blacktriangledown} = \Lambda^{\blacktriangledown\blacktriangle\blacktriangledown}$. To prove 4°, we observe that if $K = \Lambda^{\blacktriangledown}$, then

$$K^{\blacktriangle\blacktriangledown} = \Lambda^{\blacktriangledown\blacktriangle\blacktriangledown} = \Lambda^{\blacktriangledown} = K$$

by 3°; conversely, if $K = K^{\blacktriangle\blacktriangledown}$, then by definition K is the fixed field of the subgroup $K^{\blacktriangle}$ of Ω. Similarly, if $\Lambda = K^{\blacktriangle}$, then

$$\Lambda^{\blacktriangledown\blacktriangle} = K^{\blacktriangle\blacktriangledown\blacktriangle} = K^{\blacktriangle} = \Lambda$$

by 3°; conversely, if $\Lambda = \Lambda^{\blacktriangledown\blacktriangle}$, then by definition Λ is the automorphism group of E over $\Lambda^{\blacktriangledown}$.

Finally, to prove $5°$, we observe that by $4°$, the function $F \mapsto F^{\blacktriangle}$ is a function from $\mathscr{F}$ into $\mathscr{G}$, and $\Gamma \mapsto \Gamma^{\blacktriangledown}$ is a function from $\mathscr{G}$ into $\mathscr{F}$. Therefore, by the definition of a closed subfield and a closed subgroup and by Theorem 5.3, $F \mapsto F^{\blacktriangle}$ is a bijection from $\mathscr{F}$ onto $\mathscr{G}$, and its inverse is the function $\Gamma \mapsto \Gamma^{\blacktriangledown}$.

The following facts are useful in determining the automorphism group of E over K:

(1) *If $E = K(c_1, \ldots, c_n)$, then every K-automorphism of E is completely determined by its values at $c_1, \ldots, c_n$* (corollary of Theorem 37.3), that is, if σ and τ are K-automorphisms of E having the same value at c_i for all $i \in [1, n]$, then $\sigma = \tau$.

(2) *If f is a nonzero polynomial over K, then every K-automorphism σ of E induces a permutation on the set R of all roots of f in E.* Indeed, if $f = \sum_{k=0}^{n} \alpha_k X^k$ and if $f(c) = 0$, then

$$f(\sigma(c)) = \sum_{k=0}^{n} \alpha_k \sigma(c)^k = \sum_{k=0}^{n} \sigma(\alpha_k) \sigma(c)^k$$
$$= \sigma\left(\sum_{k=0}^{n} \alpha_k c^k \right) = \sigma(0) = 0.$$

Consequently, we may form a function σ_R by restricting the domain and codomain of σ to R. As σ is a permutation of E and as R is finite, σ_R is therefore a permutation of R.

(3) *If $E = K(c)$ where c is a root of an irreducible polynomial f over K and if c' is also a root of f belonging to E, then there exists a unique K-automorphism σ of E satisfying $\sigma(c) = c'$.* By the corollary of Theorem 37.7, we need only prove that $K(c') = E$. But by Theorem 37.5,

$$[K(c'): K] = \deg f = [K(c): K],$$

so $K(c')$ is a subspace of the K-vector space E having the same dimension as E and hence $K(c')$ is E.

Example 45.1. Let K be a field whose characteristic is not 2, let a be an element of K that is not a square of K, and let E be the field $K(c)$ obtained by adjoining a root c of $X^2 - a$ to K. Then $-c$ is also a root of $X^2 - a$, so there exists a K-automorphism σ of E satisfying $\sigma(c) = -c$. Consequently, $\sigma(\alpha + \beta c) = \alpha - \beta c$ for all $\alpha, \beta \in K$. Since c and $-c$ are the only roots of $X^2 - a$ in E, σ and the identity automorphism 1_E are the only K-automorphisms of E. If $x = \alpha + \beta c$ and if $\sigma(x) = x$, then $\alpha - \beta c = \alpha + \beta c$, so $\beta = 0$, and therefore, $x \in K$. Hence, K is a closed subfield of E.

In particular, the only R-automorphisms of C are 1_C and the conjugation automorphism $z \mapsto \bar{z}$. If b is a positive integer that is not a square, the only Q-automorphisms of $Q(\sqrt{b})$ are the identity automorphism and the automorphism $\alpha + \beta\sqrt{b} \mapsto \alpha - \beta\sqrt{b}$.

Example 45.2. Let $E = Z_2(c)$ where c is a root of $X^3 + X^2 + 1$, which is irreducible over Z_2. We saw in Example 37.1 that the automorphism group of E over Z_2 is $\{1_E, \sigma, \sigma^2\}$ where $\sigma(c) = c^2$. If $x = \alpha + \beta c + \gamma c^2$ and if $\sigma(x) = x$, then

$$(\alpha + \gamma) + \gamma c + (\beta + \gamma)c^2 = \alpha + \beta c + \gamma c^2,$$

so $\beta = \gamma = 0$, and therefore, $x \in Z_2$. Consequently, Z_2 is a closed subfield of E.

Example 45.3. Let $E = Q(\sqrt[3]{2})$. Therefore, E is obtained by adjoining to Q the root $c = \sqrt[3]{2}$ of the irreducible polynomial $X^3 - 2$. The other roots of $X^3 - 2$ in C are ωc and $\omega^2 c$ where $\omega = \frac{1}{2}(-1 + i\sqrt{3})$; in particular, as $E \subseteq R$, E contains no other roots of $X^3 - 2$. Consequently, the identity automorphism 1_E is the only Q-automorphism of E, and the closure of Q in E is E.

Example 45.4. Let E be the splitting field of $X^3 - 2$ over Q contained in C. Then $E = Q(c, \omega)$ where $c = \sqrt[3]{2}$ as we saw in Example 37.3. By Theorem 37.2, E is also $Q(c)(\omega)$ and $Q(\omega)(c)$. Since ω and ω^2 are the roots in E of $X^2 + X + 1$, which is therefore irreducible over R and hence over $Q(c)$, there exists a $Q(c)$-automorphism ρ of E satisfying $\rho(\omega) = \omega^2$. Now $X^3 - 2$ has no roots in $Q(\omega)$, for a root of $X^3 - 2$ has degree 3 over Q, whereas an element of $Q(\omega)$ has either degree 1 or degree 2 over Q. Therefore, $X^3 - 2$ is irreducible over $Q(\omega)$, and its roots in E are c, ωc, and $\omega^2 c$. Hence, there exists a $Q(\omega)$-automorphism σ of E satisfying $\sigma(c) = \omega c$, and we may verify at once that $\sigma^2(c) = \omega^2 c$. The following table summarizes the behavior of the Q-automorphisms 1_E, σ, σ^2, ρ, $\rho\sigma$, $\rho\sigma^2$ on the set of generators $\{\omega, c\}$ for the extension field E of Q.

	1_E	σ	σ^2	ρ	$\rho\sigma$	$\rho\sigma^2$
ω	ω	ω	ω	ω^2	ω^2	ω^2
c	c	ωc	$\omega^2 c$	c	$\omega^2 c$	ωc

By (2), every Q-automorphism of E takes ω into one of the two roots ω and ω^2 of $X^2 + X + 1$ and also takes c into one of the three roots c, ωc, $\omega^2 c$ of $X^3 - 2$, and by (1) any Q-automorphism of E is completely determined by its values at ω and c. Therefore, $\{1_E, \sigma, \sigma^2, \rho, \rho\sigma, \rho\sigma^2\}$ is the entire group Γ of Q automorphisms of E. An examination of values at c and ω shows that $\rho^2 = \sigma^3 = 1_E$ and that $\rho\sigma\rho^{-1} = \sigma^{-1}$. Hence, Γ is isomorphic to the dihedral group of order 6, the group of symmetries of an equilateral triangle. Expressing an element x of E as a linear combination of $\{1, c, c^2, \omega, \omega c, \omega c^2\}$, which is a basis of the Q-vector space E by Theorem 37.4, one may easily show that if $\rho(x) = \sigma(x) = x$, then $x \in Q$. Hence, Q is a closed subfield of E. Figure 18 pairs subgroups of Γ with their corresponding fixed subfields of E.

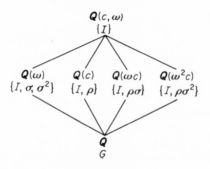

Figure 18

Theorem 45.2. (Dedekind) If $(H, \cdot)$ is a semigroup and if $(E, +, \cdot)$ is a field, the set of all nonzero homomorphisms from $(H, \cdot)$ into $(E, \cdot)$ is a linearly independent subset of the E-vector space E^H of all functions from H into E.

Proof. Let S be the set of all positive integers n such that every sequence $(\sigma_1, \ldots, \sigma_n)$ of n distinct nonzero homomorphisms is linearly independent. Clearly, $1 \in S$. Suppose that $n \in S$, let $(\sigma_1, \ldots, \sigma_{n+1})$ be a sequence of $n + 1$ distinct nonzero homomorphisms, and let $(\alpha_k)_{1 \le k \le n+1}$ be a sequence of elements of E such that

$$\sum_{k=1}^{n+1} \alpha_k \sigma_k = 0.$$

Since $\sigma_{n+1} \neq \sigma_1$, there exists $a \in H$ such that $\sigma_{n+1}(a) \neq \sigma_1(a)$. For each $x \in H$,

$$0 = \left(\sum_{k=1}^{n+1} \alpha_k \sigma_k\right)(ax) = \sum_{k=1}^{n+1} \alpha_k \sigma_k(ax) = \sum_{k=1}^{n+1} \alpha_k \sigma_k(a)\sigma_k(x),$$

and also

$$0 = \sigma_{n+1}(a)\sum_{k=1}^{n+1}\alpha_k\sigma_k(x) = \sum_{k=1}^{n+1}\alpha_k\sigma_{n+1}(a)\sigma_k(x).$$

Subtracting, we obtain

$$0 = \sum_{k=1}^{n}\alpha_k(\sigma_k(a) - \sigma_{n+1}(a))\sigma_k(x)$$

for all $x \in H$, so as $n \in S$,

$$\alpha_k(\sigma_k(a) - \sigma_{n+1}(a)) = 0$$

for each $k \in [1, n]$. Since $\sigma_1(a) \neq \sigma_{n+1}(a)$, therefore, $\alpha_1 = 0$. Consequently,

$$\sum_{k=2}^{n+1}\alpha_k\sigma_k = 0,$$

so again, since $n \in S$, we conclude that $\alpha_2 = \ldots = \alpha_{n+1} = 0$. By induction, therefore, $S = N^*$, and the proof is complete.

Theorem 45.3. If Λ is a set of n monomorphisms from a field E into a field E' and if E is a finite extension of the fixed field K of Λ, then $[E: K] \geq n$.

Proof. We shall obtain a contradiction from the assumption that $[E: K] = m < n$. Let $\Lambda = \{\sigma_1, \ldots, \sigma_n\}$, and let $(b_1, \ldots, b_m)$ be an ordered basis of the K-vector space E. By Theorem 29.8, as $m < n$, there exists a nonzero n-tuple $(a_1, \ldots, a_n)$ of elements of E' such that

$$\sigma_1(b_1)a_1 + \sigma_2(b_1)a_2 + \ldots + \sigma_n(b_1)a_n = 0$$
$$\sigma_1(b_2)a_1 + \sigma_2(b_2)a_2 + \ldots + \sigma_n(b_2)a_n = 0$$
$$\cdot \qquad \cdot \qquad \qquad \cdot \qquad \cdot$$
$$\cdot \qquad \cdot \qquad \qquad \cdot \qquad \cdot$$
$$\sigma_1(b_m)a_1 + \sigma_2(b_m)a_2 + \ldots + \sigma_n(b_m)a_n = 0.$$

We shall show that

$$a_1\sigma_1 + \ldots + a_n\sigma_n = 0.$$

Let $x \in E$, and let $\beta_1, \ldots, \beta_m \in K$ be such that $x = \sum_{k=1}^{m}\beta_k b_k$. As $\beta_i \in K$, $\sigma_j(\beta_i) = \sigma_i(\beta_i)$ for all $i \in [1, n]$. Multiplying both sides of the ith equation by

$\sigma_i(\beta_i)$, therefore, we obtain

$$\sigma_1(\beta_1 b_1)a_1 \quad + \quad \sigma_2(\beta_1 b_1)a_2 + \ldots + \sigma_n(\beta_1 b_1)a_n \qquad = 0$$

$$\sigma_1(\beta_2 b_2)a_1 \quad + \quad \sigma_2(\beta_2 b_2)a_2 + \ldots + \sigma_n(\beta_2 b_2)a_n \qquad = 0$$

$$\sigma_1(\beta_m b_m)a_1 + \sigma_2(\beta_m b_m)a_2 + \ldots + \sigma_m(\beta_m b_m)a_n \quad = 0,$$

since

$$\sigma_i(\beta_i)\sigma_j(b_i)a_j = \sigma_j(\beta_i)\sigma_j(b_i)a_j = \sigma_j(\beta_i b_i)a_j.$$

Adding, we obtain

$$\sigma_1(x)a_1 + \sigma_2(x)a_2 + \ldots + \sigma_n(x)a_n = 0.$$

Therefore, $a_1\sigma_1 + \ldots + a_n\sigma_n = 0$ and $a_i \neq 0$ for some $i \in [1, n]$, a contradiction of Theorem 45.2. Thus, $[E:K] \geq n$.

Theorem 45.4. If Γ is a subgroup of the group of all automorphisms of a field E and if K is the fixed field $\Gamma^\blacktriangledown$ of Γ, then E is a finite extension of K if and only if Γ is a finite group, in which case $[E:K]$ is the order of Γ.

Proof. If E were a finite extension of K but Γ an infinite group, then Γ would contain a finite subset Λ having $[E:K] + 1$ elements; the fixed field L of Λ would then contain K, so

$$[E:K] \geq [E:L] \geq [E:K] + 1$$

by Theorem 45.3, a contradiction. Consequently, if E is a finite extension of K, then Γ is finite. We shall assume, henceforth, that $\Gamma = \{\sigma_1, \ldots, \sigma_n\}$ is a group of n elements.

For each $x \in E$, we define $T(x)$ by

$$T(x) = \sigma_1(x) + \ldots + \sigma_n(x).$$

We shall first prove that $T(x) \in K$ for all $x \in E$ and that $T(b) \neq 0$ for some $b \in E$. For each $j \in [1, n]$, the function $\sigma \mapsto \sigma_j \circ \sigma$ is a permutation of the group Γ, so if $J(k)$ is the integer in $[1, n]$ such that $\sigma_{J(k)} = \sigma_j \circ \sigma_k$ for each $k \in [1, n]$, then J is a permutation of $[1, n]$, whence

$$\sigma_j(T(x)) = \sum_{k=1}^{n} \sigma_{J(k)}(x) = \sum_{k=1}^{n} \sigma_k(x) = T(x).$$

Therefore, since K is the fixed field of Γ, $T(x) \in K$ for all $x \in E$. If $T(x) = 0$

for all $x \in E$, then $\sigma_1 + \ldots + \sigma_n = 0$, a contradiction of Theorem 45.2. Hence, there exists $b \in E$ such that $T(b) \neq 0$.

By Theorem 45.3, we need only show that E is a finite extension of K and that $[E:K] \leq n$. But for this, it suffices to show that every finite-dimensional subspace of the K-vector space E has dimension $\leq n$, for a vector space that is not finite-dimensional has finite-dimensional subspaces of arbitrarily high dimension by an inductive argument based on Theorem 28.11. To show that every finite-dimensional subspace of the K-vector space E has dimension $\leq n$, it suffices to show that every set of $n+1$ elements of E is linearly dependent.

Let $(c_1, \ldots, c_{n+1})$ be a sequence of $n+1$ elements of E. By Theorem 29.8, there exists a nonzero $(n+1)$-tuple $(a_1, \ldots, a_{n+1})$ of elements of E such that

$$\sigma_1^{\leftarrow}(c_1)a_1 + \sigma_1^{\leftarrow}(c_2)a_2 + \ldots + \sigma_1^{\leftarrow}(c_{n+1})a_{n+1} = 0$$
$$\sigma_2^{\leftarrow}(c_1)a_1 + \sigma_2^{\leftarrow}(c_2)a_2 + \ldots + \sigma_2^{\leftarrow}(c_{n+1})a_{n+1} = 0$$
$$\vdots$$
$$\sigma_n^{\leftarrow}(c_1)a_1 + \sigma_n^{\leftarrow}(c_2)a_2 \quad \ldots + \sigma_n^{\leftarrow}(c_{n+1})a_{n+1} = 0.$$

Let $r \in [1, n+1]$ be such that $a_r \neq 0$, and let $b \in E$ satisfy $T(b) \neq 0$. Multiplying both sides of each equation by $a_r^{-1}b$ and letting $b_k = a_k a_r^{-1}b$, we obtain

$$\sigma_1^{\leftarrow}(c_1)b_1 + \sigma_1^{\leftarrow}(c_2)b_2 + \ldots + \sigma_1^{\leftarrow}(c_{n+1})b_{n+1} = 0$$
$$\sigma_2^{\leftarrow}(c_1)b_1 + \sigma_2^{\leftarrow}(c_2)b_2 + \ldots + \sigma_2^{\leftarrow}(c_{n+1})b_{n+1} = 0$$
$$\vdots$$
$$\sigma_n^{\leftarrow}(c_1)b_1 + \sigma_n^{\leftarrow}(c_2)b_2 + \ldots + \sigma_n^{\leftarrow}(c_{n+1})b_{n+1} = 0.$$

Consequently,

$$0 = \sigma_1(0) = c_1\sigma_1(b_1) + c_2\sigma_1(b_2) + \ldots + c_{n+1}\sigma_1(b_{n+1})$$
$$0 = \sigma_2(0) = c_1\sigma_2(b_1) + c_2\sigma_2(b_2) + \ldots + c_{n+1}\sigma_2(b_{n+1})$$
$$\vdots$$
$$0 = \sigma_n(0) = c_1\sigma_n(b_1) + c_2\sigma_n(b_2) + \ldots + c_{n+1}\sigma_n(b_{n+1}).$$

Adding, we obtain

$$0 = c_1 T(b_1) + c_2 T(b_2) + \ldots + c_{n+1} T(b_{n+1}).$$

But $T(b_k) \in K$ for each $k \in [1, n+1]$, and $T(b_r) = T(b) \neq 0$. Therefore,

$(c_1, \ldots, c_{n+1})$ is a linearly dependent sequence of elements of the K-vector space E, and the proof is complete.

Corollary. Every finite subgroup Λ of the group of all automorphisms of a field E is closed.

Proof. By Theorems 45.1 and 45.4,

$$\text{order } \Lambda = [E: \Lambda^{\blacktriangledown}] = [E: \Lambda^{\blacktriangledown\blacktriangle\blacktriangledown}],$$

so by Theorem 45.4, $\Lambda^{\blacktriangledown\blacktriangle}$ is a finite group that has the same order as its subgroup Λ, whence $\Lambda = \Lambda^{\blacktriangledown\blacktriangle}$.

Definition. An extension E of a field K is a **Galois extension** of K if E is an algebraic extension of K and if K is closed in E. If E is a Galois extension of K, the group $K^{\blacktriangle}$ of all K-automorphisms of E is called the **Galois group of E over K**.

Examples 45.1, 45.2, and 45.4 illustrate Galois extensions; Example 45.3 shows that a finite extension need not be a Galois extension.

Theorem 45.5. If E is a finite extension of a field K, then E is a Galois extension of K if and only if the automorphism group Γ of E over K is finite and has order $[E: K]$.

Proof. Since $\Gamma = K^{\blacktriangle}$, Γ is finite and its order is $[E: K^{\blacktriangle\blacktriangledown}]$ by Theorem 45.4 applied to $K^{\blacktriangle\blacktriangledown} = \Gamma^{\blacktriangledown}$. By the corollary of Theorem 37.4,

$$[E: K] = [E: K^{\blacktriangle\blacktriangledown}][K^{\blacktriangle\blacktriangledown}: K].$$

Therefore, the order of Γ is $[E: K]$ if and only if $[K^{\blacktriangle\blacktriangledown}: K] = 1$, or equivalently, if and only if $K^{\blacktriangle\blacktriangledown} = K$.

Lemma. If L and K are subfields of a field E such that $K \subseteq L$ and if Γ is the automorphism group of E over K, then the function

$$\sigma \circ L^{\blacktriangle} \mapsto \sigma_L$$

is a well-defined bijection from $\Gamma/L^{\blacktriangle}$ onto the set Γ_L of restrictions to L of all K-automorphisms of E.

Proof. The function is well-defined, for if $\sigma \circ L^{\blacktriangle} = \tau \circ L^{\blacktriangle}$, then $\tau^{\leftarrow} \circ \sigma \in L^{\blacktriangle}$, so $\tau^{\leftarrow}(\sigma(x)) = x$, and hence, $\sigma(x) = \tau(x)$ for all $x \in L$, whence $\sigma_L = \tau_L$. The function is clearly surjective, and it is also injective, for if $\sigma, \tau \in \Gamma$ and if

$\sigma_L = \tau_L$, then $\sigma(x) = \tau(x)$, and hence $\tau^{\leftarrow}(\sigma(x)) = x$, for all $x \in L$, so $\tau^{\leftarrow} \circ \sigma \in L^{\blacktriangle}$, and therefore, $\sigma \circ L^{\blacktriangle} = \tau \circ L^{\blacktriangle}$.

Theorem 45.6. (Fundamental Theorem of Galois Theory) Let E be a finite Galois extension of K, let Γ be the Galois group of E over K, and let L be a subfield of E containing K. Then

  1° E is a Galois extension of L,
  2° $[E: L]$ is the order of $L^{\blacktriangle}$,
  3° $[L: K]$ is the index $(\Gamma: L^{\blacktriangle})$ of $L^{\blacktriangle}$ in Γ.

Furthermore, the function $F \mapsto F^{\blacktriangle}$ is a bijection from the set $\mathscr{F}_K$ of all subfields of E containing K onto the set $\mathscr{G}_\Gamma$ of all subgroups of Γ, and its inverse is the function $\Lambda \to \Lambda^{\blacktriangledown}$.

Proof. By Theorem 45.5, Γ is a finite group and its order is $[E: K]$. Let r be the order of $L^{\blacktriangle}$ and s the index $(\Gamma: L^{\blacktriangle})$ of $L^{\blacktriangle}$ in Γ. The order of Γ is then rs by Lagrange's Theorem. By the lemma, the set Γ_L of restrictions to L of all K-automorphisms of E has s elements. The fixed field of Γ_L is by definition the set of all $x \in L$ such that $\sigma(x) = x$ for all $\sigma \in \Gamma$, which is simply K, the fixed field of Γ. Therefore, $[L: K] \geq s$ by Theorem 45.3. By Theorem 45.4 applied to the finite group $L^{\blacktriangle}$, we have $[E: L^{\blacktriangle\blacktriangledown}] = r$. Hence,

$$[E: L^{\blacktriangle\blacktriangledown}][L^{\blacktriangle\blacktriangledown}: L][L: K] = [E: K] = \text{order } \Gamma = rs$$

$$\leq r[L: K] = [E: L^{\blacktriangle\blacktriangledown}][L: K]$$

$$\leq [E: L^{\blacktriangle\blacktriangledown}][L^{\blacktriangle\blacktriangledown}: L][L: K].$$

Consequently, $[L^{\blacktriangle\blacktriangledown}: L] = 1$, so $L = L^{\blacktriangle\blacktriangledown}$, and hence, also $[E: L] = [E: L^{\blacktriangle\blacktriangledown}] = r$; thus, 1° and 2° hold. Moreover, $rs = r[L: K]$, so $[L: K] = s$; thus, 3° holds.

Every subfield of E containing K is, therefore, closed in E, and by the corollary of Theorem 45.4, every subgroup of Γ is closed. The final assertion, therefore, follows from 5° of Theorem 45.1, since the bijection $F \mapsto F^{\blacktriangle}$ from the set $\mathscr{F}$ of all closed subfields of E onto the set $\mathscr{G}$ of all closed subgroups of the group of all automorphisms of E takes $\mathscr{F}_K$ onto $\mathscr{G}_\Gamma$.

Denoting by Δ the subgroup of Γ containing only the identity automorphism of E, we may summarize Theorem 45.6 in Figure 19.

Our final example concerns finite fields.

Theorem 45.7. Let E be an extension of degree n of a finite field K having q elements. Then E is a Galois extension of K, and the Galois group Γ of E over K is cyclic of order n. Moreover, Γ is generated by the automorphism

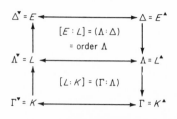

Figure 19

$\sigma: x \to x^q$, and Γ consists of the n K-automorphisms $\sigma^k: x \mapsto x^{q^k}$ where $k \in [0, n-1]$.

Proof. Since q is a power of the characteristic p of K by Theorem 38.1, σ is an automorphism of E by Corollary 38.2.2. Then $\{x \in E: x^q = x\}$ has at most q members by Corollary 35.8.1 but contains K by Theorem 38.3 and therefore is K. Consequently, σ is a K-automorphism of E, and K is closed in E; i.e., E is a Galois extension of K. An inductive argument establishes that $\sigma^k(x) = x^{q^k}$ for all $k \in N$ and all $x \in E$. By the corollary of Theorem 38.7, the multiplicative group E^* possesses a generator z. Hence, if $0 \leq j < k \leq n - 1$, then $1 \leq q^j < q^k < q^n - 1$, whence $\sigma^j(z) = z^{q^j} \neq z^{q^k} = \sigma^k(z)$. Therefore, $\{1_E, \sigma, \sigma^2, \ldots, \sigma^{n-1}\}$ is a set of n K-automorphisms of E. By Theorem 45.5, Γ has only n members, so $\Gamma = \{1_E, \sigma, \sigma^2, \ldots, \sigma^{n-1}\}$. Thus, Γ is cyclic of order n, and σ is a generator of Γ.

To illustrate the use of Theorem 45.6, we shall derive from it and Theorem 45.7 certain facts concerning subfields of finite fields.

Theorem 45.8. Let E be an extension of degree n of a finite field K having q elements. For each divisor m of n there is one and only one subfield F_{q^m} of E having q^m elements, and F_{q^m} contains K. The fields F_{q^m} where $m \mid n$ are the only subfields of E that contain K, and $F_{q^m} \subseteq F_{q^r}$ if and only if $m \mid r$.

Proof. By Theorem 45.7, the automorphism $\sigma: z \mapsto z^q$ of E is a generator of the Galois group Γ of E over K, which is cyclic of order n. Since $m \to n/m$ is a permutation of the set of positive divisors of n, by Theorem 23.9, the only subgroups of Γ are the cyclic subgroups $[\sigma^m]$ where $m \mid n$, and $[\sigma^m]$ has order n/m. Let F_{q^m} be the fixed field of $[\sigma^m]$. By Theorem 45.6, the only subfields of E containing K are the fields F_{q^m}, and

$$[F_{q^m}: K] = (\Gamma: [\sigma^m]) = n/(n/m) = m,$$

so F_{q^m} has q^m elements. By Theorem 38.3, the only subfield of E having q^m elements is $\{x \in E: x^{q^m} = x\}$, so F_{q^m} is the only subfield of E having q^m elements.

Replacing E by F_{q^r} where $r \mid n$ and observing that $[F_{q^r}: K] = r$, we see that if F_{q^r} contains F_{q^m}, then $m \mid r$, and conversely, if $m \mid r$, then F_{q^r} contains a subfield of q^m elements, which must be F_{q^m} as that is the only subfield of E having q^m elements.

Applying Theorem 45.8 to the case where K is a prime field, we obtain the following corollary:

Corollary. Let E be a finite field having p^n elements, where p is a prime. For each positive divisor m of n there is one and only one subfield F_{p^m} of E having p^m elements. The fields F_{p^m} where $m \mid n$ are the only subfields of E, and $F_{p^m} \subseteq F_{p^r}$ if and only if $m \mid r$.

EXERCISES

45.1. Prove $2°$ of Theorem 45.1.

45.2. What is the closure of Q in R? [Use Exercise 41.4.]

45.3. Let E be a finite Galois extension of K, let L_1 and L_2 be subfields of E containing K, and let Λ_1 and Λ_2 be subgroups of the Galois group of E over K.
  (a) Describe $L_1(L_2)^{\blacktriangle}$ and $(L_1 \cap L_2)^{\blacktriangle}$ in terms of $L_1^{\blacktriangle}$ and $L_2^{\blacktriangle}$.
  (b) Describe $(\Lambda_1 \cap \Lambda_2)^{\blacktriangledown}$ and the fixed field of the subgroup generated by $\Lambda_1 \cup \Lambda_2$ in terms of $\Lambda_1^{\blacktriangledown}$ and $\Lambda_2^{\blacktriangledown}$.

45.4. Let E be a finite field having p^n elements where p is a prime, and let K and L be subfields of E having p^r and p^s elements respectively. How many elements does $K \cap L$ have? How many elements does $K(L)$ have?

45.5. Let a and b be positive integers such that none of $\sqrt{a}$, $\sqrt{b}$, and $\sqrt{ab}$ is rational, and let E be the splitting field in C of $(X^2 - a)(X^2 - b)$. Describe the automorphism group Γ of E over Q. [Use Exercise 43.8(a).] Construct a diagram like that of Figure 18 pairing subgroups of Γ with their corresponding fixed fields.

*45.6. Let E be the splitting field in C of $X^4 - 2$ over Q. Describe the automorphism group Γ of E over Q. Construct a diagram like that of Figure 18 pairing subgroups of Γ with their corresponding fixed fields. [Observe that $E = Q(\sqrt[4]{2}, i)$, and model your discussion after that of Example 45.4.]

45.7. If a is an integer that is not the cube of an integer, then $Q(\sqrt[3]{a})$ is not a Galois extension of Q.

45.8. Let E be a finite Galois extension of K of degree n, and let $\{\sigma_1, \ldots, \sigma_n\}$

be the Galois group of E over K. For each $x \in E$, we define the **trace** $Tr_{E/K}(x)$ of x and the **norm** $N_{E/K}(x)$ of x over K by

$$Tr_{E/K}(x) = \sigma_1(x) + \ldots + \sigma_n(x),$$
$$N_{E/K}(x) = \sigma_1(x) \ldots \sigma_n(x).$$

(a) For all $x \in K$, $Tr_{E/K}(x)$ and $N_{E/K}(x)$ belong to K.
(b) For all $x, y \in E$, $Tr_{E/K}(x + y) = Tr_{E/K}(x) + Tr_{E/K}(y)$ and $N_{E/K}(xy) = N_{E/K}(x)N_{E/K}(y)$.
(c) The function $Tr_{E/K}$ is a nonzero linear form on the K-vector space E.
(d) What are the trace and norm of a complex number $a + bi$ over $\mathbf{R}$? of an element $a + b\sqrt{2}$ of $Q(\sqrt{2})$ over Q? of an element $a + b\sqrt{2} + c\sqrt{3} + d\sqrt{6}$ of $Q(\sqrt{2}, \sqrt{3})$ over Q?

46. Separable and Normal Extensions

Here we shall characterize finite Galois extensions of fields and determine further relationships between subfields of a finite Galois extension and subgroups of its Galois group which will prove useful in discussing polynomial equations.

Definition. A prime polynomial f over a field K is **separable** over K if there is an extension field E of K such that for some sequence $(\alpha_k)_{1 \le k \le n}$ of *distinct* elements of E,

$$f = \prod_{k=1}^{n} (X - \alpha_k).$$

A nonconstant polynomial g over K is **separable** over K if every prime factor of g in $K[X]$ is separable over K.

Thus, a prime polynomial over K of degree n is separable over K if and only if it has n roots in some extension field of K. *If g is a nonconstant polynomial over K all of whose roots in a splitting field E of g are simple, then g is separable over K*; indeed, if f is a prime factor of g in $K[X]$, then f divides g in $E[X]$ and so f is the product of distinct linear polynomials in $E[X]$. However, a polynomial may be separable over K and yet have multiple roots in an extension field; for example, $(X^2 + 1)^2$ is separable over Q. Clearly, *if g is a separable polynomial over K and if h is a nonconstant polynomial over K dividing g in $K[X]$, then h is separable over K*.

Theorem 46.1. Let f be a prime polynomial over a field K. The following conditions are equivalent:

$1°$ f is separable over K.

$2°$ $Df \neq 0$.

$3°$ Every root of f in any extension field of K is a simple root.

Proof. Condition $1°$ implies $2°$, for if Df were the zero polynomial, then every root of f in any extension field of K would be a multiple root by Theorem 35.11. To show that $2°$ implies $3°$, let c be a root of f in an extension field E of K. If c were a multiple root of f, then c would also be a root of Df by Theorem 35.11, and hence, either $Df = 0$ or $\deg f \leq \deg Df = \deg f - 1$ by Theorem 43.1, a contradiction. To see that $3°$ implies $1°$, we need only consider a splitting field of f over K.

Of importance later is the following fact: *If L is an extension of a field K and if g is a nonconstant polynomial separable over K, then g is also separable over L.* For let E be a splitting field of g over L, and let h be a prime factor of g in $L[X]$. Then in $L[X]$, h divides some prime factor f of g in $K[X]$ by Theorem 22.9 and (UFD 3'). Consequently, as every root of f in E is simple by Theorem 46.1, every root of h in E is also simple. Therefore, as E is a splitting field of g over L, h is a product of distinct linear polynomials in $E[X]$ and hence is separable over L.

Theorem 46.2. Every nonconstant polynomial over a field K of characteristic zero is separable over K.

Proof. By Theorem 46.1, it suffices to prove that every monic nonconstant polynomial over K has a nonzero derivative. If

$$f = X^n + \sum_{k=0}^{n-1} \alpha_k X^k$$

where $n \geq 1$, then

$$Df = n.X^{n-1} + \sum_{k=1}^{n-1} (k.\alpha_k) X^{k-1},$$

which is not the zero polynomial since $n.1 \neq 0$.

If the characteristic of K is a prime p, however, there may well exist a prime polynomial over K whose derivative is the zero polynomial. For example, let $K = Z_2(X)$, the field of rational fractions over Z_2. The polynomial $h = Y^2 - X \in K[Y]$ has no roots in K, for if f and g were nonzero polynomials over Z_2 such that $(f/g)^2 - X = 0$, then $f^2 = Xg^2$, but the degree of f^2 is even and that of Xg^2 is odd, a contradiction. Consequently, h is irreducible over K, but $Dh = 2.Y = 0$ as the characteristic of K is 2.

In general, when does a polynomial f of degree n over a field K whose characteristic is a prime p have a zero derivative? By definition, if

$$f = \sum_{k=0}^{n} \alpha_k X^k,$$

then

$$Df = \sum_{k=1}^{n} (k.\alpha_k) X^{k-1},$$

so $Df = 0$ if and only if $k.\alpha_k = 0$ for all $k \in [1, n]$. If $\alpha \neq 0$ and if $k \in Z$, then $k.\alpha = 0$ if and only if $p \mid k$ by Theorem 21.7. Consequently, if $Df = 0$, then $\alpha_k = 0$ whenever $p \nmid k$, so

$$f = \alpha_{rp} X^{rp} + \alpha_{(r-1)p} X^{(r-1)p} + \ldots + \alpha_p X^p + \alpha_0$$

where $rp = n$. Conversely, if

$$f = \sum_{k=0}^{r} \beta_k X^{kp},$$

then

$$Df = \sum_{k=1}^{r} p.(k.\beta_k) X^{kp-1} = 0.$$

In sum, $Df = 0$ *if and only if f belongs to the subdomain $K[X^p]$ of $K[X]$.*

Definition. An element a of an extension field E of a field K is **separable** over K if a is algebraic over K and if the minimal polynomial of a over K is separable over K. The extension field E is a **separable extension** of K if every element of E is separable over K.

By a previous observation, *if a is separable over K and if L is an extension field of K contained in E, then a is separable over L,* since by Theorem 43.3 the minimal polynomial of a over L divides the minimal polynomial of a over K in $L[X]$.

By Theorem 46.2, every algebraic extension of a field of characteristic zero is a separable extension. The following two theorems show that every algebraic extension of a finite field is also a separable extension.

Theorem 46.3. If K is a field whose characteristic is a prime p, then every algebraic extension of K is a separable extension of K if and only if the function $\sigma: \alpha \mapsto \alpha^p$ is an automorphism of K.

Proof. By Theorem 38.2, σ is a monomorphism from K into K. Therefore, we shall show that every algebraic extension of K is separable if and only if σ

is surjective. Necessity: Let $\beta \in K$, let E be a splitting field of $X^p - \beta$ over K, and let α be a root of $X^p - \beta$ in E. Then

$$X^p - \beta = X^p - \alpha^p = (X - \alpha)^p.$$

Consequently, if h is a prime factor of $X^p - \beta$ in $K[X]$, then h divides $(X - \alpha)^p$ in $E[X]$; hence, $h = (X - \alpha)^k$ for some $k \in [1, p]$, but, since every root of h in E is simple by Theorem 46.1, we have $k = 1$ and $h = X - \alpha$. As $h \in K[X]$, therefore, $\alpha \in K$. Hence, σ is surjective.

Sufficiency: We shall show that if f is a nonconstant polynomial over K whose derivative is the zero polynomial, then f is reducible over K. Since $Df = 0$, there exist $\beta_0, \ldots, \beta_r \in K$ such that

$$f = \sum_{k=0}^{r} \beta_k X^{kp}$$

as we saw above. Because σ is surjective, there exists $\alpha_k \in K$ such that $\alpha_k{}^p = \beta_k$ for each $k \in [0, r]$. Therefore,

$$f = \sum_{k=0}^{r} \alpha_k{}^p X^{kp} = \left(\sum_{k=0}^{r} \alpha_k X^k \right)^p,$$

so f is reducible over K.

Definition. A field K is **perfect** if every algebraic extension of K is separable.

Theorem 46.4. All fields of characteristic zero, all finite fields, and all algebraically closed fields are perfect.

Proof. By Theorem 46.2, all fields of characteristic zero are perfect. If K is a finite field whose characteristic is a prime p, then $\sigma: \alpha \mapsto \alpha^p$ is an automorphism of K by Corollary 38.2.2, so K is perfect by Theorem 46.3. The only algebraic extension of an algebraically closed field K is K itself by Theorem 43.10, so an algebraically closed field is perfect.

Definition. An extension field E of a field K is a **normal extension** of K if E is an algebraic extension of K and if every prime polynomial over K that has a root in E is a product of linear polynomials in $E[X]$.

For example, $Q(\sqrt[3]{2})$ is not a normal extension of Q, for $X^3 - 2$ is a prime polynomial over Q that has a root in $Q(\sqrt[3]{2})$ but is not a product of linear polynomials over $Q(\sqrt[3]{2})$, as we saw in Example 45.3.

Theorem 46.5. Let E be a finite extension of a field K. The following statements are equivalent:

1° E is a normal extension of K.

2° E is a splitting field over K of some polynomial in $K[X]$.

3° If Ω is any extension field of E and if L is any subfield of E containing K, then for every K-monomorphism φ from L into Ω there is a K-automorphism $\bar{\varphi}$ of E such that $\bar{\varphi}(x) = \varphi(x)$ for all $x \in L$.

Proof. To show that 1° implies 2°, let $\{c_1, \ldots, c_n\}$ be a basis of the K-vector space E, let g_k be the minimal polynomial of c_k over K for each $k \in [1, n]$, and let $g = g_1 g_2 \ldots g_n$. Since each g_k is a product of linear polynomials in $E[X]$ by 1°, g is also a product of linear polynomials in $E[X]$. Also E is the field generated by the union of K and the set C of roots of g in E, for C contains the basis $\{c_1, \ldots, c_n\}$ of E. Therefore, E is a splitting field of g over K.

To show that 2° implies 3°, let φ be a K-monomorphism from L into Ω where Ω is an extension field of E and where L is a subfield of E containing K, and let $L_1 = \varphi_*(L)$. By 2°, E is a splitting field over K of a polynomial $g \in K[X]$, and hence, $E = K(C)$ where C is the set of roots of g in E. Since $L_1 \supseteq K$, the subfield $K(L_1)$ of Ω generated by $K \cup L_1$ is simply L_1; therefore,

$$L_1(C) = [K(L_1)](C) = K(L_1 \cup C) = [K(C)](L_1) = E(L_1)$$

by Theorem 37.2. Consequently, $E(L_1)$ is a splitting field of g over L_1. Also E is clearly a splitting field of g over L. Therefore, by Theorem 37.9, there is an isomorphism $\bar{\varphi}$ from E onto $E(L_1)$ extending the function obtained by restricting the codomain of φ to its range, so $\bar{\varphi}(x) = \varphi(x)$ for all $x \in L$. Since φ is a K-monomorphism, $\bar{\varphi}$ is a K-isomorphism, and therefore, E is a finite-dimensional subspace of the K-vector space $E(L_1)$ that is isomorphic to $E(L_1)$. Consequently, $E(L_1)$ is finite-dimensional over K, $\dim_K E = \dim_K E(L_1)$, and $E \subseteq E(L_1)$, whence $E = E(L_1)$ by Theorem 28.13. Thus, $\bar{\varphi}$ is a K-automorphism of E satisfying $\bar{\varphi}(x) = \varphi(x)$ for all $x \in L$.

To show that 3° implies 1°, let g be a prime polynomial over K that has a root c in E, and let Ω be a splitting field of g over E. We shall show that $\Omega = E$, from which we may conclude that g is a product of linear polynomials in $E[X]$. To show that $\Omega = E$, it suffices to show that every root c' of g in Ω belongs to E. By the corollary of Theorem 37.7, there is a K-isomorphism φ from $K(c)$ onto $K(c')$ such that $\varphi(c) = c'$. The function obtained by extending the codomain of φ to Ω is then a K-monomorphism from $K(c)$ into Ω, so by 3°, there is a K-automorphism $\bar{\varphi}$ of E such that $\bar{\varphi}(x) = \varphi(x)$ for all $x \in K(c)$. In particular, $c' = \varphi(c) = \bar{\varphi}(c) \in E$. Thus, g is a product of linear polynomials in $E[X]$, and hence, E is a normal extension of K.

Corollary 46.5.1. If E is a finite extension of a field K, there is an extension Ω of E that is a finite normal extension of K.

Proof. Let $\{c_1, \ldots, c_n\}$ be a basis of the K-vector space E, let g_k be the minimal polynomial of c_k over K for each $k \in [1, n]$, let $g = g_1 g_2 \cdots g_n$, and let Ω be a splitting field of g over E. If C is the set of roots of g in Ω, then $K(C) \supseteq K(c_1, \ldots, c_n) = E$ as $\{c_1, \ldots, c_n\}$ is a basis of E, so $\Omega = E(C) = K(C)$. Consequently, Ω is a splitting field of g over K, and therefore, Ω is a finite normal extension of K by the corollary of Theorem 43.7 and Theorem 46.5.

From 3° of Theorem 46.5 we obtain the following corollary:

Corollary 46.5.2. If E is a finite normal extension of a field K, then the range of every K-monomorphism from E into an extension field Ω of E is E.

Theorem 46.6. Let E be a finite extension of a field K. The following statements are equivalent:

1° E is a Galois extension of K.
2° E is a normal separable extension of K.
3° E is a splitting field over K of a separable polynomial in $K[X]$.

Proof. Condition 2° is equivalent to the statement that for every positive integer n, the minimal polynomial over K of every element of E of degree n over K has n roots in E. To show that 1° implies 2°, therefore, let f be the minimal polynomial over K of an element c of E of degree n, and let Γ be the Galois group of E over K. By Theorem 45.6,

$$n = [K(c): K] = (\Gamma: K(c)^{\blacktriangle}).$$

Consequently, since $\Gamma / K(c)^{\blacktriangle}$ has n members, by the lemma of §45, there exist n K-automorphisms $\sigma_1, \ldots, \sigma_n$ of E no two of which have the same restriction to $K(c)$. Because σ_k is a K-automorphism, $\sigma_k(c)$ is a root of f for each $k \in [1, n]$. But $\sigma_k(c) \neq \sigma_j(c)$ if $k \neq j$, for otherwise the restrictions to $K(c)$ of σ_k and σ_j would be the same function. Hence, f has n roots in E.

Condition 2° implies 3°: By Theorem 46.5, E is a splitting field over K of a polynomial $g \in K[X]$. Every prime factor h of g in $K[X]$, therefore, has a root in E and hence is the minimal polynomial over K of an element of E. But by 2°, the minimal polynomial over K of any element of E is separable over K. Therefore, g is separable over K.

To show that 3° implies 1°, let E be the splitting field of a separable polynomial $f \in K[X]$. Since $K[X]$ is a principal ideal domain and hence a unique factorization domain, $f = \beta p_1^{n_1} \ldots p_m^{n_m}$ where $\beta \in K$ and $p_1, \ldots, p_m$ are distinct prime polynomials of $K[X]$. If $i \neq j$, then p_i and p_j are relatively prime elements of $K[X]$, so by Bezout's Identity (corollary of Theorem 22.5), there exist $a, b \in K[X]$ such that $ap_i + bp_j = 1$; hence, for each $c \in E$,

$a(c)p_i(c) + b(c)p_j(c) = 1$, so p_i and p_j have no roots in common in E. Since f is separable and since E is a splitting field of f, each p_j is the product of distinct linear polynomials in $E[X]$. Hence, the polynomial $g = p_1 \ldots p_m$ is the product of distinct linear polynomials in $E[X]$. The roots of f and g in E coincide, so E is a splitting field of g over K. By 3° of Theorem 37.9 applied to the identity automorphism of K, there are $[E:K]$ K-automorphisms of E. Therefore, E is a Galois extension of K by Theorem 45.5.

The special case where K is a field of characteristic zero, or more generally, where K is a perfect field, is worthy of notice:

Corollary. Let E be a finite extension of a perfect field K. The following statements are equivalent:

    1° E is a Galois extension of K.
    2° E is a normal extension of K.
    3° E is a splitting field over K of some polynomial in $K[X]$.

The following theorem provides an answer to a natural question: If E is a finite Galois extension of K, how may one characterize those subfields of E containing K that correspond under the bijection $L \mapsto L^{\blacktriangle}$ to normal subgroups of the Galois group over K? The answer is suggested by the terminology: They are precisely the subfields of E that are normal extensions of K. Before proving this, we observe that if $K \subseteq L \subseteq E$ and if E is a finite Galois extension of K, then E is a Galois extension of L by Theorem 45.6, but L need not be a Galois extension of K. For example, $Q(\sqrt[3]{2}, \omega)$ is a Galois extension of Q but $Q(\sqrt[3]{2})$ is not, as we saw in Examples 45.3 and 45.4.

Theorem 46.7. Let E be a finite Galois extension of K, let L be a subfield of E containing K, and let Γ be the Galois group of E over K. The following statements are equivalent:

    1° $L^{\blacktriangle}$ is a normal subgroup of Γ.
    2° $\sigma_*(L) = L$ for all $\sigma \in \Gamma$.
    3° L is a Galois extension of K.
    4° L is a normal extension of K.
    5° The order of the automorphism group Δ of L over K is $(\Gamma : L^{\blacktriangle})$.

Furthermore, under these conditions, the function $\rho: \sigma \mapsto \sigma_L$, where σ_L is the function obtained by restricting the domain and codomain of σ to L, is an epimorphism from Γ onto Δ with kernel $L^{\blacktriangle}$, so $\Gamma/L^{\blacktriangle}$ and Δ are isomorphic.

Proof. First, we shall show that for every $\sigma \in \Gamma$,

$$\sigma_*(L)^{\blacktriangle} = \sigma \circ L^{\blacktriangle} \circ \sigma^{\leftarrow}.$$

Indeed, $\tau \in \sigma_*(L)^{\blacktriangle}$ if and only if $\tau(\sigma(x)) = \sigma(x)$, or equivalently $(\sigma^{\leftarrow} \circ \tau \circ \sigma)(x)$ $= x$ for all $x \in L$. But $(\sigma^{\leftarrow} \circ \tau \circ \sigma)(x) = x$ for all $x \in L$ if and only if $\sigma^{\leftarrow} \circ \tau \circ \sigma \in L^{\blacktriangle}$, or equivalently, if and only if $\tau \in \sigma \circ L^{\blacktriangle} \circ \sigma^{\leftarrow}$.

Consequently, if $L^{\blacktriangle}$ is a normal subgroup of Γ, then $\sigma_*(L)^{\blacktriangle} = L^{\blacktriangle}$, and hence, $\sigma_*(L) = \sigma_*(L)^{\blacktriangle\blacktriangledown} = L^{\blacktriangle\blacktriangledown} = L$ for all $\sigma \in \Gamma$ as every subfield of E containing K is closed in E by Theorem 45.6. Conversely, if $\sigma_*(L) = L$ for all $\sigma \in \Gamma$, then $L^{\blacktriangle} = \sigma \circ L^{\blacktriangle} \circ \sigma^{\leftarrow}$ for all $\sigma \in \Gamma$, so $L^{\blacktriangle}$ is a normal subgroup of Γ. Thus, 1° and 2° are equivalent.

By Theorem 46.6, 3° and 4° are equivalent, since E is a separable extension of K and hence L is also. In addition, 3° and 5° are equivalent, for L is a Galois extension of K if and only if the order of Δ is $[L:K]$ by Theorem 45.5, and $[L:K] = (\Gamma:L^{\blacktriangle})$ by Theorem 45.6.

Furthermore, 2° implies 5°, for by 2°, the lemma of §45, Theorem 45.6, and Theorem 45.4, we have

$$\text{order } \Delta \geq (\Gamma:L^{\blacktriangle}) = [L:K] = [L:\Delta^{\blacktriangledown}][\Delta^{\blacktriangledown}:K] \geq [L:\Delta^{\blacktriangledown}] = \text{order } \Delta,$$

whence $(\Gamma:L^{\blacktriangle})$ is the order of Δ. Also, 4° implies 2° by Corollary 46.5.2.

By Theorem 46.5, condition 4° implies that ρ is surjective. It is easy to verify that ρ is a homomorphism with kernel $L^{\blacktriangle}$.

We conclude by establishing certain properties of separable extensions.

Theorem 46.8. Let E be an extension field of a field K.

1° If $c_1, \ldots, c_n$ are elements of E separable over K, then $K(c_1, \ldots, c_n)$ is a separable extension of K.

2° If E is a finite separable extension of K, there is a finite Galois extension of K containing E.

Proof. We shall first prove that if $c_1, \ldots, c_n$ are elements of an extension field of K that are separable over K, then there is a finite Galois extension of K containing $K(c_1, \ldots, c_n)$. Let g_k be the minimal polynomial of c_k over K for each $k \in [1, n]$. Then g_k is a separable prime polynomial over K, so $g = g_1 g_2 \ldots g_n$ is a separable polynomial over K. As in the proof of Corollary 46.5.1, a splitting field Ω of g over $K(c_1, \ldots, c_n)$ is a splitting field of g over K, and hence, Ω is a finite Galois extension of K by Theorem 46.6. Consequently, Ω is a separable extension of K by Theorem 46.6, and therefore, $K(c_1, \ldots, c_n)$ is also, so the first statement is proved. The second also follows, for if E is a finite separable extension of K, then $E = K(c_1, \ldots, c_n)$ where $\{c_1, \ldots, c_n\}$ is a basis of the K-vector space E, and by hypothesis, each c_k is separable over K.

Corollary. If E is an extension field of a field K, the set L of all elements of E separable over K is a subfield of E.

Proof. Let x and y be nonzero elements of L. Then $K(x, y)$ is a separable extension of K by Theorem 46.8 and contains $x - y$, xy, and x^{-1}, so those elements are separable over K and hence belong to L.

If E is a simple extension of K, any element c of E satisfying $K(c) = E$ is called a **primitive element** of E over K. Our final theorem, sometimes called the *theorem of the primitive element*, asserts that every finite separable extension is simple:

Theorem 46.9. A finite separable extension of a field K is a simple extension of K.

Proof. Let E be a finite separable extension of K. By Theorem 46.8, there exists a finite Galois extension Ω of K containing E. By Theorem 45.6, there are only a finite number of fields between K and Ω and *a fortiori* between K and E since the Galois group of Ω over K is finite and hence has only a finite number of subgroups. Therefore, E is a simple extension of K by Theorem 43.6.

EXERCISES

46.1. (a) If E is an extension of K of degree 2, then E is a normal extension of K.
(b) If E is a normal extension of a field K and if L is a subfield of E containing K, then E is a normal extension of L.
(c) The field $Q(\sqrt[4]{2})$ is a normal extension of $Q(\sqrt{2})$, and $Q(\sqrt{2})$ is a normal extension of Q, but $Q(\sqrt[4]{2})$ is not a normal extension of Q.

46.2. If c is a rational number whose cube root $\sqrt[3]{c}$ is irrational, then $Q(\sqrt[3]{c})$ is not a normal extension of Q.

*46.3. (a) An algebraic extension of a perfect field is perfect.
(b) A finite extension E of an imperfect field K is imperfect. [Given $b \in K$, if $a_k \in E$ satisfies $a_k^{p^k} = b$ for all $k \geq 1$ where p is the characteristic of K and if $(a_k)_{k \geq 1}$ is a sequence of distinct terms, consider a_m where m is the smallest integer such that $(a_k)_{1 \leq k \leq m}$ is linearly dependent over K.]

46.4. If Ω is an algebraically closed field whose characteristic is a prime p and if K is a subfield of Ω, then the set $E = \{x \in \Omega: x^{p^n} \in K \text{ for some } n \in N\}$ is the smallest perfect subfield of Ω containing K.

46.5. Let E be a finite extension of a field K. A field Ω is a **normal extension of K generated by E** if Ω is a normal extension of K containing E and if no proper subfield of Ω containing E is a normal extension of K.
(a) There is a polynomial $g \in K[X]$ such that an extension Ω of E is a

normal extension of K generated by E if and only if Ω is a splitting field of g over K.

(b) Two normal extensions of K generated by E are E-isomorphic.

(c) If E is a separable extension of K of degree n and if Ω is a normal extension of K generated by E, then $[\Omega : K]$ divides $n!$. [Use Theorem 46.9 and Exercise 43.6.]

46.6. Let f be a prime polynomial over a field K whose characteristic is a prime p.

(a) If $f \in K[X^q]$, then $q \mid \deg f$.

(b) There is one and only one natural number e such that $f \in K[X^{p^e}]$ but $f \notin K[X^{p^{e+1}}]$. The number e is called the **exponential degree** of f over K, and $p^{-e}(\deg f)$ is called the **reduced degree** of f over K.

*46.7. Let f be a prime polynomial over a field K whose characteristic is a prime p, and let e and m be respectively the exponential degree and the reduced degree of f over K.

(a) If $f = g(X^{p^e})$, then g is a separable prime polynomial over K of degree m.

(b) If E is a splitting field of f over K, then each root of f in E has multiplicity p^e, and consequently, f has m roots in E.

(c) The polynomial f is separable if and only if $e = 0$.

*46.8. Let K be a field whose characteristic is a prime p. If $b \in K$ and if $X^p - b$ has no root in K, then for every natural number e, the polynomial $X^{p^e} - b$ is irreducible over K. [If g is a prime factor of $X^{p^e} - b$ and c a root of g in an extension field, show that $g = (X - c)^{rp^d}$ where $p \nmid r$ and $d \leq e$; infer by Bezout's Identity that $c^{p^d} \in K$.]

*46.9. Let E be an extension of a field K whose characteristic is a prime p. An element a of E is **purely inseparable** over K if $a^{p^e} \in K$ for some natural number e. The extension E is a **purely inseparable** extension of K if every element of E is purely inseparable over K.

(a) If a is purely inseparable over K, then the minimal polynomial of a over K is $X^{p^e} - b$ for some $b \in K$ and some natural number e. [Use Exercise 46.8.]

(b) An element a of E is both separable and purely inseparable over K if and only if $a \in K$.

(c) If E is a finite purely inseparable extension of K, then $[E : K]$ is a power of p.

46.10. Let E be an algebraic extension of a field K.

(a) If L is the subfield of E consisting of all elements of E separable over K and if the characteristic of K is a prime p, then E is a purely inseparable extension of L. [Use Exercise 46.7(a).]

(b) If F is a subfield of E containing K such that E is separable over F and F is separable over K, then E is separable over K.

*46.11. Let E be a finite extension of a field K, and let L be the subfield of E consisting of all elements of E separable over K. The **separable factor** of the degree of E over K is defined to be $[L:K]$ and the **inseparable factor** of the degree of E over K is defined to be $[E:L]$. If Ω is a finite normal extension of K containing E and if n_0 is the separable degree of E over K, then there are exactly n_0 K-monomorphisms from E into Ω. [Use Theorem 46.9, $3°$ of Theorem 46.5, and Exercise 46.10(a).]

*46.12. If Ω is a finite normal extension of K, then a subfield E of Ω containing K is a normal extension of K if and only if the range of every K-monomorphism from E into Ω is E.

*46.13. If E is a finite extension of K and if n_0 is the separable factor of the degree of E over K, then there are at most n_0 K-automorphisms of E, and there are exactly n_0 K-automorphisms of E if and only if E is a normal extension of K. [Use Exercises 46.5, 46.11, and 46.12.] Infer from this and Theorem 45.5 that a finite extension of a field K is a Galois extension of K if and only if it is a normal separable extension (Theorem 46.6).

47. The Euler-Lagrange Theorem*

If K is a field over which $X^2 + 1$ is irreducible, we shall denote by i a root of $X^2 + 1$ in a stem field $K(i)$ of $X^2 + 1$ over K.

Here we shall use Galois theory, Sylow's First Theorem, and the real closure of R to prove that C is algebraically closed.

Theorem 47.1. Let K be a perfect field over which $X^2 + 1$ is irreducible. If every polynomial of odd degree over K has a root in K and if every quadratic polynomial over $K(i)$ has a root in $K(i)$, then $K(i)$ is algebraically closed.

Proof. First we shall prove that if f is a nonconstant polynomial over K, then f has a root in $K(i)$. Let E be a splitting field of f over $K(i)$. Then E is a splitting field of $(X^2 + 1)f$ over K, and hence, since K is perfect, E is a Galois extension of K by the corollary of Theorem 46.6. Let $[E:K] = 2^n r$ where r is odd. By Sylow's First Theorem, the Galois group Γ of E over K contains a subgroup Λ of order 2^n. Let L be the fixed field of Λ; then

$$[L:K] = (\Gamma:L^{\blacktriangle}) = (\Gamma:\Lambda) = r$$

by Theorem 45.6. Suppose that $r > 1$, and let h be the minimal polynomial over K of an element belonging to L but not to K. Then h is irreducible over K, $\deg h > 1$, and $\deg h$ divides $[L:K]$ and hence is odd by Theorem 43.5. By hypothesis, therefore, h has a root in K, in contradiction to the fact that h is irreducible of degree > 1. Therefore, $r = 1$.

---

* Tacit use is made in §50 of the fact that C is algebraically closed; except for this, readers will not need the contents of this section to understand the remainder of this chapter.

Consequently, $[E: K] = 2^n$, so $[E: K(i)] = 2^{n-1}$. Suppose that $n > 1$. By 1° of Theorem 45.6, E is a Galois extension of $K(i)$. By Sylow's First Theorem, the Galois group Γ_1 of E over $K(i)$ contains a subgroup Λ_1 of order 2^{n-2}. Let L_1 be the fixed field of Λ_1. Then

$$[L_1: K(i)] = (\Gamma_1: L_1^{\blacktriangle}) = (\Gamma_1: \Lambda_1) = 2$$

by Theorem 45.6. Let h_1 be the minimal polynomial over $K(i)$ of an element belonging to L_1 but not to $K(i)$. Then h_1 is an irreducible quadratic polynomial over $K(i)$. By hypothesis, h_1 has a root in $K(i)$, a contradiction. Therefore, $n = 1$. Hence, $[E: K(i)] = 1$, so $E = K(i)$, and consequently, f has a root in $K(i)$.

To complete the proof, let g be a nonconstant monic polynomial over $K(i)$, and let $\bar{g}$ be the polynomial such that for each $j \in N$ the coefficient of X^j in $\bar{g}$ is the conjugate of the coefficient a_j of X^j in g. Then $g\bar{g}$ is a polynomial over K, for if $a_j = \alpha_j + \beta_j i$ for each $j \in N$, then the coefficient of X^m in $g\bar{g}$ is

$$\sum_{j+k=m} (\alpha_j + \beta_j i)(\alpha_k - \beta_k i) = \left(\sum_{j+k=m} \alpha_j \alpha_k + \beta_j \beta_k\right) + \left(\sum_{j+k=m} \beta_j \alpha_k - \alpha_j \beta_k\right) i$$

for all $m \in N$, and

$$\sum_{j+k=m} \beta_j \alpha_k - \alpha_j \beta_k = \sum_{j=0}^{m} \beta_j \alpha_{m-j} - \sum_{k=0}^{m} \alpha_{m-k} \beta_k = 0.$$

By what we have just proved, therefore, $g\bar{g}$ has a root z in $K(i)$, and hence, z is a root either of g or of $\bar{g}$. In the latter case, $\bar{z}$ is a root of g, for

$$g(\bar{z}) = \sum_{k=0}^{n} a_k \bar{z}^k = \sum_{k=0}^{n} \bar{\bar{a}}_k \bar{z}^k = \left(\sum_{k=0}^{n} \bar{a}_k z^k\right)^{-}$$

$$= \overline{\bar{g}(z)} = \bar{0} = 0.$$

Every nonconstant monic polynomial over $K(i)$ thus has a root in $K(i)$, so $K(i)$ is algebraically closed.

An element a of a field K is a **square** of K if there exists $x \in K$ such that $x^2 = a$, and any such element x is called a **square root** of a. By 10° of Theorem 17.6, if a is a square of a totally ordered field K, then $a \geq 0$. Consequently, there is no ordering on the field $K(i)$ converting $K(i)$ into a totally ordered field, for -1 is a square of $K(i)$. In particular, the field C of complex numbers cannot be made into a totally ordered field.

If K is a totally ordered field and if $c^2 = a$, then c and $-c$ are the only square roots of a, for $X^2 - a$ has at most two roots in K. Since exactly one of c and $-c$ is positive, therefore, *every square of a totally ordered field has exactly one positive square root*.

If K is a totally ordered field, then $X^2 + 1$ is irreducible over K, as we observed in §42.

Theorem 47.2. If every positive element of a totally ordered field K is a square of K, then every element of $K(i)$ is a square of $K(i)$ and every quadratic polynomial over $K(i)$ has a root in $K(i)$.

Proof. We shall first show that every element of K is a square of $K(i)$. Indeed, if $\alpha \geq 0$, then α is a square of K and *a fortiori* of $K(i)$; if $\alpha < 0$, then $-\alpha > 0$, so there exists $\gamma \in K$ such that $\gamma^2 = -\alpha$, and consequently, $(\gamma i)^2 = \alpha$.

Next, let $z = \alpha + \beta i$ where α, $\beta \in K$ and $\beta \neq 0$. Then $\alpha^2 + \beta^2 > 0$, so there exists $\gamma > 0$ such that $\gamma^2 = \alpha^2 + \beta^2$. Since $\gamma^2 > \alpha^2$, we have $\gamma > |\alpha|$, for otherwise, we would have $\gamma \leq |\alpha|$ and hence $\gamma^2 \leq \gamma |\alpha| \leq |\alpha|^2 = \alpha^2$, a contradiction. Therefore,

$$\gamma + \alpha \geq \gamma - |\alpha| > 0,$$

so by hypothesis there exists $\lambda > 0$ such that $\lambda^2 = \frac{1}{2}(\alpha + \gamma)$. An easy calculation then establishes that $(\lambda + \mu i)^2 = z$ where $\mu = \beta/2\lambda$.

Let $aX^2 + bX + c$ be a quadratic polynomial over $K(i)$. By the preceding paragraph, there exists $d \in K(i)$ such that $d^2 = b^2 - 4ac$. An easy calculation shows that $(d - b)/2a$ is then a root of $aX^2 + bX + c$.

Theorem 47.3. If K is a totally ordered field and if $K(i)$ is algebraically closed, then every positive element of K is a square of K, and the irreducible polynomials over K are the linear polynomials and the quadratic polynomials $\alpha X^2 + \beta X + \gamma$ where $\beta^2 - 4\alpha\gamma < 0$.

Proof. Let $\alpha > 0$. As $K(i)$ is algebraically closed, there exist λ, $\mu \in K$ such that $\lambda + \mu i$ is a root of $X^2 - \alpha$. Then

$$\lambda^2 - \mu^2 = \alpha,$$
$$2\lambda\mu = 0,$$

since $(1, i)$ is an ordered basis of $K(i)$ over K. Because the characteristic of K is zero, either $\lambda = 0$ or $\mu = 0$; if $\lambda = 0$, then $\alpha = -\mu^2$, which is impossible as $\alpha > 0$; therefore, $\mu = 0$, so $\alpha = \lambda^2$ and hence is a square of K.

Let f be an irreducible polynomial over K. Since $K(i)$ is algebraically closed, there is a root c of f in $K(i)$, so

$$[K(c): K] \leq [K(i): K] = 2,$$

and hence, $[K(c):K]$ is either 1 or 2. But $[K(c):K]$ is the degree of f by Theorem 37.5, so f is either linear or quadratic.

It remains for us to show that a quadratic polynomial $\alpha X^2 + \beta X + \gamma$ has a root in K if and only if $\beta^2 - 4\alpha\gamma \geq 0$. If $\beta^2 - 4\alpha\gamma \geq 0$, there exists $\lambda \in K$ satisfying $\lambda^2 = \beta^2 - 4\alpha\gamma$ by what we have just proved, and $(\lambda - \beta)/2\alpha$ is easily seen to be a root of $\alpha X^2 + \beta X + \gamma$. Conversely, if x is a root of $\alpha X^2 + \beta X + \gamma$ in K, then

$$\beta^2 - 4\alpha\gamma = (2\alpha x + \beta)^2 \geq 0.$$

Theorem 47.4. (Euler-Lagrange) Let K be a totally ordered field. The following conditions are equivalent:

1° $K(i)$ is algebraically closed.
2° K is real-closed.
3° Every positive element of K is a square of K, and every polynomial over K of odd degree has a root in K.

Proof. To prove that 1° implies 2°, let f be a monic polynomial over K satisfying $f(a)f(b) < 0$ where $a < b$. Since f is a product of irreducible polynomials, there is a prime polynomial h over K dividing f such that $h(a)h(b) < 0$, for if a product of elements is < 0, at least one of the elements is < 0. If h were quadratic, then $h = X^2 + pX + q$ where $p^2 - 4q < 0$ by Theorem 47.3, and hence

$$h(x) = \left(x + \frac{p}{2}\right)^2 + \left(q - \frac{p^2}{4}\right) > 0$$

for all $x \in K$, whence in particular $h(a)h(b) > 0$, a contradiction. Therefore, by Theorem 47.3, $h = X - c$ for some $c \in K$. Thus, c is a root of h and hence of f. Also, $(a - c)(b - c) = h(a)h(b) < 0$ and $a - c < b - c$, so $a - c < 0 < b - c$ by 9° of Theorem 17.6, and therefore, $a < c < b$.

Next we shall prove that 2° implies 3°. Let $a > 0$, and let $f = X^2 - a$. If $a < 1$, then $f(0) < 0 < f(1)$; if $a > 1$, then $a^2 > a$, so $f(0) < 0 < f(a)$. Hence, in either case, there exists $c \in K$ such that

$$0 = f(c) = a - c^2$$

by 2°. Every positive element of K is, therefore, a square of K. Let $f = \sum_{k=0}^{n} \alpha_k X^k$ be a polynomial of odd degree $n = 2m + 1$. For every $x \in K^*$,

$$f(x) = (\alpha_n x)[x^{2m}(1 + \alpha_{n-1}\alpha_n^{-1}x^{-1} + \ldots + \alpha_0\alpha_n^{-1}x^{-n})].$$

Let

$$M = n \cdot \max\{1, |\alpha_{n-1}\alpha_n^{-1}|, \ldots, |\alpha_0\alpha_n^{-1}|\}.$$

If $|x| > M$, then

$$|\alpha_{n-k}\alpha_n^{-1}x^{-k}| < |\alpha_{n-k}\alpha_n^{-1}|\, M^{-k} \leq |\alpha_{n-k}\alpha_n^{-1}|\, M^{-1} \leq \frac{1}{n}$$

for all $k \in [1, n]$, so

$$x^{2m}(1 + \alpha_{n-1}\alpha_n^{-1}x^{-1} + \ldots + \alpha_0\alpha_n^{-1}x^{-n})$$
$$\geq x^{2m}(1 - |\alpha_{n-1}\alpha_n^{-1}x^{-1}| - \ldots - |\alpha_0\alpha_n^{-1}x^{-n}|) > 0,$$

and hence, $f(x) > 0$ if and only if $\alpha_n x > 0$. Let $c = M + 1$. Then $f(-c) < 0 < f(c)$ if $\alpha_n > 0$, and $f(c) < 0 < f(-c)$ if $\alpha_n < 0$, so by 2°, f has a root in K.

Finally, assume that 3° holds. By Theorem 39.4, the characteristic of K is zero, and hence, K is a perfect field by Theorem 46.4. Therefore, by Theorems 47.2 and 47.1, $K(i)$ is algebraically closed.

Theorem 47.5. (d'Alembert) The field C of complex numbers is an algebraically closed field.

The assertion is an immediate consequence of Theorems 42.2 and 47.4.

EXERCISES

47.1. If E is an extension of a field K, the subfield A of all elements of E algebraic over K is called the **algebraic closure of K in E**. If E is an extension of K and if E is an algebraically closed field, then the algebraic closure A of K in E is also an algebraically closed field.

47.2. An extension field E of a field K is an **algebraic closure** of K if E is an algebraic extension of K and if E is an algebraically closed field. The algebraic closure A of Q in C is an algebraic closure of Q by Exercise 47.1; elements of A are called **algebraic numbers**. Let $A_R = A \cap R$; elements of the field A_R are called **real algebraic numbers**.
(a) Show that A_R is a proper subfield of R, and conclude that A is a proper subfield of C. [Use Exercise 43.12.]
(b) Prove that A_R is a real-closed field.
(c) Conclude that $A = A_R(i)$.

47.3. If E is an algebraic closure of K, then E is an algebraic closure of every subfield of E containing K.

47.4. Any algebraic closure of R is R-isomorphic to C.

47.5. Let E be an extension of K, and let $(L_\alpha)_{\alpha \in A}$ be a family of subfields of E containing K. If F_α is the algebraic closure of K in L_α for each $\alpha \in A$, then $\bigcap_\alpha F_\alpha$ is the algebraic closure of K in $\bigcap_\alpha L_\alpha$.

*47.6. If E is an algebraic extension of a field K such that every nonconstant polynomial over K is a product of linear polynomials in $E[X]$, then E is an algebraic closure of K. [Use Theorems 43.8 and 43.10.]

47.7. If K is a real-closed, totally ordered field and if E is an algebraic extension of K, then either $E = K$ or $E = K(i)$, where i is a root of $X^2 + 1$, and consequently, $[E: K] \leq 2$. [Apply Theorem 43.10 to $E(i)$.]

*47.8. Let K be a field whose characteristic is not 2, and let A be a noncommutative division algebra over K such that every commutative division subalgebra of A has dimension ≤ 2 over K. We identify K with the subfield $K.1$ of A.
(a) There exists $a \in A$ such that $a \notin K$ but $a^2 \in K$. [Complete the square of the minimal polynomial of $a_1 \notin K$.]
(b) Let $A_+ = \{x \in A: axa^{-1} = x\}$, and let $A_- = \{x \in A: axa^{-1} = -x\}$. Show that A_+ and A_- are subspaces of the K-vector space A and that A is the direct sum of the subspaces A_+ and A_-.
(c) Show that A_+ is a division subring of A and that A_- is a one-dimensional vector space over A_+.
(d) Show that $A_+ = K[a]$, and conclude that $\dim_K A = 4$.
(e) Assume that K is a real-closed, totally ordered field. Show that for every $a \in A$ such that $a \notin K$ there exists $j \in A$ such that $j^2 = -1$ and $K[a] = K[j]$. Conclude that A is the division algebra of quaternions over K (Exercise 33.21).

47.9. (Frobenius) If K is a real-closed, totally ordered field, then a finite-dimensional division algebra over K is isomorphic either to the one-dimensional commutative division algebra K, or to the two-dimensional commutative division algebra obtained by adjoining a root of $X^2 + 1$ to K, or to the four-dimensional division algebra of quaternions. [Use Exercises 47.7 and 47.8.]

48. Roots of Unity

Finite Galois extensions of a field K are identical with splitting fields of separable polynomials over K, as we saw in §46. Here we shall investigate in detail particularly simple polynomials, namely, the polynomials $X^n - 1$ and, more generally, $X^n - b$.

Theorem 48.1. Let p be the characteristic of a field K, let b be a nonzero element of K, and let $n \in N^*$. If either $p = 0$ or $p \nmid n$, then every root of

$X^n - b$ in any extension field E of K is simple, and in particular, $X^n - b$ is separable over K.

Proof. By the corollary of Theorem 35.10, $D(X^n - b) = n.X^{n-1}$. The hypothesis concerning p implies, therefore, that the only root of $D(X^n - b)$ in E is zero. Consequently, every root of $X^n - b$ in E is simple by Theorem 35.11, and in particular, $X^n - b$ is separable over K.

An element ζ of a field K is an **nth root of unity** if $\zeta^n = 1$, i.e., if ζ is a root of $X^n - 1$. The set of nth roots of unity in K, therefore, has at most n members and is, furthermore, a subgroup of the multiplicative group K^*, for if $\zeta^n = \xi^n = 1$, then $(\zeta\xi)^n = \zeta^n\xi^n = 1$ and $(\zeta^{-1})^n = (\zeta^n)^{-1} = 1$. Consequently, by Theorem 38.7, *the group of nth roots of unity in K is a cyclic group.*

Definition. A **primitive nth root of unity** in a field K is any nth root of unity whose order (in the group of nth roots of unity) is n.

Since the group of nth roots of unity in K is cyclic and has at most n members, *an nth root of unity ζ in K is primitive if and only if the group of nth roots of unity in K has n elements and ζ is a generator of that group.* For example, Q has no primitive cube root of unity, since 1 is the only rational cube root of unity. On the other hand, if K is algebraically closed and if the characteristic of K is either zero or a prime not dividing n, then K contains n nth roots of unity of Theorem 48.1. We have an analytic expression for the n nth roots of unity of the algebraically closed field C: They are the complex numbers

$$\cos\frac{2\pi k}{n} + i\sin\frac{2\pi k}{n}$$

where $k \in [0, n - 1]$, and the primitive nth roots of unity in C are those numbers for which k is relatively prime to n.

If K possesses a primitive nth root of unity, then $X^n - 1$ splits over K, and moreover,

$$X^n - 1 = \prod_{\zeta \in G}(X - \zeta)$$

where G is the group of all nth roots of unity in K, since G contains n members and since there are at most n roots of $X^n - 1$ in K.

Definition. Let K be a field possessing a primitive nth root of unity. The **cyclotomic polynomial** over K of index n is the polynomial Φ_n defined by

$$\Phi_n = \prod_{\zeta \in S}(X - \zeta)$$

where S is the set of all primitive nth roots of unity in K.

For example, if K is any field, then 1 is the only primitive first root of unity in K, so

$$\Phi_1 = X - 1.$$

If K is a field whose characteristic is not 2, then -1 is the only primitive square root of unity in K, so

$$\Phi_2 = X + 1.$$

Let K be a field possessing a primitive nth root of unity ζ_0, let G be the multiplicative group of all nth roots of unity in K, and for each positive divisor m of n let S_m be the set of all nth roots of unity in K or order m. If $m \mid n$, then $\zeta_0^{n/m}$ is clearly a primitive mth root of unity, and every primitive mth root of unity in K is an nth root of unity and hence belongs to S_m, so

$$\Phi_m = \prod_{\zeta \in S_m} (X - \zeta).$$

By Theorem 23.8, $\{S_m : m \mid n\}$ is a partition of G, and consequently,

$$\prod_{\zeta \in G} (X - \zeta) = \prod_{m \mid n} \left(\prod_{\zeta \in S_m} (X - \zeta) \right).$$

Therefore,

(1) $$X^n - 1 = \prod_{m \mid n} \Phi_m.$$

This equality enables one recursively to calculate the coefficients of the cyclotomic polynomials. For example, if n is a prime and if K possesses a primitive nth root of unity, then $X^n - 1 = \Phi_1 \Phi_n = (X - 1)\Phi_n$, so

$$\Phi_n = X^{n-1} + X^{n-2} + \ldots + X + 1.$$

If K possesses a primitive sixth root of unity, to calculate Φ_6 we observe that $X^6 - 1 = \Phi_1 \Phi_2 \Phi_3 \Phi_6$ and $X^3 - 1 = \Phi_1 \Phi_3$, so $X^6 - 1 = (X^3 - 1)(X + 1)\Phi_6$, and hence,

$$\Phi_6 = X^2 - X + 1.$$

An easy inductive argument utilizing (1) and Theorem 34.8 establishes that *if K is a field possessing a primitive nth root of unity, then the coefficients of the cyclotomic polynomials of index n over K belong to the prime subfield of K.*

Let K be a field whose characteristic is either zero or not a divisor of n, and let L and L' be splitting fields of $X^n - 1$ over K. By Theorem 48.1, the groups of nth roots of unity in L and in L' both have n elements, and hence, both L and L' possess a primitive nth root of unity. Let S and S' be respectively the sets of primitive nth roots of unity in L and in L', and let Φ_n and Φ'_n be respectively the cyclotomic polynomials of index n over L and over L'. By the corollary of Theorem 37.9, there is a K-isomorphism σ from L onto L'; let $\bar{\sigma}$ be the isomorphism from the ring $L[X]$ onto the ring $L'[X]$ induced by σ. Clearly, ζ' is a primitive nth root of unity in L' if and only if there is a primitive nth root of unity ζ in L such that $\zeta' = \sigma(\zeta)$; therefore,

$$\Phi'_n = \prod_{\zeta' \in S'} (X - \zeta') = \prod_{\zeta \in S} (X - \sigma(\zeta))$$

$$= \bar{\sigma}\left(\prod_{\zeta \in S} (X - \zeta)\right) = \bar{\sigma}(\Phi_n).$$

But since the coefficients of Φ_n lie in the prime subfield of L and hence in K and since σ is a K-isomorphism, $\bar{\sigma}(\Phi_n) = \Phi_n$, and therefore,

$$\Phi'_n = \Phi_n.$$

Consequently, we may unambiguously make the following definition:

Definition. Let K be a field whose characteristic is either zero or not a divisor of n. The **cyclotomic polynomial** of index n over K is the cyclotomic polynomial of index n over a splitting field of $X^n - 1$ over K.

Thus, the cyclotomic polynomial of index n is defined over any field K whose characteristic is either zero or not a divisor of n and is identical with the cyclotomic polynomial of index n over the prime subfield of K. On the other hand, if L is a splitting field of $X^n - 1$ over a field K whose characteristic is a prime p dividing n and if $n = mp$, then L contains no primitive nth roots of unity, for if $\zeta^n = 1$, then

$$0 = \zeta^n - 1 = \zeta^{mp} - 1^p = (\zeta^m - 1)^p,$$

so $\zeta^m = 1$, and hence, the order of ζ is less than n.

An easy inductive argument utilizing (1) and Theorem 34.8 establishes that *the coefficients of the cyclotomic polynomials over Q are all integers.* Moreover, if p is a prime not dividing n, the cyclotomic polynomial of index n over Z_p is closely related to that of index n over Q; indeed, if φ_p is the canonical epimorphism from the ring Z onto the field Z_p and if $\bar{\varphi}_p$ is the epimorphism from the ring $Z[X]$ onto the ring $Z_p[X]$ induced by φ_p, an inductive argument utilizing (1) establishes that *the cyclotomic polynomial of index n over* $\boldsymbol{Z}_p$ *is the image under* $\bar{\varphi}_p$ *of the cyclotomic polynomial of index n over* $\boldsymbol{Q}$.

If the characteristic of a field K is either zero or not a divisor of n, the number of primitive nth roots of unity in a splitting field of $X^n - 1$ over K is the number of generators of a cyclic group of order n; that is, it is $\varphi(n)$ where φ is the Euler φ-function, since the group of nth roots of unity is cyclic of order n. Consequently,

$$\deg \Phi_n = \varphi(n).$$

The examples of cyclotomic polynomials given above suggest that each of the coefficients of Φ_n is either 1, 0, or -1; actually, this is true if $n < 105$, but the coefficient of X^{41} in Φ_{105} is -2. Moreover, a theorem of I. Schur asserts that the coefficients of the cyclotomic polynomials over $\boldsymbol{Q}$ can be arbitrarily large; i.e., for every $k \in N^*$, there exists $n \in N^*$ such that the absolute value of some coefficient of Φ_n is greater than k.

For every field K and every strictly positive integer n, we shall denote by $R_n(K)$ a splitting field of $X^n - 1$ over K. Since C is algebraically closed, C contains a splitting field of $X^n - 1$ over any subfield of itself, so if K is a subfield of C, we shall tacitly assume that $R_n(K)$ is the splitting field of $X^n - 1$ over K contained in C. As we observed above, if the characteristic of K is either zero or not a divisor of n, then $R_n(K)$ contains a primitive nth root of unity.

Theorem 48.2. If the characteristic of K is either zero or not a divisor of n, then $R_n(K)$ is a finite Galois extension of K, and the Galois group Γ of $R_n(K)$ over K is isomorphic to a subgroup of the multiplicative group of invertible elements of the ring $\boldsymbol{Z}_n$.

Proof. Since $X^n - 1$ is separable over K by Theorem 48.1, the first assertion follows from Theorem 46.6. Let ζ be a primitive nth root of unity in $R_n(K)$. Then $R_n(K) = K(\zeta)$ since every nth root of unity in $R_n(K)$ is a power of ζ. For each $\sigma \in \Gamma$, let $\chi(\sigma)$ be defined by

$$\chi(\sigma) = \{k \in \boldsymbol{Z} : \sigma(\zeta) = \zeta^k\}.$$

If $\sigma \in \Gamma$, then $\sigma(\zeta)$ is also an nth root of unity and hence is a power of ζ, so $\chi(\sigma) \neq \emptyset$; also, as the order of ζ is n, $\zeta^k = \zeta^j$ if and only if $k \equiv j \pmod{n}$, so $\chi(\sigma)$ is an element of the ring $\boldsymbol{Z}_n$, namely, the element $k + (n)$ where k is any integer such that $\sigma(\zeta) = \zeta^k$. If $\sigma, \tau \in \Gamma$ and if j and k are integers such that $\sigma(\zeta) = \zeta^j$ and $\tau(\zeta) = \zeta^k$, then

$$(\sigma \circ \tau)(\zeta) = \sigma(\zeta^k) = \sigma(\zeta)^k = \zeta^{jk},$$

and therefore,

$$\chi(\sigma \circ \tau) = \chi(\sigma)\chi(\tau).$$

Clearly, χ takes the identity automorphism of $R_n(K)$ into the identity element of the ring Z_n; conversely, if $\chi(\sigma)$ is the multiplicative identity of Z_n, then $\sigma(\zeta) = \zeta$, and hence, σ is the identity automorphism, since $R_n(K) = K(\zeta)$. Therefore, χ is an isomorphism from Γ onto a subgroup H of the multiplicative semigroup Z_n that contains the multiplicative identity of Z_n. Since H is a group, every element of H is, consequently, an invertible element of the ring Z_n.

Theorem 48.3. For every $n \in N^*$, Φ_n is irreducible over Q.

Proof. Let g be a prime factor of Φ_n in $Q[X]$. As we observed above, $\Phi_n \in Z[X]$. By Theorem 36.2, there is a scalar multiple g_0 of g with integral coefficients such that g_0 divides Φ_n in $Z[X]$. Since Φ_n is monic, the leading coefficient of g_0 is either 1 or -1, so g_0 is either g or $-g$, and therefore, g is actually a polynomial over Z. Since Φ_n divides $X^n - 1$ in $Q[X]$, g does also, so there is a polynomial $h \in Q[X]$ such that $X^n - 1 = gh$. By Theorem 34.8, $h \in Z[X]$.

Let ζ be any root of g in $R_n(Q)$. Since g divides Φ_n, ζ is a primitive nth root of unity. We shall prove that if p is a prime not dividing n, then ζ^p is also a root of g. Suppose, on the contrary, that $g(\zeta^p) \neq 0$. Then $h(\zeta^p) = 0$, since ζ^p is an nth root of unity and since $X^n - 1 = gh$. As g is a prime polynomial, g is the minimal polynomial of ζ over Q and hence divides $h(X^p)$ by Theorem 43.1. Let $u \in Q[X]$ be such that

$$h(X^p) = gu.$$

Then $u \in Z[X]$ by Theorem 34.8. For each $\alpha \in Z$, let $\bar{\alpha}$ be the coset $\alpha + (p)$ in Z_p, and let φ be the canonical epimorphism from the ring Z onto the field Z_p defined by $\varphi(\alpha) = \bar{\alpha}$. By Theorem 34.2,

$$\bar{\varphi}: \sum_{k=0}^{r} \alpha_k X^k \mapsto \sum_{k=0}^{r} \bar{\alpha}_k Y^k$$

is an epimorphism from the ring $Z[X]$ onto the ring $Z_p[Y]$; for every $v \in Z[X]$, we shall denote $\bar{\varphi}(v)$ simply by $\bar{v}$. Since $Y^n - \bar{1} = \bar{\varphi}(X^n - 1)$ and $X^n - 1 = gh$, we have

(2) $$Y^n - \bar{1} = \bar{g}\bar{h}.$$

If $h = \sum\limits_{k=0}^{r} \beta_k X^k$, then since $\bar{\beta}^p = \bar{\beta}$ for each $\bar{\beta} \in Z_p$ by Theorem 38.3, we have

$$\bar{h}^p = \left(\sum_{k=0}^{r} \bar{\beta}_k Y^k\right)^p = \sum_{k=0}^{r} \bar{\beta}_k^p Y^{kp} = \bar{\varphi}\left(\sum_{k=0}^{r} \bar{\beta}_k X^{pk}\right)$$

$$= \bar{\varphi}(h(X^p)) = \bar{\varphi}(gu) = \bar{\varphi}(g)\bar{\varphi}(u)$$

by Theorem 38.2, so

(3) $$\bar{h}^p = \bar{g}\bar{u}.$$

Since g is a monic polynomial, $\bar{g}$ has the same degree as g and, in particular, is not a constant polynomial. Let $\bar{v}$ be an irreducible factor of $\bar{g}$ in $Z_p[Y]$. By (3), Theorem 22.9, and (UFD 3'), $\bar{v}$ is also an irreducible factor of $\bar{h}$, so by (2), $\bar{v}^2$ is a factor of $Y^n - \bar{1}$. Consequently, a root of $\bar{v}$ in any splitting field of $Y^n - \bar{1}$ is a multiple root of $Y^n - \bar{1}$, a contradiction of Theorem 48.1 as $p \nmid n$. Therefore, $g(\zeta^p) = 0$.

Because g is a nonconstant factor of Φ_n, some primitive nth root of unity ζ is a root of g. We shall show that if ξ is any other primitive nth root of unity in $R_n(Q)$, then ξ is also a root of g. Since ζ is primitive, $\xi = \zeta^r$ for some integer $r > 1$. Let $r = p_1 p_2 \ldots p_m$ where $(p_k)_{1 \le k \le m}$ is a sequence of (not necessarily distinct) primes. For each $k \in [1, m]$, the prime p_k does not divide n, for if $p_k s = n$ and if $q = p_1 \ldots p_{k-1} p_{k+1} \ldots p_m$, then $\xi^s = \zeta^{nq} = 1$ and $s < n$, a contradiction of the primitivity of ξ. By what we have just proved, therefore, $\zeta, \zeta^{p_1}, \zeta^{p_1 p_2}, \ldots, \zeta^{p_1 p_2 \ldots p_m} = \xi$ are all roots of g. Hence, every primitive nth root of unity in $R_n(Q)$ is a root of g. Consequently, $\Phi_n = g$ by the definition of Φ_n and Theorem 35.8. Therefore, Φ_n is irreducible over Q.

Corollary. The field $R_n(Q)$ is a Galois extension of Q of degree $\varphi(n)$ (where φ is the Euler φ-function), and the Galois group Γ of $R_n(Q)$ over Q is isomorphic to the multiplicative group of invertible elements of the ring Z_n.

Proof. Since Φ_n is irreducible over Q and since $R_n(Q) = Q(\zeta)$ where ζ is any primitive nth root of unity,

$$[R_n(Q): Q] = \deg \Phi_n = \varphi(n)$$

by Theorem 37.5. By Theorem 48.2, Γ is isomorphic to a subgroup of the multiplicative group of invertible elements of Z_n; the latter group has $\varphi(n)$ elements, however, and the order of Γ is $\varphi(n)$ by Theorem 45.5; therefore, Γ is isomorphic to the group of all invertible elements of Z_n.

In contrast with Theorem 48.3, a cyclotomic polynomial over Z_p need not be irreducible. For example, $\Phi_6(3) = 0$ in Z_7.

Our discussion of cyclotomic polynomials enables us to resolve the classical geometric problem of constructing a regular polygon by ruler and compass. Indeed, if a regular polygon of n sides is constructible by ruler and compass, so is the angle subtended at its center by a side, an angle of $2\pi/n$ radians. Conversely, if the angle of $2\pi/n$ radians whose vertex is the origin and whose initial side is the positive half of the X-axis is constructible by ruler and compass, then the unit circle Q whose center is the origin intersects the sides of the angle at constructible points $P_0 = (1, 0)$ and $P_1 = (\cos 2\pi/n, \sin 2\pi/n)$; the circle of center P_1 passing through P_0 intersects Q at another constructible point $P_2 = (\cos 4\pi/n, \sin 4\pi/n)$; and continuing in this way we may construct by ruler and compass the vertices of a regular polygon of n sides. As we saw in §44, an angle of $2\pi/n$ radians is constructible if and only if the point $(\cos 2\pi/n, \sin 2\pi/n)$ is constructible, or equivalently, if and only if the complex number $\cos 2\pi/n + i \sin 2\pi/n$ is constructible. But this number is a primitive nth root of unity. Of course, if one primitive nth root of unity ζ is constructible, then all nth roots of unity are constructible since they are powers of ζ. Therefore, a regular polygon of n sides is constructible by ruler and compass if and only if the nth roots of unity in C are constructible complex numbers.

Theorem 48.4. The nth roots of unity in C are constructible complex numbers if and only if $\varphi(n)$ is a power of 2.

Proof. By Theorem 48.3, the degree of a primitive nth root of unity over Q is $\varphi(n)$, so the condition is necessary by Theorem 44.5.

Sufficiency: Let $\varphi(n) = 2^m$. By Theorem 44.4, it suffices to show that there is a sequence $(K_j)_{0 \le j \le m}$ of subfields of C satisfying

$$Q = K_0 \subset K_1 \subset \ldots \subset K_m = R_n(Q),$$

$$[K_j : K_{j-1}] = 2 \text{ for all } j \in [1, m].$$

Let Γ be the Galois group of $R_n(Q)$ over Q. Then Γ is an abelian group of order $\varphi(n) = 2^m$ by the corollary of Theorem 48.3. By Sylow's First Theorem, every group of order 2^n where $n > 0$ contains a subgroup of order 2^{n-1}. (Here is a sketch of a direct proof of the fact that an abelian group of order 2^n, where $n > 0$, has a subgroup of order 2^{n-1}: We proceed by induction on n. The assertion is clear if $n = 1$. Suppose that every abelian group of order 2^r has a subgroup of order 2^{r-1}, and let G be an abelian group of order 2^{r+1}. Let $a \in G$ be other than the neutral element; the order of a is then 2^k for some $k \in [1, r + 1]$; let $b = a^{2^{k-1}}$. Then the subgroup $[b]$ generated by b has order 2; let φ be the canonical epimorphism from G

onto $G/[b]$. Since $G/[b]$ has order 2^r, $G/[b]$ has a subgroup H of order 2^{r-1}; consequently, $\varphi^*(H)$ is a subgroup of G of order 2^r.) Hence, by induction, there exists a sequence $(\Gamma_j)_{0 \leq j \leq m}$ of subgroups of Γ such that

$$\Gamma = \Gamma_0 \supset \Gamma_1 \supset \ldots \supset \Gamma_m = \{1_{R_n(Q)}\}$$

and the order of $\Gamma_j = 2^{m-j}$ for each $j \in [0, m]$. Let $K_j = \Gamma_j^{\blacktriangledown}$ for each $j \in [0, m]$. By the Fundamental Theorem of Galois Theory,

$$Q = K_0 \subset K_1 \subset \ldots \subset K_m = R_n(Q),$$

$$[R_n(Q): K_j] = \text{order } \Gamma_j = 2^{m-j}$$

for all $j \in [0, m]$, whence

$$[K_j: K_{j-1}] = \frac{[R_n(Q): K_{j-1}]}{[R_n(Q): K_j]} = \frac{2^{m-j+1}}{2^{m-j}} = 2$$

for all $j \in [1, m]$. This completes the proof.

Let p be a prime. Then $\varphi(p) = p - 1$ by Theorem 23.11. Hence, $\varphi(p) = 2^m$ if and only if $p = 2^m + 1$. Primes of the form $2^m + 1$ are called **Fermat primes**. If $m = 2^h s$ where s is odd, then $2^{2^h} + 1$ divides $2^m + 1$ since

$$2^m + 1 = (2^{2^h} + 1) \sum_{k=1}^{s} (-1)^{k-1} 2^{2^h(s-k)}.$$

Therefore, if $2^m + 1$ is a prime, then m itself is a power of 2. In sum:

Theorem 48.5. If p is a prime, then a regular polygon of p sides is constructible by ruler and compass if and only if $p = 2^{2^h} + 1$ for some natural number h.

For each natural number h let $F_h = 2^{2^h} + 1$. If h is respectively 0, 1, 2, 3, 4, then F_h is respectively the prime 3, 5, 17, 257, and 65,537. No more Fermat primes are known, and it is known that F_h is not prime if $h \in [5, 12]$. Therefore, for no prime $p < 2^{8192}$ (a number of 2,467 digits, the first seven of which are 1090748) is a regular polygon of p sides constructible by ruler and compass, except for the primes 3, 5, 17, 257, and 65,537.

Theorem 48.6. If $n \geq 3$, then a regular polygon of n sides is constructible by ruler and compass if and only if either

$$n = 2^m,$$

where $m \geq 2$, or

$$n = 2^m p_1 p_2 \cdots p_r,$$

where $m \geq 0$ and $(p_k)_{1 \leq k \leq r}$ is a sequence of distinct Fermat primes.

Proof. Let p be a prime, $m \geq 1$. By Theorem 23.11, $\varphi(p^m) = p^m - p^{m-1} = p^{m-1}(p - 1)$. Hence, $\varphi(2^m)$ is a power of 2, and if $p \neq 2$, $\varphi(p^m)$ is a power of 2 if and only if $m = 1$ and p is a Fermat prime. The assertion, therefore, follows from Theorem 48.4 and Corollary 24.8.3.

A field E is a **cyclic extension** of K if E is a finite Galois extension of K and if the Galois group of E over K is cyclic. The following theorems relate cyclic extensions with splitting fields of polynomials of the form $X^n - b$, which are sometimes called **pure polynomials.**

Theorem 48.7. Let K be a field containing a primitive nth root of unity ζ. If $b \in K^*$ and if E is a splitting field of $X^n - b$ over K, then E is a finite Galois extension of K, $E = K(c)$ for any root c of $X^n - b$, and the Galois group Γ of E over K is isomorphic to a subgroup of the additive cyclic group $\mathbf{Z}_n$ and hence is, in particular, cyclic.

Proof. By Theorems 48.1 and 46.6, E is a finite Galois extension of K. Let c be a root in E of $X^n - b$. Then $c, \zeta c, \zeta^2 c, \ldots, \zeta^{n-1}c$ are n distinct roots of $X^n - b$, so every root of $X^n - b$ in E is among them, whence $E = K(c)$. For each $\sigma \in \Gamma$, let $\chi(\sigma)$ be defined by

$$\chi(\sigma) = \{k \in \mathbf{Z} : \sigma(c) = \zeta^k c\}.$$

Then $\chi(\sigma) \neq \emptyset$, since $\sigma(c)$ is also a root of $X^n - b$. Moreover, $\chi(\sigma) \in \mathbf{Z}_n$, for $\zeta^k c = \zeta^j c$ if and only if $k \equiv j \pmod n$. If $\sigma, \tau \in \Gamma$ and if $\sigma(c) = \zeta^k c$ and $\tau(c) = \zeta^j c$, then

$$(\sigma \circ \tau)(c) = \sigma(\zeta^j c) = \zeta^j \sigma(c) = \zeta^{j+k} c,$$

so

$$\chi(\sigma \circ \tau) = \chi(\sigma) + \chi(\tau).$$

If $\chi(\sigma) = 0$, then $\sigma(c) = c$, so $\sigma = 1_E$ as $E = K(c)$. Consequently, χ is an isomorphism from Γ onto a subgroup of $(\mathbf{Z}_n, +)$.

Theorem 48.8. Let K be a field containing a primitive nth root of unity ζ, and let E be an extension field of K. The following statements are equivalent:

1° There exists $b \in K$ such that the polynomial $X^n - b$ is irreducible over K and E is a splitting field of $X^n - b$ over K.

2° E is a Galois extension of K, and the Galois group Γ of E over K is cyclic of order n.

Proof. Condition 1° implies 2°: By Theorem 48.7, $E = K(c)$ where c is a root of $X^n - b$ in E, and Γ is isomorphic to a subgroup of the additive cyclic group $\mathbf{Z}_n$. Thus, since $X^n - b$ is irreducible over K,

$$n = [E: K] = \text{order } \Gamma$$

by Theorems 37.5 and 45.5. Therefore, G is isomorphic to $\mathbf{Z}_n$.

Condition 2° implies 1°: Let σ be a generator of Γ. Then Γ consists of the n distinct K-automorphisms $\sigma, \sigma^2, \ldots, \sigma^{n-1}, \sigma^n = 1_E$. By Theorem 45.2, the endomorphism

$$1_E + \zeta^{-1}\sigma + \zeta^{-2}\sigma^2 + \ldots + \zeta^{-n+2}\sigma^{n-2} + \zeta^{-n+1}\sigma^{n-1}$$

of the K-vector space E is not the zero endomorphism. Consequently, there exists $a \in E$ such that the element c defined by

$$c = a + \zeta^{-1}\sigma(a) + \zeta^{-2}\sigma^2(a) + \ldots + \zeta^{-n+2}\sigma^{n-2}(a) + \zeta^{-n+1}\sigma^{n-1}(a)$$

is not zero. Now

$$\begin{aligned}
\sigma(c) &= \sigma(a) + \zeta^{-1}\sigma^2(a) + \zeta^{-2}\sigma^3(a) + \ldots + \zeta^{-n+2}\sigma^{n-1}(a) + \zeta^{-n+1}a \\
&= \zeta[\zeta^{-1}\sigma(a) + \zeta^{-2}\sigma^2(a) + \zeta^{-3}\sigma^3(a) + \ldots + \zeta^{-n+1}\sigma^{n-1}(a) + a] \\
&= \zeta c.
\end{aligned}$$

From this, it is easy to see by an inductive argument that for all $k \in [1, n]$,

(4) $$\sigma^k(c) = \zeta^k c,$$

whence

$$\sigma^k(c^j) = \sigma^k(c)^j = \zeta^{kj}c^j$$

for all natural numbers j, and in particular,

(5) $$\sigma^k(c^n) = c^n.$$

Let $b = c^n$. Since E is a Galois extension of K, $b \in K$ by (5) and hence $X^n - b$ is a polynomial over K. We shall show that $E = K(c)$. By (4) and since $c \neq 0$, the only member of Γ leaving each element of $K(c)$ fixed is the identity automorphism, so $K(c)^{\blacktriangle} = \{1_E\}$. Consequently,

$$K(c) = K(c)^{\blacktriangle\blacktriangledown} = \{1_E\}^{\blacktriangle} = E$$

by the Fundamental Theorem of Galois Theory. Hence, E is a splitting field of $X^n - b$, since $E = K(c)$ and since $c, \zeta c, \ldots, \zeta^{n-1}c$ are n distinct roots of $X^n - b$ in E. The minimal polynomial g of c over K divides $X^n - b$ and in addition satisfies

$$\deg g = [K(c) : K] = [E : K] = \text{order } \Gamma = n,$$

so $g = X^n - b$, and therefore, $X^n - b$ is irreducible over K.

We saw in Example 45.4, in contrast, that the Galois group of a splitting field of $X^3 - 2$ over Q is not cyclic.

EXERCISES

48.1. What are the coefficients of Φ_4? Φ_8? Φ_{16}? Φ_{2^n}?

48.2. What are the coefficients of Φ_6? Φ_{10}? Φ_{14}? Φ_{22}? Φ_{2p} where p is an odd prime?

*__48.3.__ (a) If m is an odd number > 1, then $\Phi_{2m} = \Phi_m(-X)$. [Use induction.]
(b) If p is a prime and if $m \in N^*$, then $\Phi_{p^m} = \Phi_p(X^{p^{m-1}})$.
(c) If $m \in N^*$ and if p is a prime that does not divide m, then $\Phi_{pm}\Phi_m = \Phi_m(X^p)$. [Use induction.]

48.4. Exhibit Φ_n for each $n \in [1, 16]$.

48.5. (a) Prove that the coefficients of the cyclotomic polynomial of index n over a field K possessing a primitive nth root of unity belong to the prime subfield of K.
(b) Prove that the coefficients of the cyclotomic polynomials over Q are integers.
(c) Let p be a prime, and let φ_p be the canonical epimorphism from Z onto Z_p. Prove that if $\bar{\varphi}_p$ is the epimorphism from $Z[X]$ onto $Z_p[X]$ induced by φ_p and if p does not divide n, then the cyclotomic polynomial of index n over Z_p is the image under $\bar{\varphi}_p$ of the cyclotomic polynomial of index n over Q.

*__48.6.__ Let K be a field whose characteristic is either zero or not a divisor of n.
(a) The degree of every prime factor in $K[X]$ of Φ_n is $[R_n(K) : K]$.
(b) If K has q elements, then $[R_n(K) : K]$ is the smallest of the strictly positive integers m such that $n \mid q^m - 1$.

(c) If K has q elements, then Φ_n is irreducible over K if and only if the order of the coset $q + (n)$ in the group of invertible elements of Z_n is $\varphi(n)$.

48.7. (a) If p is a prime other than 2 or 3, then $[R_{12}(Z_p): Z_p] \leq 2$, Φ_{12} is reducible over Z_p, and Z_p contains a 12th root of unity other than 1. [Use Exercise 48.6.]
(b) The field Z_3 contains no 11th root of unity other than 1, but Φ_{11} is reducible over Z_3. [Use Exercise 48.6.]

48.8. Let K be a field whose characteristic is either zero or not a divisor of n. If $n \neq 2$, the product of the primitive nth roots of unity in $R_n(K)$ is 1, and if $n \neq 1$, the constant coefficient of Φ_n is 1. [Pair ζ^{-1} with ζ.]

*48.9. Let $n \in N^*$.
(a) If p is a prime not dividing n but if p divides $\Phi_n(a)$ for some integer a, then $p \equiv 1 \pmod{n}$.
(b) There are infinitely many primes of the form $nk + 1$ where $k \in N$. [If there were finitely many and if b were their product, apply (a) and Exercise 48.8 in showing that for every integer r, $\Phi_n(nbr)$ would be -1 or 1.]
(c) There are infinitely many primes p not dividing n such that Φ_n has a root in Z_p. [Argue as in (b).]

48.10. If a field E is a finite extension of its prime subfield, then E contains only finitely many roots of unity. [Use Exercise 24.12.]

48.11. Let K be a field, and let $n \in N$. For each nonzero element b of an extension field of K, the set (b, n, K) defined by

$$(b, n, K) = \{r \in Z: \text{there exists } x \in K \text{ such that } b^r = x^n\}$$

is an ideal of Z.

*48.12. Let K be a field possessing a primitive nth root of unity ζ, and let $b \in K^*$. There exist $r, s \in N^*$ and an element $c \in K$ such that $rs = n$, $c^s = b$,

$$X^n - b = (X^r - c)(X^r - \zeta^r c)(X^r - \zeta^{2r} c) \ldots (X^r - \zeta^{(s-1)r} c),$$

and $X^r - \zeta^{kr} c$ is irreducible over K for each $k \in [0, s - 1]$. [Let a be a root of $X^n - b$ in a splitting field E, and let r be the positive generator of $(a, 1, K)$. Factor $X^r - \zeta^{kr} c$ in E, and to show its irreducibility over K, consider the constant coefficient of an irreducible factor.]

48.13. If q is a prime and if K is a field possessing a primitive qth root of unity, then for every $b \in K^*$, the polynomial $X^q - b$ either is irreducible over K or has q distinct roots in K. [Use Exercise 48.12.]

48.14. Let K be a field whose characteristic is either zero or not a divisor of n, and let $b \in K^*$. If h is a prime factor of $X^n - b$ in $K[X]$, then $\deg h \in (b, n, K)$. [Factor $X^n - b$ in an extension field containing a primitive nth root of unity and a root of $X^n - b$, and compute the nth power of the constant coefficient of h.]

48.15. Let q be a prime, and let K be a field whose characteristic is not q. If $b \in K^*$, then $X^q - b$ either is irreducible over K or has a root in K. [Use Exercise 48.14.]

*48.16. Let q be an odd prime, let K be a field whose characteristic is not q, let $k \in N^*$, and let $b \in K^*$ be such that the polynomial $f = X^{q^{k-1}} - b$ is irreducible over K. Let E be a field obtained by adjoining to K a primitive q^kth root of unity ζ and a root d of the polynomial $g = X^{q^k} - b$, and let $c = d^q$, $v = \zeta^q$. Finally, let h be a prime factor of g in $K[X]$, and let s be the number of roots in E that h has in common with $X^q - c$.

(a) Express f as a product of linear polynomials over $K(v, c)$.

(b) Express g as a product of polynomials of degree q over $K(v, c)$. [Observe that $g = f(X^q)$.]

(c) Express each of the factors occurring in (b) as a product of linear polynomials over E.

(d) h has exactly s roots in E in common with each of the factors occurring in the factorization of (b), and $\deg h = sq^{k-1}$. [Use 3° of Theorem 46.5 and the irreducibility of f.]

(e) The ideal (b, q, K) contains s. [Compute the qth power of the constant coefficient of h.]

(f) $s = q$, and hence, g is irreducible over K.

(g) Show that if $b = -64$, $q = 2$, $k = 2$, then $X^{q^{k-1}} - b$ is irreducible but $X^{q^k} - b$ is reducible over Q. At what point does the preceding argument fail if q is 2 instead of an odd prime?

48.17. Let q be an odd prime, and let K be a field whose characteristic is not q. If b is an element of K^* that is not the qth power of any element of K, then $X^{q^k} - b$ is irreducible over K for all $k \in N$. [Use Exercises 48.15 and 48.16.]

*48.18. Let K be a field whose characteristic is a prime p, let $b \in K$, and let E be a splitting field of $X^p - X - b$ over K.

(a) If a is a root of $f = X^p - X - b$ in E, then $a, a + 1, a + 2, \ldots, a + (p - 1)$ are all the roots of f in E, and $E = K(a)$.

(b) The minimal polynomials over K of the roots of f all have the same degree.

(c) Either f is irreducible over K, or f has p distinct roots in K.

(d) If f is irreducible over K, then E is a Galois extension of K, and hence, the Galois group of E over K is a (cyclic) group of order p.

*48.19. Let K be a field whose characteristic is a prime p, let $b \in K$, and let $f = X^p - X - b$. If f is irreducible over K and if a is a root of f in an extension field of K, then $X^p - X - ba^{p-1}$ is irreducible over $K(a)$. [Use Exercise 48.18(c) and 4° of Theorem 37.5.]

48.20. If b is an integer and if p is a prime not dividing b, then $X^p - X - b$ is irreducible over Q. [Apply Exercise 48.18 to Z_p.]

*48.21. If E is a Galois extension of a field K whose characteristic is a prime p and if the Galois group Γ of E over K has order p, then there exists $b \in K$ such that $X^p - X - b$ is irreducible over K and E is a splitting field of $X^p - X - b$. [Let σ be a generator of Γ, let c be an element of E such that $Tr_{E/K}(c) = -1$ (Exercise 45.8), and let $u = c + 2\sigma(c) + 3\sigma^2(c) + \ldots + (p-1)\sigma^{p-2}(c)$; show that $\sigma(u) = u + 1$ and that $u^p - u \in K$.]

*48.22. Let E be a finite Galois extension of a field K such that the Galois group Γ of E over K is cyclic of order n, and let σ be a generator of Γ. If $b \in E$, then $Tr_{E/K}(b) = 0$ (Exercise 45.8) if and only if there exists $z \in E$ such that $b = z - \sigma(z)$. [Consider

$$z = \sum_{k=0}^{n-1} \sigma^k(c)[b + \sigma(b) + \ldots + \sigma^k(b)]$$

where $Tr_{E/K}(c) = 1$.]

48.23. Let q be a power of a prime p, and let K be a field having q^n elements.
(a) If $b \in K$, then $X^q - X - b$ has a root in K if and only if $b + b^q + b^{q^2} + \ldots + b^{q^{n-1}} = 0$. [Use Exercise 48.22.]
(b) If $q = p$ and if $b \in K$, then $X^p - X - b$ is irreducible over K if and only if $b + b^p + b^{p^2} + \ldots + b^{p^{n-1}} \neq 0$. [Use (a) and Exercise 48.18.]

The remaining two exercises outline a proof of Wedderburn's Theorem that a finite division ring is a field. Throughout, E is a finite division ring, C is the center of E, q is the number of elements in C, and n is the dimension of E over C.

*48.24. (a) If N is a division subring of E containing C, then the dimension of N over C divides n.
(b) If $a \in E$ but $a \notin C$, then the number of conjugates of a in the multiplicative group E^* is of the form $(q^n - 1)/(q^d - 1)$ where $d \mid n$ and $d < n$. [Show that $N = \{x \in E : ax = xa\}$ is a division subring. Use 2° of Theorem 26.1.]
(c) $\Phi_n(q)$ divides $q - 1$. [Use (1) and the equality concerning conjugate classes occurring before Theorem 26.3.]

48.25. (a) If ζ is a primitive nth root of unity in C and if $n > 1$, then $|q - \zeta| > q - 1$.
(b) Conclude that $n = 1$ and hence that $E = C$. [Use (a) and Exercise 48.24(c).]

49. Permutation Groups

To apply the concepts and methods of Galois theory to the study of polynomials, we need some preliminary information about permutation groups. We recall that $\mathfrak{G}$ is a **permutation group** on E if $\mathfrak{G}$ is a subgroup of the group $(\mathfrak{S}_E, \circ)$ of all permutations of E.

If f is a bijection from E onto F, the function Ψ'_f defined by

$$\Psi'_f: \sigma \mapsto f \circ \sigma \circ f^{\leftarrow}$$

is easily seen to be an isomorphism from the group $\mathfrak{S}_E$ onto the group $\mathfrak{S}_F$. If $\tau = \Psi'_f(\sigma)$, then $\tau \circ f = f \circ \sigma$, and hence, for every $x \in E$,

$$\tau(f(x)) = f(\sigma(x)).$$

Thus, for each $x \in E$, τ takes the element of F corresponding under f to x into the element of F corresponding under f to $\sigma(x)$; in this sense, the permutation τ of F acts on elements of F in "essentially" the same way that σ acts on elements of E. For this reason, we make the following definition:

Definition. Let $\mathfrak{G}$ and $\mathfrak{H}$ be permutation groups on sets E and F respectively. A bijection Ψ' from $\mathfrak{G}$ onto $\mathfrak{H}$ is a **permutation group isomorphism** if there exists a bijection f from E onto F such that

(1) $$\Psi'(\sigma) = f \circ \sigma \circ f^{\leftarrow}$$

for all $\sigma \in \mathfrak{G}$. We shall say that $\mathfrak{G}$ and $\mathfrak{H}$ are **isomorphic as permutation groups** if there is a permutation group isomorphism from $\mathfrak{G}$ onto $\mathfrak{H}$.

Thus, Ψ' is a permutation group isomorphism from $\mathfrak{G}$ onto $\mathfrak{H}$ if and only if Ψ' is obtained from the isomorphism Ψ'_f from $\mathfrak{S}_E$ onto $\mathfrak{S}_F$ defined by a bijection f from E onto F by restricting the domain of Ψ'_f to $\mathfrak{G}$ and its codomain to $\mathfrak{H}$.

We shall see shortly that two permutation groups may be isomorphic as groups but yet fail to be isomorphic as permutation groups. By our definition of a permutation group isomorphism, *subgroups $\mathfrak{H}$ and $\mathfrak{K}$ of $\mathfrak{S}_E$ are isomorphic as permutation groups if and only if they are conjugate subgroups of $\mathfrak{S}_E$.*

If σ and τ are permutations of E, we shall say that σ and τ are **disjoint permutations** if no element x of E satisfies both $\sigma(x) \neq x$ and $\tau(x) \neq x$.

Theorem 49.1. Let σ and τ be disjoint permutations of E. Then $\sigma\tau = \tau\sigma$. Moreover, if σ and τ have finite order, then the order of $\sigma\tau$ is the least common multiple of the orders of σ and τ.

Proof. If $\sigma(x) \neq x$, then $\sigma(\sigma(x)) \neq \sigma(x)$ as σ is injective, so $\tau(x) = x$ and $\tau(\sigma(x)) = \sigma(x)$, whence

$$\sigma(\tau(x)) = \sigma(x) = \tau(\sigma(x));$$

similarly, if $\tau(x) \neq x$, then $\tau(\sigma(x)) = \sigma(\tau(x))$; finally, if $\sigma(x) = x$ and $\tau(x) = x$, then

$$\sigma(\tau(x)) = \sigma(x) = x = \tau(x) = \tau(\sigma(x)).$$

Thus, $\sigma\tau = \tau\sigma$. Suppose further that σ and τ have respectively orders n and m. Let s be the least common multiple of n and m, and let j and k be the integers such that $nj = s = mk$. Then as σ and τ commute,

$$(\sigma\tau)^s = \sigma^s\tau^s = (\sigma^n)^j(\tau^m)^k = 1_E.$$

Suppose that $(\sigma\tau)^r = 1_E$ where $r \geq 1$. If $\sigma(x) \neq x$, then $\tau(x) = x$ and hence $\tau^r(x) = x$, so

$$x = (\sigma\tau)^r(x) = \sigma^r(\tau^r(x)) = \sigma^r(x),$$

and if $\sigma(x) = x$, then again $\sigma^r(x) = x$; hence, $\sigma^r = 1_E$. Similarly, $\tau^r = 1_E$. Therefore, $n \mid r$ and $m \mid r$, so $s \mid r$. Thus, s is the order of $\sigma\tau$.

By induction, we obtain the following corollary:

Corollary. If $\sigma_1, \ldots, \sigma_n$ is a sequence of mutually disjoint permutations of E of finite order, then the order of $\sigma_1 \ldots \sigma_n$ is the least common multiple of the orders of $\sigma_1, \ldots, \sigma_n$.

If f is a bijection from E onto F and if $(a_1, \ldots, a_n)$ is an n-cycle in $\mathfrak{S}_E$, then, as we observed in the proof of Theorem 26.7, an easy calculation establishes that

(2) $$f \circ (a_1, \ldots, a_n) \circ f^{\leftarrow} = (f(a_1), \ldots, f(a_n)).$$

Theorem 49.2. If E is a set, the order of an n-cycle belonging to the group $\mathfrak{S}_E$ is n.

Proof. Let $\sigma = (a_1, \ldots, a_n)$. An easy inductive argument establishes that $\sigma^{k-1}(a_1) = a_k$ for all $k \in [1, n]$. Hence, $\sigma^j \neq 1_E$ if $j \in [1, n-1]$, and for every $k \in [1, n]$,

$$\sigma^{n-k}(a_k) = \sigma^{n-k}(\sigma^{k-1}(a_1)) = \sigma^{n-1}(a_1) = a_n,$$

so

$$\sigma^n(a_k) = \sigma^{k-1}(\sigma(\sigma^{n-k}(a_k))) = \sigma^{k-1}(\sigma(a_n))$$
$$= \sigma^{k-1}(a_1) = a_k,$$

and therefore $\sigma^n = 1_E$. Thus, the order of σ is n.

We shall denote by 1_n the identity permutation of $[1, n]$.

Example 49.1. Let $\mathfrak{D}_2 = \{1_4, (1, 3)(2, 4), (1, 2)(3, 4), (1, 4)(2, 3)\}$, and let $\mathfrak{R} = \{1_4, (1, 2), (3, 4), (1, 2)(3, 4)\}$. Clearly, $\mathfrak{D}_2$ and $\mathfrak{R}$ are both subgroups of $\mathfrak{S}_4$ isomorphic to the four-group $Z_2 \times Z_2$. Since $\mathfrak{R}$ contains a transposition, any permutation group isomorphic as a permutation group to $\mathfrak{R}$ contains a transposition by (2). Consequently, $\mathfrak{D}_2$ and $\mathfrak{R}$ are isomorphic as groups but are not isomorphic as permutation groups.

Theorem 49.3. If $\tau \in \mathfrak{S}_n$, then τ is a product of mutually disjoint cycles.

Proof. Let R_τ be the relation on $[1, n]$ satisfying $x \, R_\tau \, y$ if and only if there exists $k \in Z$ such that $\tau^k(x) = y$. Since $\tau^0(x) = x$ for all $x \in [1, n]$, R_τ is reflexive. If $\tau^k(x) = y$, then $\tau^{-k}(y) = x$, so R_τ is symmetric. If $\tau^k(x) = y$ and if $\tau^j(y) = z$, then $\tau^{k+j}(x) = z$, so R_τ is transitive. Let $a_1, \ldots, a_s$ be numbers in $[1, n]$ such that the equivalence classes for R_τ determined by a_i and a_j are distinct if $i \neq j$ and every equivalence class for R_τ is determined by one of the numbers $a_1, \ldots, a_s$. Since $\mathfrak{S}_n$ is a finite group, τ has finite order, and therefore, for each $r \in [1, s]$ there is a smallest strictly positive integer m_r such that $\tau^{m_r}(a_r) = a_r$. It is easy to verify that $a_r, \tau(a_r), \ldots, \tau^{m_r-1}(a_r)$ is a sequence of distinct elements constituting the equivalence class determined by a_r, and consequently that

$$\tau = (a_1, \tau(a_1), \ldots, \tau^{m_1-1}(a_1)) \ldots (a_s, \tau(a_s), \ldots, \tau^{m_s-1}(a_s)),$$

a product of mutually disjoint cycles.

Theorem 49.4. If $n \geq 2$, every permutation of $[1, n]$ is a product of transpositions.

Proof. We observed that every m-cycle is a product of transpositions in the proof of Theorem 20.11. By Theorem 49.3, therefore, every permutation of $[1, n]$ is a product of transpositions.

Corollary. If $\mathfrak{H}$ is a normal subgroup of $\mathfrak{S}_n$ containing a transposition, then $\mathfrak{H} = \mathfrak{S}_n$.

Proof. By (2), $\mathfrak{H}$ contains all transpositions of $\mathfrak{S}_n$, and hence, $\mathfrak{H} = \mathfrak{S}_n$ by Theorem 49.4.

We next wish to examine in detail the groups $\mathfrak{S}_n$ and $\mathfrak{A}_n$ and, in particular, to determine all the normal subgroups of either. We shall denote the normal subgroup of $\mathfrak{S}_n$ consisting only of the identity permutation by $\mathfrak{I}_n$. Clearly, $\mathfrak{A}_1 = \mathfrak{I}_1$, $\mathfrak{A}_2 = \mathfrak{I}_2$, and $\mathfrak{A}_3 = \{1_3, (1, 2, 3), (1, 3, 2)\}$. The alternating group

$\mathfrak{A}_4$ differs in certain respects from the other alternating groups, and we shall not consider it in detail here.*

Theorem 49.5. If $n \geq 3$ and if $\mathfrak{H}$ is a normal subgroup of $\mathfrak{S}_n$ such that $\mathfrak{H} \cap \mathfrak{A}_n = \mathfrak{J}_n$, then $\mathfrak{H} = \mathfrak{J}_n$.

Proof. We shall first prove that a subgroup $\mathfrak{K}$ of $\mathfrak{S}_n$ of order 2 is not a normal subgroup of $\mathfrak{S}_n$. Let $\mathfrak{K} = \{1_n, \sigma\}$ where the order of σ is 2. Then there exists $j \in [1, n]$ such that $\sigma(j) \neq j$, and as $n \geq 3$, there exists $k \in [1, n]$ distinct from j and $\sigma(j)$. Let $\tau = (\sigma(j), k)$. Then

$$(\tau\sigma\tau^{\leftarrow})(j) = \tau\sigma(j) = k,$$

so $\tau\sigma\tau^{\leftarrow}$ is neither 1_n nor σ and hence does not belong to $\mathfrak{K}$. Consequently, $\mathfrak{K}$ is not a normal subgroup of $\mathfrak{S}_n$.

Suppose that $\mathfrak{H} \neq \mathfrak{J}_n$. Then by what we have just proved, there exist distinct permutations σ and ρ belonging to $\mathfrak{H}$ neither of which is the identity permutation. One of σ, ρ, $\sigma\rho^{\leftarrow}$ is even by Theorem 20.10, and none of them is the identity permutation, so $\mathfrak{H} \cap \mathfrak{A}_n \neq \mathfrak{J}_n$.

Theorem 49.6. If $n \geq 3$, then every even permutation of $[1, n]$ is a product of 3-cycles.

Proof. An even permutation is a product of an even number of transpositions by Theorem 49.4 and 20.10. It suffices, therefore, to show that a product of two transpositions is a product of 3-cycles. If (a, b) and (c, d) are disjoint transpositions, then

$$(a, b)(c, d) = (a, c, b)(a, c, d),$$

and if $b \neq d$,

$$(a, b)(a, d) = (a, d, b).$$

Theorem 49.7. If $n \neq 4$ and if a normal subgroup $\mathfrak{H}$ of $\mathfrak{A}_n$ contains a 3-cycle, then $\mathfrak{H} = \mathfrak{A}_n$.

Proof. From our discussion of $\mathfrak{A}_3$, the assertion is evident if $n < 4$. We shall assume, therefore, that $n \geq 5$. Let $(a_1, a_2, a_3) \in \mathfrak{H}$, and let (b_1, b_2, b_3) be any 3-cycle. Surely there is a permutation τ_1 of $[1, n]$ such that $\tau_1(a_1) = b_1$, $\tau_1(a_2) = b_2$, and $\tau_1(a_3) = b_3$. Let $\tau = \tau_1$ if τ_1 is even, and let $\tau = (b_4, b_5)\tau_1$ if τ_1 is odd, where b_4 and b_5 are distinct integers of $[1, n]$ not among b_1, b_2, b_3. Then τ is even, and therefore by (2),

$$(b_1, b_2, b_3) = \tau(a_1, a_2, a_3)\tau^{\leftarrow} \in \mathfrak{H}.$$

* A discussion is presented in §52 of the author's *Modern Algebra*, Vol. II.

Thus, $\mathfrak{H}$ contains every 3-cycle and hence is $\mathfrak{A}_n$ by Theorem 49.6.

Actually, the requirement that $n \neq 4$ is unnecessary; the only normal subgroup of $\mathfrak{A}_4$ containing a 3-cycle is $\mathfrak{A}_4$.

Definition. A group G is **simple** if the only normal subgroups of G are G itself and the subgroup containing only the neutral element.

For example, $\mathfrak{S}_n$ is not simple if $n \geq 3$, since $\mathfrak{A}_n$ is a normal subgroup of $\mathfrak{S}_n$. The subgroup $\mathfrak{D}_2$ (Example 49.1) is a normal subgroup of $\mathfrak{S}_4$ and hence of $\mathfrak{A}_4$ by (2), so $\mathfrak{A}_4$ is not simple.

The following three theorems are fundamental to our later discussion of polynomials.

Theorem 49.8. If $n \neq 4$, then $\mathfrak{A}_n$ is a simple group.

Proof. The assertion is evident if $n \leq 3$, so we shall assume that $n \geq 5$. Let $\mathfrak{H}$ be a normal subgroup of $\mathfrak{A}_n$ containing more than one element, and let m be the largest of those integers k such that $\mathfrak{H}$ contains a product of mutually disjoint cycles, one of which is a k-cycle. Since disjoint cycles commute, $\mathfrak{H}$ then contains a permutation σ satisfying $\sigma = (a_1, \ldots, a_m)\tau$, where τ is either the identity permutation or a product of mutually disjoint cycles each of which is also disjoint from $(a_1, \ldots, a_m)$.

Case 1: $m > 3$. Then the inverse (a_1, a_3, a_2) of (a_1, a_2, a_3) is disjoint from τ, so $\mathfrak{H}$ contains

$$
\begin{aligned}
[(a_1, a_2, a_3)\sigma(a_1, a_2, a_3)^{\leftarrow}]\sigma^{\leftarrow} &= (a_1, a_2, a_3)(a_1, \ldots, a_m)\tau(a_1, a_2, a_3)^{\leftarrow}\sigma^{\leftarrow} \\
&= (a_1, a_2, a_3)(a_1, \ldots, a_m)(a_1, a_2, a_3)^{\leftarrow}\tau\sigma^{\leftarrow} \\
&= (a_2, a_3, a_1, a_4, \ldots, a_m)(a_1, \ldots, a_m)^{\leftarrow} \\
&= (a_2, a_3, a_1, a_4, \ldots, a_m)(a_m, \ldots, a_4, a_3, a_2, a_1) \\
&= (a_1, a_2, a_4)
\end{aligned}
$$

by (2), and hence, $\mathfrak{H} = \mathfrak{A}_n$ by Theorem 49.7.

Case 2: $m = 3$. Then each factor of τ is either a transposition or a 3-cycle. But if one factor of τ were a 3-cycle (a_4, a_5, a_6), then $\tau = (a_4, a_5, a_6)\rho$, where ρ is either the identity permutation or a product of mutually disjoint cycles, each of which is also disjoint from (a_1, a_2, a_3) and from (a_4, a_5, a_6), and since the inverse (a_2, a_4, a_3) of (a_2, a_3, a_4) would be disjoint from ρ,

$\mathfrak{H}$ would contain

$$[(a_2, a_3, a_4)\sigma(a_2, a_3, a_4)^{\leftarrow}]\sigma^{\leftarrow}$$

$$= (a_2, a_3, a_4)(a_1, a_2, a_3)(a_4, a_5, a_6)\rho(a_2, a_3, a_4)^{\leftarrow}\rho^{\leftarrow}(a_4, a_5, a_6)^{\leftarrow}(a_1, a_2, a_3)^{\leftarrow}$$

$$= (a_2, a_3, a_4)(a_1, a_2, a_3)(a_4, a_5, a_6)(a_2, a_4, a_3)\rho\rho^{\leftarrow}(a_4, a_6, a_5)(a_1, a_3, a_2)$$

$$= (a_1, a_4, a_2, a_3, a_5),$$

a contradiction of our assumption that $m = 3$. Consequently, each factor of τ is a transposition, so $\tau^2 = 1_n$. Therefore, since τ commutes with (a_1, a_2, a_3), $\mathfrak{H}$ contains

$$\sigma^2 = (a_1, a_2, a_3)^2\tau^2 = (a_1, a_3, a_2),$$

and hence, $\mathfrak{H} = \mathfrak{A}_n$ by Theorem 49.7.

Case 3: $m = 2$. We shall see that our assumption that $n > 4$ implies that this case cannot occur. For if $m = 2$, then σ is a product of an even number of transpositions, so $\sigma = (a_1, a_2)(a_3, a_4)\rho$, where ρ is either the identity permutation or a product of mutually disjoint transpositions, each of which is also disjoint from (a_1, a_2) and from (a_3, a_4). Then the inverse (a_2, a_4, a_3) of (a_2, a_3, a_4) is disjoint from ρ, so $\mathfrak{H}$ contains

$$[(a_2, a_3, a_4)\sigma(a_2, a_3, a_4)^{\leftarrow}]\sigma^{\leftarrow}$$

$$= (a_2, a_3, a_4)(a_1, a_2)(a_3, a_4)\rho(a_2, a_3, a_4)^{\leftarrow}\rho^{\leftarrow}(a_1, a_2)(a_3, a_4)$$

$$= (a_2, a_3, a_4)(a_1, a_2)(a_3, a_4)(a_2, a_4, a_3)\rho\rho^{\leftarrow}(a_1, a_2)(a_3, a_4)$$

$$= (a_1, a_4)(a_2, a_3).$$

Since $n \geq 5$, there exists an integer $a_5 \in [1, n]$ distinct from a_1, a_2, a_3, a_4. Consequently, $\mathfrak{H}$ contains

$$\{(a_1, a_4, a_5)[(a_1, a_4)(a_2, a_3)](a_1, a_4, a_5)^{\leftarrow}\}(a_1, a_4)(a_2, a_3)$$

$$= (a_1, a_4, a_5)(a_1, a_4)(a_2, a_3)(a_1, a_5, a_4)(a_1, a_4)(a_2, a_3)$$

$$= (a_1, a_5, a_4),$$

a contradiction of our assumption that $m = 2$.

Theorem 49.9. If $n \neq 4$, then the only normal subgroups of $\mathfrak{S}_n$ are $\mathfrak{S}_n$, $\mathfrak{A}_n$, and $\mathfrak{I}_n$.

Proof. The assertion is easy to prove if $n \leq 3$, so we shall assume that $n \geq 5$. Let $\mathfrak{H}$ be a normal subgroup of $\mathfrak{S}_n$. Then $\mathfrak{H} \cap \mathfrak{A}_n$ is a normal subgroup of $\mathfrak{A}_n$ and hence is either $\mathfrak{A}_n$ or $\mathfrak{J}_n$ by Theorem 49.8. If $\mathfrak{H} \cap \mathfrak{A}_n = \mathfrak{J}_n$, then $\mathfrak{H} = \mathfrak{J}_n$ by Theorem 49.5. If $\mathfrak{H} \cap \mathfrak{A}_n = \mathfrak{A}_n$, then $\mathfrak{H} \supseteq \mathfrak{A}_n$, so the order m of $\mathfrak{H}$ divides $n!$ and is a multiple of $n!/2$, whence m is either $n!$ or $n!/2$, and consequently, $\mathfrak{H}$ is either $\mathfrak{S}_n$ or $\mathfrak{A}_n$.

Definition. A permutation group $\mathfrak{G}$ on E is a **transitive permutation group** if for all $x, y \in E$ there exists $\sigma \in \mathfrak{G}$ such that $\sigma(x) = y$.

It is easy to see that permutation groups that are isomorphic as permutation groups are either both transitive or both intransitive. The subgroup $\mathfrak{R}$ (Example 49.1) of $\mathfrak{S}_4$ is intransitive, though $\mathfrak{D}_2$, which is isomorphic as a group to $\mathfrak{R}$, is transitive.

Theorem 49.10. If q is a prime number and if $\mathfrak{G}$ is a transitive group of permutations of $[1, q]$ that contains a transposition (a, b), then $\mathfrak{G} = \mathfrak{S}_q$.

Proof. Let $M = \{j \in [1, q] : \text{either } j = a \text{ or } (a, j) \in \mathfrak{G}\}$, and let $\sigma \in \mathfrak{G}$. We shall prove that if $\sigma_*(M) \cap M \neq \emptyset$, then $\sigma_*(M) = M$. To do so, suppose that there exists $i \in M$ such that $\sigma(i) \in M$. We shall first show that $\sigma(a) \in M$. If either $a = i$ or $\sigma(a) = a$, then clearly $\sigma(a) \in M$. Consequently, we shall assume that $\sigma(a)$ is neither $\sigma(i)$ nor a. Then since (a, i) and $(a, \sigma(i))$ belong to $\mathfrak{G}$, we also have $\tau(a, i)\tau^{\leftarrow} \in \mathfrak{G}$ where $\tau = (a, \sigma(i))\sigma$. But by (2),

$$\tau(a, i)\tau^{\leftarrow} = (\tau(a), \tau(i)) = (\sigma(a), a) = (a, \sigma(a)),$$

so $\sigma(a) \in M$. Now let j be any element of M; we shall show that $\sigma(j)$ belongs to M. If either $j = a$ or $\sigma(j) = a$, then $\sigma(j) \in M$ by what we have just proved. Consequently, we shall assume that $\sigma(j)$ is neither $\sigma(a)$ nor a. Then since $(a, j) \in \mathfrak{G}$, we also have $\rho(a, j)\rho^{\leftarrow} \in \mathfrak{G}$ where $\rho = (a, \sigma(a))\sigma$ or $\rho = \sigma$ according as $a \neq \sigma(a)$ or $a = \sigma(a)$. But by (2),

$$\rho(a, j)\rho^{\leftarrow} = (\rho(a), \rho(j)) = (a, \sigma(j)),$$

so $\sigma(j) \in M$. Therefore, $\sigma_*(M) \subseteq M$. But since M is finite and since σ is a permutation of $[1, q]$, we conclude that $\sigma_*(M) = M$.

Thus, for each $\sigma \in \mathfrak{G}$, either $\sigma_*(M) = M$ or $\sigma_*(M) \cap M = \emptyset$. From this we may conclude that $\{\sigma_*(M) : \sigma \in \mathfrak{G}\}$ is a partition of $[1, q]$; indeed, as $\mathfrak{G}$ is transitive, for every $k \in [1, q]$ there exists $\sigma \in \mathfrak{G}$ such that $\sigma(a) = k$, whence $k \in \sigma^*(M)$; moreover, if $j \in \tau_*(M) \cap \rho_*(M)$ where $\tau, \rho \in \mathfrak{G}$, then $\tau^{\leftarrow}(j) \in M \cap (\tau^{\leftarrow}\rho)_*(M)$, so $M = \tau^{\leftarrow}_*(\rho_*(M))$ and therefore $\tau_*(M) = \rho_*(M)$ by what we have just proved. All members of the partition clearly have the same number of elements, so the number m of elements in M divides q. But q is a prime and $m \geq 2$, since a and b belong to M. Therefore, $m = q$, so $M = [1, q]$, and hence, $(a, s) \in \mathfrak{G}$ for all $s \in [1, q]$. From this we may conclude that $\mathfrak{G}$ contains all transpositions, for if (i, j) is a transposition in $\mathfrak{S}_q$ where $i \neq a, j \neq a$, then

$$(i, j) = (a, i)(a, j)(a, i) \in \mathfrak{G}.$$

By Theorem 49.4, therefore, $\mathfrak{G} = \mathfrak{S}_q$.

EXERCISES

49.1. Into how many different conjugate classes do the subgroups of order 2 of $\mathfrak{S}_5$ divide? of $\mathfrak{S}_6$? of $\mathfrak{S}_n$? Into how many different conjugate classes do the subgroups of a given prime order p of $\mathfrak{S}_n$ divide?

49.2. Show that if $\mathfrak{G}$ is a transitive permutation group on E and if $\mathfrak{H}$ is a permutation group on F isomorphic as a permutation group to $\mathfrak{G}$, then $\mathfrak{H}$ is a transitive permutation group.

49.3. (a) The group $\mathfrak{S}_n$ is generated by the $n - 1$ transpositions $(1, 2), (2, 3), \ldots,$ $(n - 1, n)$. [Use (2).]
(b) The group $\mathfrak{S}_n$ is generated by $(1, 2)$ and $(1, 2, \ldots, n)$. [Use (2) and (a).]

49.4. (a) The group $\mathfrak{A}_n$ is generated by the $n - 2$ 3-cycles $(1, 2, 3), (1, 2, 4), \ldots,$ $(1, 2, n)$. [Use Theorem 49.6 and (2).]
(b) If n is odd, then $\mathfrak{A}_n$ is generated by $(1, 2, 3)$ and $(1, 2, \ldots, n)$; if n is even, then $\mathfrak{A}_n$ is generated by $(1, 2, 3)$ and $(2, 3, \ldots, n)$.

49.5. Let $\mathfrak{G}$ be a permutation group on E, and let T be the relation on E satisfying $x \, T \, y$ if and only if there exists $\sigma \in \mathfrak{G}$ such that $\sigma(x) = y$. Show that T is an equivalence relation. (The equivalence class determined by x for T is called the **transitivity class** of x determined by $\mathfrak{G}$.)

49.6. If $\mathfrak{G}$ is a transitive permutation group on E and if $\mathfrak{H}$ is a normal subgroup of $\mathfrak{G}$, then any two transitivity classes determined by $\mathfrak{H}$ have the same number of elements.

49.7. Let q be a prime, let E be a set having q elements, and let $\mathfrak{G}$ be a transitive permutation group on E. If $\mathfrak{H}$ is a normal subgroup of $\mathfrak{G}$ containing at least two elements, then $\mathfrak{H}$ is a transitive permutation group on E. [Use Exercise 49.6.]

*49.8. A permutation σ of $[1, q]$ is **linear** if there exist $b, c \in Z$ such that

(3) $\sigma(t) \equiv bt + c$ $(\bmod \, q)$

for all $t \in [1, q]$. A subgroup $\mathfrak{G}$ of $\mathfrak{S}_q$ is **linear** if every member of $\mathfrak{G}$ is a linear permutation. Let $\mathfrak{G}$ be a linear subgroup of $\mathfrak{S}_q$, for each $\sigma \in \mathfrak{G}$ let

$$B_\sigma = \{b \in Z: \text{there exists } c \in Z \text{ such that (3) holds for all } t \in [1, q]\},$$

and let $B: \sigma \mapsto B_\sigma$.

(a) For each $\sigma \in \mathfrak{G}$, B_σ is an invertible element of the ring Z_q of integers modulo q.

(b) B is a homomorphism from $\mathfrak{G}$ into the multiplicative group of invertible elements of Z_q^*.

(c) Let K be the kernel of B. For each $\sigma \in K$, the set C_σ defined by

$$C_\sigma = \{c \in Z: \sigma(t) \equiv t + c \pmod{q} \text{ for all } t \in [1, q]\}$$

is an element of Z_q, and the function $C: \sigma \mapsto C_\sigma$ is a monomorphism from $(K, +)$ into the additive group Z_q.

(d) If q is a prime and if σ is a permutation of $[1, q]$ other than the identity permutation that satisfies (3) for all $t \in [1, q]$, then $\sigma(t) \neq t$ for all $t \in [1, q]$ if and only if $b \equiv 1 \pmod{q}$.

*49.9. If f is a homomorphism from $\mathfrak{S}_n$ into $(R, \cdot)$, then either $f(\sigma) = 0$ for all $\sigma \in \mathfrak{S}_n$, or $f(\sigma) = 1$ for all $\sigma \in \mathfrak{S}_n$, or $f = \text{sgn}$. [What are the finite multiplicative subsemigroups of R? Use (2).]

50. Solving Polynomials by Radicals

As every beginning algebra student knows, the roots of the quadratic $X^2 + bX + c$, where $b, c \in Q$, lie in the field $Q(\sqrt{D})$ obtained by adjoining to Q a square root $\sqrt{D}$ (i.e., a root of the polynomial $X^2 - D$) of the quantity $D = b^2 - 4c$; the roots are then $\frac{1}{2}(-b + \sqrt{D})$ and $\frac{1}{2}(-b - \sqrt{D})$. Actually, this result also holds if Q is replaced by any field whose characteristic is not 2.

The general solution of the quadratic equation was known to antiquity, but it was not until 1530 that a similar solution of the general cubic equation was obtained by Tartaglia. It is easily shown that if $g = X^3 + aX^2 + bX + c \in Q[X]$, then the roots of g are obtained by subtracting $\frac{1}{3}a$ from those of $f = X^3 + pX + q$, where $p = b - a^2/3$, $q = c - ab/3 + 2a^3/27$. Explicit formulas for the roots of f (and hence of g) may be given; they lie in the field obtained by first adjoining to Q a square root $\sqrt{-3}$ of -3 and a square root $\sqrt{D}$ of $D = -4p^3 - 27q^2$, and then adjoining to that field a cube root of the quantity

$$-\tfrac{27}{2}q + \tfrac{3}{2}\sqrt{-3}\sqrt{D}.$$

This procedure works equally well if Q is replaced by any field whose characteristic is neither 2 nor 3. In 1545, Ferrari similarly obtained explicit formulas for the roots of a quartic.

Needless to say, mathematicians then began work to find explicit formulas for the roots of a quintic. In the second quarter of the nineteenth century, however, Abel and Galois made the astounding discovery that similar formulas for the roots of a quintic simply did not exist, in general, for they do not always lie in a field obtained by successive adjunctions to Q of nth roots.

Here we shall establish the celebrated result that a polynomial over a field of characteristic zero is solvable "by radicals" if and only if the Galois group of its splitting field has a certain property called "solvability." Using this theorem, we may construct explicit quintics over Q that cannot be solved by radicals.*

Definition. An extension field F of a field K is a **radical extension** of K if there exist a sequence $(a_j)_{1 \leq j \leq m}$ of elements of F and a sequence $(s_j)_{1 \leq j \leq m}$ of integers >1 such that

(1) $a_1^{s_1} \in K,$

(2) $a_j^{s_j} \in K(a_1, \ldots, a_{j-1})$ for each $j \in [2, m],$

(3) $F = K(a_1, \ldots, a_m).$

Thus, if we adjoin to K an s_1th root of some element of K to obtain an extension K_1 of K, then adjoin to K_1 an s_2th root of some element of K_1 to obtain an extension K_2, and so forth for m steps, the extensions $K_1, K_2, \ldots,$ K_m are all radical extensions of K.

Actually, in the definition we may assume that each s_j is a prime:

Theorem 50.1. If F is a radical extension of a field K, then F is a finite extension of K, and there exist a sequence $(b_k)_{1 \leq k \leq n}$ of elements of F and a sequence $(r_k)_{1 \leq k \leq n}$ of primes such that

$$b_1^{r_1} \in K,$$

$$b_k^{r_k} \in K(b_1, \ldots, b_{k-1}) \text{ for each } k \in [2, n],$$

$$F = K(b_1, \ldots, b_n).$$

Proof. Let $(a_j)_{1 \leq j \leq m}$ be a sequence of elements of F and $(s_j)_{1 \leq j \leq m}$ a sequence

---

* A discussion of quadratics, cubics, and quartics and the Galois groups of their splitting fields occurs in §51 and §52 of the author's *Modern Algebra*, Vol. II.

of integers > 1 such that (1), (2), and (3) hold. Then $[K(a_1): K] \le s_1$ and $[K(a_1, \ldots, a_j): K(a_1, \ldots, a_{j-1})] \le s_j$ for each $j \in [2, m]$, so

$$[F: K] = [K(a_1, \ldots, a_m): K(a_1, \ldots, a_{m-1})] \ldots [K(a_1): K]$$

$$\le s_m \ldots s_1.$$

For each $j \in [1, m]$, let $(p_{j,k})_{1 \le k \le t(j)}$ be a sequence of (not necessarily distinct) primes such that

$$s_j = p_{j,1} \ldots p_{j,t(j)},$$

and let

$$c_{j,k} = \begin{cases} a_j^{p_{j,k+1}p_{j,k+2}\cdots p_{j,t(j)}} & \text{if } k \in [1, t(j) - 1,] \\ a_j & \text{if } k = t(j). \end{cases}$$

Then

$$c_{j,1}^{p_{j,1}} = a_j^{s_j}$$

and for each $k \in [2, t(j)]$,

$$c_{j,k}^{p_{j,k}} = c_{j,k-1}.$$

Thus, as $K(c_{i,1}, \ldots, c_{i,t(i)}) = K(a_i)$ for each $i \in [1, m]$,

$$c_{j,1}^{p_{j,1}} = a_j^{s_j} \in K(a_1, \ldots, a_{j-1})$$

$$\subseteq K(c_{1,1}, \ldots, c_{1,t(1)}, \ldots, c_{j-1,1}, \ldots, c_{j-1,t(j-1)})$$

and for each $k \in [2, t(j)]$,

$$c_{j,k}^{p_{j,k}} \in K(c_{j,k-1})$$

$$\subseteq K(c_{1,1}, \ldots, c_{1,t(1)}, \ldots, c_{j,1}, \ldots, c_{j,k-1}).$$

Moreover,

$$F = K(a_1, \ldots, a_m)$$

$$= K(c_{1,1}, \ldots, c_{1,t(1)}, \ldots, c_{m,1}, \ldots, c_{m,t(m)}).$$

Consequently, the sequences $(b_k)_{1 \le k \le n}$ and $(r_k)_{1 \le k \le n}$ have the desired properties where

$$n = t(1) + \ldots + t(m),$$

where for all $k \in [1, t(1)]$,

$$b_k = c_{1,k}$$

$$r_k = p_{1,k},$$

and where for each $j \in [2, m]$ and each $k \in [1, t(j)]$,

$$b_{t(1)+...+t(j-1)+k} = c_{j,k},$$

$$r_{t(1)+...+t(j-1)+k} = p_{j,k}.$$

Definition. A nonconstant polynomial f over a field K is **solvable by radicals** over K if there is a radical extension of K that contains a splitting field of f over K.

The importance of permutation groups in the study of polynomials arises from the fact that the Galois group of a splitting field of a separable polynomial is isomorphic in a "natural" way to a permutation group. Indeed, let f be a separable polynomial over a field K, let E be a splitting field of f over K, and let c be a sequence $(c_1, \ldots, c_n)$ of distinct terms consisting of all the roots of f in E. A K-automorphism σ of E takes roots of f into roots of f. For each root c_j of f, let $\sigma_c(j)$ be the unique integer $k \in [1, n]$ such that $\sigma(c_j) = c_k$, so that

$$\sigma(c_j) = c_{\sigma_c(j)}.$$

Since σ is a permutation of E, σ_c is a permutation of $[1, n]$, for if $\sigma_c(j) = \sigma_c(k)$, then $\sigma(c_j) = \sigma(c_k)$, whence $c_j = c_k$ and therefore $j = k$. Furthermore, if σ and τ are K-automorphisms of E, then $(\sigma\tau)_c = \sigma_c \tau_c$, for

$$c_{(\sigma\tau)_c(j)} = (\sigma\tau)(c_j) = \sigma(\tau(c_j))$$

$$= \sigma(c_{\tau_c(j)}) = c_{\sigma_c(\tau_c(j))}.$$

Therefore, $\sigma \mapsto \sigma_c$ is a monomorphism from the Galois group G of E over K into $\mathfrak{S}_n$, for if σ_c is the identity element of $\mathfrak{S}_n$, then $\sigma(c_j) = c_j$ for all $j \in [1, n]$, whence $\sigma = 1_E$ since $E = K(c_1, \ldots, c_n)$.

Definition. Let f be a separable polynomial over a field K, and let c be a sequence $(c_1, \ldots, c_n)$ of distinct terms consisting of all the roots of f in a splitting field E of f over K. The **Galois group of f over K defined by** c is the range of the monomorphism $\sigma \mapsto \sigma_c$ from the Galois group of E over K into $\mathfrak{S}_n$.

If f has n roots in E, there may be many different Galois groups of f, corresponding to the $n!$ different ways of arranging the roots of f in a sequence. But these groups are all isomorphic as permutation groups. Indeed, if $c = (c_1, \ldots, c_n)$ and $d = (d_1, \ldots, d_n)$ are sequences of distinct terms each consisting of all the roots of f in E, then there is a permutation

h of $[1, n]$ such that
$$c_k = d_{h(k)}$$
for all $k \in [1, n]$; since
$$d_{\sigma_d(h(k))} = \sigma(d_{h(k)}) = \sigma(c_k) = c_{\sigma_c(k)} = d_{h(\sigma_c(k))}$$
for all $k \in [1, n]$, we infer that
$$\sigma_d \circ h = h \circ \sigma_c;$$

therefore, $\sigma_c \mapsto \sigma_d$ is a permutation group isomorphism from the Galois group of f defined by c onto the Galois group of f defined by d. We shall denote by $\mathfrak{G}_K(f)$ the Galois group of a separable polynomial $f \in K[X]$ defined by some sequence of the roots of f in a splitting field E of f over K.

Example 50.1. Let $E = Q(\sqrt{2}, \sqrt{3})$, let $f = (X^2 - 2)(X^2 - 3)$, and let $g = X^4 - 10X^2 + 1$. We observed in Example 37.2 that E is the splitting field of both f and g over Q. We also saw that $X^2 - 3$ is irreducible over $Q(\sqrt{2})$ and similarly that $X^2 - 2$ is irreducible over $Q(\sqrt{3})$; hence, there exist a $Q(\sqrt{2})$-automorphism σ of E such that $\sigma(\sqrt{3}) = -\sqrt{3}$ and a $Q(\sqrt{3})$-automorphism τ of E such that $\tau(\sqrt{2}) = -\sqrt{2}$. Then σ, τ, $\sigma\tau$, and the identity automorphism 1_E are all the Q-automorphisms of E as $[E : Q] = 4$. The Galois group of f defined by the sequence $(\sqrt{3}, \sqrt{2}, -\sqrt{3}, -\sqrt{2})$ is, therefore, $\mathfrak{R} = \{1_E, (1, 3), (2, 4), (1, 3)(2, 4)\}$. The roots of g are $\sqrt{2} + \sqrt{3}$, $-\sqrt{2} + \sqrt{3}$, $\sqrt{2} - \sqrt{3}$, and $-\sqrt{2} - \sqrt{3}$. Computing the values of σ, τ, and $\sigma\tau$ at those roots, we see that the Galois group of g defined by that sequence is $\mathfrak{D}_2 = \{1_E, (1, 3)(2, 4), (1, 2)(3, 4), (1, 4)(2, 3)\}$. As we saw in Example 49.1, the Galois groups of f and g are isomorphic as groups but not as permutation groups.

Our next goal is to obtain a condition on $\mathfrak{G}_K(f)$ equivalent to the solvability of f by radicals where K is a field whose characteristic is zero. For this we need some preliminary definitions and theorems.

Definition. A group $(G, \cdot)$ with neutral element e is **solvable** if there exists a sequence $(G_i)_{0 \leq i \leq n}$ of subgroups of G satisfying the following three conditions:

1° $G_0 = G$ and $G_n = \{e\}$.
2° For each $i \in [1, n]$, G_i is a normal subgroup of G_{i-1}.
3° For each $i \in [1, n]$, the group G_{i-1}/G_i is abelian.

A sequence of subgroups $(G_i)_{0 \le i \le n}$ satisfying these three conditions is called a **solvable sequence** for G.

As we shall see in Theorem 50.2, for finite groups, condition 3° may be replaced by the condition that G_{i-1}/G_i be cyclic (and hence abelian) of prime order.

Lemma 50.1. Let f be an epimorphism from a group G onto a group G', let H' and K' be subgroups of G' such that K' is a normal subgroup of H', and let $H = f^*(H')$ and $K = f^*(K')$. Then H and K are subgroups of G, K is a normal subgroup of H, and the group H/K is isomorphic to the group H'/K'.

Proof. If $x, y \in H$, then $f(x), f(y) \in H'$, so $f(xy^{-1}) = f(x)f(y)^{-1} \in H'$ and therefore $xy^{-1} \in H$; hence, H is a subgroup of G by Theorem 8.3. Also, $f_*(H) = H'$, for if $y \in H'$, then there exists $x \in G$ such that $f(x) = y$ as f is surjective, whence $x \in f^*(H') = H$. Similarly, K is a subgroup of G and $f_*(K) = K'$. Moreover, K is a normal subgroup of H, for if $x \in H$, then

$$f_*(xKx^{-1}) = f(x)f_*(K)f(x)^{-1} = f(x)K'f(x)^{-1} = K',$$

whence $xKx^{-1} \subseteq f^*(K') = K$.

The function $f_H \colon H \to H'$ defined by $f_H(x) = f(x)$ for all $x \in H$ is an epimorphism from H onto H', and $\varphi \colon x' \mapsto x'K'$ is an epimorphism from H' onto H'/K'. Consequently, $\varphi \circ f_H$ is an epimorphism from H onto H'/K' whose kernel is K, for $x \in K$ if and only if $f(x) \in K'$. Therefore, H/K is isomorphic to H'/K' by Theorem 20.9.

Definition. Let H and K be subgroups of a group G such that $H \supseteq K$. A **cyclic sequence from H to K** is a sequence $(J_k)_{0 \le k \le r}$ of subgroups of H satisfying the following three conditions:

1° $J_0 = H$ and $J_r = K$.
2° For each $k \in [1, r]$, J_k is a normal subgroup of J_{k-1}.
3° For each $k \in [1, r]$, the group J_{k-1}/J_k is a cyclic group whose order is a prime.

Theorem 50.2. If G is a finite solvable group, then there is a cyclic sequence from G to $\{e\}$.

Proof. Let $(G_i)_{0 \le i \le n}$ be a solvable sequence for G, and let S be the set of all strictly positive integers m such that for any subgroups H and L of G, if L is a normal subgroup of H and if the group H/L is an abelian group of order $\le m$, then there is a cyclic sequence from H to L. It suffices to prove that $S = N^*$, for then there exists a cyclic sequence from G_{i-1} to G_i for each $i \in [1, n]$, and by stringing such sequences together, we obtain a cyclic sequence from G to $\{e\}$.

Clearly, $1 \in S$, for if the order of H/L is 1, then $L = H$, so the sequence whose only term is H is the desired cyclic sequence. Suppose that $m \in S$, and let L be a normal subgroup of H such that H/L is an abelian group of order $m + 1$. Let $H' = H/L$, let a be an element of H not belonging to L, and let s be the order of the element aL of H'. Then $s > 1$, so there exist a prime p and a positive integer r such that $s = pr$. Let $b = a^r$, let K' be the cyclic subgroup of H' generated by bL, and let $K = \varphi_L^*(K')$ where φ_L is the canonical epimorphism from H onto H'. The order of bL is clearly p, so K' is a subgroup of H' of order p. By Lemma 50.1, K is a normal subgroup of H as $H = \varphi_L^*(H')$, and H/K is isomorphic to H'/K'; in particular, H/K is abelian since H' is. Let L' be the subgroup of K' consisting only of the neutral element. Then $\varphi_L^*(L') = L$, so K/L is isomorphic to K'/L' by Lemma 50.1, and K'/L' is clearly isomorphic to K'. Therefore, K/L is a cyclic group of order p. Since

$$m + 1 = \text{order } H' = (\text{order } K')(\text{order } H'/K') = p \cdot (\text{order } H/K),$$

the order of H/K is less than m. Therefore, as $m \in S$, there exists a cyclic sequence $(J_k)_{0 \le k \le r}$ from H to K. Then $(J_k)_{0 \le k \le r+1}$ is a cyclic sequence from H to L where $J_{r+1} = L$. Consequently, $m + 1 \in S$, so by induction $S = N^*$, and the proof is complete.

Example 50.2. Clearly any abelian group G is solvable, for if G is abelian, the sequence $G, \{e\}$ is a solvable sequence for G. In particular, $\mathfrak{S}_1$ and $\mathfrak{S}_2$ are solvable groups. Examples of nonabelian solvable groups are furnished by $\mathfrak{S}_3$ and $\mathfrak{S}_4$. Indeed, the sequence $\mathfrak{S}_3, \mathfrak{A}_3, \{1_3\}$ is a solvable sequence for $\mathfrak{S}_3$, and the sequence $\mathfrak{S}_4, \mathfrak{A}_4, \mathfrak{D}_2, \{1_4\}$ is a solvable sequence for $\mathfrak{S}_4$ (as noted before Theorem 49.8, $\mathfrak{D}_2$ (Example 49.1) is a normal subgroup of $\mathfrak{S}_4$).

By the following theorem, every subgroup of these groups is also solvable.

Theorem 50.3. A subgroup H of a solvable group G is solvable.

Proof. Let $(G_i)_{0 \le i \le n}$ be a solvable sequence for G, and let $H_i = G_i \cap H$ for each $i \in [0, n]$. Then $H_0 = H$ and $H_n = \{e\}$. Also, H_i is a normal subgroup of H_{i-1} for each $i \in [1, n]$ by Theorem 18.5, for if $x \in H_{i-1}$, then

$$xH_ix^{-1} = (xH_ix^{-1}) \cap H \subseteq (xG_ix^{-1}) \cap H \subseteq G_i \cap H = H_i.$$

Let f_i be the restriction to H_{i-1} of the canonical epimorphism from G_{i-1}

onto G_{i-1}/G_i. Then x belongs to the kernel of f_i if and only if

$$x \in H_{i-1} \cap G_i = H \cap G_{i-1} \cap G_i = H_i.$$

Consequently, by Theorem 20.9, the group H_{i-1}/H_i is isomorphic to a subgroup of G_{i-1}/G_i and hence is abelian. Therefore, $(H_i)_{0 \leq i \leq n}$ is a solvable sequence for H.

Theorem 50.4. If f is an epimorphism from a group G onto a group G' and if K is the kernel of f, then G is a solvable group if and only if both G' and K are solvable groups.

Proof. Necessity: K is a solvable group by Theorem 50.3. Let $(G_i)_{0 \leq i \leq n}$ be a solvable sequence for G, and let $G'_i = f_*(G_i)$ for each $i \in [0, n]$. Then G'_i is a subgroup of G' for all $i \in [0, n]$, $G'_0 = G'$, and $G'_n = \{e'\}$, where e' is the neutral element of G'. Also G'_i is a normal subgroup of G'_{i-1} for each $i \in [1, n]$, for if $y \in G'_{i-1}$, then there exists $x \in G_{i-1}$ such that $y = f(x)$, whence

$$y G'_i y^{-1} = f(x) f_*(G_i) f(x)^{-1} = f_*(x G_i x^{-1}) \subseteq f_*(G_i) = G'_i.$$

For each $i \in [1, n]$, let f_{i-1} be the function obtained by restricting the domain of f to G_{i-1} and the codomain of f to G'_{i-1}, and let φ_{i-1} be the canonical epimorphism from G'_{i-1} onto G'_{i-1}/G'_i. Then $\varphi_{i-1} \circ f_{i-1}$ is clearly an epimorphism from G_{i-1} onto G'_{i-1}/G'_i whose kernel is G_i. Thus, by Theorem 20.9, G'_{i-1}/G'_i is isomorphic to G_{i-1}/G_i and hence is abelian. Therefore, $(G'_i)_{0 \leq i \leq n}$ is a solvable sequence for G'.

Sufficiency: Let $(G'_i)_{0 \leq i \leq m}$ and $(K_j)_{0 \leq j \leq n}$ be solvable sequences for G' and K respectively. Let $G_i = f^*(G'_i)$ for each $i \in [0, m]$, and let $G_{m+j} = K_j$ for each $j \in [1, n]$. For each $i \in [1, m]$, G_i is a normal subgroup of G_{i-1} and G_{i-1}/G_i is isomorphic to G'_{i-1}/G'_i and hence is abelian by Lemma 50.1. Also $G_m = f^*(\{e'\}) = K_0$, so $G_{m+j-1}/G_{m+j} = K_{j-1}/K_j$ for each $j \in [1, n]$. Therefore, $(G_k)_{0 \leq k \leq m+n}$ is a solvable sequence for G.

Corollary. If H is a normal subgroup of G, then G is a solvable group if and only if both H and G/H are solvable groups.

Theorem 50.5. If E is a radical extension of a field K, then there exists an extension F of E that is a normal radical extension of K.

Proof. Let $(a_i)_{1 \leq i \leq m}$ be a sequence of elements of E and $(s_i)_{1 \leq i \leq m}$ a sequence of integers > 1 satisfying (1)–(3). For each $i \in [1, m]$, let g_i be the minimal polynomial of a_i over K, let $g = g_1 \ldots g_m$, and let F be a splitting field of g over E. Then F is also a splitting field of g over K, since $g \in K[X]$ and since E is an extension of K generated by certain roots of g. Therefore, F is a finite normal extension of K by Theorem 46.5. Consequently, F is a finite extension of the closure of K in F, so there are only a finite number of K-automorphisms of F by Theorem 45.4; let them be $\sigma_0, \sigma_1, \ldots, \sigma_r$. For each $k \in [0, r]$ and for each $i \in [1, m]$ let

$$b_{km+i} = \sigma_k(a_i).$$

Then $F = K(b_1, b_2, \ldots, b_{rm+m})$; indeed, if c is a root of g in F, then c is a root of g_i for some $i \in [1, m]$; by the corollary of Theorem 37.7 there exists a K-isomorphism τ from $K(a_i)$ onto $K(c)$ such that $\tau(a_i) = c$ as g_i is irreducible over K; and by Theorem 46.5, there exists $k \in [0, r]$ such that $\tau(x) = \sigma_k(x)$ for all $x \in K(a_i)$, whence

$$c = \sigma_k(a_i) = b_{km+i}.$$

Thus, as F is generated by K and the roots of g, $F = K(b_1, \ldots, b_{rm+m})$.

For each $j \in [1, m]$ and each $k \in [0, r]$, we have $K(b_{km+1}, \ldots, b_{km+j}) = (\sigma_k)_*(K(a_1, \ldots, a_j))$; indeed, as $b_{km+i} = \sigma_k(a_i)$ for each $i \in [1, m]$,

$$K(b_{km+1}, \ldots, b_{km+j}) \subseteq (\sigma_k)_*(K(a_1, \ldots, a_j));$$

since $a_i = \sigma_k^{\leftarrow}(b_{km+i})$ for each $i \in [1, m]$,

$$K(a_1, \ldots, a_j) \subseteq \sigma_k^*(K(b_{km+1}, \ldots, b_{km+j})),$$

whence

$$(\sigma_k)_*(K(a_1, \ldots, a_j)) \subseteq (\sigma_k)_*(\sigma_k^*(K(b_{km+1}, \ldots, b_{km+j})))$$
$$= K(b_{km+1}, \ldots, b_{km+j}).$$

For each $k \in [0, r]$,

$$b_{km+1}^{s_1} = \sigma_k(a_1^{s_1}) \in (\sigma_k)_*(K) = K \subseteq K(b_1, \ldots, b_{km});$$

and for each $k \in [0, r]$ and each $i \in [2, m]$,

$$b_{km+i}^{s_i} = \sigma_k(a_i^{s_i}) \in (\sigma_k)_*(K(a_1, \ldots, a_{i-1}))$$
$$= K(b_{km+1}, \ldots, b_{km+i-1}) \subseteq K(b_1, \ldots, b_{km+i-1}).$$

Therefore, F is a normal radical extension of K.

Theorem 50.6. If f is a nonconstant monic polynomial over a field K whose characteristic is zero, then f is solvable by radicals over K if and only if $\mathfrak{G}_K(f)$ is a solvable group.

Proof. Necessity: By Theorem 50.5, there is a normal radical extension F of K that contains a splitting field E of f over K (Figure 20). Let $(a_j)_{1 \leq j \leq m}$ be a sequence of elements of F and $(s_j)_{1 \leq j \leq m}$ a sequence of integers > 1 such that (1)–(3) hold, and let $s = s_1 s_2 \ldots s_m$. As the characteristic of F is zero, there exists an extension N of F generated by a primi-tive sth root of unity ζ. If g is a polynomial over K such that F is a splitting field of g over K, then N is a splitting field over K of the polynomial $(X^s - 1)g$, since $1, \zeta, \zeta^2, \ldots, \zeta^{s-1}$ are s distinct roots of $X^s - 1$ in N; consequently, N is a Galois extension of K by the corollary of Theorem 46.6. Also by that corollary, the field $L = K(\zeta)$ is a Galois extension of K as L is a splitting field of $X^s - 1$ over K. Let Γ be the Galois group of N over K. Since the splitting field E of f is a Galois extension of K, $\mathfrak{G}_K(f)$ is isomorphic to a quotient group of

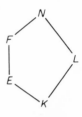

Figure 20

Γ by Theorem 46.7; hence, to prove that $\mathfrak{G}_K(f)$ is solvable, it suffices by the corollary of Theorem 50.4 to prove that Γ is solvable. Let Λ be the Galois group of N over L, and let Δ be the Galois group of L over K. Then Λ is a normal subgroup of Γ and Δ is isomorphic to Γ/Λ by Theorem 46.7. By Theorem 48.2, Δ is an abelian and hence a solvable group; to show that Γ is a solvable group, therefore, it suffices by the corollary of Theorem 50.4 to show that Λ is solvable.

Let $L_0 = L$, and for each $j \in [1, m]$ let $L_j = L(a_1, \ldots, a_j)$; also for each $j \in [0, m]$ let Λ_j be the Galois group of N over L_j. Let $j \in [1, m]$, let $\zeta_j = \zeta^{s/s_j}$, and let $b_j = a_j^{s_j}$. Then L_{j-1} contains ζ_j, which clearly is a primitive s_jth root of unity. Consequently, L_j is a splitting field of $X^{s_j} - b_j$ over L_{j-1}, for $a_j, \zeta_j a_j, \zeta_j^2 a_j, \ldots, \zeta_j^{s_j-1} a_j$ are s_j distinct roots of $X^{s_j} - b_j$ in L_j. Therefore, by Theorem 48.7, L_j is a Galois extension of L_{j-1}, and the Galois group of L_j over L_{j-1} is a subgroup of a cyclic group and hence is abelian; but by Theorem 46.7, the Galois group of L_j over L_{j-1} is isomorphic to Λ_{j-1}/Λ_j. Since $\Lambda_0 = \Lambda$ and $\Lambda_m = \{1_N\}$, therefore, $(\Lambda_j)_{0 \leq j \leq m}$ is a solvable sequence for Λ.

Sufficiency: Let E be a splitting field of f over K, and let $n = [E : K]$. As the characteristic of E is zero, there exists an extension N of E generated by a primitive $n!$th root of unity ζ (Figure 21). We shall show that N is a radical extension of K. Since N is a splitting field of $(X^{n!} - 1)f$ over K, N is a Galois extension of K. Let Γ be the Galois group of N over K, and let Δ be the Galois group of N over E. By Theorem 46.7, since E is a normal extension of K, Δ is a normal sub-group of Γ and Γ/Δ is isomorphic to the solvable group

Figure 21

$\mathfrak{G}_K(f)$. But Δ is isomorphic to a subgroup of a cyclic group and hence is abelian by Theorem 48.2. Therefore, Γ is solvable by the corollary of Theorem 50.4. Also, by Theorem 50.3, the Galois group Λ of N over the field $L = K(\zeta)$ is solvable since Λ is a subgroup of Γ. Let $(\Lambda_k)_{0 \leq k \leq m}$ be a cyclic sequence from Λ to $\{1_N\}$, let p_k be the order of Λ_{k-1}/Λ_k for each $k \in [1, m]$, and let $L_k = \Lambda_k^\blacktriangle$ for each $k \in [0, m]$. Then

$$L_0 = \Lambda^\blacktriangledown = L,$$
$$L_m = \Lambda_m^\blacktriangledown = \{1_N\}^\blacktriangledown = N.$$

Let $k \in [1, m]$. By Theorem 45.6, N is a finite Galois extension of L_{k-1} and L_k, and the Galois groups of N over L_{k-1} and L_k, respectively, are Λ_{k-1} and Λ_k. Since $\Lambda_k = L_k^\blacktriangle$ is a normal subgroup of Λ_{k-1}, L_k is a Galois extension of L_{k-1} by Theorem 46.7, and the Galois group of L_k over L_{k-1} is isomorphic to the cyclic group Λ_{k-1}/Λ_k of order p_k. Certainly

$$\deg_K \zeta \geq \deg_E \zeta,$$

so as

$$[N\colon L](\deg_K \zeta) = [N\colon L][L\colon K] = [N\colon K]$$
$$= [N\colon E][E\colon K] = (\deg_E \zeta)[E\colon K],$$

we have

$$p_k \leq [N\colon L] \leq [E\colon K] = n.$$

Therefore, L and hence also L_{k-1} contain $\zeta_k = \zeta^{n!/p_k}$, which clearly is a primitive p_kth root of unity. By Theorem 48.8, there exists $b_k \in L_{k-1}$ such that L_k is a splitting field of $X^{p_k} - b_k$ over L_{k-1}. Let a_k be a root of $X^{p_k} - b_k$ in L_k. Then $a_k, \zeta_k a_k, \zeta_k^2 a_k, \ldots, \zeta_k^{p_k-1} a_k$ are p_k distinct roots of $X^{p_k} - b_k$, so $L_k = L_{k-1}(a_k)$ and $a_k^{p_k} \in L_{k-1}$. Thus,

$$N = L_m = L(a_1, \ldots, a_m) = K(\zeta, a_1, \ldots, a_m),$$
$$\zeta^{n!} \in K,$$
$$a_1^{p_1} \in L = K(\zeta),$$

and

$$a_k^{p_k} \in L_{k-1} = K(\zeta, a_1, \ldots, a_{k-1})$$

for each $k \in [2, m]$, so N is a radical extension of K. Hence, f is solvable by radicals.

Theorem 50.7. If $n \geq 5$, then $\mathfrak{A}_n$ and $\mathfrak{S}_n$ are insolvable groups.

Proof. Since

$$(1, 2, 3)(1, 2, 4) \neq (1, 2, 4)(1, 2, 3),$$

$\mathfrak{A}_n$ is not abelian and hence is not solvable by Theorem 49.8. By Theorem 50.3, therefore, $\mathfrak{S}_n$ is also insolvable.

Theorem 50.8. If f is a separable prime polynomial over a field K, then $\mathfrak{G}_K(f)$ is a transitive permutation group.

Proof. Let $c_1, \ldots, c_n$ be the roots of f in a splitting field E of f over K; E is a Galois extension of K by Theorem 46.6. Let $i, j \in [1, n]$. As c_i and c_j are roots of f, there is a K-isomorphism σ_1 from $K(c_i)$ onto $K(c_j)$ such that $\sigma_1(c_i) = c_j$ by the corollary of Theorem 37.7. By Theorem 46.5, there is a K-automorphism σ of E satisfying $\sigma(x) = \sigma_1(x)$ for all $x \in K(c_i)$. Consequently, the permutation of $[1, n]$ corresponding to σ is an element of $\mathfrak{G}_K(f)$ taking i into j. Therefore, $\mathfrak{G}_K(f)$ is transitive.

Theorem 50.9. Let K be a subfield of $\mathbf{R}$. If f is a prime polynomial over K whose degree p is a prime and if f has exactly two nonreal roots in $\mathbf{C}$, then $\mathfrak{G}_K(f) = \mathfrak{S}_p$.

Proof. Since f is a prime polynomial, $\mathfrak{G}_K(f)$ is a transitive permutation group by Theorem 50.8. Let c_1 be a nonreal complex root of f. Since the coefficients of f are real, the complex conjugate $\bar{c}_i$ of c_i is also a root of f, and hence $\bar{c}_i = c_j$ for some $j \in [1, p]$ distinct from i. Let E be the splitting field of f over K contained in $\mathbf{C}$. Since $\sigma: z \mapsto \bar{z}$, $z \in E$, is a K-monomorphism from E into $\mathbf{C}$, by Corollary 46.5.2 $\bar{z} \in E$ for all $z \in E$, so the function obtained by restricting the codomain of σ to E is a K-automorphism of E. The permutation in $\mathfrak{G}_K(f)$ corresponding to this automorphism is (i, j). Hence, $\mathfrak{G}_K(f) = \mathfrak{S}_p$ by Theorem 49.10.

At last we are able to exhibit a quintic that is not solvable by radicals over $\mathbf{Q}$.

Example 50.3. Let f be the polynomial $X^5 - 4X + 2$ over $\mathbf{Q}$. By Eisenstein's Criterion, f is irreducible over $\mathbf{Q}$. Since $f(-2) < 0$, $f(0) > 0, f(1) < 0, f(2) > 0, f$ has at least three real roots and hence, at most, two nonreal complex roots by Theorem 42.2. But

$$Df = 5X^4 - 4 = (\sqrt{5}X^2 + 2)(\sqrt{5}X^2 - 2),$$

so Df has only two real roots, namely, $(\frac{4}{5})^{1/4}$ and $-(\frac{4}{5})^{1/4}$. Consequently, f has, at most, three real roots by the corollary of Theorem 42.3. Hence, f has exactly two nonreal complex roots. By Theorems 50.6, 50.7, and 50.9, therefore, f is not solvable by radicals over $\mathbf{Q}$.

EXERCISES

50.1. A finite group of order p^n where p is a prime is solvable.

50.2. Show that the following polynomials are not solvable by radicals over Q.

(a) $X^5 - 9X + 3$ (b) $X^5 - 5X^3 - 20X + 5$
(c) $X^5 - 6X^2 + 3$ (d) $X^5 - 10X^3 + 5$
(e) $X^5 - 8X + 6$ (f) $X^5 - 15X^4 + 6$.

50.3. Is every nonconstant polynomial with complex coefficients solvable by radicals over C? Is every nonconstant polynomial with real coefficients solvable by radicals over R?

50.4. If an irreducible polynomial f over a field K has a root in a radical extension of K, then f is solvable by radicals over K.

*50.5. Let f be a separable prime polynomial of prime degree q over a field K such that the Galois group Γ of a splitting field E of f over K is solvable, and let $(\Gamma_k)_{0 \le k \le n}$ be a solvable sequence for Γ such that Γ_{n-1} is a cyclic subgroup containing more than one element. Let c be a root of f in E, let ρ be a generator of Γ_{n-1}, and let $c_i = \rho^i(c)$ for each $i \in [1, q]$.
(a) The sequence $c = (c_1, \ldots, c_q)$ is a sequence of distinct terms consisting of all the roots of f in E. [Use Exercise 49.7.]
(b) Let $\mathfrak{G}_K(f)$ be the Galois group of f over K determined by c, and for each $k \in [0, n]$ let $\mathfrak{G}_k$ be the subgroup of $\mathfrak{G}_K(f)$ corresponding to Γ_k. Then for each $m \in [1, n]$, the permutation group $\mathfrak{G}_{n-m}$ is a linear permutation group (Exercise 49.8), and every cycle of length q belonging to $\mathfrak{G}_{n-m}$ belongs to $\mathfrak{G}_{n-1}$. [Use (2) of §49, Exercise 49.8(d), and induction in considering $\tau \rho \tau^{\leftarrow}$.]
(c) Conclude that $\mathfrak{G}_K(f)$ is a linear permutation group containing the cycle $(1, 2, \ldots, q)$.

50.6. If f is a prime polynomial of prime degree q over a field K whose characteristic is zero, then f is solvable by radicals over K if and only if there exists a sequence $c = (c_1, \ldots, c_q)$ of distinct terms consisting of all the roots of f in a splitting field of f such that the Galois group $\mathfrak{G}_K(f)$ of f determined by c is a linear permutation group containing the cycle $(1, 2, \ldots, q)$. [Use Exercises 50.5 and 49.8.]

*50.7. Let f be a prime polynomial of prime degree q over a field K whose characteristic is zero, and let c and c' be two roots of f in a splitting field E of f over K. If f is solvable by radicals over K, then $E = K(c, c')$. [Use Exercise 50.6 to show that $K(c, c')^{\blacktriangle} = \{1_E\}$.]

50.8. Let K be a subfield of R, and let f be a prime polynomial over K of prime degree q.

(a) If f is solvable by radicals over K and if f has two real roots, then every root of f in C is a real number.

(b) If $q \geq 5$ and if f has n real roots where $2 < n < q$, then f is not solvable by radicals over K. [Use Exercise 50.7.]

50.9. Let q be a prime ≥ 5.

(a) Let f be one of the following polynomials over Q:

$$X^q + aX + b$$

$$X^q + aX^2 + b$$

$$X^q + a_{q-1}X^{q-1} + a_{q-2}X^{q-2} + \ldots + a_3X^3 + a_0.$$

If f is a prime polynomial that is solvable by radicals over Q, then f has exactly one real root.

(b) Show that $X^q - 4X + 2$ is a prime polynomial that is not solvable by radicals over Q. Construct two other examples of prime polynomials of degree q that are not solvable by radicals over Q.

*50.10. If u is a constructible complex number, then (with the terminology of Theorem 44.4) there exists an admissible sequence $(L_j)_{0 \leq j \leq q}$ of subfields for u such that L_q is a normal extension of Q. [Argue as in the proof of Theorem 50.5.]

*50.11. (a) A complex number u is constructible if and only if u belongs to a finite normal extension E of Q such that $[E:Q]$ is a power of 2. [Use Exercise 50.10; for sufficiency, argue as in the proof of Theorem 48.4.] (b) A complex number u is constructible if and only if u is algebraic over Q and the degree over Q of the splitting field of its minimal polynomial is a power of 2.

APPENDIX A

INDUCED N-ARY OPERATIONS

Let $\triangle$ be a composition on E. We wish to generalize the definition of $\triangle^n a$ for every $a \in E$ and every $n \in N^*$ by giving a definition of the expression $a_1 \triangle a_2 \triangle \ldots \triangle a_{n-1} \triangle a_n$ for every n-tuple $(a_1, \ldots, a_n) \in E^n$, so that $a_1 \triangle a_2 \triangle \ldots \triangle a_{n-1} \triangle a_n$ may reasonably be described as the element obtained after n steps by starting first with a_1, forming then the composite $a_1 \triangle a_2$ of it with a_2, forming next the composite $(a_1 \triangle a_2) \triangle a_3$ of the result with a_3, forming next the composite $((a_1 \triangle a_2) \triangle a_3) \triangle a_4$ of that result with a_4, etc. Thus, our definition of $a_1 \triangle a_2 \triangle \ldots \triangle a_{n-1} \triangle a_n$ will turn out to be

$$(\ldots ((a_1 \triangle a_2) \triangle a_3) \triangle \ldots \triangle a_{n-1}) \triangle a_n$$

where all the left parentheses occur at the beginning. To make our definition precise we shall use the Principle of Recursive Definition.

Definition. Let $n \in N^*$. An ***n*-ary operation** on a set E is a function from E^n into E.

We shall, of course, use the words "unary," "binary," "ternary," etc., for "1-ary," "2-ary," "3-ary." Binary operations on E are thus just the compositions on E.

Theorem A.1. Let $\triangle$ be a composition on E. There is one and only one sequence $(\triangle_k)_{k \geq 1}$ such that for all $n \in N^*$, $\triangle_n$ is an n-ary operation on E,

(1) $$\triangle_1(a) = a$$

for every $a \in E$, and

(2) $$\triangle_{n+1}(a_1, \ldots, a_n, a_{n+1}) = \triangle_n(a_1, \ldots, a_n) \triangle a_{n+1}$$

485

for every $(n + 1)$-tuple $(a_1, \ldots, a_{n+1}) \in E^{n+1}$. In particular, $\triangle_2$ is the given composition $\triangle$.

Proof. Let

$$\mathscr{E} = \{\nabla : \text{for some } n \in N^*, \ \nabla \text{ is an } n\text{-ary operation on } E\}.$$

Let s be the function from $\mathscr{E}$ into $\mathscr{E}$ defined as follows: For each n-ary operation ∇ on E, $s(\nabla)$ is the $(n + 1)$-ary operation defined by

$$s(\nabla)(a_1, \ldots, a_n, a_{n+1}) = \nabla(a_1, \ldots, a_n)\triangle a_{n+1}$$

for all $(a_1, \ldots, a_n, a_{n+1}) \in E^{n+1}$. By the Principle of Recursive Definition, there is a unique sequence $(\triangle_k)_{k \geq 1}$ such that $\triangle_1$ is the unary operation defined by (1) and $\triangle_{n+1} = s(\triangle_n)$ for each $n \in N^*$. An easy inductive argument shows that $\triangle_n$ is an n-ary operation on E for each $n \in N^*$, and by the definition of s, (2) holds for every $(a_1, \ldots, a_n, a_{n+1}) \in E^{n+1}$.

We shall call the nth term $\triangle_n$ of the sequence $(\triangle_k)_{k \geq 1}$ the **n-ary operation defined by** $\triangle$.

Definition. If $\triangle$ is a composition on E, for each $(a_1, \ldots, a_n) \in E^n$ the **composite** of $(a_1, \ldots, a_n)$ for $\triangle$ is the value at $(a_1, \ldots, a_n)$ of the n-ary operation defined by $\triangle$. However, the composite of $(a_1, \ldots, a_n)$ for a composition denoted by a symbol similar to $+$ is called the **sum** of $(a_1, \ldots, a_n)$, and the composite of $(a_1, \ldots, a_n)$ for a composition denoted by a symbol similar to $\cdot$ is called the **product** of $(a_1, \ldots, a_n)$.

The composite of $(a_1, \ldots, a_n)$ for $\triangle$ is ordinarily denoted by

$$\mathop{\triangle}_{k=1}^{n} a_k \qquad \text{or} \qquad a_1 \triangle \ldots \triangle a_n.$$

The composite of $(a_1, \ldots, a_n)$ for a composition denoted by a symbol similar to $+$ is also denoted by

$$\sum_{k=1}^{n} a_k,$$

and the composite of $(a_1, \ldots, a_n)$ for a composition denoted by a symbol similar to $\cdot$ is also denoted by

$$\prod_{k=1}^{n} a_k.$$

If an ordered n-tuple is given by a simple formula, that formula may be used to denote its composite. For example,

$$\sum_{k=1}^{n} k^2 \quad \text{is} \quad \sum_{k=1}^{n} a_k$$

where $a_k = k^2$ for each $k \in [1, n]$.

An easy inductive argument establishes the following theorem, which shows that our definition of $\underset{k=1}{\overset{n}{\triangle}} a_k$ essentially generalizes that of $\triangle^n a$.

Theorem A.2. Let $\triangle$ be a composition on a set E, and let $a \in E$. If $(a_1, \ldots, a_n)$ is the ordered n-tuple defined by $a_k = a$ for each $k \in [1, n]$, then

$$\underset{k=1}{\overset{n}{\triangle}} a_k = \triangle^n a.$$

Definition. A **sequence** is a function whose domain is a subset of N. If the codomain of a sequence is contained in E, the sequence is said to be a **sequence of elements of** E, or a **sequence in** E. A **sequence of distinct elements** (or **terms**) **of** E is an injection from a subset of N into E. A **sequence of n terms** is a sequence whose domain has n elements, a **finite sequence** is a sequence whose domain is finite, and an **infinite sequence** is a sequence whose domain is infinite.

Custom has given the following notation for sequences: If f is a sequence and if A is the domain of f, some letter or symbol is chosen, say "a," $f(k)$ is denoted by a_k for each $k \in A$, and f itself is denoted by $(a_k)_{k \in A}$. Any expression denoting the domain of f may be used in place of "$k \in A$"; for example, if A is the set of all natural numbers $\geq n$, the sequence may be denoted by $(a_k)_{k \geq n}$, and if A is the integer interval $[p, q]$, the sequence may be denoted by $(a_k)_{p \leq k \leq q}$. If a sequence is defined by a simple formula, that formula is often used to denote the sequence as in the following examples: $(k^2)_{3 \leq k \leq 7}$ is the sequence $(a_k)_{3 \leq k \leq 7}$ where $a_k = k^2$ for all $k \in [3, 7]$, and $(\log(k^2 - 5))_{k \geq 3}$ is the sequence $(b_k)_{k \geq 3}$ where $b_k = \log(k^2 - 5)$ for all $k \geq 3$.

With this notation, a sequence $(a_k)_{k \in A}$ is a sequence of distinct elements if and only if $a_j \neq a_k$ for all $j, k \in A$ such that $j \neq k$. If $(E, \leqslant)$ is an ordered structure, a sequence $(a_k)_{k \in A}$ of elements of E is strictly increasing if and only if for all $j, k \in A$, if $j < k$, then $a_j \prec a_k$.

Let $(a_k)_{k \in A}$ be a sequence of n terms and let σ be a permutation of A. Then $(a_k) \circ \sigma$ is a sequence whose value at each $k \in A$ is $a_{\sigma(k)}$: for this reason, the sequence $(a_k) \circ \sigma$ is denoted by $(a_{\sigma(k)})_{k \in A}$. The *ordered n-tuple defined by the sequence* $(a_{\sigma(k)})_{k \in A}$ is the ordered n-tuple $(a_{\sigma(k_1)}, a_{\sigma(k_2)}, \ldots, a_{\sigma(k_n)})$ where

k_1 is the smallest element of A, k_2 the next smallest element, etc., and k_n the largest element of A. We may make this definition more precise (that is, we may eliminate the appeal to the word "etc.") by use of the following theorems.

Theorem A.3. If $(E, \leq)$ and $(F, \preccurlyeq)$ are totally ordered structures and if E and F are finite sets having the same number of elements, then there is exactly one isomorphism from the ordered structure $(E, \leq)$ onto the ordered structure $(F, \preccurlyeq)$.

Proof. It suffices to consider the case where $(F, \preccurlyeq)$ is $(N_n, \leq)$ for some natural number n. Let S be the set of all natural numbers n such that if E is any set having n elements and if $\leq$ is any total ordering on E, then there is exactly one isomorphism from $(E, \leq)$ onto $(N_n, \leq)$. Clearly, $0 \in S$. Let $n \in S$. To show that $n + 1 \in S$, let $(E, \leq)$ be a totally ordered structure such that E has $n + 1$ elements. By Theorem 12.9, E has a greatest element b. Then $E - \{b\}$ has n elements by Theorem 12.4, so since $n \in S$ there exists a unique isomorphism f from the ordered structure $E - \{b\}$ (with the total ordering induced from that of E) onto the totally ordered structure N_n. The function g defined by

$$g(x) = \begin{cases} f(x) & \text{if } x \in E - \{b\}, \\ n & \text{if } x = b \end{cases}$$

is clearly the desired isomorphism from $(E, \leq)$ onto $(N_{n+1}, \leq)$. If h is an isomorphism from $(E, \leq)$ onto $(N_{n+1}, \leq)$, then $h(b)$ is surely n, so the restriction of h to $E - \{b\}$ to a function from $E - \{b\}$ onto N_n is an isomorphism and hence is f. Thus,

$$h(x) = f(x) = g(x)$$

for all $x \in E - \{b\}$, and

$$h(b) = n = g(b),$$

so $h = g$. Therefore, $n + 1 \in S$. Consequently, $S = N$ by induction, and the proof is complete.

To illustrate Theorem A.3, we mention explicitly what the unique isomorphism is if $E = N_{n-m}$ and $F = [m + 1, n]$ where $m < n$:

Theorem A.4. If m and n are natural numbers such that $m < n$, then $[m + 1, n]$ has $n - m$ elements, and the function h defined by

$$h(x) = x + m + 1$$

for all $x \in N_{n-m}$ is the unique isomorphism from the totally ordered structure N_{n-m} onto the totally ordered structure $[m+1, n]$, where the orderings of both N_{n-m} and $[m+1, n]$ are those induced by the ordering of N.

Definition. If $(a_k)_{k \in A}$ is a sequence of n terms and if σ is a permutation of A, the **ordered n-tuple defined by the sequence** $(a_{\sigma(k)})_{k \in A}$ is the ordered n-tuple $(a_{\sigma(\tau(j))})_{1 \leq j \leq n}$, where τ is the unique isomorphism (guaranteed by Theorem A.3) from the totally ordered structure $[1, n]$ onto the totally ordered structure A.

For example, if $A = \{0, 3, 4, 9\}$, the ordered quadruple defined by $(a_k)_{k \in A}$ (where σ is the identity permutation) is (a_0, a_3, a_4, a_9), and if σ is the permutation $(0, 9, 3)$ of A, expressed in cyclic notation, the ordered quadruple defined by $(a_{\sigma(k)})_{k \in A}$ is (a_9, a_0, a_4, a_3).

Let $\triangle$ be a composition on E. If $(a_k)_{k \in A}$ is a sequence of n terms of E and if σ is a permutation of A, we shall denote by

$$\underset{k \in A}{\triangle} a_{\sigma(k)}$$

the composite of the ordered n-tuple defined by $(a_{\sigma(k)})_{k \in A}$. If A is the integer interval $[m, n]$, then $\underset{k \in A}{\triangle} a_{\sigma(k)}$ is often denoted by

$$a_{\sigma(m)} \triangle \cdots \triangle a_{\sigma(n)}.$$

Thus, by definition, if A has at least two elements and if q is the largest, then

$$\underset{k \in A}{\triangle} a_{\sigma(k)} = \left(\underset{k \in A-\{q\}}{\triangle} a_{\sigma(k)} \right) \triangle a_{\sigma(q)}.$$

In general, there are many ways of meaningfully inserting parentheses in the expression $a_1 \triangle \cdots \triangle a_n$; for example, there are five ways if $n = 4$, namely, $((a_1 \triangle a_2) \triangle a_3) \triangle a_4$, which by definition is $a_1 \triangle a_2 \triangle a_3 \triangle a_4$, $(a_1 \triangle (a_2 \triangle a_3)) \triangle a_4$, $(a_1 \triangle a_2) \triangle (a_3 \triangle a_4)$, $a_1 \triangle ((a_2 \triangle a_3) \triangle a_4)$, and $a_1 \triangle (a_2 \triangle (a_3 \triangle a_4))$. Our next theorem allows us to conclude that all such ways yield the same element if $\triangle$ is associative.

Theorem A.5. (General Associativity Theorem) Let $\triangle$ be an associative composition on E, let $(a_k)_{p+1 \leq k \leq p+n}$ be a sequence of elements of E, and let $(r_k)_{0 \leq k \leq s}$ be a strictly increasing sequence of natural numbers such that $r_0 = p$ and $r_s = p + n$. If

$$b_k = \underset{j=r_{k-1}+1}{\overset{r_k}{\triangle}} a_j$$

for each $k \in [1, s]$, then

(3)
$$\underset{k=1}{\overset{s}{\triangle}} b_k = \underset{k=p+1}{\overset{p+n}{\triangle}} a_k,$$

that is,

$$\underset{k=1}{\overset{s}{\triangle}} (a_{r_{k-1}+1} \triangle a_{r_{k-1}+2} \triangle \ldots \triangle a_{r_k}) = a_{p+1} \triangle \ldots \triangle a_{p+n}.$$

Proof. Let S be the set of all $n \in N^*$ such that (3) holds for every sequence $(a_k)_{p+1 \le k \le p+n}$ of elements of E and every strictly increasing sequence $(r_k)_{0 \le k \le s}$ of natural numbers such that $r_0 = p$ and $r_s = p + n$. Then $1 \in S$, for if $n = 1$, then $r_s = r_0 + 1$, so $s = 1$, and hence,

$$\underset{k=1}{\overset{s}{\triangle}} b_k = b_1 = a_{p+1} = \underset{k=p+1}{\overset{p+n}{\triangle}} a_k.$$

Let $n \in S$. To show that $n + 1 \in S$, let $(a_k)_{p+1 \le k \le p+n+1}$ be a sequence of elements of E, and let $(r_k)_{0 \le k \le s}$ be a strictly increasing sequence of natural numbers such that $r_0 = p$ and $r_s = p + n + 1$. Then $r_{s-1} \le p + n$.

Case 1: $r_{s-1} = p + n$. Then $b_s = a_{p+n+1}$. Since $n \in S$,

$$a_{p+1} \triangle \ldots \triangle a_{p+n} = b_1 \triangle \ldots \triangle b_{s-1}.$$

Hence, by the definition of composite,

$$
\begin{aligned}
a_{p+1} \triangle \ldots \triangle a_{p+n+1} &= (a_{p+1} \triangle \ldots \triangle a_{p+n}) \triangle a_{p+n+1} \\
&= (b_1 \triangle \ldots \triangle b_{s-1}) \triangle b_s \\
&= b_1 \triangle \ldots \triangle b_s.
\end{aligned}
$$

Case 2: $r_{s-1} < p + n$. Let

$$b'_s = a_{r_{s-1}+1} \triangle \ldots \triangle a_{r_s-1}.$$

Then, by the definition of composite,

$$b_s = b'_s \triangle a_{p+n+1}.$$

Since $n \in S$,

$$a_{p+1} \triangle \ldots \triangle a_{p+n} = b_1 \triangle \ldots \triangle b_{s-1} \triangle b'_s.$$

Therefore, by the associativity of $\triangle$ and by the definition of composite,

$$b_1\triangle \ldots \triangle b_s = (b_1\triangle \ldots \triangle b_{s-1})\triangle(b_s'\triangle a_{p+n+1})$$
$$= ((b_1\triangle \ldots \triangle b_{s-1})\triangle b_s')\triangle a_{p+n+1}$$
$$= (b_1\triangle \ldots \triangle b_{s-1}\triangle b_s')\triangle a_{p+n+1}$$
$$= (a_{p+1}\triangle \ldots \triangle a_{p+n})\triangle a_{p+n+1}$$
$$= a_{p+1}\triangle \ldots \triangle a_{p+n+1}.$$

Thus, $n+1 \in S$. Therefore, $S = N^*$ by induction, and the proof is complete.

If $\triangle$ is associative, no matter how parentheses are meaningfully inserted in the expression $a_{p+1}\triangle \ldots \triangle a_{p+n}$, one may, by the General Associativity Theorem, remove at least one pair at a time by working from the inside out. For example, by letting the sequence $(r_k)_{0\leq k\leq s}$ of Theorem A.5 be first the sequence $(2, 3, 5)$, then $(1, 2, 5, 6)$, and finally $(0, 1, 6)$, we see that

$$a_1\triangle(a_2\triangle(a_3\triangle(a_4\triangle a_5))\triangle a_6) = a_1\triangle(a_2\triangle(a_3\triangle a_4\triangle a_5)\triangle a_6)$$
$$= a_1\triangle(a_2\triangle a_3\triangle a_4\triangle a_5\triangle a_6)$$
$$= a_1\triangle a_2\triangle a_3\triangle a_4\triangle a_5\triangle a_6.$$

Theorem A.6. Let $\triangle$ be an associative composition on E, let $(a_k)_{1\leq k\leq n}$ be a sequence of elements of E, and let $b \in E$. If b commutes with a_k for each $k \in [1, n]$, then b commutes with $a_1\triangle \ldots \triangle a_n$.

The assertion is easily established by means of induction and Theorem 4.5.

Theorem A.7. (General Commutativity Theorem) Let $\triangle$ be an associative composition on E, and let $(a_k)_{1\leq k\leq n}$ be a sequence of elements of E. If $a_i\triangle a_j = a_j\triangle a_i$ for all $i, j \in [1, n]$, then for every permutation σ of $[1, n]$,

(4) $$a_{\sigma(1)}\triangle \ldots \triangle a_{\sigma(n)} = a_1\triangle \ldots \triangle a_n.$$

Proof. Let S be the set of all $n \in N^*$ such that (4) holds for every sequence $(a_k)_{1\leq k\leq n}$ of n terms of E satisfying $a_i\triangle a_j = a_j\triangle a_i$ for all $i, j \in [1, n]$ and every permutation σ of $[1, n]$. Clearly, $1 \in S$. Let $n \in S$. To show that $n+1 \in S$, let $(a_k)_{1\leq k\leq n+1}$ be a sequence of $n+1$ terms of E satisfying $a_i\triangle a_j = a_j\triangle a_i$ for all $i, j \in [1, n+1]$, and let σ be a permutation of $[1, n+1]$.

Case 1: $\sigma(n+1) = n+1$. The function obtained by restricting the domain and codomain of σ to $[1, n]$ is then a permutation of $[1, n]$, so

$$a_{\sigma(1)}\triangle \ldots \triangle a_{\sigma(n)} = a_1\triangle \ldots \triangle a_n$$

since $n \in S$, whence

$$a_{\sigma(1)}\triangle \ldots \triangle a_{\sigma(n+1)} = (a_{\sigma(1)}\triangle \ldots \triangle a_{\sigma(n)})\triangle a_{\sigma(n+1)}$$
$$= (a_1\triangle \ldots \triangle a_n)\triangle a_{n+1}$$
$$= a_1\triangle \ldots \triangle a_{n+1}.$$

Case 2: $\sigma(1) = n + 1$. Let $\tau: k \mapsto \sigma(k + 1)$, $k \in [1, n]$. Since

$$[1, n + 1] = [1, n] \cup \{n + 1\}$$

by Corollary 11.4.2, τ is clearly a permutation of $[1, n]$, so

$$a_{\tau(1)}\triangle \ldots \triangle a_{\tau(n)} = a_1\triangle \ldots \triangle a_n$$

as $n \in S$. Therefore, by Theorems A.5 and A.6,

$$a_{\sigma(1)}\triangle \ldots \triangle a_{\sigma(n+1)} = a_{\sigma(1)}\triangle(a_{\sigma(2)}\triangle \ldots \triangle a_{\sigma(n+1)})$$
$$= a_{n+1}\triangle(a_{\tau(1)}\triangle \ldots \triangle a_{\tau(n)})$$
$$= a_{n+1}\triangle(a_1\triangle \ldots \triangle a_n)$$
$$= (a_1\triangle \ldots \triangle a_n)\triangle a_{n+1}$$
$$= a_1\triangle \ldots \triangle a_{n+1}.$$

Case 3: $\sigma(m) = n + 1$ for some $m \in [2, n]$. Let τ be the function from $[1, n + 1]$ into $[1, n + 1]$ defined by

$$\tau(k) = \begin{cases} \sigma(k) & \text{if } k \in [1, m - 1], \\ \sigma(k + 1) & \text{if } k \in [m, n], \\ n + 1 & \text{if } k = n + 1. \end{cases}$$

Clearly, τ is a permutation of $[1, n + 1]$, so by Case 1,

$$a_{\tau(1)}\triangle \ldots \triangle a_{\tau(n+1)} = a_1\triangle \ldots \triangle a_{n+1}.$$

Therefore, by Theorems A.5 and A.6,

$$a_{\sigma(1)}\triangle \ldots \triangle a_{\sigma(n+1)} = (a_{\sigma(1)}\triangle \ldots \triangle a_{\sigma(m-1)})\triangle(a_{\sigma(m)}\triangle(a_{\sigma(m+1)}\triangle \ldots \triangle a_{\sigma(n+1)}))$$
$$= (a_{\tau(1)}\triangle \ldots \triangle a_{\tau(m-1)})\triangle(a_{\tau(n+1)}\triangle(a_{\tau(m)}\triangle \ldots \triangle a_{\tau(n)}))$$
$$= (a_{\tau(1)}\triangle \ldots \triangle a_{\tau(m-1)})\triangle((a_{\tau(m)}\triangle \ldots \triangle a_{\tau(n)})\triangle a_{\tau(n+1)})$$
$$= a_{\tau(1)}\triangle \ldots \triangle a_{\tau(n+1)}$$
$$= a_1\triangle \ldots \triangle a_{n+1}.$$

Thus, $n + 1 \in S$. Therefore, $S = N^*$ by induction, and the proof is complete.

Another application of induction yields the following theorem:

Theorem A.8. (General Distributivity Theorem) If $+$ and $\cdot$ are compositions on E such that $\cdot$ is distributive over $+$, then for every sequence $(a_k)_{1 \leq k \leq n}$ of elements of E and every $b \in E$,

$$(a_1 + \ldots + a_n)b = a_1 b + \ldots + a_n b,$$
$$b(a_1 + \ldots + a_n) = ba_1 + \ldots + ba_n.$$

Occasionally, the special notation for sequences is also employed for functions that are not sequences. If f is a function from A into E, some letter or symbol is chosen, say "x," and $f(\alpha)$ is denoted by x_α for all $\alpha \in A$ and f itself by $(x_\alpha)_{\alpha \in A}$. When this notation is used, the domain A of f is called the set of **indices** of $(x_\alpha)_{\alpha \in A}$, and $(x_\alpha)_{\alpha \in A}$ is called a **family of elements of E indexed by** A instead of a function from A into E. If the function $(x_\alpha)_{\alpha \in A}$ is injective, or equivalently, if $x_\alpha \neq x_\beta$ whenever $\alpha \neq \beta$, then $(x_\alpha)_{\alpha \in A}$ is said to be a **family of distinct elements** of E.

If $\triangle$ is an associative commutative composition on E and if $(x_\alpha)_{\alpha \in A}$ is a family of elements of E indexed by a finite nonempty set A, the General Commutativity Theorem permits us to define $\underset{\alpha \in A}{\triangle} x_\alpha$ in a natural way; indeed, if A has n elements, if σ is a bijection from $[1, n]$ onto A, and if $y_k = x_{\sigma(k)}$ for each $k \in [1, n]$, then $(y_1, \ldots, y_n)$ is an ordered n-tuple of elements of E, and it is natural to define $\underset{\alpha \in A}{\triangle} x_\alpha$ to be $y_1 \triangle \ldots \triangle y_n$. The only thing we must check is that our definition does not depend on our choice of σ; that is, if τ is also a bijection from $[1, n]$ onto A and if $z_k = x_{\tau(k)}$ for each $k \in [1, n]$, then

$$z_1 \triangle \ldots \triangle z_n = y_1 \triangle \ldots \triangle y_n.$$

Let $\rho = \tau^{\leftarrow} \circ \sigma$. Then ρ is a permutation of $[1, n]$, and $z_{\rho(k)} = y_k$ for each $k \in [1, n]$, so the desired equality holds by Theorem A.7. We may, therefore, unambiguously define $\underset{\alpha \in A}{\triangle} x_\alpha$ by

$$\underset{\alpha \in A}{\triangle} x_\alpha = \overset{n}{\underset{k=1}{\triangle}} x_{\sigma(k)}$$

where σ is any bijection whatever from $[1, n]$ onto A.

To prove an analogue of the General Associativity Theorem for families of elements indexed by arbitrary finite sets, we need the following intuitively obvious theorem.

Theorem A.9. If $(r_k)_{0 \leq k \leq n}$ is a strictly increasing sequence of natural numbers and if

$$A_k = [r_{k-1} + 1, r_k]$$

for each $k \in [1, n]$, then $\{A_k : k \in [1, n]\}$ is a partition of $[r_0 + 1, r_n]$.

Proof. Let $j \in [1, n]$. Since $(r_k)_{0 \leq k \leq n}$ is strictly increasing and since $0 \leq j - 1 < j \leq n$,

$$r_0 \leq r_{j-1} < r_j \leq r_n,$$

and hence

$$r_0 + 1 \leq r_{j-1} + 1 \leq r_j \leq r_n$$

by Theorem 11.4; consequently, A_j is a nonempty subset of $[r_0 + 1, r_n]$. Also, since $(r_k)_{0 \leq k \leq n}$ is strictly increasing, $A_j \cap A_k = \emptyset$ whenever $1 \leq j < k \leq n$. It remains for us to show that if $m \in [r_0 + 1, r_n]$, then $m \in A_k$ for some $k \in [1, n]$. The set

$$J = \{j \in [0, n] : m \leq r_j\}$$

is not empty since $n \in J$. Let k be the smallest member of J. Then $k \neq 0$ since $r_0 < m$. Hence, $k \in [1, n]$, and by its definition, $r_{k-1} < m \leq r_k$, whence $r_{k-1} + 1 \leq m \leq r_k$ by Theorem 11.4. Therefore, $m \in A_k$.

Theorem A.10. Let $\triangle$ be an associative commutative composition on E, let $(x_\alpha)_{\alpha \in A}$ be a family of elements of E indexed by a finite nonempty set A, and let $(B_k)_{1 \leq k \leq n}$ be a sequence of distinct subsets of A forming a partition of A. Then

$$\mathop{\triangle}_{k=1}^{n} \left(\mathop{\triangle}_{\alpha \in B_k} x_\alpha \right) = \mathop{\triangle}_{\alpha \in A} x_\alpha.$$

Proof. Let p_k be the number of elements in B_k for each $k \in [1, n]$, let $r_0 = 0$, let $r_k = \sum_{j=1}^{k} p_j$ for each $k \in [1, n]$, and let $p = r_n$. Then $r_k - r_{k-1} = p_k$, so by Theorem A.4, both $[1, p_k]$ and $[r_{k-1} + 1, r_k]$ have p_k elements. By Theorem A.3, therefore, there is a unique isomorphism τ_k from the totally ordered structure $[1, p_k]$ onto the totally ordered structure $[r_{k-1} + 1, r_k]$, where the orderings on both integer intervals are those induced by the ordering on N (it is easy to see that $\tau_k(j) = r_{k-1} + j$ for all $j \in [1, p_k]$). For each $k \in [1, n]$, let ρ_k be a bijection from $[1, p_k]$ onto B_k. By Theorem A.9, the function σ defined by

$$\sigma(j) = \rho_k(\tau_k^{\leftarrow}(j))$$

for all $j \in [r_{k-1} + 1, r_k]$ and all $k \in [1, n]$ is a bijection from $[1, p]$ onto A. Let

$$y_j = x_{\sigma(j)}$$

for each $j \in [1, p]$. By definition,

$$\underset{\alpha \in A}{\overset{p}{\triangle}} x_\alpha = \underset{j=1}{\overset{p}{\triangle}} x_{\sigma(j)} = \underset{j=1}{\overset{p}{\triangle}} y_j,$$

and for each $k \in [1, n]$,

$$\underset{\alpha \in B_k}{\triangle} x_\alpha = \underset{i=1}{\overset{p_k}{\triangle}} x_{\rho_k(i)}.$$

Also by definition,

$$\underset{j=r_{k-1}+1}{\overset{r_k}{\triangle}} y_j = \underset{i=1}{\overset{p_k}{\triangle}} y_{r_k(i)} = \underset{i=1}{\overset{p_k}{\triangle}} x_{\sigma(r_k(i))} = \underset{i=1}{\overset{p_k}{\triangle}} x_{\rho_k(i)}.$$

Therefore, by the General Associativity Theorem,

$$\underset{\alpha \in A}{\overset{p}{\triangle}} x_\alpha = \underset{j=1}{\overset{p}{\triangle}} y_j = \underset{k=1}{\overset{n}{\triangle}} \left(\underset{j=r_{k-1}+1}{\overset{r_k}{\triangle}} y_j \right)$$
$$= \underset{k=1}{\overset{n}{\triangle}} \left(\underset{i=1}{\overset{p_k}{\triangle}} x_{\rho_k(i)} \right) = \underset{k=1}{\overset{n}{\triangle}} \left(\underset{\alpha \in B_k}{\triangle} x_\alpha \right).$$

In practice, indices are often drawn from the cartesian product of two or more sets. As an illustration of Theorem A.10, let $(x_{ij})_{(i,j) \in A}$ be a family of elements of E indexed by $A = [1, n] \times [1, m]$. If $\triangle$ is an associative commutative composition on E, then

$$\underset{i=1}{\overset{n}{\triangle}} \left(\underset{j=1}{\overset{m}{\triangle}} x_{ij} \right) = \underset{(i,j) \in A}{\triangle} x_{ij} = \underset{j=1}{\overset{m}{\triangle}} \left(\underset{i=1}{\overset{n}{\triangle}} x_{ij} \right),$$

for we need only apply Theorem A.10 to the partition $\{B_1, \ldots, B_n\}$ of A where $B_i = \{i\} \times [1, m]$ to obtain the first equality, and to the partition $\{C_1, \ldots, C_m\}$ of A where $C_j = [1, n] \times \{j\}$ to obtain the second.

EXERCISES

A.1. Let $\triangle$ be an associative composition on E, and let $(a_1, \ldots, a_8) \in E^8$. Use the General Associativity Theorem to show that each of the following is $\underset{k=1}{\overset{8}{\triangle}} a_k$:

$$(a_1 \triangle ((a_2 \triangle a_3) \triangle a_4)) \triangle ((a_5 \triangle a_6) \triangle (a_7 \triangle a_8))$$
$$a_1 \triangle (a_2 \triangle (a_3 \triangle (a_4 \triangle (a_5 \triangle (a_6 \triangle (a_7 \triangle a_8))))))$$
$$(a_1 \triangle (a_2 \triangle a_3)) \triangle ((a_4 \triangle (a_5 \triangle a_6)) \triangle (a_7 \triangle a_8))$$
$$(((a_1 \triangle a_2) \triangle (a_3 \triangle a_4)) \triangle a_5) \triangle ((a_6 \triangle a_7) \triangle a_8).$$

List at each step the sequence $(r_k)_{0 \le k \le s}$ of natural numbers used.

A.2. Let $\triangle$ be an associative commutative composition on E, and let $(a_1, \ldots, a_8) \in E^8$. Use the General Commutativity Theorem to show that each of the following is $\underset{k=1}{\overset{8}{\triangle}} a_k$:

$$a_5 \triangle a_4 \triangle a_8 \triangle a_1 \triangle a_3 \triangle a_7 \triangle a_6 \triangle a_2$$

$$a_8 \triangle a_7 \triangle a_6 \triangle a_5 \triangle a_4 \triangle a_3 \triangle a_2 \triangle a_1$$

$$a_1 \triangle a_6 \triangle a_5 \triangle a_4 \triangle a_8 \triangle a_3 \triangle a_7 \triangle a_2$$

$$a_3 \triangle a_2 \triangle a_8 \triangle a_4 \triangle a_5 \triangle a_6 \triangle a_7 \triangle a_1.$$

Give for each the permutation of $[1, 8]$ used.

A.3. Prove Theorem A.2.

A.4. Prove Theorem A.4.

A.5. Prove Theorem A.6.

A.6. Prove Theorem A.8.

*A.7. Let $\triangle$ be a composition on E, and assume that the composite of a finite sequence of elements of E for $\triangle$ has not yet been defined. For each $n \in N^*$, let $\lambda(n)$ be the number of ways of inserting parentheses in the expression $a_1 \triangle \ldots \triangle a_n$ meaningfully and without redundance (we adopt the convention that $\lambda(1) = \lambda(2) = 1$). Thus, $\lambda(3) = 2$, $\lambda(4) = 5$, and $\lambda(5) = 14$. Give a formula for $\lambda(n + 1)$ in terms of $\lambda(1), \ldots, \lambda(n)$, and use it to calculate $\lambda(m)$ for each $m \in [1, 10]$.

A.8. Prove the following equalities:

(a) $\displaystyle\sum_{k=1}^{n} k = \frac{n(n + 1)}{2}$.

(b) $\displaystyle\sum_{k=1}^{n} k^2 = \frac{n(n + 1)(2n + 1)}{6}$.

(c) $\displaystyle\sum_{k=1}^{n} k^3 = \left(\sum_{k=1}^{n} k\right)^2$.

(d) $\displaystyle\sum_{k=1}^{n} k^5 + \sum_{k=1}^{n} k^7 = 2\left(\sum_{k=1}^{n} k\right)^4$.

A.9. Let $\cdot$ be a composition on E distributive over an associative commutative composition $+$ on E. If $(x_i)_{1 \le i \le n}$ and $(y_j)_{1 \le j \le m}$ are sequences of elements of E and if $A = [1, n] \times [1, m]$, then

$$\sum_{i=1}^{n} x_i \left(\sum_{j=1}^{n} y_j\right) = \sum_{(i,j) \in A} x_j y_j = \sum_{j=1}^{n} \left(\sum_{i=1}^{n} x_i\right) y_j.$$

A.10. If $(x_k)_{1 \le k \le n}$ is a sequence of invertible elements of a semigroup $(E, \cdot)$, then $\displaystyle\prod_{k=1}^{n} x_k$ is invertible, and

$$\left(\prod_{k=1}^{n} x_k\right)^{-1} = \prod_{k=1}^{n} x_{n+1-k}^{-1}.$$

*A.11. The function f from $N \times N$ into N defined by

$$f(m, n) = \left(\sum_{k=0}^{m+n} k \right) + n$$

for all $(m, n) \in N \times N$ is known as **Cantor's diagonal mapping.** (To see why, list all ordered couples (m, n) such that $m + n \leq 6$ in a triangular table so that those ordered couples having m as first term lie in the $(m + 1)$st row and those having n as second term lie in the $(n + 1)$st column, and over each entry (m, n) write the value of f at (m, n).) Prove that f is a bijection from $N \times N$ onto N. [If $f(p, q) = f(m, n)$, first prove that $p + q = m + n$; to show that a given natural number s belongs to the range of f, first define r so that $r + 1$ is the smallest of those numbers k satisfying $s < \sum_{j=0}^{k} j$.]

A.12. If E and F are denumerable (countable) sets (Exercise 12.7), then $E \times F$ is denumerable (countable). The cartesian product of a finite number of countable sets is countable.

A.13. The union of a finite number of countable sets is countable. [If $\{A_1, \ldots, A_n\}$ is a partition of A and if each A_k is countable, construct an injection from A into $N \times [1, n]$.]

A.14. If E is infinite, then for every natural number n, E contains a subset having n elements.

APPENDIX B

COMBINATORIAL ANALYSIS

Combinatorial analysis is concerned with the problem of counting the elements of a set. We present here certain theorems of combinatorial analysis used in the text, most of which are either intuitively evident or learned in the study either of probability or of combinations and permutations in elementary algebra. The most fundamental (and obvious) result of combinatorial analysis is that if a set A is partitioned into a finite number of finite subsets, then the number of elements of A is the sum of the numbers of elements in the members of the partition (Theorem B.1).

Lemma. If $(r_k)_{p \leq k \leq q}$ is a sequence of elements of an ordered structure $(E, \leqslant)$ whose domain is an integer interval $[p, q]$, then $(r_k)_{p \leq k \leq q}$ is strictly increasing if and only if $r_{k-1} \prec r_k$ for all $k \in [p + 1, q]$.

Proof. The condition is clearly necessary since $k - 1 < k$ for all $k \in N^*$. Sufficiency: If $(r_k)_{p \leq k \leq q}$ were not strictly increasing, then the set K of all natural numbers $k \in [p, q]$ such that there exists $j \in [p, q]$ satisfying $j < k$ and $r_j \geqslant r_k$ would not be empty and hence would contain a smallest member m. Consequently, there would exist $j \in [p, q]$ such that $j < m$ and $r_j \geqslant r_m$; then $j \leq m - 1$, and $m - 1 \notin K$ as $m - 1 < m$, so

$$r_j \leqslant r_{m-1} \prec r_m \leqslant r_j,$$

a contradiction.

Theorem B.1. (Fundamental Principle of Counting) Let A be a set, and let $(B_k)_{1 \leq k \leq n}$ be a sequence of distinct finite subsets of A forming a partition of A. If p_k is the number of elements in B_k for each $k \in [1, n]$, then A is finite and has $\sum_{k=1}^{n} p_k$ elements.

Proof. Let $r_0 = 0$, and let

$$r_k = \sum_{j=1}^{k} p_j$$

for each $k \in [1, n]$. Then $r_{k-1} + p_k = r_k$, so $r_{k-1} < r_k$ and $[r_{k-1} + 1, r_k]$ has $r_k - r_{k-1} = p_k$ elements by Theorem A.4. Consequently, there is a bijection σ_k from B_k onto $[r_{k-1} + 1, r_k]$ for each $k \in [1, n]$. Let σ be the function from A into N satisfying

$$\sigma(x) = \sigma_k(x)$$

for each $x \in B_k$ and each $k \in [1, n]$. By the Lemma, $(r_k)_{0 \leq k \leq n}$ is a strictly increasing sequence of natural numbers, so by Theorem A.9, the function obtained by restricting the codomain of σ to its range is a bijection from A onto $[1, r_n]$, which has r_n members by Theorem A.4. Hence, A has $r_n = \sum_{k=1}^{n} p_k$ members.

Theorem B.2. If there is a partition of E consisting of n subsets, each subset having m elements, then E is finite and has nm elements.

Proof. Let $p_k = m$ for each $k \in [1, n]$. By Theorem A.2 and the definition of multiplication,

$$\sum_{k=1}^{n} p_k = nm.$$

Therefore, E has nm elements by Theorem B.1.

Theorem B.3. If E has n elements and if A is a subset of E having m elements, then $E - A$ has $n - m$ elements.

Proof. The assertion is evident if either $A = E$ or $A = \emptyset$. Otherwise, $\{A, E - A\}$ is a partition of E, so if p is the number of elements in $E - A$, we have $m + p = n$ by Theorem B.1, whence $p = n - m$.

Theorem B.4. If E has n elements and if F has m elements, then $E \times F$ has nm elements.

Proof. The assertion is clear if either $n = 0$ or $m = 0$, for then $E \times F$ is the empty set and so has $0 = nm$ elements. Hence, we may assume that $n > 0$ and that $m > 0$. For each $a \in E$, the function g_a defined by

$$g_a(y) = (a, y)$$

for all $y \in F$ is a bijection from F onto $\{a\} \times F$, so $\{a\} \times F$ has m elements.

Moreover, the set

$$\mathscr{F} = \{\{a\} \times F : a \in E\}$$

is a partition of $E \times F$ consisting of n sets, for the function h defined by

$$h(a) = \{a\} \times F$$

is clearly a bijection from E onto $\mathscr{F}$. Hence, $E \times F$ has nm elements by Theorem B.2.

Theorem B.5. If E has n elements and if F has m elements, then F^E has m^n elements.

Proof. Let $S = \{k \in N$: if A is a set of k elements, then F^A has m^k members$\}$. Clearly, $0 \in S$, for $F^\emptyset$ has only one member, namely, the unique function whose domain is $\emptyset$ and whose codomain is F. Assume that $k \in S$, and let A be a set of $k + 1$ elements. Let $a \in A$, and let $B = A - \{a\}$; then B has $(k + 1) - 1 = k$ elements by Theorem B.3. For each $g \in F^B$ and each $z \in F$, let f_z be the function from A into F defined by

$$f_z(x) = g(x) \quad \text{if } x \in B,$$

$$f_z(a) = z.$$

Clearly, $z \mapsto f_z$ is a bijection from F onto $\mathscr{F}_g = \{f \in F^A$: the restriction of f to B is $g\}$. Hence, $\mathscr{F}_g$ has m elements. Moreover, $\{\mathscr{F}_g : g \in F^B\}$ is a partition of F^A. Since F^B has m^k elements because $k \in S$, therefore, F^A has $m \cdot m^k = m^{k+1}$ elements by Theorem B.2. Thus, $k + 1 \in S$, and by induction the proof is complete.

Theorem B.6. If $k \in [0, n]$, a set of n elements contains $\binom{n}{k}$ subsets of k elements.

Proof. Let $S = \left\{ n \in N$: for each $k \in [0, n]$, a set of n elements has $\binom{n}{k}$ subsets of k elements$\right\}$. Clearly, $0 \in S$, since the empty set has only one subset, namely itself, and $\binom{0}{0} = 1$. Suppose that $n \in S$, and let A be a set of $n + 1$ elements. If $k = 0$ or if $k = n + 1$, then A has $\binom{n + 1}{k}$ subsets of k elements; indeed, the only subset of A having no elements is $\emptyset$, and $\binom{n + 1}{0} = 1$; and the only subset of A having $n + 1$ elements is A itself by Theorem 12.5, and $\binom{n + 1}{n + 1} = 1$. Assume, therefore, that $k \in [1, n]$. Let $a \in A$, and let $B = A - \{a\}$, a set of n elements. If X is any subset of A of k elements, $X \cap B$

has either $k - 1$ or k elements by Theorem B.3, since $X - \{a\} = X \cap B \subseteq X$. For each subset Y of B of $k - 1$ elements, there is exactly one subset X of A having k elements such that $X \cap B = Y$, namely, $X = Y \cup \{a\}$. As $n \in S$, there are $\binom{n}{k-1}$ subsets of B having $k - 1$ elements; therefore, there are $\binom{n}{k-1}$ subsets X of A having k elements such that $X \cap B$ has $k - 1$ elements. As $n \in S$, there are $\binom{n}{k}$ subsets of B having k elements. Hence, the number of subsets of A having k elements is

$$\binom{n}{k-1} + \binom{n}{k} = \binom{n+1}{k}$$

by Theorems B.1 and 15.8. Therefore, $n + 1 \in S$, and the proof is complete by induction.

Theorem B.7. If E is a set having n elements, then $\mathfrak{P}(E)$ has 2^n members.

Proof. Let $F = \{0, 1\}$, and for each subset A of E, let χ_A be the member of F^E defined by

$$\chi_A(x) = \begin{cases} 1 & \text{if } x \in A, \\ 0 & \text{if } x \notin A. \end{cases}$$

Clearly, $\chi: A \mapsto \chi_A$ is a bijection from $\mathfrak{P}(E)$ onto F^E, so $\mathfrak{P}(E)$ has 2^n members by Theorem B.5.

Corollary.

$$\sum_{m=0}^{n} \binom{n}{m} = 2^n.$$

The assertion follows from Theorems B.1, B.6, and B.7.

Theorem B.8. If E and F have n elements, there are $n!$ bijections from E onto F.

Proof. Let $S = \{n \in N$: the number of bijections from a set of n elements onto a set of n elements is $n!\}$. Then $0 \in S$ since there is only one function whose domain and codomain are both the empty set and since $0! = 1$. Assume that $n \in S$, let E and F be sets of $n + 1$ elements, and let $\mathscr{B}$ be the set of all bijections from E onto F. Let $a \in E$, and let $E' = E - \{a\}$, a set of n elements. If Y is a subset of F of n elements, for each bijection g from

E' onto Y there is exactly one bijection $f \in \mathscr{B}$ extending g, namely, the function f defined by

$$f(x) = g(x) \quad \text{if } x \in E',$$

$$f(a) = b,$$

where $\{b\} = F - Y$. Therefore, as $n \in S$, the set $\mathscr{B}_Y$ of all bijections f from E onto F such that $f_*(E') = Y$ has $n!$ members. Clearly, $\{\mathscr{B}_Y : Y \text{ is a subset of } F \text{ of } n \text{ elements}\}$ is a partition of $\mathscr{B}$. By Theorem B.6, there are $\binom{n+1}{n} =$ $n+1$ subsets Y of F having n elements; therefore, $\mathscr{B}$ has $(n+1)n! = (n+1)!$ members by Theorem B.2.

Corollary. If E has n elements, the group $\mathfrak{S}_E$ of all permutations of E has $n!$ members.

EXERCISES

In the following exercises, the number of elements in a finite set X is denoted by $n(X)$.

B.1. If A and B are finite subsets of E, then $A \cup B$ is finite, and

$$n(A \cup B) + n(A \cap B) = n(A) + n(B).$$

[Note that $A \cup B = (A - (A \cap B)) \cup B$.]

B.2. If A and B are nonempty finite subsets of E, then $\{A, B\}$ is a partition of $A \cup B$ if and only if

$$n(A \cup B) = n(A) + n(B).$$

B.3. Let $(A_k)_{1 \le k \le m}$ be a sequence of distinct nonempty finite subsets of E, and let $A = \bigcup_{k=1}^{m} A_k$. Then A is finite, $n(A) \le \sum_{k=1}^{m} n(A_k)$, and furthermore, $n(A) = \sum_{k=1}^{m} n(A_k)$ if and only if $\{A_1, \ldots, A_m\}$ is a partition of A.

B.4. If $(A_k)_{1 \le k \le m}$ is a sequence of finite sets, then $\prod_{k=1}^{m} A_k$ is finite and has $\prod_{k=1}^{m} n(A_k)$ elements.

B.5. If $k \in [0, n]$, there are $n!/(n-k)!$ injections from a set of k elements into a set of n elements. [Use Theorems B.6 and B.8.]

B.6. Let $g_{n,k}$ be the number of k-tuples $(a_1, \ldots, a_k)$ of natural numbers such that $a_1 + a_2 + \ldots + a_k = n$.
(a) Show that $g_{n,1} = 1$ and that

$$g_{n,k} = \sum_{i=0}^{n} g_{i,k-1}$$

for all $n \in N$ and all $k \geq 2$.
(b) Show that $g_{n,k} = g_{n-1,k} + g_{n,k-1}$ for all $n \geq 1$ and all $k \geq 2$.
(c) Infer that

$$g_{n,k} = \binom{n + k - 1}{k - 1}$$

for all $n \in N$, $k \in N^*$.

*B.7. Let E be a set having n elements, and let $(p_k)_{1 \leq k \leq m}$ be a sequence of strictly positive integers such that $\sum_{k=1}^{m} p_k = n$. The number of ordered m-tuples $(A_1, \ldots, A_m)$ of subsets of E such that A_k has p_k elements for each $k \in [1, m]$ and $\{A_1, \ldots, A_m\}$ is a partition of E is $n! \left/ \prod_{k=1}^{m} p_k! \right.$.

*B.8. For every $n \in N$ and every $m \in [1, n]$, let $\sigma_n(m)$ be the number of surjections from a set having n elements onto a set having m elements. Show that for each $m \in [1, n]$,

$$\sigma_n(m) = m^n - \sum_{k=0}^{m-1} \binom{m}{k} \sigma_n(k).$$

Infer that

$$\sigma_n(m) = \sum_{k=0}^{m-1} (-1)^k \binom{m}{m - k}(m - k)^n = \sum_{j=1}^{m} (-1)^{m-j} \binom{m}{j} j^n.$$

B.9. Let $(E, \leq)$ and $(F, \leqslant)$ be totally ordered structures. If E has m elements and if F has n elements where $m \leq n$, the set of all strictly increasing functions from E into F has $\binom{n}{m}$ members.

*B.10. Let $(E, \leq)$ and $(F, \leqslant)$ be totally ordered structures. If E has n elements and if F has m elements, there are $\binom{n - 1}{m - 1}$ increasing surjections from E onto F, if $m \leq n$, and there are $\sum_{k=1}^{n} \binom{n - 1}{k - 1}\binom{m}{k}$ increasing functions from E into F.

LIST OF SYMBOLS

References are to page and exercise numbers. Decimal point numbers refer to exercises.

A^*, 123

A_+, A_+^*, 147

$A^{\leftarrow}$, 5.8

$|a|$, 372

(a), 166

$A/\mathfrak{a}$, 164

$[a]$, 214

(a, b), (a, b, c), 7, 8, 95

$[a, b]$, $]a, b[$, $[a, b[$, $]a, b]$, 379

$[a, \rightarrow[$, $]a, \rightarrow[$, $]\leftarrow, a]$, $]\leftarrow, a[$, 379

$A \cup B$, $A \cap B$, 20

$A - B$, 21

$\cup \mathscr{A}$, $\cap \mathscr{A}$, 24

$a \bigtriangleup X$, $X \bigtriangleup a$, 156

$+_m$, 14

a_k', 294

$(a_k)_{k \in A}$, 487

$(a_1, \ldots, a_n)$, 63, 94, 182

$a_1 \bigtriangleup a_2 \bigtriangleup \ldots \bigtriangleup a_n$, 486

a^n, 87

$\mathfrak{A}_n$, 184

$\rightarrow$, $\leftarrow$, 15

$\mathrm{Aut}(E)$, 62

$a.T$, 247

a^x, 396

$a.z$, 238

$B \circ A$, 5.8

B^c, 21

C, 4, 399

$[c]$, 11.14

(Canc), 54

$\subseteq$, $\supseteq$, 5

$\subset$, $\supset$, 102

$\mathrm{Conj}(S)$, $\mathrm{Conj}_H(S)$, 246

$\mathscr{C}(\boldsymbol{Q})$, 380

$\mathfrak{D}_2$, 464

$\deg f$, 323

$\delta_{\alpha\beta}$, 292

Δ_A, 59

$\Delta_{\mathfrak{P}}$, 161

505

$n \cdot a$, 87

$\sqrt[n]{a}$, 398

$n \cdot m$, 86

$n - m$, 81, 113

$\binom{n}{m}$, 5.4, 130

N_n, 14, 80

(NO 1)–(NO 4), 80

(NO 5), 11.7

(NO 6)–(NO 7), 11.10

$\mathcal{N}(Q)$, 381

$N(S)$, $N_H(S)$, 246

$\nu(u)$, 281

$\emptyset$, 5

1, 27, 80

1_E, 34

(OP), 7

(OR), 147

(OS), 76

(P 1)–(P 4), 148

$\mathfrak{P}(E)$, 10

φ, 218

φ_H, 174

φ_R, 136

Φ_n, 448

$\prod_{k=1}^{n} a_k$, 95

$\prod_{k=1}^{n} E_k$, 95

(PID), 198

$P(K)$, 261

$P_m(K)$, 261

(PS), 11.7

Q, 3, 147

$[r]$, 384

R, 3, 386

$\mathcal{R}$, 383

$\mathfrak{R}$, 464

R, $\cancel{R}$, 68

R_A, 70

$R^{\leftarrow}$, 72

$R_n(K)$, 451

$[r_1, \ldots, r_m]_b$, 108

$\mathfrak{R}z$, 180, 400

$\rho(u)$, 281

S_c, 329

$\mathfrak{S}_E$, 53

$\mathfrak{S}_m$, 54

σ_c, 473

sgn σ, 182

$\sum_{k=1}^{n} a_k$, 95

sup A, 370

T, 179

$[u; (b)_m, (a)_n]$, 305

(UFD 1)–(UFD 3), (UFD 3′), 202

u^t, 31.9

$V_f(x)$, 42.13

(VS 1)–(VS 4), 253

X, 321

x^Λ, 294

x^{-1}, $-x$, 28

x^*, 28

$|x|_R$, 136

xRy, 68

$\langle x, t' \rangle$, 293

$\{x : x$ has property $Q\}$, 5

$x \circ y$, 16

$x - y$, 114

$x \triangle y$, 12

$X \triangle Y$, 53, 155

Z, 3, 110

$\bar{z}$, 399

Z_m, 169

$Z(G)$, 61

0, 27, 80

INDEX

References are to page and exercise numbers. Decimal point numbers refer to exercises.

509